What's Next?

The Mathematical Legacy of William P. Thurston

EDITED BY
Dylan P. Thurston

PRINCETON UNIVERSITY PRESS
PRINCETON AND OXFORD
2020

Copyright © 2020 by Princeton University Press

Requests for permission to reproduce material from this work should be sent to permissions@press.princeton.edu

Published by Princeton University Press
41 William Street, Princeton, New Jersey 08540
6 Oxford Street, Woodstock, Oxfordshire OX20 1TR

press.princeton.edu

All Rights Reserved

ISBN 978-0-691-16776-3
ISBN (pbk.) 978-0-691-16777-0
ISBN (ebook) 978-0-691-18589-7

British Library Cataloging-in-Publication Data is available

Editorial: Susannah Shoemaker and Lauren Bucca
Production Editorial: Nathan Carr
Text Design: Leslie Flis
Production: Jacquie Poirier
Publicity: Matthew Taylor and Katie Lewis

This book has been composed in LATEX

Printed on acid-free paper. ∞

Printed in the United States of America

10 9 8 7 6 5 4 3 2 1

Annals of Mathematics Studies

Number 205

Contents

Preface

BILL THURSTON MADE fundamental contributions to topology, geometry, and dynamical systems. But beyond these specific accomplishments he introduced new ways of thinking about and of seeing mathematics that have had a profound influence on the entire mathematical community. He discovered connections between disciplines that led to the creation of entirely new fields. In June of 2014 mathematicians from a broad spectrum of areas came together to describe recent advances and explore future directions motivated by Thurston's transformative ideas. The conference was called "What's Next? The Mathematical Legacy of Bill Thurston."

This volume collects works by the speakers at the conference. We selected talks and papers to reflect the breadth of Bill's contributions to mathematics and his intensely geometric style. This includes two papers in dynamics based on Bill's unfinished work ("Degree-d-Invariant Laminations" and "On Balanced Planar Graphs, Following W. Thurston"), as well as a paper by Misha Gromov ("Morse Spectra"), continuing the long tradition of Gromov and Thurston building off of each other. The title of the contribution by Delecroix and Zorich, "Cries and Whispers in Wind-Tree Forests," exhibits another of Bill's refrains: look for ways to gain geometric intuition through senses other than sight. This geometric refrain continues through all of the papers, up to the last contribution by Dennis Sullivan with recollections of Bill's life, first published in the *Notices of the AMS*.

This volume would not have been possible without the efforts of many people, including the conference organizers (John Hubbard, Steve Kerckhoff, John Smillie, Karen Vogtmann, and especially David Gabai), Alan Reid, and Vickie Kearn and Susannah Shoemaker, who went above and beyond their usual duties at Princeton University Press.

What's Next?

Certifying the Thurston Norm via SL(2, ℂ)-twisted Homology

IAN AGOL AND NATHAN M. DUNFIELD

In memory of Bill Thurston: his amazing mathematics will live on, but as a collaborator, mentor, and friend he is sorely missed.

1 INTRODUCTION

For a compact orientable 3-manifold M, the Thurston norm on $H_2(M, \partial M; \mathbb{Z}) \cong H^1(M; \mathbb{Z})$ measures the minimal topological complexity of a surface representing a particular homology class. Twisted Alexander polynomials are a powerful tool for studying the Thurston norm; such a polynomial $\tau(M, \phi, \alpha)$ depends on a class $\phi \in H^1(M; \mathbb{Z})$ and a representation $\alpha \colon \pi_1(M) \to \mathrm{GL}(V)$, where V is a finite-dimensional vector space over a field K. The polynomial $\tau(M, \phi, \alpha)$ is constructed from the homology with coefficients twisted by α of the cyclic cover of M associated to ϕ. The degree of any such $\tau(M, \phi, \alpha) \in K[t^{\pm 1}]$ gives a lower bound on the Thurston norm of ϕ [FK1]. Remarkably, Friedl and Vidussi [FV2] showed that given M and ϕ one can always choose α so that this lower bound is sharp, with the possible exception of when M is a closed graph manifold; their results rely on the fact that most Haken 3-manifold groups are full of cubulated goodness [Wis, Liu, PW1, PW2] so that [A] applies.

Here, we explore whether one can get sharp bounds from just representations to $\mathrm{SL}_2\mathbb{C}$, especially those that originate in a hyperbolic structure on M. When M is the exterior of a hyperbolic knot K in S^3, there is a well-defined *hyperbolic torsion polynomial* $\mathscr{T}_K \in \mathbb{C}[t^{\pm 1}]$ which is (a refinement of) the twisted Alexander polynomial associated to a lift to $\mathrm{SL}_2\mathbb{C}$ of the holonomy representation $\pi_1(M) \to \mathrm{Isom}^+(\mathbb{H}^3) = \mathrm{PSL}_2\mathbb{C}$. The experimental evidence in [DFJ] forcefully led to

Conjecture 1.1 (DFJ). *For a hyperbolic knot in S^3, the hyperbolic torsion polynomial determines the Seifert genus $g(K)$; precisely, $\deg \mathscr{T}_K = 4g(K) - 2$.*

Here, we prove this conjecture for a large class of knots, which includes infinitely many knots whose ordinary Alexander polynomial is trivial. We call a knot $K \subset S^3$ *libroid* if there is a collection Σ of disjointly embedded minimal genus Seifert surfaces in its exterior $X = S^3 \setminus N(K)$ so that $X \setminus \Sigma$ is a union of books of I-bundles in a way that respects the structure of $X \setminus \Sigma$ as a sutured manifold; see Section 6.3 for the precise definitions. We show

Theorem 6.2. *Conjecture 1.1 holds for libroid hyperbolic knots in S^3.*

Libroid knots generalize the notion of a fibroid surface introduced in [CS], and includes all fibered knots. The class of libroid knots is closed under Murasugi sum (Lemma 6.6) and contains all *special* arborescent knots obtained from plumbing oriented bands (this includes 2-bridge knots), as well as many knots whose ordinary Alexander polynomial is trivial (Theorem 6.1). Previous to Theorem 6.2, Conjecture 1.1 was known only in the case of 2-bridge knots, by work of Morifuji and Tran [Mor, MT].

1.2   Motivation

While twisted Alexander polynomials give sharp bounds on the Thurston norm if one allows arbitrary representations to $\mathrm{GL}_n\mathbb{C}$ by [FV2], there are still compelling reasons to consider questions such as Conjecture 1.1. First, if the Thurston norm is detected by representations of uniformly bounded degree, then one should be able to use ideas from [Kup] to show that the KNOT GENUS problem of [AHT] is in **NP** ∩ **co-NP** for knots in S^3 using a finite-field version of $\tau(M, \phi, \alpha)$ as the **co-NP** certificate. (As with the results in [Kup], this would be conditional on the Generalized Riemann Hypothesis. Subsequent to our work here, Lackenby has shown that KNOT GENUS is in **NP** ∩ **co-NP** by different methods [Lac].) Second, since $\mathscr{T}_K$ is easily computable in practice, a proof of Conjecture 1.1 should lead to an effectively polynomial-time algorithm for computing $g(K)$ for knots in S^3. Finally, Conjecture 1.1 would be another beautiful Thurstonian connection between the topology and geometry of 3-manifolds.

1.3   Sutured manifolds

The Thurston norm bounds associated to twisted Alexander polynomials can be understood in the following framework of [FK2]. Throughout, see Section 2 for precise definitions. Let $M = (M, R_-, R_+, \gamma)$ be a sutured manifold. Given a representation $\alpha\colon \pi_1(M) \to \mathrm{GL}(V)$, we say that M is an α-homology product if the inclusion-induced maps

$$H_*(R_+; E_\alpha) \to H_*(M; E_\alpha) \quad \text{and} \quad H_*(R_-; E_\alpha) \to H_*(M, E_\alpha)$$

are all isomorphisms; here E_α is the system of local coefficients associated to α. An α-homology product is necessarily a taut sutured manifold (see Theorem 3.2 for the precise statement). Conversely, every taut sutured manifold is an α-homology product for *some* representation α by [FK2]. A weaker, less geometric, parallel to Conjecture 1.1 is

Conjecture 1.4. *For a taut sutured manifold M, there exists $\alpha\colon \pi_1(M) \to \mathrm{SL}_2\mathbb{C}$ for which M is a homology product.*

Theorem 6.2 will follow easily from the next result, establishing a strong version of Conjecture 1.4 for books of I-bundles (see Section 4.3 for the definitions).

Theorem 4.1. *Let M be a taut sutured manifold which is a book of I-bundles. Suppose $\alpha\colon \pi_1(M) \to \mathrm{SL}_2\mathbb{C}$ has $\mathrm{tr}\big(\alpha(\gamma)\big) \neq 2$ for every curve γ which is the core of a gluing annulus for an I-bundle page. Then M is an α-homology product.*

In trying to attack Conjecture 1.4, an intriguing aspect of Theorem 4.1 is the very weak hypothesis on the representation α. Unfortunately, for more complicated taut sutured manifolds one must put additional restrictions on α to get a homology product, as the next result shows.

Theorem 5.7. *There exists a taut sutured manifold M with a faithful discrete and purely hyperbolic representation $\alpha\colon \pi_1(M) \to \mathrm{SL}_2\mathbb{C}$ where M is not an α-homology product. The manifold M is acylindrical with respect to the pared locus consisting of the sutures.*

Another instance where we can prove Conjecture 1.4 is

Theorem 4.2. *Suppose M is a sutured manifold which is a genus 2 handlebody with suture set γ a single curve separating ∂M into two once-punctured tori. If the pared manifold (M, γ) is acylindrical and $M \setminus \gamma$ is incompressible, then M is a homology product with respect to some $\alpha\colon \pi_1(M) \to \mathrm{SL}_2\mathbb{C}$.*

With both Theorems 4.1 and 4.2, it is easy to construct sutured manifolds satisfying their hypotheses which are *not* homology products with respect to $H_*(\, \cdot \, ; \mathbb{Q})$.

1.5   Outline of contents

After reviewing the needed definitions in Section 2, we establish the basic properties of homology products in Section 3 and so relate Conjectures 1.1 and 1.4. Section 4 is devoted to proving Conjecture 1.4 in the two cases mentioned above. Section 5 studies one sutured manifold in detail, characterizing which $\mathrm{SL}_2\mathbb{C}$-representations make it a homology product (Theorem 5.5); Theorem 5.7 is an easy consequence of this. Finally, Section 6 is devoted to studying libroid knots, both showing that this is a large class of knots and also proving Theorem 6.2 follows from Theorem 4.1.

1.6   Acknowledgments

Agol was partially supported by the Simons Foundation and US NSF grants DMS-1105738 and DMS-1406301. Dunfield was partially supported by US NSF grant DMS-1106476, a Simons Fellowship, and the GEAR Network (US NSF grant DMS-1107452), and this work was partially completed while visiting ICERM (Brown University) and the University of Melbourne. We thank Stefan Friedl for several helpful discussions, and we are very grateful to the referee for an extraordinarily quick and yet very thorough review of this paper.

2   BACKGROUND

We begin with the precise definitions of the basic objects we will be working with. Throughout, all manifolds will be assumed orientable and moreover oriented.

2.1   Taut surfaces

For a connected surface, define $\chi_-(S) = \max\left(-\chi(S), 0 \right)$; extend this to all surfaces via $\chi_-(S \sqcup S') = \chi_-(S) + \chi_-(S')$. For a 3-manifold M and a (possibly empty) subsurface $A \subset \partial M$, the Thurston norm of $z \in H_2(M, A; \mathbb{Z})$ is defined by

$$\|z\| = \min\{\chi_-(S) \mid S \text{ is a properly embedded surface representing } z \text{ with } \partial S \subset A\}.$$

A properly embedded compact surface S in a 3-manifold M is *taut* if S is incompressible and realizes the Thurston norm for the class $[S, \partial S]$ in $H_2(M, N(\partial S); \mathbb{Z})$.

2.2   Sutured manifolds

A sutured manifold (M, R_+, R_-, γ) is a compact 3-manifold with a partition of ∂M into two subsurfaces R_+ and R_- along their common boundary γ. The surface R_+ is oriented by the outward-pointing normal, and R_- is oriented by the inward-pointing one. Note that the orientations of $R_\pm$ induce a common orientation on γ. A sutured manifold is *taut* if it is irreducible and the surfaces $R_\pm$ are both taut. A connected sutured manifold M is *balanced* if it is irreducible, $\chi(R_-) = \chi(R_+)$, not a solid torus with $\gamma = \varnothing$, and if any component of $R_\pm$ has positive $\chi > 0$ then M is D^3 with a single suture. A disconnected sutured manifold is balanced if each connected component is. Note that any taut sutured manifold is necessarily balanced.

2.3   Notes on conventions

We follow [Sch] in requiring taut surfaces to be incompressible; this is not universal, and the difference is just that the more restrictive definition excludes a solid torus with no sutures and a ball with more than one suture. Like [FK2] but unlike many sources, we do not allow torus sutures consisting of an entire torus component of ∂M. Our definition of balanced is slightly more restrictive than that of [FK2] and also differs from that of [Juh].

2.4   Twisted homology

Suppose X is a connected CW complex with a representation $\alpha\colon \pi_1(X) \to \mathrm{GL}(V)$, where V is a vector space over a field K. Let E_α be the system of local coefficients over X corresponding to α; precisely, $E_\alpha \to X$ is the induced vector bundle where we give each fiber the *discrete topology* so that $E_\alpha \to X$ is actually a covering map. (Alternatively, you can view E_α as an ordinary vector bundle equipped with a flat connection.) Throughout, we use the geometric definition of homology with local coefficients $H_*(X; E_\alpha)$ given in [Hat, pg. 330–336] which does not require a choice of basepoint; it is equivalent to the more algebraic definition of, e.g., [Hat, pg. 328–330]. More generally, if X is not connected, we can consider a bundle $E \to X$ with fiber V and the associated homology $H_*(X; E)$. We also use the geometrically defined cohomology $H^*(X; E)$ of [Hat, pg. 333]. Of course, both $H_*(X; E)$ and $H^*(X; E)$ satisfy all the usual properties: a relative version for (X, A), long exact sequence of a pair, Mayer-Vietoris, etc.

If X is a compact oriented n-manifold with ∂X partitioned into two submanifolds with common boundary A and B then one has Poincaré duality:

$$D_M\colon H^k(X, A; E) \xrightarrow{\cong} H_{n-k}(X, B; E) \tag{2.5}$$

where D_M is given by cap product with the ordinary relative fundamental class $[X, \partial X] \in H_n(X, \partial X; \mathbb{Z})$.

Let $E^* \to X$ denote the bundle where we have replaced each fiber with its dual vector space; for E_α, this corresponds to using the dual or contragredient representation $\alpha^*\colon \pi_1(X) \to \mathrm{GL}(V^*)$ defined by $\alpha^*(g) = \left(\rho(g^{-1})\right)^*$. When X has finitely many cells, the relevant version of universal coefficients is that $H_k(X; E) \cong H^k(X; E^*)$ as K-vector spaces.

When $E^* \cong E$ as bundles over X, we say that E is self-dual. Examples include E_α where $\alpha\colon \pi_1(X) \to \mathrm{SL}_2 K$; specifically, α^* is conjugate to α via $\left(\begin{smallmatrix} 0 & 1 \\ -1 & 0 \end{smallmatrix}\right)$. Seen another way, the action of $\mathrm{SL}_2 K$ on K^2 preserves the standard symplectic form $x_1 y_2 - x_2 y_1$ and hence E_α has a nondegenerate inner product on each fiber allowing us to identify E_α with E_α^*. Representations that are unitary with respect to some involution on K may not be self-dual, but still satisfy $H_*(X, A; E_\alpha) \cong H^*(X, A; E_\alpha)$, as K-vector spaces, for any $A \subset X$; we call such representations/bundles *homologically self-dual*.

3 BASICS OF TWISTED HOMOLOGY PRODUCTS

Throughout this section, E will be a system of local coefficients over a sutured manifold M with fiber a vector space of dimension $n \geq 1$. As in the introduction, we say that M is an *E-homology product* if the inclusion induced maps $H_*(R_\pm; E) \to H_*(M; E)$ are both isomorphisms. This is equivalent to the notion of an *E-cohomology product* where $H^*(M; E) \to H^*(R_\pm; E)$ are isomorphisms: the former is the same as $H_*(M, R_\pm; E) = 0$, the latter is the same as $H^*(M, R_\pm; E) = 0$, and by Poincaré duality one has $H_k(M, R_\pm; E) \cong H^{3-k}(M, R_\mp; E)$. These concepts are parallel to [FK2], where they consider unitary representations of balanced sutured manifolds where $H_1(M, R_-; E_\alpha) = 0$ because of:

Proposition 3.1. *Suppose M is a connected balanced sutured manifold with both $R_\pm$ nonempty. If E is homologically self-dual, then M is an E-homology product if and only if any one of the following eight groups vanish: $H_k(M, R_\pm; E)$ and $H^k(M, R_\pm; E)$ for $1 \leq k \leq 2$.*

Proof. Since both of $R_\pm$ are nonempty, it follows that $H_0(M, R_\pm; E) = H^0(M, R_\pm; E) = 0$, and so by Poincaré duality we have $H_3(M, R_\mp; E) = 0$. We focus on the case where $H_1(M, R_-; E) = 0$; the other cases are similar. Since M is balanced, we have $\chi(R_-) = \chi(M)$ and hence $\chi\big(H_*(M, R_-; E)\big) = 0$. Since we know that $H_k(M, R_-; E) = 0$ for every $k \neq 2$, this forces $H_2(M, R_-; E) = 0$ as well. By Poincaré, we have $H^*(M, R_+; E) = 0$. Since E is homologically self-dual, this gives $H_*(M, R_+; E) = 0$, and so M is an E-homology product as claimed. $\qquad\square$

Our motivation for studying twisted homology products is the following two results:

Theorem 3.2 (FK2, §3). *Suppose M is an irreducible sutured manifold which is an E-homology product and where no component of M is a solid torus without sutures. Then M is taut.*

Theorem 3.3 (FK2, §4). *Suppose X is a compact irreducible 3-manifold with ∂X a (possibly empty) union of tori. For $\phi \in H^1(X; \mathbb{Z})$ nontrivial and $\alpha\colon \pi_1(X) \to \mathrm{GL}(V)$, the torsion polynomial $\tau(X, \phi, \alpha)$ gives a sharp lower bound on the Thurston* *norm $\|\phi\|$ if and only if when S is a taut surface without nugatory tori dual to ϕ the sutured manifold M which is X cut along S is an α-homology product.*

Here, a set of torus components of a taut surface S are *nugatory* if they collectively bound a submanifold of X disjoint from ∂X. Theorem 3.2 is explicit and Theorem 3.3 is implicit in Sections 3 and 4 of [FK2] respectively; however, to make this paper more self-contained, we include proofs of both results.

Let S be a properly embedded compact surface in a 3-manifold N; we do not assume either S or N is connected, and N is allowed to be noncompact and have boundary. We say S *separates* N *into* N_+ *and* N_- if $N = N_+ \cup_S N_-$, the positive side of S is contained in N_+, the negative side of S is contained in N_-, and every component of $N_\pm$ meets S. The linchpin for Theorems 3.2 and 3.3 is the following lemma, where all homology groups are with respect to some system E of local coefficients on N, and all maps on homology are induced by inclusion:

Lemma 3.4. *Suppose S separates N into $N_\pm$. If both $H_*(N_\pm) \to H_*(N)$ are surjective then so are $H_*(S) \to H_*(N_\pm)$ and $H_*(S) \to H_*(N)$. Moreover, if for some k both $H_k(N_\pm) \to H_k(N)$ are isomorphisms then so are $H_k(S) \to H_k(N_\pm)$ and $H_k(S) \to H_k(N)$.*

Proof. Since both $H_*(N_\pm) \to H_*(N)$ are surjective, the Mayer-Vietoris sequence for $N = N_+ \cup_S N_-$ splits into short exact sequences

$$0 \to H_k(S) \xrightarrow{i_+ \oplus i_-} H_k(N_+) \oplus H_k(N_-) \xrightarrow{j_+ - j_-} H_k(N) \to 0. \qquad (3.5)$$

To see that $H_k(S) \to H_k(N_+)$ is surjective, take $c_+ \in H_k(N_+)$ and choose $c_- \in H_k(N_-)$ which maps to the same element in $H_k(N)$ as c_+; then $(c_+, c_-) \mapsto 0$ under $j_+ - j_-$ and hence c_+ is the image of some element of $H_k(S)$ by exactness of (3.5). Symmetrically, $H_k(S) \to H_k(N_-)$ is also surjective, proving the first part of the lemma.

Suppose in addition that both $H_k(N_\pm) \cong H_k(N)$. Since S is compact and $H_k(S)$ surjects $H_k(N_\pm)$ and $H_k(N)$, it follows that all four K-vector spaces are finite-dimensional. Since $H_k(N_\pm) \cong H_k(N)$, exactness of (3.5) forces $H_k(S) \cong H_k(N)$, and hence the surjections $H_k(S) \to H_k(N_\pm)$ must be isomorphisms as claimed. $\qquad \square$

We now show that a sutured manifold which is a homology product must be taut.

Proof of Theorem 3.2. We may assume that M is connected. All homology groups will have coefficients in E unless otherwise indicated, and let n be the dimension of the fiber of E. We first reduce to the case where every component of $R_\pm$ has $\chi \leq 0$. If a component of $R_\pm$ is a sphere, then M must be D^3 by irreducibility with (say) $R_+ = \partial M$ and $R_- = \varnothing$. Since E must be trivial over D^3, we get that $\dim(H_0(M)) = n$. However, $H_0(R_-) = 0$ contradicting that M is an E-homology product. If some component of $R_\pm$ is a disc, say $D \subset R_+$, then $\dim(H_0(D)) = n$ and since a connected space will have H_0 of dimension at most n, we conclude that $\dim(H_0(M)) = n$. However, then E must be the trivial bundle, since nontrivial monodromy around some loop would reduce $\dim(H_0(M))$ below n. It follows that M is also a homology product with respect to $H_*(\cdot\,; K)$, and hence $R_\pm$ are both connected and thus discs; by irreducibility, M is D^3 with one suture and hence taut. So from now on we assume that every component of $R_\pm$ has $\chi \leq 0$.

Since we have excluded M from being a solid torus with no sutures, all of the torus components of $R_\pm$ are incompressible. Thus to prove that M is taut it remains to show that $R_\pm$ realize the Thurston norm of their common class in $H_2(M, N(\gamma); \mathbb{Z})$. Note this is automatic if $R_+ = \varnothing$ since the homology product condition implies $\chi(R_-) = 0$, so from now on we assume both $R_\pm$ are nonempty. Suppose S is any other surface in that homology class. Throwing away components of S that bound submanifolds of M that are disjoint from ∂M, we can assume that S separates M into $M_\pm$, where each $M_\pm$ contains $R_\pm$ respectively. We next show that the theorem follows from:

Claim 3.6. *The maps $H_k(R_\pm) \to H_k(M_\pm)$ are isomorphisms for $k \neq 1$ and injective for $k = 1$. The maps $H_k(S) \to H_k(M_\pm)$ are isomorphisms for $k \neq 1$ and surjective for $k = 1$.*

From the claim we get that $H_k(S) \cong H_k(R_\pm)$ for $k \neq 1$ and $\dim H_1(S) \geq \dim H_1(R_\pm)$; hence

$$n \cdot \chi(S) = \chi\big(H_*(S)\big) \leq \chi\big(H_*(R_\pm)\big) = n \cdot \chi(R_\pm)$$

and so

$$\chi_-(S) \geq -\chi(S) \geq -\chi(R_\pm) = \chi_-(R_\pm).$$

Thus $R_\pm$ must realize the Thurston norm in its class, establishing the proposition modulo Claim 3.6.

To prove the claim, first note that S, $R_\pm$, and $M_\pm$ are all homotopy equivalent to 2-complexes and so we need only consider $k \leq 2$. Since $R_\pm \hookrightarrow M$ gives isomorphisms on H_*, we know $H_*(R_\pm) \to H_*(M_\pm)$ is injective and $H_*(M_\pm) \to H_*(M)$ is surjective. Since every component of $M_\pm$ meets $R_\pm$, it follows that $H_0(R_\pm) \to H_0(M_\pm)$ is onto and hence an isomorphism; consequently, so is $H_0(M_\pm) \to H_0(M)$. Since $H_*(M, R_\pm) = 0$, the long exact sequence of the triple $(M, M_\pm, R_\pm)$ gives that $H_2(M_\pm, R_\pm) \cong H_3(M, M_\pm)$; by excision and Poincaré duality, we have $H_3(M, M_\pm) \cong H_3(M_\mp, S) \cong H^0(M_\mp, R_\mp)$ and the latter vanishes since each component of $M_\mp$ meets $R_\mp$. Thus we have shown $H_2(M_\pm, R_\pm) = 0$, and hence $H_2(R_\pm) \to H_2(M_\pm)$ is an isomorphism.

By Lemma 3.4, we know that each $H_*(S) \to H_*(M_\pm)$ is surjective and moreover is an isomorphism for $* = 0$. To see that $H_2(S) \to H_2(M_\pm)$ is an injection (and hence an isomorphism), just note that $H_3(M_\pm, S) \cong H^0(M_\pm, R_\pm) = 0$. This proves the claim and thus the theorem. $\qquad\square$

The last part of this section is devoted to proving the relationship between the homology product condition and the Thurston norm bounds coming from twisted torsion/Alexander polynomials.

Proof of Theorem 3.3. Let $n = \dim V$. All homology groups will have coefficients in E_α. Let $\widetilde{X}$ denote the infinite cyclic cover of X corresponding to ϕ; it has the structure of a $\mathbb{Z}$'s worth of copies of M stacked end to end so that the R_+ on one block is glued to the R_- on the next. (Note that if ϕ is not primitive then $\widetilde{X}$ is disconnected; you can reduce to the case of ϕ primitive to avoid this issue if you prefer.) Let $\widetilde{S}$ be a lift of S to $\widetilde{X}$ corresponding to the top of a preferred copy of M in $\widetilde{X}$, and note that $\widetilde{S}$ separates $\widetilde{X}$ into $\widetilde{X}_+$ and $\widetilde{X}_-$ which consist of the blocks "above" and "below" S respectively.

Unwinding the definitions, the precise form of the lower bound given in Theorem 14 of [FV1] (which is Theorem 6.6 in the arXiv version) is equivalent to

$$\|\phi\| \geq \frac{1}{n}\left(\dim H_1(\widetilde{X}) - \dim H_0(\widetilde{X}) - \dim H_2(\widetilde{X})\right) \tag{3.7}$$

where if $H_1(\widetilde{X})$ is infinite-dimensional the convention is to declare the right-hand side as 0. (When $H_1(\widetilde{X})$ is finite-dimensional so is $H_2(\widetilde{X})$, see, e.g., [FV1].)

The only if direction is easy: if M is an α-homology product, the Mayer-Vietoris sequence and the fact that homology is compactly supported imply that $\widetilde{S} \hookrightarrow \widetilde{X}$ gives an isomorphism on H_*; thus one has

$$\|\phi\| = \chi_-(S) \geq -\chi(S) = -\frac{1}{n}\chi\big(H_*(S)\big) = -\frac{1}{n}\chi\big(H_*(\widetilde{X})\big) = \text{RHS of } (3.7) \tag{3.8}$$

where we have used that $H_3(\widetilde{X})$ must be 0 since $\widetilde{X}$ is noncompact.

Conversely, suppose that (3.7) is sharp. We will show:

Claim 3.9. *The maps* $H_*(\widetilde{S}) \to H_*(\widetilde{X}_\pm) \to H_*(\widetilde{X})$ *are all isomorphisms.*

The claim implies the theorem as follows: if we take $\widetilde{X}'_-$ to be $\widetilde{X}_-$ shifted down by one, we have $\widetilde{X} = \widetilde{X}'_- \cup_{\widetilde{S}'} M \cup_{\widetilde{S}} \widetilde{X}_+$. Applying the Mayer-Vietoris sequence to this decomposition, the claim gives that $H_*(M) \to H_*(\widetilde{X})$ is an isomorphism. Again by the claim, the inclusions of $\widetilde{S} = R_+$ and $\widetilde{S}' = R_-$ into M induce isomorphisms on H_*, and so M is a homology product.

To prove the claim, begin by noting that $H_*(\widetilde{X})$ is finitely generated, and the $\mathbb{Z}$-action on $\widetilde{X}$ can take any particular generating set to one which lies entirely in $\widetilde{X}_+$; hence $H_*(\widetilde{X}_+) \to H_*(\widetilde{X})$ is onto, as is $H_*(\widetilde{X}_-) \to H_*(\widetilde{X})$. By Lemma 3.4, we know $H_*(\widetilde{S}) \to H_*(\widetilde{X}_\pm)$ is onto and an isomorphism when $* = 2$ since $H_3(\widetilde{X}, \widetilde{X}_\pm) \cong H_3(\widetilde{X}_\mp, \widetilde{S}) \cong 0$ since (each component of) $\widetilde{X}_\mp$ is noncompact. For $* = 0$, we can build a compact subset A of $\widetilde{X}_\pm$ so that $H_0(A) \to H_0(\widetilde{X}_\pm)$ is onto and $H_0(A) \to H_0(\widetilde{X})$ is an isomorphism; consequently, $H_0(\widetilde{X}_\pm) \to H_0(\widetilde{X})$ is an isomorphism and hence so is $H_0(S) \to H_0(\widetilde{X}_\pm)$ by Lemma 3.4. Finally, from (3.8), we see that $\chi(H_*(S)) = \chi(H_*(X))$ and hence the surjection $H_1(S) \to H_1(X)$ must be an isomorphism, proving the claim and thus the theorem. $\qquad\square$

4   SOME HOMOLOGY PRODUCTS

This section is devoted to the proof of Conjecture 1.4 in two nontrivial cases, both of which include many examples which are not $\mathbb{Q}$-homology products:

Theorem 4.1. *Let M be a taut sutured manifold which is a book of I-bundles. Suppose $\alpha \colon \pi_1(M) \to \mathrm{SL}_2\mathbb{C}$ has $\mathrm{tr}\left(\alpha(\gamma)\right) \neq 2$ for every curve γ which is the core of a gluing annulus for an I-bundle page. Then M is an α-homology product.*

Theorem 4.2. *Suppose M is a sutured manifold which is a genus 2 handlebody with suture set γ a single curve separating ∂M into two once-punctured tori. If the pared manifold (M, γ) is acylindrical and $M \setminus \gamma$ is incompressible, then M is a homology product with respect to some $\alpha \colon \pi_1(M) \to \mathrm{SL}_2\mathbb{C}$.*

4.3 Books of I-bundles

Recall that a *book of I-bundles* is a 3-manifold built from solid tori (the *bindings*) and I-bundles over possibly nonorientable compact surfaces (the *pages*) glued in the following way. For a page P which is an I-bundle over a surface S, the *vertical annuli* are the components of the preimage of ∂S. One is allowed to glue such a vertical annulus to any homotopically essential annulus in the boundary of the binding. We do not require that all vertical annuli are glued; those that are not are called *free*. For a page P, the *vertical boundary* $\partial_v P$ is the union of all the vertical annuli; the *horizontal boundary* $\partial_h P$ is $\partial P \setminus \partial_v P$. We say a sutured manifold is a book of I-bundles if the underlying manifold has such a description where the sutures are exactly the cores of the free vertical annuli.

Lemma 4.4. *If M is a taut sutured manifold which is a book of I-bundles, then it has such a structure where all the pages are product I-bundles. If the base surface of a page P is not an annulus, then one component of the horizontal boundary is contained in R_+ and the other contained in R_-. The cores of the vertical annuli in the alternate description are homotopic to those in the original one.*

Proof. Suppose some page P is a twisted I-bundle over a connected nonorientable surface S. Then the horizontal boundary $\partial_h P$ is connected and hence contained entirely in one of $R_\pm$, say R_+. Then $(R_+ \setminus \partial_h P) \cup \partial_v P$ is a surface homologous to R_+ with Euler characteristic $\chi(R_+) - 2\chi(S)$. Since R_+ is taut, we must have that S is a Möbius band. The pair $(P, \partial_v P)$ is homeomorphic to a solid torus B with an annulus that represents twice a generator of $\pi_1(B)$. Thus we can replace P with a product bundle over the annulus to which we have attached a new component of the binding.

If a page P is a product I-bundle over an orientable surface S, the same argument shows that if $\partial_h P$ is contained in just one of R_+ and R_- then the base surface must be an annulus. This proves the lemma. $\square$

The proof of Theorem 4.1 rests on the following simple observation.

Lemma 4.5. *Suppose $\alpha \colon \pi_1(S^1) \to \mathrm{SL}_2\mathbb{C}$ is such that $\mathrm{tr}\left(\alpha(\gamma)\right) \neq 2$ where γ is a generator of $\pi_1(S^1)$. Then $H_*(S^1; E_\alpha) = 0$.*

Proof. As with any space, $H_0(S^1; E_\alpha)$ is the set of co-invariants of α, that is, the quotient of $\mathbb{C}^2$ by $\{\alpha(g)v - v \mid g \in \pi_1(S^1), v \in \mathbb{C}^2\}$. If $\alpha(\gamma)$ is diagonalizable, then this is 0 since neither eigenvalue of $\alpha(\gamma)$ can be 1 by the trace condition; alternatively, if $\alpha(\gamma)$ is parabolic then by the trace assumption it is conjugate to $\left(\begin{smallmatrix} -1 & 1 \\ 0 & -1 \end{smallmatrix}\right)$

and again the co-invariants vanish. Since $0 = 2\chi(S^1) = \chi\big(H_*(S^1; E_\alpha)\big)$ it follows that $H_1(S^1; E_\alpha) = 0$ as well, proving the lemma. $\qquad\square$

We next establish the first main result of this section.

Proof of Theorem 4.1. As usual, all homology will have coefficients in E_α. Consider the decomposition of M into $B \cup_A P$, where B is the binding, P is the union of all the pages, and A is the union of attaching annuli. By Lemma 4.4, we can assume that $P = (S \times [-1, 1]) \cup Y$ where $S \times \{\pm 1\} \subset R_+$ and Y is a union of (annulus) $\times I$. By our hypothesis on α, Lemma 4.5 implies that $H_*(A) = 0$ and $H_*(Y) = 0$. Moreover, $H_*(B) = 0$ since the generator of π_1(component of B) has a power which has $\operatorname{tr}(\alpha) \neq 2$ and hence must have $\operatorname{tr}(\alpha) \neq 2$ as well. Set $B' = B \cup Y$ and let $A' \subset A$ be the interface between B' and $S \times [-1, 1]$. Applying Mayer-Vietoris to the decomposition $M = B' \cup_{A'} (S \times [-1, 1])$ immediately gives that $H_*(S \times [-1, 1]) \to H_*(M)$ is an isomorphism. The same reasoning shows that $H_*(S \times \pm 1) \to H_*(R_\pm)$ are isomorphisms. Combining, we get that $H_*(R_\pm) \to H_*(M)$ are isomorphisms, and so M is an α-homology product as claimed. $\qquad\square$

4.6 Acylindrical sutured handlebodies

We turn now to the proof of Theorem 4.2. The following is an immediate consequence of the results in [MFP].

Theorem 4.7. *Suppose M is a sutured manifold where each component of $R_\pm$ is a torus. If the interior of M has a complete hyperbolic metric of finite volume, then there exists a lift $\alpha\colon \pi_1(M) \to \mathrm{SL}_2\mathbb{C}$ of its holonomy representation so that $H_*(M; E_\alpha) = 0$ and $H_*(R_\pm; E_\alpha) = 0$. In particular, M is an α-homology product.*

Proof. By Lemma 3.9 of [MFP], there is a lift α of the holonomy representation so that for each component of ∂M there is some curve c with $\operatorname{tr}(\alpha(c)) = -2$. Corollary 3.6 of [MFP] now implies that $H^*(\partial M; E_\alpha) = 0$, and Theorem 0.1 of [MFP] then gives that $H^*(M; E_\alpha) = 0$ as well. Since E_α is self-dual, it follows that $H_*(\partial M; E_\alpha) = H_*(M; E_\alpha) = 0$; since $H_*(\partial M; E_\alpha) = H_*(R_-; E_\alpha) \oplus H_*(R_+; E_\alpha)$ we are done. $\qquad\square$

Lemma 4.8. *Let M be a sutured manifold and N be the sutured manifold resulting from attaching a 2-handle to M along a component of the suture set γ. Let E be a system of local coefficients on N. If N is an E-homology product then M is an $E|_M$-homology product.*

This is a natural result since if N is taut then so is M, though the converse is not always true.

Proof. Throughout, all homology is with coefficients in E. Let $\bar{R}_+ \subset \partial N$ be the extension of R_+ to the new sutured manifold N. Note that $\bar{R}_+ = R_+ \cup D^2$ and $N = M \cup (D^2 \times I)$. Consider the associated Mayer-Vietoris sequences and natural maps:

$$\begin{array}{ccccc}
\longrightarrow & H_k(S^1) & \longrightarrow & H_k(R_+) \oplus H_k(D^2) & \longrightarrow & H_k(\bar{R}_+) & \longrightarrow \\
& \downarrow & & \downarrow & & \downarrow & \\
\longrightarrow & H_k(S^1 \times I) & \longrightarrow & H_k(M) \oplus H_k(D^2 \times I) & \longrightarrow & H_k(N) & \longrightarrow
\end{array}$$

The leftmost vertical arrow is an isomorphism since it comes from a homotopy equivalence. The rightmost vertical arrow is an isomorphism by hypothesis. By the five lemma, the middle arrow must be an isomorphism; since it is the direct sum of the maps $H_k(R_+) \to H_k(M)$ and $H_k(D^2) \to H_k(D^2 \times I)$ we conclude that $H_k(R_+) \to H_k(M)$ is an isomorphism. The symmetric argument proves that $H_k(R_-) \to H_k(M)$ is an isomorphism for every k and so M is indeed an E-homology product. $\square$

Theorem 4.9. *Suppose that M is a sutured manifold so that each component of $R_\pm$ is a (possibly) punctured torus. If adding 2-handles to M along all the sutures results in a hyperbolic manifold, then there exists $\alpha \colon \pi_1(M) \to \mathrm{SL}_2\mathbb{C}$ so that M is an α-homology product.*

Proof. Let N be the result of adding 2-handles to the sutures of M. Let $\alpha \colon \pi_1(N) \to \mathrm{SL}_2\mathbb{C}$ be the lift of the holonomy representation of the hyperbolic structure on N given by Theorem 4.7. Applying Lemma 4.8 inductively shows that M is a homology product with respect to the induced representation $\pi_1(M) \to \mathrm{SL}_2\mathbb{C}$ as needed. $\square$

We can now prove the other main result of this section.

Proof of Theorem 4.2. By Theorem 4.9 it suffices to prove that the result M_γ of attaching a 2-handle to M along γ is hyperbolic. Being a handlebody, M is irreducible and atoroidal. Since ∂M is compressible and $M \setminus \gamma$ is incompressible, Theorems A, 1, and 2 of [EM] together imply that M_γ is irreducible, acylindrical, atoroidal, and has incompressible boundary (when applying Theorems 1 and 2, note that γ is separating, which is one of the special cases mentioned in the final paragraph of the statements of these results). Thus $\mathrm{int}(M_\gamma)$ has a complete hyperbolic metric of finite volume as needed. $\square$

Remark 4.10. The representation α given in the proof of Theorem 4.2 may seem a bit unnatural since it is reducible on $\pi_1(R_\pm)$. However, it can be perturbed to β for which M is still a homology product and where β is parabolic free on $\pi_1(M)$ and hence faithful. The point is just that the set of all such β is the complement of a countable union of *proper* Zariski closed subsets in the character variety $X(M) \cong \mathbb{C}^3$, and hence is dense in $X(M)$. Specifically, as discussed in Section 5, the locus where M is not a homology product is Zariski closed, as of course is the set where a fixed nontrivial $\gamma \in \pi_1(M)$ is parabolic.

5   AN EXAMPLE

Suppose M is a balanced sutured manifold which is homeomorphic to a genus 2 handlebody. Assuming that each of $R_\pm$ is connected, then either $R_\pm$ are both tori with one boundary component or both pairs of pants. In this section, we compute

$H^1(M, R_+; E_\alpha)$ in a specific example as α varies over the $SL_2\mathbb{C}$ character variety of $\pi_1(M)$, and so characterize the α for which M is an α-homology product. This leads to the proof of Theorem 5.7 which was discussed in the introduction.

5.1   Basic setup

Both $\pi_1(R_+)$ and $\pi_1(M)$ are free groups of rank two, say generated by $\langle x, y \rangle$ and $\langle a, b \rangle$ respectively; let $i_*\colon \pi_1(M) \to \pi_1(R_+)$ be the map induced by the inclusion $i\colon R_+ \hookrightarrow M$. For $w \in \langle x, y \rangle$ we denote its Fox derivatives in $\mathbb{Z}[\langle x, y \rangle]$ by $\partial_x w$ and $\partial_y w$, where

$$\partial_x x = 1, \quad \partial_x x^{-1} = -x^{-1}, \quad \partial_x y^{\pm 1} = 0, \quad \text{and} \quad \partial_x(w_1 \cdot w_2) = \partial_x w_1 + w_1 \cdot \partial_x w_2.$$

Now fix a representation $\alpha\colon \pi_1(M) \to GL(V)$ where $\dim(V) = 2$, and extend to a ring homomorphism $\alpha\colon \mathbb{Z}[\pi_1(M)] \to \mathrm{End}(V)$.

Proposition 5.2. *The sutured manifold M is an α-homology product precisely when the 4×4 matrix*

$$\begin{pmatrix} \alpha\big(\partial_x i_*(a)\big) & \alpha\big(\partial_y i_*(a)\big) \\ \alpha\big(\partial_x i_*(b)\big) & \alpha\big(\partial_y i_*(b)\big) \end{pmatrix}$$

has nonzero determinant.

Proof. Consider the 2-complex W with one vertex v, four edges e_x, e_y, e_a, e_b, and two faces r_a, r_b with attaching maps specified by the words $i_*(a) \cdot a^{-1}$ and $i_*(b) \cdot b^{-1}$. For the subcomplex $B = e_a \cup e_b$, there is a map $j\colon (W, B) \to (M, R_+)$ which induces homotopy equivalences $W \to M$ and $B \to R_+$ corresponding to the natural maps on fundamental groups ($[e_x] \mapsto x$, $[e_a] \mapsto a$, etc.). By the long exact sequence of the pair and the five lemma, it follows that j_* induces an isomorphism $H^*(M, R_+; E_\alpha) \to H^*(W, B; E_{\alpha \circ j_*})$.

   By Proposition 3.1, to show M is an α-homology product, it remains to show $H^1(W, B; E_{\alpha \circ j_*}) = 0$. As a left module over $\Lambda = \mathbb{Z}[\langle x, y \rangle]$, the chain complex of the universal cover $\widetilde{W}$ of W has the form:

$$C_*(\widetilde{W}; \mathbb{Z})\colon \quad 0 \to \Lambda r_a \oplus \Lambda r_b \xrightarrow{\partial_2} \Lambda e_x \oplus \Lambda e_y \oplus \Lambda e_a \oplus \Lambda e_b \xrightarrow{\partial_1} \Lambda v \to 0.$$

Since Λ is noncommutative, it is most natural to write the matrices $[\partial_i]$ for the left module maps ∂_i so that they act on row vectors to their left, that is, $\partial_i(v) = v \cdot [\partial_i]$. In this form, we have the following, where we have denoted $i_*(a)$ and $i_*(b)$ in $\langle x, y \rangle$ by just a and b:

$$[\partial_1] = \begin{pmatrix} x - 1 \\ y - 1 \\ a - 1 \\ b - 1 \end{pmatrix} \qquad [\partial_2] = \begin{pmatrix} \partial_x a & \partial_y a & -1 & 0 \\ \partial_x b & \partial_y b & 0 & -1 \end{pmatrix}$$

Applying the functor $\mathrm{Hom}(\,\cdot\,, V_\alpha)$ to get $C^*(W; E_{\alpha \circ j_*})$ has the effect of replacing each copy of Λ with V, where the matrices of the coboundary maps d^i are the

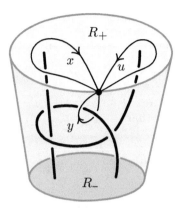

Figure 5.3: The sutured manifold M sketched at left is D^3 with open neighborhoods of the two dark arcs removed, where R_+ and R_- are the pairs of pants indicated. The manifold M is homeomorphic to a handlebody, with $\pi_1(M)$ freely generated by the loops x and y; the element u in $\pi_1(M)$ is $yxyx^{-1}y^{-1}$. These claims can be checked by a straightforward calculation starting with a Reidemeister-like presentation for $\pi_1(M)$.

result of applying $\alpha\colon \Lambda \to \operatorname{End}(V)$ entrywise to the $[\partial_i]$; here the matrices $[d^i]$ act on column vectors to their right. Restricting to the subcomplex of cochains vanishing on B gives:

$$C^*(W, B; E_{\alpha\circ j_*})\colon \quad 0 \leftarrow V^2 \xleftarrow{d^1} V^2 \leftarrow 0 \leftarrow 0$$

where d^1 is precisely the matrix in the statement of the proposition; the result follows. □

5.4 Pants example

Let M be the sutured manifold shown in Figure 5.3, where the free group $\pi_1(R_+)$ has generators

$$\pi_1(R_+) = \langle x, \, yxyx^{-1}y^{-1} \rangle.$$

Let $X(M)$ be the $\mathrm{SL}_2\mathbb{C}$ character variety of $\pi_1(M) = \langle x, y \rangle$. Now $X(M) \cong \mathbb{C}^3$ with coordinates $\{\bar{x}, \bar{y}, \bar{z}\}$ corresponding to the trace functions of $\{x, y, xy\}$. Despite the fact that M is a product with respect to ordinary $\mathbb{Z}$ homology, we will show:

Theorem 5.5. *The locus L of $[\alpha] \in X(M)$ where M is not an α-homology product is a (complex) 2-dimensional plane, namely $\{\bar{x} + \bar{y} - \bar{z} = 3\}$.*

Remark 5.6. Unlike for irreducible representations, characters $[\alpha] \in X(M)$ consisting of *reducible* representations may contain nonconjugate representations. For such classes, there is thus ambiguity in which local system E to associate with $[\alpha]$. However, it turns out that whether M is an E-homology product is independent of this choice. Similar to [DFJ, Lemma 7.1], the point is that reducible representations with the same character share the same diagonal part and one uses this with Proposition 5.2 to verify the claim; since our focus is on irreducible representations, we leave the details to the interested reader.

Proof. By Proposition 5.2, we are interested in when

$$\det \begin{pmatrix} \alpha(1) & 0 \\ \alpha(y - yxyx^{-1}) & \alpha(1 + yx - yxyx^{-1}y^{-1}) \end{pmatrix} = 0$$

or equivalently when $\det(\alpha(w)) = 0$ for $w = 1 + xy - xyx^{-1} \in \mathbb{Z}[\langle x, y \rangle]$. Any irreducible α can be conjugated so that

$$\alpha(x) = \begin{pmatrix} 0 & 1 \\ -1 & \bar{x} \end{pmatrix} \quad \text{and} \quad \alpha(y) = \begin{pmatrix} \bar{y} & -u \\ u^{-1} & 0 \end{pmatrix} \quad \text{where } u + u^{-1} = \bar{z}.$$

Applying this α to w and eliminating variables yields that $\det(\alpha(w)) = 0$ if and only if $\bar{x} + \bar{y} - \bar{z} - 3 = 0$; thus L is as claimed. $\qquad\square$

One representation in L is $(\bar{x}, \bar{y}, \bar{z}) = (4, 4, 5)$ which can be realized by

$$\alpha(x) = \begin{pmatrix} 1 & 1 \\ 2 & 3 \end{pmatrix} \quad \text{and} \quad \alpha(y) = \begin{pmatrix} 1 & -2 \\ -1 & 3 \end{pmatrix}.$$

An easy calculation shows that the axes of these hyperbolic elements cross in $\mathbb{H}^2$; since $\alpha(xyx^{-1}y^{-1})$ is also hyperbolic with negative trace, it follows that $\alpha(\langle x, y \rangle)$ is a Fuchsian Schottky group [Pur]. In particular, α is discrete, faithful, and purely hyperbolic. This proves:

Theorem 5.7. *There exists a taut sutured manifold M with a faithful discrete and purely hyperbolic representation $\alpha\colon \pi_1(M) \to \mathrm{SL}_2\mathbb{C}$ where M is not an α-homology product. The manifold M is acylindrical with respect to the pared locus consisting of the sutures.*

Remark 5.8. Representations that cover the same homomorphism $\pi_1(M) \to \mathrm{PSL}_2\mathbb{C}$ need not give rise to isomorphic cohomology. For example, the Schottky representation above covers the same $\mathrm{PSL}_2\mathbb{C}$ representation as β where $(\bar{x}, \bar{y}, \bar{z}) = (-4, 4, -5)$, which is not in L, and hence M is a β-homology product. In fact, in this example, every irreducible representation to $\mathrm{PSL}_2\mathbb{C}$ has *some* lift to $\mathrm{SL}_2\mathbb{C}$ for which M is a homology product.

Remark 5.9. For each $N \geq 2$, the group $\mathrm{SL}_2\mathbb{C}$ has a unique irreducible N-dimensional complex representation, which we denote $\iota_N\colon \mathrm{SL}_2\mathbb{C} \to \mathrm{SL}_N\mathbb{C}$. Let L_N be the locus of $[\alpha]$ in $X(M)$ where M is not an $\iota_N \circ \alpha$ homology product. A straightforward calculation with Gröbner bases finds:

$$L_3 = \left\{ 2\bar{x}\bar{y}\bar{z} - \bar{x}^2 - \bar{y}^2 - 3\bar{z}^2 + 3 = 0 \right\}$$

$$L_4 = \big\{ 3\bar{x}^2\bar{y}^2\bar{z} - 3\bar{x}^2\bar{y}\bar{z}^2 - 3\bar{x}\bar{y}^2\bar{z}^2 + \bar{x}^4 - 2\bar{x}^3\bar{y} - 2\bar{x}\bar{y}^3 + \bar{y}^4 + 2\bar{x}^3\bar{z}$$

$$+ 3\bar{x}^2\bar{y}\bar{z} + 3\bar{x}\bar{y}^2\bar{z} + 2\bar{y}^3\bar{z} - 3\bar{x}\bar{y}\bar{z}^2 + 2\bar{x}\bar{z}^3 + 2\bar{y}\bar{z}^3 + \bar{z}^4 - 3\bar{x}^3 - 3\bar{y}^3$$

$$+ 3\bar{z}^3 - 3\bar{x}^2 + 6\bar{x}\bar{y} - 3\bar{y}^2 - 6\bar{x}\bar{z} - 6\bar{y}\bar{z} - 3\bar{z}^2 + 6\bar{x} + 6\bar{y} - 6\bar{z} + 9 = 0 \big\}.$$

The intersection $L_2 \cap L_3 \cap L_4$ is zero-dimensional, as one would expect from the intersection of three (complex) surfaces in $\mathbb{C}^3$. Computing out a bit farther, we found that $\bigcap_{N=2}^5 L_N = \bigcap_{N=2}^{10} L_N$ contains a single point $(\bar{x}, \bar{y}, \bar{z}) = (2, 2, 1)$ outside the reducible representations; in particular, there are no purely hyperbolic representations in this intersection.

6   LIBROID SEIFERT SURFACES

In this last section, we study libroid knots, a notion generalizing fibered knots and fibroid surfaces which is defined in Section 6.3 below. We will show that this is a large class of knots for which Conjecture 1.1 holds:

Theorem 6.1. *All special arborescent knots, except the $(2, n)$–torus knots, are hyperbolic libroid knots. Moreover, there are infinitely many hyperbolic libroid knots whose ordinary Alexander polynomial is trivial.*

Theorem 6.2. *Conjecture 1.1 holds for libroid hyperbolic knots in S^3.*

6.3   Library sutured manifolds

We call a taut sutured manifold $(M, R_\pm, \gamma)$ a *library* if there is a taut surface $(\Sigma, \partial\Sigma) \subset (M, N(\gamma))$ such that $[\Sigma] = n[R_+] \in H_2(M, N(\gamma); \mathbb{Z})$ for some $n \geq 0$, and the sutured manifold $M \setminus \Sigma$ is a book of I-bundles in the sense of Section 4.3. Note that $M \setminus \Sigma$ has at least $n+1$ connected components, and thus is a collection of books of I-bundles, that is, a "library." We say that a taut surface $S \subset X^3$ is a *libroid surface* if $X \setminus S$ is a library sutured manifold. This generalizes the notion of a *fibroid surface* [CS], and in fact the surface $S \cup \Sigma$ is a fibroid surface. We say that a knot in S^3 is libroid if it has a minimal genus Seifert surface which is libroid. Definitions in hand, we now deduce Theorem 6.2 from Theorem 4.1.

Proof of Theorem 6.2. Let K be a libroid knot with X its exterior, and let $\alpha\colon \pi_1(X) \to \mathrm{SL}_2\mathbb{C}$ be a lift of the holonomy representation of the hyperbolic structure on X. Let S be a minimal genus Seifert surface for K which is libroid. By Theorem 3.3, we just need to show that the sutured manifold $M = X \setminus S$ is an α-homology product. This is immediate if M is a product, so we will assume from now on that X is not fibered. Let $\{\Sigma_i\}$ be disjoint minimal genus Seifert surfaces cutting M up into sutured manifolds that are each a book of I-bundles; for notational convenience, set $\Sigma_0 = S$. It is enough to show that each such book B is an α-homology product, since they are stacked one atop another to form M. To apply Theorem 4.1, we need to check that no core γ of a gluing annulus has $\mathrm{tr}\left(\alpha(\gamma)\right) = 2$. Assume γ is such a core, so in particular $\alpha(\gamma)$ is parabolic.

First note that γ is isotopic to an essential curve in some Σ_i. Since Σ_i is minimal genus and not a fiber, by Fenley [Fen] it is a quasi-Fuchsian surface in X and in particular the only embedded curve in Σ_i whose image under α is parabolic is $\partial\Sigma_i$, which is the homological longitude $\lambda \in \pi_1(\partial X)$. But by [Cal, Corollary 2.6] or [MFP, Corollary 3.11], one always has $\mathrm{tr}\left(\alpha(\lambda)\right) = -2$, which contradicts that $\alpha(\gamma)$ has trace $+2$. So we can apply Theorem 4.1 as desired, proving the theorem.   □

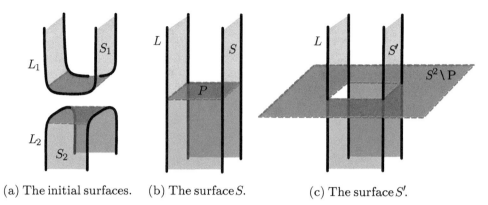

(a) The initial surfaces.   (b) The surface S.   (c) The surface S'.

Figure 6.5: The Murasugi sum with $k = 2$.

6.4   A plethora of libroid knots

We now turn to showing that there are many hyperbolic libroid knots. A key tool for this will be the notion of Murasugi sum, which we quickly review. Consider two oriented surfaces with boundary S_1 and S_2 in S^3, and let $L_i = \partial S_i \subset S^3$ be the associated links. Suppose that S_1 and S_2 intersect so that there is a sphere $S^2 \subset S^3$ with $S^3 = B_1 \cup_{S^2} B_2$, so that $S_i \subset B_i$; see Figure 6.5(a, b), where the interface between B_1 and B_2 is a horizontal plane separating S_1 and S_2, which have been pulled apart slightly for clarity. Moreover, assume that $S_1 \cap S_2 = P \subset S^2$ is a $2k$-sided polygon, where the edges of ∂P are cyclically numbered so that the odd edges lie in L_1, and the even edges lie in L_2. Also, assume that the orientations of S_1 and S_2 agree on P. Let $L = \partial(S_1 \cup S_2)$ be the link obtained as a boundary of the union of the two surfaces. Then L is said to be obtained by *Murasugi sum* from L_1 and L_2. If $k = 1$, this is connected sum, and if $k = 2$, then this operation is known as *plumbing*. There are two natural Seifert surfaces for L shown in Figure 6.5(b, c), given by $S = S_1 \cup S_2$, and $S' = \left((S_1 \cup S_2) - P\right) \cup \overline{(S^2 - P)}$. Note that S' is also a Murasugi sum of the surfaces $(S_i - P) \cup \overline{S^2 - P}$, which are isotopic to S_i.

Gabai showed that if each S_i is minimal genus, then so is S; similarly if each S_i is a fiber, then so is S [Gab1]. We generalize these results to:

Lemma 6.6. *If S_1 and S_2 are libroid surfaces, and S is obtained from S_1 and S_2 by Murasugi sum, then S is also a libroid surface.*

Proof. The Seifert surfaces S and S' for L can be disjointly embedded as sketched in Figure 6.7. In detail, take a regular neighborhood $N(L)$, and form the exterior $E(L) = \overline{S^3 - N(L)}$. We'll use the notation above for Murasugi sum. Then $S^2 \cap E(L)$ is a $2k$-punctured sphere, dividing $E(L)$ into tangle complements $T_i = E(L) \cap B_i$. Take a regular neighborhood R_{3-i} of $\overline{S_i - P} \cap T_i$ inside T_i; then the relative boundary of R_{3-i} in T_i is two parallel copies of $\overline{S_i - P}$. The union with $S^2 - (R_1 \cup R_2)$ gives our two disjointly embedded Seifert surfaces $S \cup S'$.

The complements $S^3 - S_i$, $S^3 - S$, and $S^3 - S'$ naturally admit sutured manifold structures as described in Section 4 of [Sak]. Moreover, the two complementary regions $S^3 - (S \cup S')$ may be identified with $(S^3 - S_i) \cup R_i$, where R_i is the product sutured manifold described above, and R_i is attached to $S^3 - S_i$ along k product disks in the sutures corresponding to $R_i \cap S^2$ (recall k is defined by $S_1 \cap S_2 = P$ is

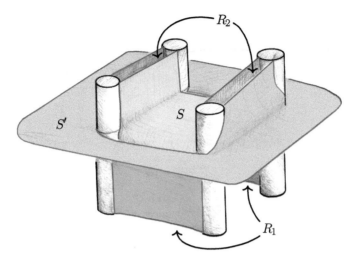

Figure 6.7: The surfaces S and S' made disjoint.

a $2k$-gon). But $S^3 - S_i$ is a library sutured manifold, which may be extended as products into R_i to obtain a library decomposition of $S^3 - (S \cup S')$. Thus S and S' are libroid Seifert surfaces for L. □

Remark 6.8. The sutured manifold decomposition in the above proof is the same as that in [Sak, Condition 4.2]; while we first decompose along $S \cup S'$ and then along the $2k$ product disks and remove the product sutured manifolds R_i, Sakuma first decomposes along S and then along the disk $S^2 - P$, resulting in the union of the sutured manifolds $S^3 - S_i$.

The class of *arborescent links* are those obtained by plumbing together twisted bands in a tree-like pattern (see, e.g., [Gab2, BS] for a definition). It is important to note that the bands are allowed to have an odd number of twists. With a few known exceptions, these links are hyperbolic (see [BS] or [FG, Theorem 1.5]). The subclass of *special arborescent links* studied by Sakuma [Sak] are those obtained by plumbing bands with even numbers of twists, and hence the plumbed surface is a Seifert surface for the link. Inductively applying Lemma 6.6 shows that all *special* arborescent knots are libroid. A famous family of non-special arborescent knots are the Kinoshita-Terasaka knots; to complete the proof of Theorem 6.1, it suffices to show:

Theorem 6.10. *The Kinoshita-Terasaka knots $KT_{2,n}$ shown in Figure 6.9(a) are libroid hyperbolic knots with trivial ordinary Alexander polynomial.*

Proof. These knots are hyperbolic since they are arborescent and not one of the exceptional cases, and their Alexander polynomials were calculated in [KT]. Minimal genus Seifert surfaces were found by Gabai [Gab2, §5]; we review his construction to verify that these knots are libroid.

Let L be the $(3, -2, 2, -3)$–pretzel link shown in Figure 6.9(b); a Seifert surface S for one orientation of L is shown in Figure 6.9(c). The surface S is a twice-punctured torus, and hence taut since $S^3 \setminus L$ is hyperbolic. The $KT_{2,n}$ knot can be obtained

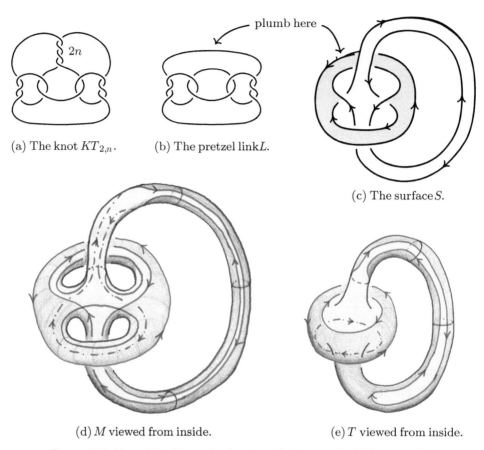

(a) The knot $KT_{2,n}$.   (b) The pretzel link L.

(c) The surface S.

(d) M viewed from inside.   (e) T viewed from inside.

Figure 6.9: Kinoshita-Terasaka knots and the proof of Theorem 6.10.

by plumbing a band with $2n$-twists onto S in the location shown, so by Lemma 6.6 it suffices to prove that the complement of S is a book of I-bundles.

Thickening S to a handlebody, we get the picture in Figure 6.9(d); the outside of this handlebody is the sutured manifold M we seek to understand. Each short red curve meets the long blue oriented sutures in two points and bounds an obvious disk in M. These are product discs in the sense of [Gab2], so we decompose along them to get the sutured manifold T which is the *exterior* of the solid torus shown in Figure 6.9(e). Note that T is a solid torus with four sutures that each wind once around in the core direction. In particular T is taut and hence so is M; moreover, thinking backwards to build M from T by reattaching the product discs shows that M is a book of I-bundles with a single binding which is basically T.   □

REFERENCES

[A]   I. Agol. Criteria for virtual fibering. *J. Topol.* **1** (2008), 269–284. arXiv: 0707.4522, MR2399130.

[AHT] I. Agol, J. Hass, and W. Thurston. The computational complexity of knot genus and spanning area. *Trans. Amer. Math. Soc.* **358** (2006), 3821–3850. arXiv:math/0205057, MR2219001.

[BS] F. Bonahon and L. C. Siebenmann. New Geometric Splittings of Classical Knots and the Classification and Symmetries of Arborescent knots. http://www-bcf.usc.edu/~fbonahon/.

[Cal] D. Calegari. Real places and torus bundles. *Geom. Dedicata* **118** (2006), 209–227. arXiv:math/0510416, MR2239457.

[CS] M. Culler and P. B. Shalen. Volumes of hyperbolic Haken manifolds. I. *Invent. Math.* **118** (1994), 285–329. MR1292114.

[DFJ] N. M. Dunfield, S. Friedl, and N. Jackson. Twisted Alexander polynomials of hyperbolic knots. *Exp. Math.* **21** (2012), 329–352. arXiv:1108.3045, MR3004250.

[EM] M. Eudave Muñoz. On nonsimple 3-manifolds and 2-handle addition. *Topology Appl.* **55** (1994), 131–152. MR1256216.

[Fen] S. R. Fenley. Quasi-Fuchsian Seifert surfaces. *Math. Z.* **228** (1998), 221–227.
MR1630563.

[FK1] S. Friedl and T. Kim. Twisted Alexander norms give lower bounds on the Thurston norm. *Trans. Amer. Math. Soc.* **360** (2008), 4597–4618. arXiv:math/0505682, MR2403698.

[FK2] S. Friedl and T. Kim. Taut sutured manifolds and twisted homology. *Math. Res. Lett.* **20** (2013), 289–303. arXiv:1209.0254, MR3151648.

[FV1] S. Friedl and S. Vidussi. A survey of twisted Alexander polynomials. In *The mathematics of knots*, volume 1 of *Contrib. Math. Comput. Sci.*, pages 45–94. Springer, Heidelberg, 2011. arXiv:0905.0591, MR2777847.

[FV2] S. Friedl and S. Vidussi. The Thurston norm and twisted Alexander polynomials. *J. Reine Angew. Math.* **707** (2015), 87–102. arXiv:1204.6456, MR3403454.

[FG] D. Futer and F. Guéritaud. Angled decompositions of arborescent link complements. *Proc. Lond. Math. Soc. (3)* **98** (2009), 325–364. arXiv:math/0610775, MR2481951.

[Gab1] D. Gabai. The Murasugi sum is a natural geometric operation. In *Low-dimensional topology (San Francisco, Calif., 1981)*, volume 20 of *Contemp. Math.*, pages 131–143. Amer. Math. Soc., Providence, RI, 1983. MR718138.

[Gab2] D. Gabai. Genera of the arborescent links. *Mem. Amer. Math. Soc.* **59** (1986), i–viii and 1–98. MR823442.

[Hat] A. Hatcher. *Algebraic topology.* Cambridge University Press, Cambridge, 2002. MR1867354.

[Juh] A. Juhasz. Knot Floer homology and Seifert surfaces. *Algebr. Geom. Topol.* **8** (2008), 603–608. arXiv:math/0702514, MR2443240.

[KT] S. Kinoshita and H. Terasaka. On unions of knots. *Osaka Math. J.* **9** (1957), 131–153. MR0098386.

[Kup] G. Kuperberg. Knottedness is in NP, modulo GRH. *Adv. Math.* **256** (2014), 493–506. arXiv:1112.0845, MR3177300.

[Lac] M. Lackenby. The efficient certification of knottedness and Thurston norm. arXiv:1604.00290.

[Liu] Y. Liu. Virtual cubulation of nonpositively curved graph manifolds. *J. Topol.* **6** (2013), 793–822. arXiv:1110.1940, MR3145140.

[MFP] P. Menal-Ferrer and J. Porti. Twisted cohomology for hyperbolic three manifolds. *Osaka J. Math.* **49** (2012), 741–769. arXiv:1001.2242, MR2993065.

[Mor] T. Morifuji. On a conjecture of Dunfield, Friedl and Jackson. *C. R. Math. Acad. Sci. Paris* **350** (2012), 921–924. arXiv:1209.4239, MR2990904.

[MT] T. Morifuji and A. T. Tran. Twisted Alexander polynomials of 2-bridge knots for parabolic representations. *Pacific J. Math.* **269** (2014), 433–451. arXiv:1301.1101, MR3238485.

[PW1] P. Przytycki and D. T. Wise. Graph manifolds with boundary are virtually special. *J. Topol.* **7** (2014), 419–435. arXiv:1110.3513, MR3217626.

[PW2] P. Przytycki and D. T. Wise. Mixed 3-manifolds are virtually special. *J. Amer. Math. Soc.* **31** (2018), 319–347. arXiv:1205.6742, MR3758147.

[Pur] N. Purzitsky. Two-generator discrete free products. *Math. Z.* **126** (1972), 209–223. MR0346070.

[Sak] M. Sakuma. Minimal genus Seifert surfaces for special arborescent links. *Osaka J. Math.* **31** (1994), 861–905. MR1315011.

[Sch] M. Scharlemann. Sutured manifolds and generalized Thurston norms. *J. Differential Geom.* **29** (1989), 557–614. MR992331.

[Wis] D. T. Wise. The structure of groups with a quasiconvex hierarchy. Preprint 2012, 189 pages.

The Profinite Completion of 3-Manifold Groups, Fiberedness and the Thurston Norm

MICHEL BOILEAU AND STEFAN FRIEDL

Dedicated to the memory of William Thurston.

1   INTRODUCTION

In this paper we will study the question: what properties of 3-manifolds are determined by the set of finite quotients of their fundamental groups? The modern reformulation of the above question (see Lemma 2.2 for details) is to ask: what properties of a 3-manifold N are determined by the profinite completion $\widehat{\pi_1(N)}$ of its fundamental group? Here, and throughout the paper, all 3-manifolds are understood to be compact, orientable, and connected with empty or toroidal boundary.

Wilton–Zalesskii [WZ17] showed that the profinite completion of the fundamental group of a closed 3-manifold can detect whether or not it is hyperbolic, and they showed that it can detect whether or not it is Seifert fibered. On the other hand it is known by work of Funar [Fu13, Proposition 5.1] and Hempel [He14] that the profinite completion of the fundamental group can not always distinguish between pairs of torus bundles and between certain pairs of Seifert manifolds. It is still an open question though whether the profinite completion can distinguish any two hyperbolic 3-manifolds.

In this paper we are mostly interested in the relation between the profinite completion, fiberedness and the Thurston norm of a 3-manifold. We quickly recall the relevant definitions. Given a surface Σ with connected components $\Sigma_1, \ldots, \Sigma_k$ its complexity is defined to be

$$\chi_-(\Sigma) := \sum_{i=1}^{d} \max\{-\chi(\Sigma_i), 0\}.$$

Given a 3-manifold N and $\phi \in H^1(N; \mathbb{Z})$ the Thurston norm is defined as

$$x_N(\phi) := \min\{\chi_-(\Sigma) \mid \Sigma \subset N \text{ properly embedded and dual to } \phi\}.$$

Thurston [Th86] showed that x_N is a seminorm on $H^1(N; \mathbb{Z})$ and that x_N extends to a seminorm on $H^1(N; \mathbb{R})$. We also recall that an integral class $\phi \in H^1(N; \mathbb{Z}) = \text{Hom}(\pi_1(N), \mathbb{Z})$ is called *fibered* if there exists a fibration $p \colon N \to S^1$ such that

$\phi = p_*\colon \pi_1(N) \to \mathbb{Z}$. More generally, we say that a class $\phi \in H^1(N;\mathbb{R})$ is *fibered* if ϕ can be represented by a nowhere-vanishing closed 1–form. By [Ti70] the two notions of being fibered coincide for integral cohomology classes.

In the following let N_1 and N_2 be two 3-manifolds and suppose there exists an isomorphism $f\colon \widehat{\pi_1(N_1)} \to \widehat{\pi_1(N_2)}$ of the profinite completions of the fundamental groups. Such an isomorphism induces an isomorphism $\widehat{H_1(N_1;\mathbb{Z})} \to \widehat{H_1(N_2;\mathbb{Z})}$ of the profinite completions of the homology groups. It is straightforward to see that this implies that $H_1(N_1;\mathbb{Z})$ and $H_1(N_2;\mathbb{Z})$ are abstractly isomorphic, but in general the isomorphism $\widehat{H_1(N_1;\mathbb{Z})} \to \widehat{H_1(N_2;\mathbb{Z})}$ is not induced by an isomorphism of the homology groups. Since we want to compare the Thurston norm and the fibered classes of N_1 and N_2 it is convenient to assume that $\widehat{H_1(N_1;\mathbb{Z})} \to \widehat{H_1(N_2;\mathbb{Z})}$ is in fact induced by an isomorphism $H_1(N_1;\mathbb{Z}) \to H_1(N_2;\mathbb{Z})$.

This leads us to the following definition: we say that an isomorphism $f\colon \widehat{\pi_1} \to \widehat{\pi_2}$ between profinite completions of two groups π_1 and π_2 is *regular* if the induced isomorphism $\widehat{H_1(\pi_1;\mathbb{Z})} \to \widehat{H_1(\pi_2;\mathbb{Z})}$ is induced by an isomorphism $H_1(\pi_1;\mathbb{Z}) \to H_1(\pi_2;\mathbb{Z})$. This isomorphism is then necessarily uniquely determined by f and by a slight abuse of notation we also denote it by f. (It is quite straightforward to see that the isomorphism provided by Funar [Fu13] is regular, whereas the isomorphisms given by Hempel [He14] are not necessarily regular.)

It follows from the Sphere Theorem that a 3-manifold, with the adjectives mentioned above, is aspherical if and only if it is prime, has infinite fundamental group and is not homeomorphic to $S^1 \times S^2$.

Now we can formulate our main theorem.

Theorem 1.1. *Let N_1 and N_2 be two aspherical 3-manifolds with empty or toroidal boundary. Suppose $f\colon \widehat{\pi_1(N_1)} \to \widehat{\pi_1(N_2)}$ is a regular isomorphism. Let $\phi \in H^1(N_2;\mathbb{R})$. Then*

$$x_{N_2}(\phi) = x_{N_1}(f^*\phi).$$

Furthermore

$$\phi \in H^1(N_2;\mathbb{R}) \text{ is fibered} \quad \Longleftrightarrow \quad f^*\phi \in H^1(N_1;\mathbb{R}) \text{ is fibered.}$$

The proof rests on the fact that 3-manifold groups are good in the sense of Serre [Se97] and that profinite completions contain enough informations on certain twisted Alexander polynomials which by [FV13, FV15, FN15] determine fiberedness and the Thurston norm. Both types of results rely on the recent work of Agol [Ag08, Ag13], Przytycki–Wise [PW18] and Wise [Wi09, Wi12a, Wi12b].

Remark. Let N_1 and N_2 be two aspherical 3-manifolds with empty or toroidal boundary. If there exists an isomorphism $f\colon \widehat{\pi_1(N_1)} \to \widehat{\pi_1(N_2)}$ that is not necessarily regular one can not compare the Thurston norms of specific classes. But one can still ask whether N_1 is fibered if and only if N_2 is fibered. The methods presented below in the knot case will answer the question in the affirmative if $b_1(N_1) = b_1(N_2) = 1$. We do not know how to address the question if $b_1(N_1) = b_1(N_2) > 1$.

Remark. Recently Jaikin-Zapirain [JZ18] proved the following variation on Theorem 1.1: If M and N are two compact orientable 3-manifolds such that the profinite completions of the fundametal groups are isomorphic, then M is fibered if and only if N is fibered.

For the remainder of this introduction we consider the special case of knot complements. Given a knot $K \subset S^3$ we denote by $X(K) = S^3 \setminus \nu K$ its exterior and we denote by $\pi(K) := \pi_1(S^3 \setminus \nu K)$ the corresponding knot group. We say that K is fibered if the knot exterior $X(K)$ is a surface bundle over S^1. Furthermore, we refer to the minimal genus of a Seifert surface for K as the genus $g(K)$ of K.

Our main theorem for knots is the following variation on Theorem 1.1.

Theorem 1.2. *If J and K are two knots such that the profinite completions of their groups are isomorphic, then $g(J) = g(K)$. Furthermore, J is fibered if and only if K is fibered.*

This theorem is not an immediate corollary to Theorem 1.1 since we do not assume that the isomorphism of the profinite completions is regular.

It is also natural to ask, to what degree does the profinite completion of knot groups distinguish knots? Here, and throughout the paper, we say that two knots J and K are *equivalent* if there exists a diffeomorphism h of S^3 with $h(J) = K$. Evidently the profinite completion of knot groups can only determine knots which are already determined by their groups. Whitten [Wh87] showed that prime knots are determined by their groups. The following question thus arises.

Question 1.3. *Let J and K be two prime knots with $\widehat{\pi(J)} \cong \widehat{\pi(K)}$. Does it follow that J and K are equivalent?*

It is straightforward to see that the profinite completion detects the unknot. For completeness' sake we provide the proof in Lemma 4.8. Also, in many real-life situations it can be quite easy to distinguish two given knots by their finite quotients.

Using Theorem 1.2 we quickly obtain the following corollary.

Corollary 1.4. *Let J be the trefoil or the figure-8 knot. If K is a knot with $\widehat{\pi(J)} \cong \widehat{\pi(K)}$, then J and K are equivalent.*

More generally we prove the following result for torus knots.

Theorem 1.5. *Let J be a torus knot. If K is a knot with $\widehat{\pi(J)} \cong \widehat{\pi(K)}$, then J and K are equivalent.*

Two knots in the 3-sphere are said commensurable if their exteriors have homeomorphic finite-sheeted covers. If these homeomorphic finite-sheeted covers are cyclic covers, then the knots are said cyclically commensurable. The following result shows that cyclically commensurable hyperbolic knots are distinguished by their finite quotients when their Alexander polynomials are not a product of cyclotomic polynomials.

Theorem 1.6. *Let K_1 and K_2 be two cyclically commensurable hyperbolic knots. If the Alexander polynomial of K_1 has at least one zero which is not a root of unity and $\widehat{\pi(K_1)} \cong \widehat{\pi(K_2)}$, then K_1 and K_2 are equivalent.*

A hyperbolic knot K has no hidden symmetries if the commensurator of $\pi(K)$ in $PSL(2,\mathbb{C})$ coincides with its normalizer. The following corollary is a straightforward consequence of Theorem 1.6 above and the fact that two hyperbolic knots in S^3 without hidden symmetries are commensurable iff they are cyclically commensurable, see [BBCW12, Theorem 1.4].

Corollary 1.7. *Let K_1 and K_2 be two commensurable hyperbolic knots in S^3 without hidden symmetries. If the Alexander polynomial of K_1 has at least one zero which is not a root of unity and $\widehat{\pi(K_1)} \cong \widehat{\pi(K_2)}$, then K_1 and K_2 are equivalent.*

Remark.

1. Theorem 1.6 applies in particular to the pretzel knot K of type $(2,3,7)$. It is known as the Fintushel-Stern knot K and admits two lens space surgeries. So its exterior is cyclically covered by the exteriors of two distinct knots. It is a fibered knot with Alexander polynomial the Lehmer polynomial $\Delta_K = 1 + t - t^3 - t^4 - t^5 - t^6 - t^7 + t^9 + t^{10}$ which is the integral polynomial of smallest known Mahler measure (see Section 4.6 for the definition) and which has two real roots, one being the Salem number $1.17628\ldots$.
2. Currently, the only hyperbolic knots known to admit hidden symmetries are the figure-8 and the two dodecahedral knots of Aitchison and Rubinstein, see [AR92]. By [Re91] the figure-8 is the only knot in S^3 with arithmetic complement, hence it is the unique knot in its commensurability class. The two dodecahedral knots are commensurable, but one is fibered and the other one not, so it follows from Theorem 1.2 that their groups cannot have the same profinite completion.

The paper is structured as follows. In Section 2 we recall the definition and some basic facts on profinite completions. In Section 3 we recall the definition of twisted homology and cohomology groups and of twisted Alexander polynomials of 3-manifolds. In Section 4 we will relate profinite completions to the degrees of certain twisted Alexander polynomials. This will then allow us to prove Theorem 1.1 and a slightly stronger version of Theorem 1.2. In Section 4.5 we prove Theorem 1.5 about torus knots and in Section 4.6 we prove Theorem 1.6 about cyclically commensurable hyperbolic knots.

Recently there has been great interest in determining which properties of 3-manifolds are determined by the profinite completion of the fundamental group. We refer to the work of Martin Bridson and Alan Reid [BR15] and Martin Bridson, Alan Reid and Henry Wilton [BRW17] for results which have a certain overlap with our work. We also refer to the work of Gareth Wilkes [Wk17, Wk18a, Wk18b, Wk19] for more results.

Convention

Unless it says specifically otherwise, all groups are assumed to be finitely generated, all modules are assumed to be finitely generated, all manifolds are assumed to be orientable, connected and compact, and all 3-manifolds are assumed to have empty or toroidal boundary. Finally, p will always be a prime and $\mathbb{F}_p$ denotes the field with p coefficients.

Acknowledgment

We wish to thank Baskar Balasubramanyam, Nathan Dunfield, Marc Lackenby, José Pedro Quintanilha, Jacob Rasmussen, Alan Reid, Henry Wilton and Pavel Zalesskii for helpful conversations and feedback. Furthermore we want to thank the referee for quickly refereeing the paper and for making lots of helpful comments. The second author is grateful for the hospitality at Keble College, IISER Pune, Université de Toulouse and Aix-Marseille Université. The second author was supported by the SFB 1085 "Higher Invariants" at the University of Regensburg, funded by the Deutsche Forschungsgemeinschaft (DFG). The first author was partially supported by ANR (projects 12-BS01-0003-01 and 12-BS01-0004-01). While working on revising the paper both authors benefited from the support and the hospitality of the Isaac Newton Institute for Mathematical Sciences during the program Homology Theories in Low Dimensional Topology supported by EPSRC Grant Number EP/KO32208/1.

2   THE PROFINITE COMPLETION OF A GROUP

In this section we recall several basic properties of profinite completions. Throughout this section we refer to [RZ10] and [Wi98] for details.

2.1   The Definition of the Profinite Completion of a Group

Given a group π we consider the inverse system $\{\pi/\Gamma\}_\Gamma$ where Γ runs over all finite index normal subgroups of π. The profinite completion $\widehat{\pi}$ of π is then defined as the inverse limit of this system, i.e.,

$$\widehat{\pi} = \varprojlim \pi/\Gamma.$$

Note that the natural map $\pi \to \widehat{\pi}$ is injective if and only if π is residually finite. It follows from [He87] and the proof of the Geometrization Conjecture that fundamental groups of 3-manifolds are residually finite.

When π is finitely generated, a deep result in [NS07] states that every finite index subgroup of $\widehat{\pi}$ is open. It means that $\widehat{\widehat{\pi}} = \widehat{\pi}$. Then the following is a consequence of the definitions, see [RZ10, Proposition 3.2.2] for details.

Lemma 2.1. *Let π be a finitely generated group. Then for any finite group G the map $\pi \to \widehat{\pi}$ induces a bijection $\mathrm{Hom}(\widehat{\pi}, G) \to \mathrm{Hom}(\pi, G)$.*

2.2   Groups with isomorphic profinite completions

In this section A and B are finitely generated groups. A group homomorphism $\varphi \colon A \to B$ induces a homomorphism $\widehat{\varphi} \colon \widehat{A} \to \widehat{B}$. Evidently, if φ is an isomorphism, then so is $\widehat{\varphi}$. On the other hand, an isomorphism $\phi \colon \widehat{A} \to \widehat{B}$ is not necessarily induced by a homomorphism $\varphi \colon A \to B$. There are even isomorphisms $\widehat{\mathbb{Z}} \to \widehat{\mathbb{Z}}$ that are not induced by an automorphism of $\mathbb{Z}$. Moreover, it follows from [NS07] that any homomorphism $\phi \colon \widehat{A} \to \widehat{B}$ is in fact continuous.

If $f \colon \widehat{A} \to \widehat{B}$ is an isomorphism, then it follows from Lemma 2.1 that for any finite group G we have bijections

$$\mathrm{Hom}(B, G) \leftarrow \mathrm{Hom}(\widehat{B}, G) \xrightarrow{f^*} \mathrm{Hom}(\widehat{A}, G) \rightarrow \mathrm{Hom}(A, G).$$

Given $\beta \in \mathrm{Hom}(B, G)$ we will, by a slight abuse of notation, denote by $\beta \circ f$ the resulting homomorphism from A to G. In particular given a representation $\beta \colon B \rightarrow \mathrm{GL}(k, \mathbb{F}_p)$ we obtain an induced representation $A \rightarrow \mathrm{GL}(k, \mathbb{F}_p)$ which we denote by $\beta \circ f$.

Given a group π we denote by $Q(\pi)$ the set of finite quotients of π. We just showed that finitely generated groups A and B with isomorphic profinite completions have the same finite quotients, i.e., the sets $Q(A)$ and $Q(B)$ are the same. Somewhat surprisingly the converse also holds, more precisely, by [RZ10, Corollary 3.2.8] the following lemma holds.

Lemma 2.2. *Two finitely generated groups A and B have isomorphic profinite completions if and only if $Q(A) = Q(B)$.*

3   TWISTED ALEXANDER POLYNOMIALS OF KNOTS

3.1   Definition of twisted homology and cohomology groups

Let X be a connected CW-complex. We write $\pi = \pi_1(X)$ and we denote by $p \colon \widetilde{X} \rightarrow X$ the universal covering of X. Note that π acts on $\widetilde{X}$ on the left via deck transformations. We can thus view $C_*(\widetilde{X})$ as a left $\mathbb{Z}[\pi]$-module. Let R be a commutative ring and let V be an R-module. Let $\alpha \colon \pi \rightarrow \mathrm{Aut}_R(V)$ be a representation. We henceforth view V as a left $\mathbb{Z}[\pi]$-module. Given any i we refer to

$$H^i_\alpha(X; V) := H_i\big(\mathrm{Hom}_{\mathbb{Z}[\pi]}\big(C_*(\widetilde{X}), V\big)\big)$$

as the *i-th twisted cohomology* of (X, α).

Using the standard involution $g \mapsto g^{-1}$ we can turn the $\mathbb{Z}[\pi]$-left module $C_*(\widetilde{X})$ into a right $\mathbb{Z}[\pi]$-module. Given any i we then refer to

$$H^\alpha_i(X; V) := H_i\big(C_*(\widetilde{X}) \otimes_{\mathbb{Z}[\pi]} V\big)$$

as the *i-th twisted homology* of (X, α).

3.2   Orders of modules

Let $\mathbb{F}$ be a field and let H be a finitely generated $\mathbb{F}[t^{\pm 1}]$-module. Since $\mathbb{F}[t^{\pm 1}]$ is a PID there exists an isomorphism $H \cong \oplus_{i=1}^n \mathbb{F}[t^{\pm 1}]/f_i(t)$ where $f_1(t), \ldots, f_n(t) \in \mathbb{F}[t^{\pm 1}]$. We refer to $\mathrm{ord}(H) := \prod_{i=1}^n f_i(t) \in \mathbb{F}[t^{\pm 1}]$ as the *order of H*. The order is well-defined up to multiplication by a unit in $\mathbb{F}[t^{\pm 1}]$. Furthermore it is non-zero if and only if H is an $\mathbb{F}[t^{\pm 1}]$-torsion module and it is a unit if and only if $H = 0$.

3.3   The definition of twisted Alexander polynomials

Let X be a CW-complex, let $\phi \in H^1(X; \mathbb{Z})$ and let $\alpha \colon \pi_1(X) \rightarrow \mathrm{GL}(k, \mathbb{F})$ be a representation over a field $\mathbb{F}$. We write $\pi = \pi_1(X)$ and $\mathbb{F}[t^{\pm 1}]^k := \mathbb{F}^k \otimes_{\mathbb{Z}} \mathbb{Z}[t^{\pm 1}]$ and we denote by $\alpha \otimes \phi$ the tensor representation

$$\alpha \otimes \phi \colon \pi \to \mathrm{Aut}_{\mathbb{F}[t^{\pm 1}]}(\mathbb{F}[t^{\pm 1}]^k)$$

$$g \mapsto \left(\begin{array}{c} \mathbb{F}[t^{\pm 1}]^k \to \mathbb{F}[t^{\pm 1}]^k \\ \sum_i v_i \otimes p_i(t) \mapsto \sum_i \alpha(g)(v_i) \otimes t^{\phi(g)} p_i(t) \end{array} \right).$$

This allows us to view $\mathbb{F}[t^{\pm 1}]^k$ as a left $\mathbb{Z}[\pi]$-module. We then consider the twisted homology groups $H_i^{\alpha \otimes \phi}(X; \mathbb{F}[t^{\pm 1}]^k)$ which are naturally $\mathbb{F}[t^{\pm 1}]$-modules. Given $i \in \mathbb{N}$ we denote by $\Delta^{\alpha}_{X,\phi,i} \in \mathbb{F}[t^{\pm 1}]$ the order of the $\mathbb{F}[t^{\pm 1}]$-module $H_i^{\alpha \otimes \phi}(X; \mathbb{F}[t^{\pm 1}]^k)$ and we refer to it as the *i-th twisted Alexander polynomial* of (X, ϕ, α). We refer to the original papers [Li01, Wa94, Ki96, KL99] and the survey papers [FV10, DFL15] for more information on twisted Alexander polynomials.

The twisted Alexander polynomials are well-defined up to multiplication by a unit in $\mathbb{F}[t^{\pm 1}]$, i.e., up to multiplication by some at^k where $a \in \mathbb{F} \setminus \{0\}$ and $k \in \mathbb{Z}$. In the following, given $p, q \in \mathbb{F}[t^{\pm 1}]$ we write $p \doteq q$ if p and q agree up to multiplication by a unit in $\mathbb{F}[t^{\pm 1}]$.

For future reference we recall two lemmas about twisted Alexander polynomials. The first lemma is [FK06, Lemma 2.4].

Lemma 3.1. *Let X be a CW-complex, let $\phi \in H^1(X; \mathbb{Z})$ be non-zero and let $\alpha \colon \pi_1(X) \to \mathrm{GL}(k, \mathbb{F})$ be a representation over a field $\mathbb{F}$. Then the zeroth twisted Alexander polynomial $\Delta^{\alpha}_{X,\phi,0}$ is non-zero.*

Given a representation $\alpha \colon \pi \to \mathrm{GL}(k, \mathbb{F}_p)$ we denote by $\alpha^* \colon \pi \to \mathrm{GL}(k, \mathbb{F}_p)$ the representation which is given by $\alpha^*(g) := \alpha(g^{-1})^t$ for $g \in \pi$. The following lemma is [FK06, Proposition 2.5].

Lemma 3.2. *Let N be a 3-manifold, let $\phi \in H^1(N; \mathbb{Z})$ be non-zero and let $\alpha \colon \pi_1(N) \to \mathrm{GL}(k, \mathbb{F})$ be a representation over a field $\mathbb{F}$. Suppose that $\Delta^{\alpha}_{N,\phi,1}$ is non-zero. Then the following hold:*

1. *If N has non-trivial boundary, then $\Delta^{\alpha}_{N,\phi,2} \doteq 1$.*
2. *If N is closed, then $\Delta^{\alpha}_{N,\phi,2} \doteq \Delta^{\alpha^*}_{N,\phi,0}$.*

3.4   Twisted Alexander polynomials, fiberedness and the Thurston norm

Given a polynomial $f(t) = \sum_{k=r}^{s} a_k t^k \in \mathbb{F}[t^{\pm 1}]$ with $a_r \neq 0$ and $a_s \neq 0$ we define $\deg(f(t)) = s - r$. We extend this definition to the zero polynomial by setting $\deg(0) := +\infty$. It is clear that the degree of a twisted Alexander polynomial is well-defined, i.e., not affected by the indeterminacy in the definition.

Theorem 3.3. *Let N be a 3-manifold and let $\phi \in H^1(N; \mathbb{Z})$ be non-zero. Then the following hold:*

1. *Pick a prime p. The class ϕ is fibered if and only if for any representation $\alpha \colon \pi_1(N) \to \mathrm{GL}(k, \mathbb{F}_p)$ we have $\Delta^{\alpha}_{N,\phi,1} \neq 0$.*
2. *Pick a prime p. Let $\alpha \colon \pi_1(N) \to \mathrm{GL}(k, \mathbb{F}_p)$ be a representation such that $\Delta^{\alpha}_{N,\phi,1}$ is non-zero. Then the twisted Alexander polynomials $\Delta^{\alpha}_{N,\phi,i}$ are non-zero*

for $i = 0, 1, 2$ and

$$x_N(\phi) \geq \max\left\{0, \tfrac{1}{k}\left(-\deg\left(\Delta^\alpha_{N,\phi,0}\right) + \deg\left(\Delta^\alpha_{N,\phi,1}\right) - \deg\left(\Delta^\alpha_{N,\phi,2}\right)\right)\right\}.$$

3. *If N is aspherical, then there exists a prime p and a representation $\alpha \colon \pi_1(N) \to$ $\mathrm{GL}(k, \mathbb{F}_p)$ such that the twisted Alexander polynomials $\Delta^\alpha_{N,\phi,i}$ are non-zero for $i = 0, 1, 2$ and*

$$x_N(\phi) = \max\left\{0, \tfrac{1}{k}\left(-\deg\left(\Delta^\alpha_{N,\phi,0}\right) + \deg\left(\Delta^\alpha_{N,\phi,1}\right) - \deg\left(\Delta^\alpha_{N,\phi,2}\right)\right)\right\}.$$

Here the "only if" direction of (1) was proven by various authors, see, e.g., [Ch03, GKM05, FK06]. The "if" direction was proven in [FV13]. The inequality in (2) was proven in [FK06, Fr14]. Finally statement (3) is proven in [FN15] building on [FV15]. Here the "if" statement of (1) and the proof of (3) build on the work of Agol [Ag08, Ag13], Przytycki–Wise [PW18] and Wise [Wi09, Wi12a, Wi12b].

3.5   Degrees of twisted Alexander polynomials

Throughout the paper, given a group π, $\phi \in H^1(\pi; \mathbb{Z}) = \mathrm{Hom}(\pi, \mathbb{Z})$ and $n \in \mathbb{N}$ we denote by $\phi_n \colon \pi \xrightarrow{\phi} \mathbb{Z} \to \mathbb{Z}_n$ the composition of ϕ with the obvious projection map. Furthermore, given a representation $\alpha \colon \pi \to \mathrm{GL}(k, \mathbb{F})$ and $n \in \mathbb{N}$ we write $\mathbb{F}[\mathbb{Z}_n]^k = \mathbb{F}^k \otimes_{\mathbb{Z}} \mathbb{Z}[\mathbb{Z}_n]$ and we denote by $\alpha \otimes \phi_n \colon \pi \to \mathrm{Aut}(\mathbb{F}[\mathbb{Z}_n]^k)$ the representation which is defined in a completely analogous way as we defined $\alpha \otimes \phi$ above. Later on we will make use of the following proposition.

Proposition 3.4. *Let X be a CW-complex, let $\phi \in H^1(X; \mathbb{Z})$ be non-trivial and let $\alpha \colon \pi_1(X) \to \mathrm{GL}(k, \mathbb{F}_p)$ be a representation. Then the following equalities hold:*

$$\deg \Delta^\alpha_{X,\phi,0}(t) = \max\left\{\dim_{\mathbb{F}_p}\left(H_0^{\alpha \otimes \phi_n}(X; \mathbb{F}_p[\mathbb{Z}_n]^k)\right) \big| n \in \mathbb{N}\right\}$$

$$\deg \Delta^\alpha_{X,\phi,1}(t) = \max\left\{\dim_{\mathbb{F}_p}\left(H_1^{\alpha \otimes \phi_n}(X; \mathbb{F}_p[\mathbb{Z}_n]^k)\right)\right.$$
$$\left. - \dim_{\mathbb{F}_p}\left(H_0^{\alpha \otimes \phi_n}(X; \mathbb{F}_p[\mathbb{Z}_n]^k)\right) \big| n \in \mathbb{N}\right\}.$$

The proof of this proposition will require the remainder of this section. For simplicity we will henceforth write $\mathbb{F} = \mathbb{F}_p$, $\Lambda = \mathbb{F}[t^{\pm 1}]$ and for each $n \in \mathbb{N}$ we write $\Lambda_n = \mathbb{F}[t^{\pm 1}]/(t^n - 1)$. We first give several elementary lemmas.

Lemma 3.5. *For any Λ-module M we have $\dim_{\mathbb{F}}(M) = \deg(\mathrm{ord}(M))$.*

Proof. We first observe that for any polynomial $q(t) \in \Lambda$, not necessarily non-zero, we have $\dim_{\mathbb{F}}(\Lambda/q(t)\Lambda) = \deg(q(t))$. Since Λ is a PID the general case of the lemma follows immediately from the cyclic case. $\qquad\qquad\square$

Lemma 3.6. *Let M be a Λ-module. Then*

$$\dim_{\mathbb{F}}(M) = \sup\left\{\dim_{\mathbb{F}}\left(M \otimes_\Lambda \Lambda_n\right) \mid n \in \mathbb{N}\right\} \in \mathbb{N} \cup \{\infty\}.$$

Proof. We write $M = \Lambda^r \oplus T$ where T is a torsion Λ-module. First we consider the case that $r > 0$. In this case $\dim_{\mathbb{F}}(M) = \infty$. On the other hand, for any n we have

$$\dim_{\mathbb{F}}(M \otimes_{\Lambda} \Lambda_n) \geq \dim_{\mathbb{F}}(\Lambda^r \otimes_{\Lambda} \Lambda_n) = \dim_{\mathbb{F}}(\Lambda_n^r) = rn.$$

Thus we showed that the claimed equality holds if $r > 0$.

Now suppose that $r = 0$. For any n the ring epimorphism $\Lambda \to \Lambda_n$ induces an epimorphism $M = M \otimes_{\Lambda} \Lambda \to M \otimes_{\Lambda} \Lambda_n$. We thus see that for any n we have

$$\dim_{\mathbb{F}}(M) \geq \dim_{\mathbb{F}}(M \otimes_{\Lambda} \Lambda_n).$$

Since M is a finite abelian group there exists an n such that multiplication by t^n acts like the identity on M. Put differently, multiplication by $t^n - 1$ is the zero map. For such n it is straightforward to see that the map

$$M \otimes_{\Lambda} \Lambda_n \to M$$
$$m \otimes [q(t)] \to mq(t)$$

is a well-defined isomorphism of $\mathbb{F}$-modules. In particular $\dim_{\mathbb{F}}(M) = \dim_{\mathbb{F}}(M \otimes_{\Lambda} \Lambda_n)$. Together with the above inequality this implies the lemma. $\square$

Lemma 3.7. *Let C_* be a chain complex of modules over Λ such that $H_0(C_*)$ is Λ-torsion. Then*

$$H_0(C_* \otimes_{\Lambda} \Lambda_n) = H_0(C_*) \otimes_{\Lambda} \Lambda_n$$
$$H_1(C_* \otimes_{\Lambda} \Lambda_n) \cong H_1(C_*) \otimes_{\Lambda} \Lambda_n \oplus H_0(C_*) \otimes_{\Lambda} \Lambda_n.$$

Proof. By the universal coefficient theorem for chain complexes of modules over the PID Λ we have for any i that

$$H_i(C_* \otimes_{\Lambda} \Lambda_n) \cong H_i(C_*) \otimes_{\Lambda} \Lambda_n \oplus \mathrm{Tor}_{\Lambda}(H_{i-1}(C_*), \Lambda_n).$$

The lemma now follows easily from the definitions and from the fact that for any Λ-module H we have

$$\mathrm{Tor}_{\Lambda}(H, \Lambda_n) \cong \mathrm{Tor}_{\Lambda}(H) \otimes_{\Lambda} \Lambda_n. \qquad \square$$

Now we are finally in a position to prove Proposition 3.4.

Proof of Proposition 3.4. Let X be a CW-complex, let $\phi \in H^1(X; \mathbb{Z})$ be non-trivial and let $\alpha \colon \pi_1(X) \to \mathrm{GL}(k, \mathbb{F}_p)$ be a representation. As usual we denote by $\widetilde{X}$ the universal cover of X. We consider the Λ-chain complex

$$C_* := C_*(\widetilde{X}) \otimes_{\mathbb{Z}[\pi_1(X)]} \Lambda^k.$$

With this notation we have

$$H_i(X; \Lambda^k) = H_i(C_*) \text{ and } H_i(X; \Lambda_n^k) = H_i(C_* \otimes_{\Lambda} \Lambda_n).$$

(Here and throughout the proof we drop the representation in the notation for twisted homology groups.) By Lemma 3.1 we know that $H_0(X; \Lambda^k)$ is Λ-torsion. Thus in the following we can apply Lemma 3.7 to the chain complex C_*.

First we consider $\deg \Delta^\alpha_{X,\phi,0}(t)$. It follows from Lemmas 3.5, 3.6 and 3.7 that

$$
\begin{aligned}
\deg \Delta^\alpha_{X,\phi,0}(t) &= \dim_\mathbb{F} \left(H_0(X; \Lambda^k) \right) \\
&= \sup \left\{ \dim_\mathbb{F} \left(H_0(X; \Lambda^k) \otimes_\Lambda \Lambda_n \right) \mid n \in \mathbb{N} \right\} \\
&= \sup \left\{ \dim_\mathbb{F} \left(H_0(X; \Lambda^k_n) \right) \mid n \in \mathbb{N} \right\}.
\end{aligned}
$$

Now we turn to the proof of the second equality. It follows from applying Lemma 3.5 and 3.6 once and from applying Lemma 3.7 twice that

$$
\begin{aligned}
\deg \Delta^\alpha_{X,\phi,1}(t) &= \dim_\mathbb{F} \left(H_1(X; \Lambda^k) \right) \\
&= \sup \left\{ \dim_\mathbb{F} \left(H_1(X; \Lambda^k) \otimes_\Lambda \Lambda_n \right) \mid n \in \mathbb{N} \right\} \\
&= \sup \left\{ \dim_\mathbb{F} \left(H_1(X; \Lambda^k_n) \right) - \dim_\mathbb{F} \left(H_0(X; \Lambda^k) \otimes_\Lambda \Lambda_n \right) \mid n \in \mathbb{N} \right\} \\
&= \sup \left\{ \dim_\mathbb{F} \left(H_1(X; \Lambda^k_n) \right) - \dim_\mathbb{F} \left(H_0(X; \Lambda^k_n) \right) \mid n \in \mathbb{N} \right\}. \qquad \square
\end{aligned}
$$

4   THE PROFINITE COMPLETION AND TWISTED ALEXANDER POLYNOMIALS

4.1   Twisted homology groups and profinite completions

Given a group π we denote by π_i, $i \in I$ the inverse system given by all finite index normal subgroups. Given a discrete abelian group A and a representation $\alpha \colon \pi \to \mathrm{Aut}_\mathbb{Z}(A)$ we define

$$
H^i_\alpha(\widehat{\pi}; A) \; = \; \varprojlim H^i_\alpha(\pi/\pi_i; A^{\pi_i}),
$$

where $H^i_\alpha(\pi/\pi_i; A^{\pi_i})$ is the usual twisted cohomology of the finite group π/π_i. Following Serre [Se97, D.2.6 Exercise 2] we say that a group π is *good* if the following holds: for any representation $\alpha \colon \pi \to \mathrm{Aut}_\mathbb{Z}(A)$, where A is a finite abelian group, the inclusion $\iota \colon \pi \to \widehat{\pi}$ induces for any i an isomorphism $\iota^* \colon H^i_\alpha(\widehat{\pi}; A) \to H^i_\alpha(\pi; A)$.

The following theorem was first proved by Cavendish [Ca12, Section 3.5, Lemma 3.7.1]. We also refer to [AFW15, (H.26)] for an alternative approach which builds on [WZ10] and the work of Agol [Ag08], Przytycki–Wise [PW18] and Wise [Wi09, Wi12a, Wi12b].

Theorem 4.1. *All 3-manifold groups are good.*

The fact that 3-manifolds are good gives us the following useful corollary.

Corollary 4.2. *Let N_1, N_2 be aspherical 3-manifolds such that $\widehat{\pi_1(N_1)} \cong \widehat{\pi_1(N_2)}$. Then N_1 is closed if and only if N_2 is closed.*

Proof. By goodness, by the hypothesis that the manifolds N_1, N_2 are aspherical and by the hypothesis that $\widehat{\pi_1(N_1)} \cong \widehat{\pi_1(N_2)}$ we have isomorphisms

$$H^3(\pi_1(N_1); \mathbb{Z}_2) \xleftarrow{\cong} H^3(\widehat{\pi_1(N_1)}; \mathbb{Z}_2) \cong H^3(\widehat{\pi_1(N_2)}; \mathbb{Z}_2) \xrightarrow{\cong} H^3(\pi_1(N_2); \mathbb{Z}_2).$$

The corollary follows from the fact that a 3-manifold is closed if and only if the third cohomology group with $\mathbb{Z}_2$-coefficients is non-zero. □

The main goal of this section is to prove the following proposition.

Proposition 4.3. *Let π_1 and π_2 be good groups. Let $f\colon \widehat{\pi_1} \xrightarrow{\cong} \widehat{\pi_2}$ be an isomorphism. Let $\beta\colon \pi_2 \to \mathrm{GL}(k, \mathbb{F}_p)$ be a representation. Then for any i we have an isomorphism*

$$H_i^{\beta \circ f}(\pi_1; \mathbb{F}_p^k) \cong H_i^{\beta}(\pi_2; \mathbb{F}_p^k).$$

Before we give a proof of the proposition we need to formulate a lemma which relates twisted cohomology groups to twisted homology groups. Here we recall that given a representation $\alpha\colon \pi \to \mathrm{GL}(k, \mathbb{F})$ over a field $\mathbb{F}$ we denote by $\alpha^*\colon \pi \to \mathrm{GL}(k, \mathbb{F})$ the representation which is given by $\alpha^*(g) := \alpha(g^{-1})^t$ for $g \in \pi$.

Lemma 4.4. *Let π be a group and let $\gamma\colon \pi \to \mathrm{GL}(k, \mathbb{F})$ be a representation over a field $\mathbb{F}$. Then $H^i_{\gamma^*}(\pi; \mathbb{F}^k) \cong H_i^{\gamma}(\pi; \mathbb{F}^k)$ for any i.*

Proof. Let Y be a $K(\pi, 1)$. As usual we denote by $\widetilde{Y}$ the universal cover of Y. In the following we denote by $\mathbb{F}^k_{\gamma}$, respectively $\mathbb{F}^k_{\gamma^*}$, the vector space $\mathbb{F}^k$ together with the structure as a left $\mathbb{Z}[\pi]$-module induced by the representation γ, respectively γ^*. A direct calculation shows that

$$\mathrm{Hom}_{\mathbb{Z}[\pi]}\left(C_*(\widetilde{Y}), \mathbb{F}^k_{\gamma^*}\right) \to \mathrm{Hom}_{\mathbb{F}}\left(C_*(\widetilde{Y}) \otimes_{\mathbb{Z}[\pi]} \mathbb{F}^k_{\gamma}, \mathbb{F}\right)$$
$$(\varphi\colon C_i(\widetilde{Y}) \to \mathbb{F}^k) \mapsto \begin{pmatrix} C_i(\widetilde{Y}) \otimes_{\mathbb{Z}[\pi]} \mathbb{F}^k \to \mathbb{F} \\ \sigma \otimes v \mapsto \varphi(\sigma)^t v \end{pmatrix}$$

is an isomorphism of chain complexes of $\mathbb{F}$-vector spaces. It follows from this isomorphism and from the universal coefficient theorem applied to the chain complex $C_*(\widetilde{Y}) \otimes_{\mathbb{F}[\pi]} \mathbb{F}^k_{\gamma}$ and to the field $\mathbb{F}$ that for any i we have an isomorphism

$$H^i_{\gamma^*}(Y; \mathbb{F}^k) \cong \mathrm{Hom}_{\mathbb{F}}\left(H_i^{\gamma}(Y; \mathbb{F}^k), \mathbb{F}\right).$$

Now the claim follows from the fact that a finite-dimensional vector space is isomorphic to its dual vector space. □

Now we are in a position to prove Proposition 4.3.

Proof of Proposition 4.3. Let π_1 and π_2 be good groups. Let $f\colon \widehat{\pi_1} \xrightarrow{\cong} \widehat{\pi_2}$ be an isomorphism. Let $\beta\colon \pi_2 \to \mathrm{GL}(k, \mathbb{F}_p)$ be a representation. Since π_1 and π_2 are good we know that the inclusion maps $\pi_j \to \widehat{\pi}_j$, $j = 1, 2$ and the map f give us for any i isomorphisms

$$H^i_{\beta*}(\pi_2; \mathbb{F}^k_p) \xleftarrow{\cong} H^i_{\beta*}(\widehat{\pi}_2; \mathbb{F}^k_p) \xrightarrow{f^*} H^i_{\beta* \circ f}(\widehat{\pi}_1; \mathbb{F}^k_p) \xrightarrow{\cong} H^i_{\beta* \circ f}(\pi_1; \mathbb{F}^k_p).$$

But by Lemma 4.4 we also have

$$H^i_{\beta*}(\pi_2; \mathbb{F}^k_p) \cong H^\beta_i(\pi_2; \mathbb{F}^k_p) \text{ and } H^i_{\beta* \circ f}(\pi_1; \mathbb{F}^k_p) \cong H^{\beta \circ f}_i(\pi_1; \mathbb{F}^k_p).$$

The proposition follows from combining all these isomorphisms. □

4.2   Proof of Theorem 1.1

For the reader's convenience we recall the statement of Theorem 1.1.

Theorem 1.1. *Let N_1 and N_2 be two aspherical 3-manifolds. Suppose $f \colon \widehat{\pi_1(N_1)} \to \widehat{\pi_1(N_2)}$ is a regular isomorphism. Let $\phi \in H^1(N_2; \mathbb{Z})$. Then*

$$x_{N_2}(\phi) = x_{N_1}(f^*\phi).$$

Furthermore

$$(N_2, \phi) \text{ is fibered} \quad \Longleftrightarrow \quad (N_1, f^*\phi_1) \text{ is fibered}.$$

In the proof of the theorem we will need the following lemma.

Lemma 4.5. *Let N_1 and N_2 be two 3-manifolds. Suppose $f \colon \widehat{\pi_1(N_1)} \to \widehat{\pi_1(N_2)}$ is a regular isomorphism. Then for any non-trivial $\phi \in H^1(N_2)$ and any representation $\alpha \colon \pi_1(N_2) \to \mathrm{GL}(k, \mathbb{F}_p)$ and $i \in \{0, 1\}$ we have*

$$\deg\left(\Delta^{\alpha \circ f}_{N_1, \phi \circ f, i}\right) = \deg\left(\Delta^\alpha_{N_2, \phi, i}\right).$$

Furthermore, if N_1 and N_2 are aspherical and if $\Delta^{\alpha \circ f}_{N_1, \phi \circ f, 1}$ is non-zero, then we also have

$$\deg\left(\Delta^{\alpha \circ f}_{N_1, \phi \circ f, 2}\right) = \deg\left(\Delta^\alpha_{N_2, \phi, 2}\right).$$

Proof. We first point out that for any n we have $(\alpha \circ f) \otimes (\phi_n \circ f) = (\alpha \otimes \phi_n) \circ f$. Also, as usual the twisted homology groups in dimensions 0 and 1 only depend on the fundamental group. Together with Theorem 4.1 and Proposition 4.3 this implies that for any n and $i \in \{0, 1\}$ we have

$$\dim_{\mathbb{F}_p}\left(H^{\alpha \otimes \phi_n}_i(N_2; \mathbb{F}_p[\mathbb{Z}_n]^k)\right) = \dim_{\mathbb{F}_p}\left(H^{(\alpha \circ f) \otimes (\phi_n \circ f)}_i(N_1; \mathbb{F}_p[\mathbb{Z}_n]^k)\right).$$

For $i = 0, 1$ the equality of the lemma is now an immediate consequence of Proposition 3.4.

Now we assume that N_1 and N_2 are aspherical and that $\Delta^{\alpha \circ f}_{N_1, \phi \circ f, 1}$ is non-zero. By the equality of the degrees of the first twisted Alexander polynomials we also obtain that $\Delta^{\alpha \circ f}_{N_2, \phi \circ f, 2}$ is non-zero. The argument above also shows that

$$\deg \left(\Delta^{\alpha^* \circ f}_{N_1, \phi \circ f, 0} \right) = \deg \left(\Delta^{\alpha^*}_{N_2, \phi, 0} \right).$$

The desired equality of degrees now follows from Lemma 3.2 together with Corollary 4.2. □

Now we are finally in a position to prove Theorem 1.1.

Proof of Theorem 1.1. Let N_1 and N_2 be two aspherical 3-manifolds and suppose that we are given a regular isomorphism $f \colon \widehat{\pi_1(N_1)} \to \widehat{\pi_1(N_2)}$. Let $\phi \in H^1(N_2; \mathbb{R})$. We need to show the following two statements:

1. $x_{N_2}(\phi) = x_{N_1}(f^* \phi)$.
2. The class $\phi \in H^1(N_2; \mathbb{R})$ is fibered if and only if $f^* \phi \in H^1(N_1; \mathbb{R})$ is fibered.

We first prove (1) and (2) in the case that $\phi \in H^1(N_2; \mathbb{R})$ is an integral cohomology class.

1. By Theorem 3.3 (3) there exists a prime p and a representation $\alpha \colon \pi_1(N_2) \to \mathrm{GL}(k, \mathbb{F}_p)$ such that the twisted Alexander polynomials $\Delta^\alpha_{N_2, \phi, i}$, $i = 0, 1, 2$ are non-zero and such that

$$x_{N_2}(\phi) = \frac{1}{k} \left(- \deg \left(\Delta^\alpha_{N_2, \phi, 0} \right) + \deg \left(\Delta^\alpha_{N_2, \phi, 1} \right) - \deg \left(\Delta^\alpha_{N_2, \phi, 2} \right) \right).$$

By Lemma 4.5 the degrees on the right-hand side are the same for the twisted Alexander polynomials $\Delta^{\alpha \circ f}_{N_1, \phi \circ f, i}$, $i = 0, 1, 2$. If we combine this observation with the above equality and with Theorem 3.3 (2) we obtain that $x_{N_1}(\phi \circ f) \geq x_{N_2}(\phi)$. If we run through this argument with the roles of N_1 and N_2 switched we see that $x_{N_1}(\phi \circ f) = x_{N_2}(\phi)$. We thus obtained the desired equality.

2. We suppose that $\phi \in H^1(N_2; \mathbb{Z})$ is non-fibered. By Theorem 3.3 (1) there exists a representation $\alpha \colon \pi_1(N_2) \to \mathrm{GL}(k, \mathbb{F}_p)$ such that $\deg \left(\Delta^\alpha_{N_2, \phi, 1} \right) = \infty$. By Lemma 4.5 we have $\deg \left(\Delta^{\alpha \circ f}_{N_1, \phi \circ f, 1} \right) = \infty$. But by Theorem 3.3 (1) this implies that $f^* \phi \in H^1(N_1; \mathbb{Z})$ is non-fibered. Running the same argument backwards we see that ϕ is fibered if and only if $f^* \phi$ is fibered.

Summarizing, we just showed that $f^* \colon H^1(N_2; \mathbb{Z}) \to H^1(N_1; \mathbb{Z})$ is an isometry with respect to the Thurston norms and it defines a bijection of the fibered classes. Since the norms are homogeneous and continuous it follows that $f^* \colon H^1(N_2; \mathbb{R}) \to H^1(N_1; \mathbb{R})$ is also an isometry with respect to the Thurston norms. Furthermore, Thurston [Th86] showed that the set of fibered classes of a 3-manifold is given by the union on cones on open top-dimensional faces of the Thurston norm ball of the 3-manifolds. The fact that f^* defines an isometry of Thurston norms and that it defines a bijection of integral fibered classes thus also implies that f^* defines a bijection of real fibered classes. □

4.3   Proof of Theorem 1.2

In the following, recall that given a knot $K \subset S^3$ we denote by $X(K) := S^3 \setminus \nu K$ its exterior and we denote by $\pi(K) := \pi_1(S^3 \setminus \nu K)$ the corresponding knot group. We say that K *is fibered* if the knot exterior $X(K)$ is a surface bundle over S^1. Furthermore, we refer to the minimal genus of a Seifert surface for K as the *genus* $g(K)$ *of* K.

If $\phi \in H^1(X(K); \mathbb{Z})$ is a generator, then K is fibered if and only if ϕ is a fibered class. Furthermore, it is straightforward to see that $x_{X(K)}(\phi) = \max\{0, 2g(K) - 1\}$. Finally it is a well-known consequence of the Sphere Theorem that $X(K)$ is aspherical. We thus see that Theorem 1.2 is an immediate consequence of the following theorem.

Theorem 4.6. *Let N_1 and N_2 be two aspherical 3-manifolds with $H_1(N_1; \mathbb{Z}) \cong H_1(N_2; \mathbb{Z}) \cong \mathbb{Z}$. Suppose there exists an isomorphism $f : \widehat{\pi_1(N_1)} \to \widehat{\pi_1(N_2)}$. Let $\phi_i \in H^1(N_i; \mathbb{Z})$, $i = 1, 2$ be generators. Then $x_{N_1}(\phi_1) = x_{N_2}(\phi_2)$. Furthermore $\phi_1 \in H^1(N_1; \mathbb{Z})$ is fibered if and only if $\phi_2 \in H^1(N_2; \mathbb{Z})$ is fibered.*

We point out that Theorem 4.6 is not an immediate consequence of Theorem 1.1 since we do not assume that there exists a *regular* isomorphism between the profinite completions of $\pi_1(N_1)$ and $\pi_1(N_2)$.

In the proof of Theorem 4.6 we will need one more lemma.

Lemma 4.7. *Let N be a 3-manifold with $H_1(N; \mathbb{Z}) \cong \mathbb{Z}$. Let $\beta : \pi_1(N) \to \mathrm{GL}(k, \mathbb{F}_p)$ be a representation and let $\phi_n : \pi_1(N) \to \mathbb{Z}_n$ and $\psi_n : \pi_1(N) \to \mathbb{Z}_n$ be two epimorphisms. Then given any i there exists an isomorphism*

$$H_i^{\beta \otimes \phi_n}(N; \mathbb{F}_p[\mathbb{Z}_n]^k) \cong H_i^{\beta \otimes \psi_n}(N; \mathbb{F}_p[\mathbb{Z}_n]^k).$$

Proof. We denote by $\widetilde{N}$ the universal cover of N. Since $H_1(N; \mathbb{Z}) \cong \mathbb{Z}$ there exists an $r \in \mathbb{N}$, coprime to n, such that $\psi_n = r \cdot \phi_n$. Therefore it suffices to show that for all i we have

$$H_i^{\beta \otimes \phi_n}(N; \mathbb{F}_p[\mathbb{Z}_n]^k) \cong H_i^{\beta \otimes r \cdot \phi_n}(N; \mathbb{F}_p[\mathbb{Z}_n]^k).$$

This in turn follows from the observation that

$$C_*(\widetilde{N}) \otimes_{\mathbb{Z}[\pi_1(N)]} \left(\mathbb{F}_p^k \otimes \mathbb{Z}[\mathbb{Z}_n] \right)_{\beta \otimes r \cdot \phi_n} \to C_*(\widetilde{N}) \otimes_{\mathbb{Z}[\pi_1(N)]} \left(\mathbb{F}_p^k \otimes \mathbb{Z}[\mathbb{Z}_n] \right)_{\beta \otimes \phi_n}$$
$$\sigma \otimes \left(v \otimes \textstyle\sum_{g \in \mathbb{Z}_n} a_g g \right) \mapsto \sigma \otimes \left(v \otimes \textstyle\sum_{g \in \mathbb{Z}_n} a_g (r \cdot g) \right)$$

is an isomorphism of chain complexes. $\square$

Now we are ready to provide the proof of Theorem 4.6.

Proof of Theorem 4.6. Let M and N be two aspherical 3-manifolds with $H_1(M; \mathbb{Z}) \cong H_1(N; \mathbb{Z}) \cong \mathbb{Z}$ and let $f : \widehat{\pi_1(M)} \to \widehat{\pi_1(N)}$ be an isomorphism. Let $\phi \in H^1(M; \mathbb{Z})$ and $\psi \in H^1(N; \mathbb{Z})$ be generators.

It follows from Lemma 4.7, Theorem 4.1 and Proposition 4.3 that for any n and any $i \in \{0, 1\}$ we have

$$\dim_{\mathbb{F}_p}\left(H_i^{\alpha\otimes\phi_n}(M;\mathbb{F}_p[\mathbb{Z}_n]^k)\right) \;=\; \dim_{\mathbb{F}_p}\left(H_i^{(\alpha\circ f)\otimes\psi_n}(N;\mathbb{F}_p[\mathbb{Z}_n]^k)\right).$$

It follows from Lemma 4.5 that for any representation $\alpha\colon \pi_1(N)\to \mathrm{GL}(k,\mathbb{F}_p)$ we have

$$\deg\left(\Delta_{M,\phi,i}^{\alpha\circ f}\right) \;=\; \deg\left(\Delta_{N,\psi,i}^{\alpha}\right), \quad i=0,1$$

and the equality also holds for $i=2$ if the first twisted Alexander polynomials are non-zero. The argument of the proof of Theorem 1.1 now carries over to prove the desired statements. $\qquad\Box$

4.4 The profinite completion of the unknot, the trefoil and the figure-8 knot

As promised in the introduction we now recall the argument that the profinite completion detects the unknot.

Lemma 4.8. *Let U be the unknot. If K is a knot with $\widehat{\pi(U)}\cong\widehat{\pi(K)}$, then K is the unknot.*

Proof. It is a well-known consequence of Dehn's lemma that a knot K is the unknot if and only if $\pi(K)\cong\mathbb{Z}$. Since $H_1(\pi(K);\mathbb{Z})\cong\mathbb{Z}$ it follows that a knot is the unknot if and only if $\pi(K)$ is abelian. By [He87] knot groups are residually finite. It thus follows that a knot is the unknot if and only if all finite quotients are abelian. Since the finite quotients of a group are the same as the finite quotient of its profinite completion it now follows that a knot K is trivial if and only if all finite quotients of $\widehat{\pi(K)}$ are abelian. We thus showed that the profinite completion of a knot group determines whether or not the knot is trivial. $\qquad\Box$

Given a knot K we denote by $X(K)_n$ the n-fold cyclic cover of $X(K)$. We have the following elementary lemma.

Lemma 4.9. *Let J and K be two knots such that the profinite completions of $\pi(J)$ and $\pi(K)$ are isomorphic. Then for any n we have $H_1(X(J)_n;\mathbb{Z})\cong H_1(X(K)_n;\mathbb{Z})$.*

Proof. The isomorphism $f\colon \widehat{\pi(J)}\to\widehat{\pi(K)}$ induces an isomorphism of the profinite completions of $\ker\{\pi(J)\to\mathbb{Z}_n\}$ and $\ker\{\pi(K)\to\mathbb{Z}_n\}$. But if two groups have isomorphic profinite completions the abelianizations have to agree. We thus see that

$$H_1(X(J)_n;\mathbb{Z})\cong H_1(\ker\{\pi(J)\to\mathbb{Z}_n\};\mathbb{Z})\cong H_1(\ker\{\pi(K)\to\mathbb{Z}_n\};\mathbb{Z})$$
$$\cong H_1(X(K)_n;\mathbb{Z}). \qquad\Box$$

This lemma allows us to prove the following proposition.

Proposition 4.10. *Let J and K be two knots such that the profinite completions of $\pi(J)$ and $\pi(K)$ are isomorphic. Then the following hold:*

1. *The Alexander polynomial Δ_J has a zero that is an n-th root of unity if and only if Δ_K has a zero that is an n-th root of unity.*

2. *If neither Alexander polynomial has a zero that is a root of unity, then $\Delta_J = \pm \Delta_K$.*

Proof. Given a set S we write $|S| = 0$ if S is infinite and otherwise we denote by $|S|$ the number of elements. Let K be a knot and let $n \in \mathbb{N}$. Fox [Fo56], see also [We79, Tu86], showed that

$$H_1(X(K)_n; \mathbb{Z}) \cong \mathbb{Z} \oplus A$$

where A is a group with

$$|A| = \left| \prod_{k=1}^{n} \Delta_K \left(e^{2\pi i k/n} \right) \right|.$$

In particular we have $b_1(X(K)_n) = 1$ if and only if no n-th root of unity is a zero of $\Delta_K(t)$.

The first statement of the proposition is now an immediate consequence of Lemma 4.9. The second statement follows from combining the above formula with Lemma 4.9 and a deep result of Fried [Fr88]. $\qquad \square$

Now we can also prove the following corollary which we already mentioned in the introduction.

Corollary 1.4. *Let J be the trefoil or the figure-8 knot. If K is a knot with $\widehat{\pi(J)} \cong \widehat{\pi(K)}$, then J and K are equivalent.*

Proof. Let J be the trefoil or the figure-8 knot and let K be another knot with $\widehat{\pi(J)} \cong \widehat{\pi(K)}$. It is well-known that J is a fibered knot with $g(J) = 1$. It follows from Theorem 1.2 that K is also a fibered knot with $g(K) = g(J) = 1$. From [BZ85, Proposition 5.14] we deduce that K is either the trefoil or the figure-8 knot. Thus it suffices to show that the profinite completion can distinguish the trefoil from the figure-8 knot. But this is a consequence of Proposition 4.10 and the fact that the Alexander polynomial of the trefoil is the cyclotomic polynomial $t^{-1} - 1 + t$ whereas the Alexander polynomial of the figure-8 knot is $t^{-1} - 3t + t = t^{-1}(t - \frac{3+i\sqrt{5}}{2})$ $(t - \frac{3-i\sqrt{5}}{2})$. $\qquad \square$

4.5   Torus knots

In this section we prove Theorem 1.5 stating that each torus knot is distinguished, among knots, by the profinite completion of its group. First we prove that the profinite completion detects torus knots.

Proposition 4.11. *Let J be a torus knot. If K is a knot with $\widehat{\pi(J)} \cong \widehat{\pi(K)}$, then K is a torus knot.*

Wilton–Zalesskii [WZ17] showed that the profinite completion of the fundamental group determines whether a *closed* 3-manifold is Seifert fibered. Proposition 4.11 proves the same result for knot complements. Our proof is quite different from the proof provided by Wilton–Zalesskii since our main tool for dealing with hyperbolic JSJ-components is the paper by Long–Reid [LR98].

Proof. We argue by contradiction by assuming that either K is a hyperbolic knot or a satellite knot.

In the first case, by [LR98] the group $\pi(K)$ is residually a non-abelian finite simple group. (In [LR98] the group authors only state that $\pi(K)$ is residually finite simple, but it is clear from the proof that they prove the slightly stronger statement that the group $\pi(K)$ is in fact residually a non-abelian finite simple group.) Therefore $\pi(K)$, and so $\pi(J)$, admit co-final towers of finite regular coverings with non-abelian simple covering groups. But this is impossible for $\pi(J)$, since it has a non-trivial cyclic center. So we can assume that K is a satellite knot.

Since $\pi(J)$ has a non-trivial infinite cyclic center and since abelian subgroups of 3-manifold groups are separable [Ha01], $\widehat{\pi(J)}$ has a procyclic center $\widehat{Z} \cong \widehat{\mathbb{Z}}$, and the quotient is the profinite completion of the free product of two finite cyclic groups of pairwise prime order, which is centerless by [ZM88, Theorem 3.16].

Since K is a satellite knot, the exterior $X(K) = S^3 \setminus \nu K$ has a non-trivial JSJ-splitting. Such a splitting induces a graph-of-groups decomposition of $\pi(K)$. The profinite topology on $\pi(K)$ is efficient for this decomposition which means that the vertex and edge groups are closed and that the profinite topology on $\pi(K)$ induces the full profinite topologies on each vertex and edge group (see [WZ10, Theorem A]). Therefore $\widehat{\pi(K)}$ is a profinite graph of profinite completions of the corresponding vertex and edge groups. Since $\widehat{\pi(K)} \cong \widehat{\pi(J)}$, the non-trivial procyclic center $\widehat{Z}$ must belong to each edge group of this graph-of-profinite groups decomposition for $\widehat{\pi(K)}$: this follows from [ZM88, Theorem 3.16], since the graph-of-groups decomposition of $\pi(K)$ is not of dihedral type, otherwise $\pi(K)$ would be solvable, which is not possible for a non-trivial knot.

Therefore $\widehat{Z}$ belongs to the profinite completion of each vertex group of the graph-of-groups decomposition of $\pi(K)$ induced by the JSJ-splitting. Since each vertex corresponds to a hyperbolic or a Seifert piece in the geometric decomposition of $S^3 \setminus K$, it follows from the first step that all the pieces are Seifert fibered and thus $X(K) = S^3 \setminus \nu K$ is a graph 3-manifold.

Let $\widehat{\mathbb{Z} \times \mathbb{Z}}$ be an edge group of the graph-of-profinite groups decomposition of $\widehat{\pi(K)}$. It corresponds to the profinite completion of the corresponding edge group $\mathbb{Z} \times \mathbb{Z}$ of the JSJ-graph-of-groups decomposition of $\pi_1(M)$. The two vertex groups G_1 and G_2 containing this edge group correspond to the fundamental groups of Seifert pieces of M. Each group $G_i, i = 1, 2$, is an extension:

$$1 \to \mathbb{Z} = Z_i \to G_i \to \Gamma_i \to 1,$$

where $\Gamma_i, i = 1, 2$, is a free product of finite cyclic groups. The profinite completion $\widehat{\Gamma}_i$ is the free product (in the profinite category) of finite cyclic groups (see [RZ10, Exercice 9.2.7]), and hence is centerless [ZM88, Theorem 3.16], except perhaps if it is $\widehat{\mathbb{Z}_2 \star \mathbb{Z}_2}$. But in this last case by [GZ11, Proposition 4.3] Γ_i would be isomorphic to $\mathbb{Z}_2 \star \mathbb{Z}_2$ and the Seifert piece would contain an embedded Klein bottle, which is impossible in S^3. Therefore each profinite completion $\widehat{G_i}, i = 1, 2$, has a procyclic center $\widehat{Z_i}$ which must contain the procyclic center $\widehat{Z}$ of $\widehat{\pi(K)}$, since the quotient $\widehat{G_i}/\widehat{Z_i}$ is centerless.

For each prime p, the p-Sylow subgroup $\widehat{Z}_{(p)}$ of $\widehat{Z}$ is of finite index in each p-Sylow subgroup $\widehat{Z}_{i\,(p)} \subset \widehat{Z}_i$, by [RZ10, Proposition 2.7.1]. Since the subgroup generated by Z_1 and Z_2 is of finite index in the edge group $\mathbb{Z} \times \mathbb{Z}$, the subgroup generated by $\widehat{Z}_1$ and $\widehat{Z}_2$ is of finite index, say n, in the profinite edge group $\widehat{\mathbb{Z} \times \mathbb{Z}} \cong \widehat{\mathbb{Z}} \times \widehat{\mathbb{Z}}$. It follows that the p-Sylow subgroup $\widehat{Z}_{(p)} \cong \widehat{\mathbb{Z}}_{(p)}$ of $\widehat{Z}$ is of finite index in the p-Sylow subgroup $\widehat{\mathbb{Z}}_{(p)} \times \widehat{\mathbb{Z}}_{(p)}$ of the edge group for a prime p which does not divide n. This gives the desired contradiction, since $\widehat{\mathbb{Z}}_{(p)} \times \widehat{\mathbb{Z}}_{(p)}$ is a free $\widehat{\mathbb{Z}}_{(p)}$-module of rank 2. □

We give now the proof of Theorem 1.5.

By Proposition 4.11 K is a torus knot. If two torus knots $T_{p,q}$ and $T_{r,s}$ have isomorphic profinite completions, their quotients $\widehat{\mathbb{Z}_p \star \mathbb{Z}_q}$ and $\widehat{\mathbb{Z}_r \star \mathbb{Z}_s}$ by the pro-cyclic center are also isomorphic. Then it follows again from [GZ11, Proposition 4.3] that the groups $\mathbb{Z}_p \star \mathbb{Z}_q$ and $\mathbb{Z}_r \star \mathbb{Z}_s$ are isomorphic, and so the knots K and J are equivalent.

Another more topological argument is that the two torus knots $T_{p,q}$ and $T_{r,s}$ have the same genus by Theorem 1.2. So $(p-1)(q-1) = (r-1)(s-1)$. Moreover the first cyclic covers of the knot complements with maximal Betti number correspond to the pq-cover for $T_{p,q}$ and the rs-covers for $T_{r,s}$. It follows that $pq = rs$ and thus $p+q = r+s$ by the genus equality. Therefore $(p,q) = (r,s)$ or $(p,q) = (s,r)$, in both cases the torus knots $T_{p,q}$ and $T_{r,s}$ are equivalent.

Remark. Torus knots are all commensurable. Moreover they are cyclically commensurable if they have the same genus. However they are distinguished by the profinite completions of their groups. In the next section we discuss the case of hyperbolic knots.

4.6 Commensurable knots

In this section we prove Theorem 1.6 which shows that two cyclically commensurable hyperbolic knots can be distinguished by the profinite completions of their groups, provided that their Alexander polynomials are not a product of cyclotomic polynomials. The proof relies on the following proposition:

Proposition 4.12. *Let K_1 and K_2 be two distinct and cyclically commensurable hyperbolic knots. Then there is a compact orientable 3-manifold Y with $H_1(Y, \mathbb{Z}) = \mathbb{Z}$ and two coprime integers p_1 and p_2 such that $X(K_1)$ is a p_1-cyclic cover of Y and $X(K_2)$ is a p_2-cyclic cover of Y.*

Proof. For a hyperbolic knot K, the positive solution of the Smith conjecture implies that the subgroup $Z(K)$ of $\mathrm{Isom}^+(X(K))$ which acts freely on $\partial X(K)$ is cyclic, and that the action of $Z(K)$ extends to a finite cyclic action on S^3 which is conjugated to an orthogonal action. Therefore the orientable orbifold $\mathcal{Z}_K = X(K)/Z(K)$ is a knot exterior in the quotient $\mathcal{L} = S^3/Z(K)$ which is an orbi-lens space. The notion of orbi-lens space was introduced in [BBCW12, Section 3]. An orbi-lens space $\mathcal{L}$ is a 3-orbifold whose underlying space $|\mathcal{L}|$ is a lens space and the singular locus $\Sigma(\mathcal{L})$ is a closed submanifold of the union of the cores $C_1 \cup C_2$ of a genus one Heegaard splitting $V_1 \cup V_2$ of the underlying space $|\mathcal{L}|$. In particular, there are coprime positive integers $a_1, a_2 \geq 1$ such that a point of C_j has isotropy group $\mathbb{Z}/a_j$, and thus the orbifold fundamental group $\pi_1^{orb}(\mathcal{L}) \cong \mathbb{Z}/(a_1 a_2 |\pi_1(|\mathcal{L}|)|)$. We use the notation $\mathcal{L}(p, q; a)$ to denote such an orbi-lens space with $a_1 = a$ and $a_2 = 1$. When $a = 1$, $\mathcal{L}(p, q; a)$ is the lens space $L(p, q)$.

By [BBCW12, Proposition 4.7] if K_1 and K_2 are two distinct cyclically commensurable hyperbolic knots, then up to orientation preserving homeomorphism, $\mathcal{Z}_{K_1} = \mathcal{Z}_{K_2}$. The orbifold $\mathcal{Z} = \mathcal{Z}_{K_1} = \mathcal{Z}_{K_2}$ embeds as a knot exterior in both orbi-lens spaces $\mathcal{L}(p_1, q_1; a) = S^3 / Z(K_1)$ and $\mathcal{L}(p_2, q_2; a) = S^3 / Z(K_2)$, with $ap_1 = |Z(K_1)|$, $ap_2 = |Z(K_2)|$ and p_1 coprime to p_2, by [BBCW12, Propositions 5.7 and 5.8]. In particular the ap_2 cyclic cover M of $X(K_1)$ coincides with the ap_1 cyclic cover of $X(K_2)$.

By [BBCW12, Theorem 1.5] the orbifold $\mathcal{Z}$ admits a fibration by 2-orbifolds with base the circle. This fibration lifts to fibrations by surfaces over the circle in the exteriors $X(K_1)$, $X(K_2)$ and in their common cyclic covering M. So the knots K_1 and K_2 are fibered knots, and their fibrations lift to the same fibration over the circle in M. It follows that the fiber for these three fibrations is the same surface F, and the monodromy for the fibration on M is $\phi = \phi_1^{ap_2} = \phi_2^{ap_1} : F \to F$, where ϕ_1 is the monodromy of K_1 and ϕ_2 the monodromy of K_2. Since K_1 and K_2 are hyperbolic knots, the monodromies ϕ_1 and ϕ_2 are pseudo-Anosov homeomorphisms of F. The uniqueness of the root of a pseudo-Anosov element in the mapping class group of a surface with boundary, [BoPa09, Theorem 4.5], implies that $\phi_1^{p_2} = \phi_2^{p_1}$. Therefore the p_2-cyclic cover N of $X(K_1)$ coincides with the p_1-cyclic cover of $X(K_2)$. Since p_1 and p_2 are coprime, the deck transformations of the cyclic covers $N \to X(K_1)$ and $N \to X(K_2)$ generate a cyclic subgroup C of $\mathrm{Isom}^+(N)$ of order $p_1 p_2$. The quotient $Y = N/C$ is cyclically covered by $X(K_1)$ with order p_1 and by $X(K_2)$ with order p_2. It follows that Y is a manifold, since the order of the isotropy group of a point of Y must divide p_1 and p_2.

Therefore Y is the exterior of a primitive knot in the lens space $L(p_1, q_1)$, since its preimage in S^3 is the knot K_1. In particular $H_1(Y, \mathbb{Z}) = \mathbb{Z}$. $\square$

We give now the proof of Theorem 1.6.

Let K_1 and K_2 be two distinct and cyclically commensurable hyperbolic knots. Let Y be the compact orientable 3-manifold provided by Proposition 4.12. Then $H_1(Y, \mathbb{Z}) = \mathbb{Z}$ and there are two coprime integers p_1 and p_2 such that $X(K_1)$ is a p_1-cyclic cover of Y, while $X(K_2)$ is a p_2-cyclic cover of Y. Let $\tilde{Y}$ be the maximal free abelian cover of Y, then its first homology group $H_1(\tilde{Y}, \mathbb{Z})$ is a torsion module over $\mathbb{Z}[t, t^{-1}]$, whose order is $\Delta_Y \in \mathbb{Z}[t, t^{-1}]$.

Given a non-zero polynomial $p \in \mathbb{C}[t^{\pm 1}]$ with top coefficient c and zeros $z_1, \ldots, z_m$ we denote by

$$m(p) := |c| \cdot \prod_{i=1}^{m} \max\{1, |Z_i|\}$$

its Mahler measure. Furthermore, given a topological space Z we write $t(Z) := |\mathrm{Tor}(H_1(Z; \mathbb{Z}))|$. Now let Z be a 3-manifold with $H_1(Z; \mathbb{Z}) \cong \mathbb{Z}$. We denote by $\Delta_Z(t) \in \mathbb{Z}[t^{\pm 1}]$ the corresponding Alexander polynomial and given $n \in \mathbb{N}$ we denote by Z_n the cover corresponding to the epimorphism $\pi_1(Z) \to H_1(Z, \mathbb{Z}) \cong \mathbb{Z} \to \mathbb{Z}/n\mathbb{Z}$.

Silver–Williams [SW02, Theorem 2.1] proved that

$$\lim_{n \to \infty} \frac{\log(t(Z_n))}{n} = \ln(m(\Delta_Z(t))).$$

By applying this equation to $X(K_1), X(K_2)$ and Y it follows, see [Fr13, Section 3.4.1], that:

$$\ln(m(\Delta_{K_1})) = \lim_{k \to \infty} \frac{\log(t(X(K_1)_k))}{k} = p_1 \lim_{k \to \infty} \frac{\log(t(Y_{kp_1}))}{kp_1} = p_1 \ln(m(\Delta_Y(t))),$$

and

$$\ln(m(\Delta_{K_2})) = \lim_{k \to \infty} \frac{\log(t(X(K_2)_k))}{k} = p_2 \lim_{k \to \infty} \frac{\log(t(Y_{kp_2}))}{kp_2} = p_2 \ln(m(\Delta_Y(t))).$$

If $\widehat{\pi(K_1)} \cong \widehat{\pi(K_2)}$, then $t(X(K_1)_k) = t(X(K_2)_k)$ for all $k \geq 1$ and thus $p_1 = p_2$ if $\ln(m(\Delta_Y)) \neq 0$. This is the case if the Alexander polynomial Δ_{K_1} has a zero that is not a root of unity, because then by Kronecker's theorem, see [Pr10, Theorem 4.5.4], Δ_{K_1} has a zero z with $|z| > 1$, hence the Mahler measure $m(\Delta_{K_1}) \neq 1$ and $\ln(m(\Delta_{K_1})) \neq 0$. But, in this case this contradicts the fact that p_1 and p_2 are coprime.

Remark. It follows from the proof of Theorem 1.6 that the Mahler measures of the Alexander polynomials of two distinct and cyclically commensurable hyperbolic knots cannot be equal, except if it is equal to 1.

A knot group is bi-orderable if it admits a strict total ordering of its elements which is invariant under multiplication on both sides. By [CR12, Theorem 1.1], if the group of a fibered knot K is bi-orderable, then the Alexander polynomial of K has a positive real root. Hence, we get the following corollary:

Corollary 4.13. *Let K_1 and K_2 be two cyclically commensurable hyperbolic knots. If $\pi(K_1)$ is bi-orderable and $\widehat{\pi(K_1)} \cong \widehat{\pi(K_2)}$, then K_1 and K_2 are equivalent.*

Remark. A strong conjecture (see [Lü13, Conjecture 1.12(2)]) predicts that the volume of the complement of a hyperbolic knot is determined by the growth of homology torsions of its finite covers. By [Ka18, Theorem 1] this conjecture implies that two hyperbolic knots such that the profinite completions of the fundamental groups are isomorphic, have the same volume. Since cyclically commensurable distinct hyperbolic knots have distinct volumes by [BBCW12, Theorem 1.7], this would imply that cyclically commensurable hyperbolic knots are distinguished by the profinite completion of their group. It would imply furthermore that given a knot K there are only finitely many hyperbolic knots whose groups have a profinite completion isomorphic to $\widehat{\pi(K)}$ by Jorgensen-Thurston's results on volumes of complete hyperbolic 3-manifolds (namely the properness of the volume function on the set of complete hyperbolic 3-manifolds with finite volume, endowed with the geometric topology, see [BP92, Chapter E]).

REFERENCES

[Ag08]   I. Agol, *Criteria for virtual fibering*, J. Topol. 1 (2008), no. 2, 269–284.

[Ag13]   I. Agol, *The virtual Haken conjecture*, with an appendix by I. Agol, D. Groves and J. Manning, Documenta Math. 18 (2013), 1045–1087.

[AR92]   I. R. Aitchison and J. H. Rubinstein, *Combinatorial cubings, cusps, and the dodecahedral knots*, in Topology '90, Ohio State Univ. Math. Res. Inst. Publ. 1, 17–26, de Gruyter (1992).

[AFW15]   M. Aschenbrenner, S. Friedl and H. Wilton, *3-manifold groups*, European Mathematical Society Series of Lectures in Mathematics (2015).

[BP92]   R. Benedetti and C. Petronio, *Lectures on hyperbolic geometry*, Universitext, Springer, Berlin (1992).

[BBCW12] M. Boileau, S. Boyer, R. Cebanu and G. S. Walsh, *Knot commensurability and the Berge conjecture*, Geom. Top. 16 (2012), 625–664.

[BoPa09] C. Bonatti and L. Paris, *Roots in the mapping class groups*, Proc. Lond. Math. Soc. (3) 98 (2009), no. 2, 471–503.

[BR15]   M. Bridson and A. Reid, *Profinite rigidity, fibering, and the figure-eight knot*, preprint, this volume.

[BRW17]   M. R. Bridson, A. W. Reid and H. Wilton, *Profinite rigidity and surface bundles over the circle*, Bull. London Math. Soc. 49 (2017), 831–841.

[BZ85]   G. Burde and H. Zieschang, *Knots,* de Gruyter Studies in Mathematics, 5, Walter de Gruyter & Co., Berlin, 1985.

[Ca12]   W. Cavendish, *Finite-sheeted covering spaces and solenoids over 3-manifolds*, thesis, Princeton University (2012).

[Ch03]   J. Cha, *Fibred knots and twisted Alexander invariants*, Trans. Amer. Math. Soc. 355 (2003), 4187–4200.

[CR12]   A. Clay and D. Rolfsen, *Ordered groups, eigenvalues, knots, surgery and L-spaces*, Math. Proc. Cambridge Philos. Soc. 152 (2012), no. 1, 115–129.

[DFL15]   J. Dubois, S. Friedl and W. Lück, *Three flavors of twisted invariants of knots*, Introduction to Modern Mathematics, Advanced Lectures in Mathematics 33 (2015), 143–170.

[Fo56]   R. H. Fox, *Free differential calculus II. Subgroups*, Ann. of Math. (2) 64 (1956), 407–419.

[Fr88]   D. Fried, *Cyclic resultants of reciprocal polynomials*, Holomorphic dynamics (Mexico, 1986), 124–128, Lecture Notes in Math., 1345, Springer, Berlin, 1988.

[Fr13] S. Friedl, *Commensurability of knots and L^2-invariants*, Geometry and topology down under, 263–279, Contemp. Math. 597, Amer. Math. Soc., Providence, RI, 2013.

[Fr14] S. Friedl, *Twisted Reidemeister torsion, the Thurston norm and fibered manifolds*, Geom. Dedicata 172 (2014), 135–145.

[FK06] S. Friedl and T. Kim, *The Thurston norm, fibered manifolds and twisted Alexander polynomials*, Topology 45 (2006), 929–953.

[FN15] S. Friedl and M. Nagel, *Twisted Reidemeister torsion and the Thurston norm: graph manifolds and finite representations*, Ill. J. Math. 59 (2015), 691–705.

[FV10] S. Friedl and S. Vidussi, *A survey of twisted Alexander polynomials*, The Mathematics of Knots: Theory and Application (Contributions in Mathematical and Computational Sciences), editors: Markus Banagl and Denis Vogel (2010), 45–94.

[FV13] S. Friedl and S. Vidussi, *A vanishing theorem for twisted Alexander polynomials with applications to symplectic 4-manifolds*, J. Eur. Math. Soc. 15 (2013), no. 6, 2127–2041.

[FV15] S. Friedl and S. Vidussi, *The Thurston norm and twisted Alexander polynomials*, J. für reine und angewandte Math. 707 (2015), 87–102.

[Fu13] L. Funar, *Torus bundles not distinguished by TQFT invariants*, Geom. Top. 17 (2013), 2289–2344.

[GKM05] H. Goda, T. Kitano and T. Morifuji, *Reidemeister torsion, twisted Alexander polynomial and fibred knots*, Comment. Math. Helv. 80 (2005), 51–61.

[GZ11] F. Grunewald and P. Zalesskii, *Genus of groups*, Journal of Algebra 326 (2011), 130–168.

[Ha01] E. Hamilton, *Abelian subgroup separability of Haken 3-manifolds and closed hyperbolic n-orbifolds*, Proc. London Math. Soc. 83 (2001), no. 3, 626–646.

[He87] J. Hempel, *Residual finiteness for 3-manifolds*, Combinatorial group theory and topology (Alta, Utah, 1984), 379–396, Ann. of Math. Stud. 111, Princeton Univ. Press, Princeton, NJ, 1987.

[He14] J. Hempel, *Some 3-manifold groups with the same finite quotients*, preprint, 2014.

[JZ18] A. Jaikin-Zapirain, *Recognition of being fibred for compact 3-manifolds*, preprint (2018) http://verso.mat.uam.es/~andrei.jaikin/preprints/fibering.pdf.

[Ka18] H. Kammeyer, *A remark on torsion growth in homology and volume of 3-manifolds*, preprint (2018), arXiv:1802.09244.

[KL99] P. Kirk and C. Livingston, *Twisted Alexander invariants, Reidemeister torsion and Casson–Gordon invariants*, Topology 38 (1999), no. 3, 635–661.

[Ki96] T. Kitano, *Twisted Alexander polynomials and Reidemeister torsion*, Pacific J. Math. 174 (1996), no. 2, 431–442.

[Li01] X. S. Lin, *Representations of knot groups and twisted Alexander polynomials*, Acta Math. Sin. (Engl. Ser.) 17, no. 3 (2001), 361–380.

[LR98] D. Long and A. Reid, *Simple quotients of hyperbolic 3-manifold groups*, Proc. Amer. Math. Soc. 126, no. 3 (1998), 877–880.

[Lü13] W. Lück, *Approximating L^2-invariants and homology growth*, Geom. Funct. Anal. 23 (2013), 622–663.

[NS07] N. Nikolov and D. Segal, *On finitely generated profinite groups I: Strong completeness and uniform bounds*, Ann. of Math. 165 (2007), 171–236.

[Pr10] V. Prasolov, *Polynomials*, Translated from the 2001 Russian second edition by Dimitry Leites. 2nd printing. Algorithms and Computation in Mathematics 11. Berlin: Springer (2010).

[PW18] P. Przytycki and D. Wise, *Mixed 3-manifolds are virtually special*, J. Amer. Math. Soc. 31 (2018), no. 2, 319–347.

[Re91] A. W. Reid, *Arithmeticity of knot complements*, J. London Math. Soc. 43 (1991), 171–184.

[RZ10] L. Ribes and P. Zalesskii, *Profinite groups. Second edition*, Ergebnisse der Mathematik und ihrer Grenzgebiete. 3. Folge. A Series of Modern Surveys in Mathematics 40. Springer-Verlag, Berlin, 2010.

[Se97] J.-P. Serre, *Galois cohomology*, Springer-Verlag, Berlin, 1997.

[SW02] D. Silver and S. Williams, *Mahler measure, links and homology growth*, Topology 41 (2002), 979–991.

[Th86] W. Thurston, *A norm for the homology of 3-manifolds*, Mem. Amer. Math. Soc. 59 (1986), no. 339, 99–130.

[Ti70] D. Tischler, *On fibering certain foliated manifolds over S^1*, Topology 9 (1970), 153–154.

[Tu86] V. Turaev, *Reidemeister torsion in knot theory*, Russian Math. Surveys 41 (1986), no. 1, 119–182.

[Wa94] M. Wada, *Twisted Alexander polynomial for finitely presentable groups*, Topology 33, no. 2 (1994), 241–256.

[Wi98] J. Wilson, *Profinite groups*, London Mathematical Society Monographs, New Series, vol. 19, Clarendon Press, Oxford University Press, New York, 1998.

[We79]   C. Weber, *Sur une formule de R. H. Fox concernant l'homologie des revêtements cycliques*, Enseign. Math. 25 (1979), 261–272.

[Wh87]   W. Whitten, *Knot complements and groups*, Topology 26 (1987), no. 1, 41–44.

[Wk17]   G. Wilkes, *Profinite rigidity for Seifert fibre spaces*, Geom. Dedicata 188 (2017), 141–163.

[Wk18a] G. Wilkes, *Profinite rigidity of graph manifolds and JSJ decompositions of 3-manifolds*, Journal of Algebra 502 (2018), 538–587.

[Wk18b] G. Wilkes, *Profinite completions, cohomology and JSJ decompositions of compact 3-manifolds*, New Zealand J. Math. 48 (2018), 101–113.

[Wk19]   G. Wilkes, *Profinite rigidity of graph manifolds, II: knots and mapping classes*, preprint (2018), Israel J. Math. 233 (2019), no. 1, 351–378.

[WZ10]   H. Wilton and P. Zalesskii, *Profinite properties of graph manifolds*, Geom. Dedicata 147 (2010), 29–45.

[WZ17]   H. Wilton and P. Zalesskii, *Distinguishing geometries using finite quotients*, Geom. Top. 21 (2017), 345–384.

[Wi09]   D. Wise, *The structure of groups with a quasi-convex hierarchy*, Electronic Res. Ann. Math. Sci. 16 (2009), 44–55.

[Wi12a] D. Wise, *The structure of groups with a quasi-convex hierarchy*, 189 pages, preprint (2012), downloaded on October 29, 2012 from http://www.math.mcgill.ca/wise/papers.html.

[Wi12b] D. Wise, *From riches to RAAGs: 3-manifolds, right–angled Artin groups, and cubical geometry*, CBMS Regional Conference Series in Mathematics, 117, Amer. Math. Soc. (2012).

[ZM88]   P. Zalesskii and O. V. Melńikov, *Subgroups of profinite groups acting on trees*, (Russian) Mat. Sb. (N.S.) 135 (177) (1988), no. 4, 419–439, 559; translation in Math. USSR-Sb. 63 (1989), no. 2, 405—424.

Profinite Rigidity, Fibering, and the Figure-Eight Knot

MARTIN R. BRIDSON AND ALAN W. REID

1 INTRODUCTION

We are interested in the extent to which the set of finite quotients of a 3-manifold group $\Gamma = \pi_1(M)$ determines Γ and M. In this article we shall prove that fibered 3-manifolds that have non-empty boundary and first betti number 1 can be distinguished from all other compact 3-manifolds by examining the finite quotients of $\pi_1(M)$. A case where our methods are particularly well suited is when M is the complement of the figure-eight knot. Throughout, we denote this knot by $\mathcal{K}$ and write $\Pi = \pi_1(\mathbb{S}^3 \setminus \mathcal{K})$.

The manifold $\mathbb{S}^3 \setminus \mathcal{K}$ and the group Π hold a special place in the interplay of 3-manifold topology and hyperbolic geometry. $\mathcal{K}$ became the first example of a *hyperbolic knot* when Riley [29] constructed a discrete faithful representation of Π into $\mathrm{PSL}(2, \mathbb{Z}[\omega])$, where ω is a cube root of unity. Subsequently, Thurston [34] showed that the hyperbolic structure on $\mathbb{S}^3 \setminus \mathcal{K}$ could be obtained by gluing two copies of the regular ideal tetrahedron from $\mathbb{H}^3$. The insights gained from understanding the complement of the figure-eight knot provided crucial underpinning for many of Thurston's great discoveries concerning the geometry and topology of 3-manifolds, such as the nature of geometric structures on manifolds obtained by Dehn surgery, and the question of when a surface bundle admits a hyperbolic structure. In our exploration of the finite quotients of 3-manifold groups, the figure-eight knot stands out once again as a special example.

Throughout this article, we allow manifolds to have non-empty boundary, but we assume that any spherical boundary components have been removed by capping-off with a 3-ball. We write $\mathcal{C}(G)$ for the set of isomorphism classes of the finite quotients of a group G.

Theorem A. *Let M be a compact connected 3-manifold. If $\mathcal{C}(\pi_1(M)) = \mathcal{C}(\Pi)$, then the interior of M is homeomorphic to $\mathbb{S}^3 \setminus \mathcal{K}$.*

Corollary 1.1. *Let $J \subset \mathbb{S}^3$ be a knot and let $\Lambda = \pi_1(S^3 \setminus J)$. If $\mathcal{C}(\Lambda) = \mathcal{C}(\Pi)$, then J is isotopic to $\mathcal{K}$.*

Bridson was supported in part by grants from the EPSRC and a Royal Society Wolfson Merit Award. Reid was supported in part by an NSF grant.

It is useful to arrange the finite quotients of a group Γ into a directed system, and to replace $\mathcal{C}(\Gamma)$ by the *profinite completion* $\widehat{\Gamma}$: the normal subgroups of finite index $N \lhd \Gamma$ are ordered by reverse inclusion and $\widehat{\Gamma} := \varprojlim \Gamma/N$. A standard argument shows that for finitely generated groups, $\mathcal{C}(\Gamma_1) = \mathcal{C}(\Gamma_2)$ if and only if $\widehat{\Gamma}_1 \cong \widehat{\Gamma}_2$ (see [11]). Henceforth, we shall express our results in terms of $\widehat{\Gamma}$ rather than $\mathcal{C}(\Gamma)$.

$\mathbb{S}^3 \smallsetminus \mathcal{K}$ has the structure of a once-punctured torus bundle over the circle (see §3), and hence Π is an extension of a free group of rank 2 by $\mathbb{Z}$. We shall use Theorem A to frame a broader investigation into 3-manifolds whose fundamental groups have the same profinite completion as a free-by-cyclic group. To that end, we fix some notation.

We write F_r to denote the free group of rank $r \geq 2$ and $b_1(X)$ to denote the first betti number rk $\mathrm{H}^1(X, \mathbb{Z})$, where X is a space or a finitely generated group. Note that for knot complements $b_1(X) = 1$. Stallings's Fibering Theorem [31] shows that a compact 3-manifold N with non-empty boundary fibers over the circle, with compact fiber, if and only if $\pi_1 N$ is of the form $F_r \rtimes \mathbb{Z}$.

In Section 3 we shall elucidate the nature of compact 3-manifolds whose fundamental groups have the same profinite completion as a group of the form $F_r \rtimes \mathbb{Z}$. In particular we shall prove:

Theorem B. *Let M be a compact connected 3-manifold and let $\Gamma = F_r \rtimes_\phi \mathbb{Z}$. If $b_1(M) = 1$ and $\widehat{\Gamma} \cong \widehat{\pi_1(M)}$, then M has non-empty boundary, fibers over the circle with compact fiber, and $\pi_1(M) \cong F_r \rtimes_\psi \mathbb{Z}$ for some $\psi \in Out(F_r)$.*

Corollary 1.2. *Let M and N be compact connected 3-manifolds with $\widehat{\pi_1(N)} \cong \widehat{\pi_1(M)}$. If M has non-empty incompressible boundary and fibers over the circle, and $b_1(M) = 1$, then N has non-empty incompressible boundary and fibers over the circle, and $b_1(N) = 1$.*

One can remove the hypothesis $b_1(M) = 1$ from Theorem B at the expense of demanding more from the isomorphism $\widehat{\pi_1 M} \to \widehat{F_r \rtimes \mathbb{Z}}$: it is enough to require that $\pi_1 M$ has cyclic image under the composition

$$\pi_1 M \to \widehat{\pi_1 M} \to \widehat{F_r \rtimes \mathbb{Z}} \to \widehat{\mathbb{Z}},$$

or else that the isomorphism $\widehat{\pi_1 M} \to \widehat{F_r \rtimes \mathbb{Z}}$ is regular in the sense of [5]; see remark 5.2. Corollary 1.2 can be strengthened in the same manner.

We shall pay particular attention to the case $r = 2$. In that setting we prove that for each Γ with $b_1(\Gamma) = 1$, there are only finitely many possibilities for M, and either they are all hyperbolic or else none of them are. (We refer the reader to Theorem D at the end of the paper for a compilation of related results.) These considerations apply to the complement of the figure-eight knot, because it is a once-punctured torus bundle with monodromy

$$\phi = \begin{pmatrix} 2 & 1 \\ 1 & 1 \end{pmatrix},$$

whence $\Pi \cong F_2 \rtimes_\phi \mathbb{Z}$. In this case, a refinement of the argument which shows that there are only finitely many possibilities for M shows that in fact there is a unique possibility. Similar arguments show that the trefoil knot complement and the

Gieseking manifold are uniquely defined by the finite quotients of their fundamental groups (see §3 and §6 for details).

The most difficult step in the proof of Theorem B is achieved by appealing to the cohomological criterion provided by the following theorem.

Theorem C. *Let M be a compact, irreducible, 3-manifold with non-empty boundary which is a non-empty union of incompressible tori and Klein bottles. Let $f \colon \pi_1(M) \to \mathbb{Z}$ be an epimorphism. Then, either $\ker f$ is finitely generated and free (and M is fibered) or else the closure of $\ker f$ in $\widehat{\pi_1 M}$ has cohomological dimension at least 2.*

Our proof of this theorem relies on the breakthroughs of Agol [1] and Wise [37] concerning cubical structures on 3-manifolds and the subsequent work of Przytycki-Wise [24] on the separability of embedded surface groups, as refined by Liu [19, 20]. We isolate from this body of work a technical result of independent interest: *if M is a closed 3-manifold and $S \subset M$ is a closed embedded incompressible surface, then the closure of $\pi_1 S$ in $\widehat{\pi_1 M}$ is isomorphic to $\widehat{\pi_1 S}$.* The surfaces to which we apply this result are obtained using a construction of Freedman and Freedman [12]. Our argument also exploits the calculation of L^2-betti numbers by Lott and Lück [21], and the Cyclic Surgery Theorem [9]. Results concerning the goodness and cohomological dimension of profinite completions also play an important role in the proof of Theorem B.

Our results here, and similar results in [5], are the first to give credence to the possibility, raised in [2] and [27], that Kleinian groups of finite co-volume might be distinguished from each other and from all other 3-manifold groups by their profinite completions, i.e., that they are *profinitely rigid* in the sense of [27]. In contrast, the fundamental groups of 3-manifolds modelled on the geometries Sol and $\mathbb{H}^2 \times \mathbb{R}$ are not profinitely rigid in general; see [13], [17] and Remark 3.7. There are also lattices in higher-dimensional semi-simple Lie groups that are not determined by their profinite completions (see [3]). We refer the reader to [27, §10] for a wider discussion of profinite rigidity for 3-manifold groups, and note that the recent work of Wilton and Zalesskii [35] provides an important advance in this area.

Theorem D prompts the question: *to what extent are free-by-cyclic groups profinitely rigid?* We shall return to this question in a future article (cf. Remark 3.7). Note that one cannot hope to distinguish a free-by-cyclic group from an arbitrary finitely presented, residually finite group by means of its finite quotients without first resolving the deep question of whether finitely generated free groups themselves can be distinguished.

Remarks

During the writing of this paper we became aware that M. Boileau and S. Friedl were working on similar problems [5]. Their results, which appear elsewhere in this volume, overlap with ours but the methods of proof are very different. Both papers appeared in preprint form in May 2015. The present text retains its original form, but we should comment on subsequent advances concerning profinite rigidity in low-dimensional topology.

In joint work with Wilton [7] that builds on the present paper, we prove that the fundamental group of every once-punctured torus bundle is profinitely rigid in the class of fundamental groups of compact 3-manifolds. We also extended the fibering criterion in Corollary C to cover closed manifolds, retaining the hypothesis $b_1 = 1$. Subsequently, Jaikin-Zapirain [18] was able to remove the hypothesis $b_1 = 1$, thereby establishing that fibering is a profinite invariant of compact 3-manifolds. Meanwhile, Wilton and Zalesskii proved a profinite recognition theorem for finite volume hyperbolic manifolds [36, Corollary C] in the context of their work on JSJ recognition. This can be used to simplify the proof of profinite rigidity for the figure-eight knot complement: the arguments we give in subsection 5.1 are unnecessary in the hyperbolic case.

Acknowledgment

We thank Yi Liu for helpful conversations and correspondence, and Cameron Gordon for Dehn surgery advice. We would also like to thank Stefan Friedl for helpful comments concerning the issue in Remark 5.2.

2   PRELIMINARIES CONCERNING PROFINITE GROUPS

In this section we gather the results about profinite groups that we shall need.

2.1   Basics

Let Γ be a finitely generated group. We have

$$\widehat{\Gamma} = \varprojlim \Gamma/N$$

where the inverse limit is taken over the normal subgroups of finite index $N \triangleleft \Gamma$ ordered by reverse inclusion. $\widehat{\Gamma}$ is a compact topological group.

The natural homomorphism $i : \Gamma \to \widehat{\Gamma}$ is injective if and only if Γ is residually finite. The image of i is dense regardless of whether Γ is residually finite or not, so the restriction to Γ of any continuous epimorphism from $\widehat{\Gamma}$ to a finite group is onto. A deep theorem of Nikolov and Segal [23] implies that if Γ is finitely generated then *every* homomorphism from $\widehat{\Gamma}$ to a finite group is continuous.

For every finite group Q, the restriction $f \mapsto f \circ i$ gives a bijection from the set of epimorphisms $\widehat{\Gamma} \twoheadrightarrow Q$ to the set of epimorphisms $\Gamma \twoheadrightarrow Q$. From consideration of the finite abelian quotients, the following useful lemma is easily deduced.

Lemma 2.1. *Let Γ_1 and Γ_2 be finitely generated groups. If $\widehat{\Gamma}_1 \cong \widehat{\Gamma}_2$, then $H_1(\Gamma_1, \mathbb{Z}) \cong H_1(\Gamma_2, \mathbb{Z})$.*

Notation and terminology

Given a subset X of a profinite group G, we write $\overline{X}$ to denote the closure of X in G.

Let Γ be a finitely generated, residually finite group and let Δ be a subgroup. The inclusion $\Delta \hookrightarrow \Gamma$ induces a continuous homomorphism $\widehat{\Delta} \to \widehat{\Gamma}$ whose image is $\overline{\Delta}$. The map $\widehat{\Delta} \to \widehat{\Gamma}$ is injective if and only if $\overline{\Delta} \cong \widehat{\Delta}$; and in this circumstance one says that Γ *induces the full profinite topology on* Δ. For finitely generated groups, this is equivalent to the statement that for every subgroup of finite index $I < \Delta$ there is a subgroup of finite index $S < \Gamma$ such that $S \cap \Delta \subset I$.

Note that if $\Delta < \Gamma$ is of finite index, then Γ induces the full profinite topology on Δ.

There are two other situations of interest to us where Γ induces the full profinite topology on a subgroup. First, suppose that Γ is a group and H a subgroup of Γ, then Γ is called H-separable if for every $g \in G \smallsetminus H$, there is a subgroup K of finite index in Γ such that $H \subset K$ but $g \notin K$; equivalently, the intersection of all finite index subgroups in Γ containing H is precisely H. The group Γ is called *LERF* (or *subgroup separable*) if it is H-separable for every finitely generated subgroup H, or equivalently, if every finitely generated subgroup is a closed subset in the profinite topology.

It is important to note that even if a finitely generated subgroup $H < \Gamma$ is separable, it need not be the case that Γ induces the full profinite topology on H. However, LERF does guarantee this: each subgroup of finite index $H_0 < H$ is finitely generated, hence closed in Γ, so we can write H as a union of cosets $h_1 H_0 \cup \cdots \cup h_n H_0$ and find subgroups of finite index $K_i < \Gamma$ such that $H_0 \subset K_i$, but $h_i \notin K_i$; the intersection of the K_i is then a subgroup of finite index in Γ with $K \cap H = H_0$.

The second situation that is important for us is the following.

Lemma 2.2. *If N is finitely generated, then any semidirect product of the form $\Gamma = N \rtimes Q$ induces the full profinite topology on N.*

Proof. The characteristic subgroups $C < N$ of finite index are a fundamental system of neighborhoods of 1 defining the profinite topology on N, so it is enough to construct a subgroup of finite index $S < \Gamma$ such that $S \cap N = C$. As C is characteristic in N, it is invariant under the action of Q, so the action $\Psi : Q \to \mathrm{Aut}(N)$ implicit in the semidirect product descends to an action $Q \to \mathrm{Aut}(N/C)$, with image Q_C say. Define S to be the kernel of $\Gamma \to (N/C) \rtimes Q_C$. □

2.2 On the cohomology of profinite groups

We recall some facts about the cohomology of profinite groups that we shall use. We refer the reader to [28, Chapter 6] and [30] for details.

Let G be a profinite group, M a discrete G-module (i.e., an abelian group M equipped with the discrete topology on which G acts continuously) and let $C^n(G, M)$ be the set of all continuous maps $G^n \to M$. When equipped with the usual coboundary operator $d : C^n(G, M) \to C^{n+1}(G, M)$, this defines a chain complex $C^*(G, M)$ whose cohomology groups $H^q(G; M)$ are the *continuous cohomology groups* of G with coefficients in M.

Now let Γ be a finitely generated group. Following Serre [30], we say that a group Γ is *good* if for all $q \geq 0$ and for every finite Γ-module M, the homomorphism of cohomology groups

$$H^q(\widehat{\Gamma}; M) \to H^q(\Gamma; M)$$

induced by the natural map $\Gamma \to \widehat{\Gamma}$ is an isomorphism between the cohomology of Γ and the continuous cohomology of $\widehat{\Gamma}$.

Returning to the general setting again, let G be a profinite group, and p a prime number. The *p-cohomological dimension* of G is the least integer n such that for every discrete G-module M and for every $q > n$, the p-primary component of $H^q(G; M)$ is zero. This is denoted by $cd_p(G)$. The cohomological dimension of G is defined as the supremum of $cd_p(G)$ over all primes p, and this is denoted by $cd(G)$.

We also retain the standard notation $cd(\Gamma)$ for the cohomological dimension (over $\mathbb{Z}$) of a discrete group Γ.

Lemma 2.3. *If the discrete group Γ is good, then $cd(\widehat{\Gamma}) \leq cd(\Gamma)$. If, in addition, Γ is the fundamental group of a closed aspherical manifold, then $cd(\widehat{\Gamma}) = cd(\Gamma)$.*

Proof. If $cd(\Gamma) \leq n$ then $H^q(\Gamma, M) = 0$ for every Γ-module M and every $q > n$. By goodness, this vanishing transfers to the profinite setting in the context of finite modules.

If Γ is the fundamental group of a closed aspherical d-manifold M, then $cd(\Gamma) = d$. And by Poincaré duality, $H^d(\Gamma, \mathbb{F}_2) \neq 0$, whence $H^d(\widehat{\Gamma}, \mathbb{F}_2) \neq 0$. $\square$

We also recall the following result (see [30, Chapter 1 §3.3]).

Proposition 2.4. *Let p be a prime, let G be a profinite group, and H a closed subgroup of G. Then $cd_p(H) \leq cd_p(G)$.*

Fundamental groups of closed surfaces are good and (excluding $\mathbb{S}^2$ and $\mathbb{RP}^2$) have cohomological dimension 2. Free groups are good and have cohomological dimension 1.

Corollary 2.5. *If F is a non-abelian free group and Σ is the fundamental group of a closed surface other than $\mathbb{S}^2$, then $\widehat{F}$ does not contain $\widehat{\Sigma}$.*

If one has a short exact sequence $1 \to N \to \Gamma \to Q \to 1$ where both N and Q are good, and $H^q(N; M)$ is finite for every finite Γ-module M then Γ is good (see [30, Chapter 1 §2.6]). Hence we have:

Lemma 2.6. *If $F \neq 1$ is finitely generated and free, then every group of the form $F \rtimes \mathbb{Z}$ is good and $cd(\widehat{F \rtimes \mathbb{Z}}) = 2$.*

2.3 L^2-betti numbers

The standard reference for this material is Lück's treatise [22]. L^2-betti numbers of groups are defined analytically, but Lück's Approximation Theorem provides a purely algebraic surrogate definition of the first L^2-betti number for finitely presented, residually finite groups: if (N_i) is a sequence of finite-index normal subgroups in Γ with $\cap_i N_i = 1$, then $b_1^{(2)}(\Gamma) = \lim_i b_1(N_i)/[\Gamma : N_i]$. The fundamental group of every compact 3-manifold X is residually finite (see, for example, [4]), so we can use the surrogate definition $b_1^{(2)}(X) = b_1^{(2)}(\pi_1(X))$.

We shall require the following result [6, Theorem 3.2].

Proposition 2.7. *Let Λ and Γ be finitely presented residually finite groups and suppose that Λ is a dense subgroup of $\widehat{\Gamma}$. Then $b_1^{(2)}(\Gamma) \leq b_1^{(2)}(\Lambda)$.*

In particular, if $\widehat{\Lambda} \cong \widehat{\Gamma}$ then $b_1^{(2)}(\Gamma) = b_1^{(2)}(\Lambda)$.

The following special case of a result of Gaboriau [14] will also be useful.

Proposition 2.8. *Let $1 \to N \to \Gamma \to Q \to 1$ be a short exact sequence of groups. If Γ is finitely presented, N finitely generated and Q is infinite, then $b_1^{(2)}(\Gamma) = 0$.*

3   GROUPS OF THE FORM $F \times \mathbb{Z}$

In this section we explore the extent to which free-by-cyclic groups are determined by their profinite completions, paying particular attention to the figure-eight knot. We remind the reader of our notation: $b_1(G)$ is the rank of $H^1(G, \mathbb{Z})$ and F_r is a free group of rank r.

Let Γ be a finitely generated group. If $b_1(\Gamma) = 1$, then there is a unique normal subgroup $N \subset \Gamma$ such that $\Gamma/N \cong \mathbb{Z}$ and the kernel of the induced map $\widehat{\Gamma} \to \widehat{\mathbb{Z}}$ is the closure $\overline{N}$ of N. We saw in Lemma 2.2 that if N is finitely generated then $\overline{N} \cong \widehat{N}$, in which case we have a short exact sequence:

$$1 \to \widehat{N} \to \widehat{\Gamma} \to \widehat{\mathbb{Z}} \to 1. \tag{†}$$

Lemma 3.1. *Let $\Gamma_1 = N_1 \rtimes \mathbb{Z}$ and $\Gamma_2 = N_2 \rtimes \mathbb{Z}$, with N_1 and N_2 finitely generated. If $b_1(\Gamma_1) = 1$ and $\widehat{\Gamma}_1 \cong \widehat{\Gamma}_2$, then $\widehat{N}_1 \cong \widehat{N}_2$.*

Proof. For finitely generated groups, $\widehat{\Gamma}_1 \cong \widehat{\Gamma}_2$ implies $H_1(\Gamma_1, \mathbb{Z}) \cong H_1(\Gamma_2, \mathbb{Z})$. Thus $b_1(\Gamma_2) = 1$. We fix an identification $\widehat{\Gamma}_1 = \widehat{\Gamma}_2$. Then N_1 and N_2 are dense in the kernel of the canonical epimorphism $\widehat{\Gamma}_1 \to \widehat{\mathbb{Z}}$ described in (†). And by Lemma 2.2, this kernel is isomorphic to $\widehat{N}_i$ for $i = 1, 2$. □

3.1   The groups $\Gamma_\phi = F_2 \rtimes_\phi \mathbb{Z}$

The action of $\mathrm{Aut}(F_2)$ on $H_1(F_2, \mathbb{Z})$ gives an epimorphism $\mathrm{Aut}(F_2) \to \mathrm{GL}(2, \mathbb{Z})$ whose kernel is the group of inner automorphisms. The isomorphism type of Γ_ϕ depends only on the conjugacy class of the image of ϕ in $\mathrm{Out}(F_2) = \mathrm{GL}(2, \mathbb{Z})$, and we often regard ϕ as an element of $\mathrm{GL}(2, \mathbb{Z})$. We remind the reader that finite-order elements of $\mathrm{GL}(2, \mathbb{Z})$ are termed *elliptic*, infinite-order elements with an eigenvalue of absolute value bigger than 1 are *hyperbolic*, and the other infinite-order elements are *parabolic*. Note that an element of infinite order in $\mathrm{SL}(2, \mathbb{Z})$ is hyperbolic if and only if $|\mathrm{tr}(\phi)| > 2$.

Every automorphism ϕ of F_2 sends $[a, b]$ to a conjugate of $[a, b]^{\pm 1}$ and hence can be realized as an automorphism of the once-punctured torus (orientation preserving or reversing according to whether $\det \phi = \pm 1$).

3.2   Features distinguished by $\widehat{\Gamma}_\phi$

Proposition 3.2. *Suppose $\widehat{\Gamma}_{\phi_1} \cong \widehat{\Gamma}_{\phi_2}$. If ϕ_1 is hyperbolic then ϕ_2 is hyperbolic and has the same eigenvalues as ϕ_1 (equivalently, $\det \phi_1 = \det \phi_2$ and $\mathrm{tr}\, \phi_1 = \mathrm{tr}\, \phi_2$).*

We break the proof of this proposition into three lemmas. First, for the assertion about $\det \phi$, we use the short exact sequence (†).

Lemma 3.3. *If $b_1(\Gamma_{\phi_1}) = 1$ and $\widehat{\Gamma}_{\phi_1} \cong \widehat{\Gamma}_{\phi_2}$, then $\det \phi_1 = \det \phi_2$.*

Proof. We use the shorthand $\Gamma_i = \Gamma_{\phi_i}$, fix an identification $\widehat{\Gamma}_1 = \widehat{\Gamma}_2$, and write $N = \widehat{F}_2$ for the kernel of the canonical map $\widehat{\Gamma}_i \to \widehat{\mathbb{Z}}$. The canonical map $F_2 \to H_1(F_2, \mathbb{Z}/3)$ induces an epimorphism $N \to (\mathbb{Z}/3)^2$ and the action of $\widehat{\Gamma}_i$ by conjugation on N induces a map $\widehat{\mathbb{Z}} \to \mathrm{GL}(2, \mathbb{Z}/3)$ with cyclic image generated by the reduction of ϕ_i for $i = 1, 2$. Thus $\det \phi_1 = \det \phi_2$ is determined by whether the image of $\widehat{\mathbb{Z}}$ lies in $\mathrm{SL}(2, \mathbb{Z}/3)$ or not. $\square$

Lemma 3.4. *If $b_1(\Gamma_{\phi_1}) = 1$ and $\widehat{\Gamma}_{\phi_1} \cong \widehat{\Gamma}_{\phi_2}$, then $\widehat{\Gamma}_{\phi_1^r} \cong \widehat{\Gamma}_{\phi_2^r}$ for all $r \neq 0$.*

Proof. $\Gamma_{\phi^r} \cong \Gamma_{\phi^{-r}}$, so assume $r > 0$. We again consider the canonical map from $\widehat{\Gamma}_1 = \widehat{\Gamma}_2$ to $\widehat{\mathbb{Z}}$. Let N_r be the kernel of the composition of this map and $\widehat{\mathbb{Z}} \to \mathbb{Z}/r$. Then N_r is the closure of $F_2 \rtimes_{\phi_i} r\mathbb{Z} < \Gamma_{\phi_i}$ for $i = 1, 2$. Thus $\widehat{\Gamma}_{\phi_1^r} \cong N_r \cong \widehat{\Gamma}_{\phi_2^r}$. (Here we have used the fact that the profinite topology on any finitely generated group induces the full profinite topology on any subgroup of finite index.) $\square$

Lemma 3.5.

1. $b_1(\Gamma_\phi) = 1$ if and only if $1 + \det \phi \neq \mathrm{tr}\,\phi$.
2. If $b_1(\Gamma_\phi) = 1$ then $H_1(\Gamma_\phi, \mathbb{Z}) \cong \mathbb{Z} \oplus T$, where $|T| = |1 + \det \phi - \mathrm{tr}\,\phi|$.

Proof. By choosing a representative $\phi_* \in \mathrm{Aut}(F_2)$, we get a presentation for Γ_ϕ,

$$\langle a, b, t \mid tat^{-1} = \phi_*(a),\ tbt^{-1} = \phi_*(b) \rangle.$$

By abelianizing, we see that $H_1(\Gamma_\phi, \mathbb{Z})$ is the direct sum of $\mathbb{Z}$ (generated by the image of t) and $\mathbb{Z}^2$ modulo the image of $\phi - I$. The image of $\phi - I$ has finite index if and only if $\det(\phi - I)$ is non-zero, and a trivial calculation shows that this determinant is $1 - \mathrm{tr}\,\phi + \det \phi$. If the index is finite, then the quotient has order $|\det(\phi - I)|$. $\square$

Corollary 3.6. *$b_1(\Gamma_{\phi^r}) = 1$ for all $r \neq 0$ if and only if ϕ is hyperbolic.*

Proof. If ϕ is hyperbolic, then ϕ^r is hyperbolic for all $r \neq 0$; in particular $\mathrm{tr}\,\phi^r \neq 0$, and $|\mathrm{tr}\,\phi^r| > 2$ if $\det \phi^r = 1$. Thus the result follows from Lemma 3.5(1), in the hyperbolic case. If ϕ is elliptic, then $\phi^r = I$ for some r, whence $b_1(\Gamma_{\phi^r}) = 3$. If ϕ is parabolic, then there exists $r > 0$ so that ϕ^r had determinant 1 and is conjugate in $\mathrm{GL}(2, \mathbb{Z})$ to $\begin{pmatrix} 1 & n \\ 0 & 1 \end{pmatrix}$ with $n > 0$. In particular, $b_1(\Gamma_{\phi^r}) = 2$. $\square$

Proof of Proposition 3.2. Lemma 3.4 and Corollary 3.6 imply that Γ_{ϕ_2} is hyperbolic, Lemma 3.3 shows that $\det \phi_1 = \det \phi_2$, and then Lemma 3.5(2) implies that $\mathrm{tr}\,\phi_1 = \mathrm{tr}\,\phi_2$ (since $H_1(\Gamma_{\phi_1}, \mathbb{Z}) \cong H_1(\Gamma_{\phi_2}, \mathbb{Z})$). $\square$

Remark 3.7. So far, the only quotients that we have used to explore the profinite completion of Γ_ϕ are the abelian quotients of $F_2 \rtimes_{\phi^r} \mathbb{Z}$. Since these all factor through $A_{\phi^r} = \mathbb{Z}^2 \rtimes_{\phi^r} \mathbb{Z}$, we were actually extracting information about ϕ from the groups $\widehat{A}_{\phi^r}$. Up to isomorphism, such a completion $\widehat{A}_\psi$ is determined by the *local* conjugacy class of ψ, meaning $\widehat{A}_\psi \cong \widehat{A}_{\psi'}$ if and only if the image of ψ is conjugate to the image of ψ' in $\mathrm{GL}(2, \mathbb{Z}/m)$ for all integers $m > 1$. Local conjugacy does not imply that ψ and ψ' are conjugate in $\mathrm{GL}(2, \mathbb{Z})$: Stebe [32] proved that

$$\begin{pmatrix} 188 & 275 \\ 121 & 177 \end{pmatrix} \quad \text{and} \quad \begin{pmatrix} 188 & 11 \\ 3025 & 177 \end{pmatrix}$$

have this property and Funar [13] described infinitely many such pairs. The corresponding torus bundles over the circle provide pairs of Sol manifolds whose fundamental groups are not isomorphic but have the same profinite completion [13].

Remark 3.8. The conclusion of Proposition 3.2 could also be phrased as saying that ϕ_1 and ϕ_2 lie in the same conjugacy class in $\mathrm{GL}(2, \mathbb{Q})$. When intersected with $\mathrm{GL}(2, \mathbb{Z})$, this will break into a finite number of conjugacy classes; how many can be determined using class field theory [33].

3.3 The figure-eight knot

For small examples, instead of using class field theory, one can calculate the conjugacy classes in $\mathrm{GL}(2, \mathbb{Z})$ with a given trace and determinant by hand. For example, up to conjugacy in $\mathrm{GL}(2, \mathbb{Z})$ the only matrix with trace 3 and determinant 1 is $\begin{pmatrix} 2 & 1 \\ 1 & 1 \end{pmatrix}$. This calculation yields the following consequence of Proposition 3.2 and Lemma 3.1.

We retain the notation established in the introduction: $\mathcal{K}$ is the figure-eight knot and $\Pi = \pi_1(\mathbb{S}^3 \smallsetminus \mathcal{K})$.

Proposition 3.9. *Let $\Gamma = F \rtimes_\phi \mathbb{Z}$, where F is finitely generated and free. If $\widehat{\Gamma} \cong \widehat{\Pi}$, then F has rank two, $\Gamma \cong \Pi$, and ϕ is conjugate to $\begin{pmatrix} 2 & 1 \\ 1 & 1 \end{pmatrix}$ in any identification of $\mathrm{Out}(F)$ with $\mathrm{GL}(2, \mathbb{Z})$.*

3.4 Uniqueness for the trefoil knot and the Gieseking manifold

From Lemma 3.5 we know that the only monodromies $\phi \in \mathrm{GL}(2, \mathbb{Z})$ for which $H_1(\Gamma_\phi, \mathbb{Z}) \cong \mathbb{Z}$ are those for which $(\mathrm{tr}\,(\phi), \det \phi)$ is one of $(1, 1)$, $(1, -1)$, $(3, 1)$. We have already discussed how the last possibility determines the figure-eight knot. Each of the other possibilities also determines a unique conjugacy class in $\mathrm{GL}(2, \mathbb{Z})$, represented by

$$\begin{pmatrix} 1 & -1 \\ 1 & 0 \end{pmatrix} \quad \text{and} \quad \begin{pmatrix} 1 & 1 \\ 1 & 0 \end{pmatrix},$$

respectively. The punctured-torus bundle with the first monodromy is the complement of the trefoil knot, and the second monodromy produces the Gieseking manifold, which is the unique non-orientable 3-manifold whose oriented double cover is the complement of the figure-eight knot.

Note that the first matrix is elliptic while the second is hyperbolic. The following proposition records two consequences of this discussion. For item (2) we need to appeal to Lemma 3.1.

Proposition 3.10.

1. *The only groups of the form $\Lambda = F_2 \rtimes \mathbb{Z}$ with $H_1(\Lambda, \mathbb{Z}) = \mathbb{Z}$ are the fundamental groups of (i) the figure-eight knot complement, (ii) the trefoil knot, and (iii) the Gieseking manifold.*
2. *Let Λ be one of these three groups and let F be a free group. If $\Gamma = F \rtimes \mathbb{Z}$ and $\widehat{\Gamma} \cong \widehat{\Lambda}$, then $\Gamma \cong \Lambda$.*

Example 3.11. Instead of appealing to the general considerations in this section, one can distinguish the profinite completions of the groups for the figure-eight knot and the trefoil knot directly. Indeed, setting $y^2 = 1$ in the standard presentation

$$\Pi = \langle x, y \mid yxy^{-1}xy = xyx^{-1}yx \rangle$$

one sees that Π maps onto the dihedral group D_{10}. But T, the fundamental group of the trefoil knot, cannot map onto D_{10}, because D_{10} is centerless and has no elements of order 3, whereas $T = \langle a, b \mid a^2 = b^3 \rangle$ is a central extension of $\mathbb{Z}/2 * \mathbb{Z}/3$.

3.5 Finite ambiguity

We do not know if all free-by-cyclic groups can be distinguished from one another by their profinite completions (cf. Remark 3.7), but the elementary analysis in the previous subsection enables one to show that the ambiguity in the case when the free group has rank 2 is at worst finite. (A more sophisticated argument allows one to remove this finite ambiguity [7].)

Proposition 3.12. *For every $\phi \in \mathrm{GL}(2, \mathbb{Z})$, there exist only finitely many conjugacy classes $[\psi]$ in $\mathrm{GL}(2, \mathbb{Z})$ such that $\widehat{\Gamma}_\phi \cong \widehat{\Gamma}_\psi$. Moreover, all such ψ are of the same type: hyperbolic, parabolic, or elliptic.*

Proof. If ϕ is hyperbolic, this follows from Proposition 3.2. If ϕ is parabolic then either it has trace 2 and $b_1(\Gamma_\phi) = 2$, or trace -2 in which case $b_1(\Gamma_\phi) = 1$. In the former case, ϕ is conjugate in $\mathrm{GL}(2, \mathbb{Z})$ to $\begin{pmatrix} 1 & n \\ 0 & 1 \end{pmatrix}$ with $n > 0$, and $H_1(\Gamma_\phi, \mathbb{Z}) \cong \mathbb{Z}^2 \oplus \mathbb{Z}/n$, so n is determined by $H_1(\Gamma_\phi, \mathbb{Z})$, hence by $\widehat{\Gamma}_\phi$. In the case where the trace is -2, we have $b_1(\Gamma_\phi) = 1$ and Lemma 3.4 reduces us to the previous case.

For the elliptic case, since the possible orders are 2, 3, 4 or 6 and there are only finitely many conjugacy classes in $\mathrm{GL}(2, \mathbb{Z})$ of elements of such orders, we are reduced to distinguishing elliptics from non-elliptics by means of $\widehat{\Gamma}_\phi$. If ϕ is elliptic and $b_1(\Gamma_\phi) = 1$, then Lemma 3.4 completes the proof, because for non-elliptics $b_1(\Gamma_{\phi^r})$ is never greater than 2, whereas for elliptics it becomes 3. The only other possibility, up to conjugacy, is $\begin{pmatrix} 0 & 1 \\ 1 & 0 \end{pmatrix}$, which cannot be distinguished from parabolics such as $\begin{pmatrix} 1 & 1 \\ 0 & 1 \end{pmatrix}$ by means of $H_1(\Gamma_\phi, \mathbb{Z})$. However, these can be distinguished on passage to subgroups of finite index, since for $\phi = \begin{pmatrix} 0 & 1 \\ 1 & 0 \end{pmatrix}$, $b_1(\Gamma_{\phi^2}) = 3$, while for parabolics, $b_1(\Gamma_{\phi^2}) = 2$. □

3.6 Passing to finite-sheeted covers

It is easy to see that a subgroup of finite index in a free-by-cyclic group is itself free-by-cyclic. Finite extensions of $F_r \rtimes \mathbb{Z}$, even if they are torsion-free, will not be free-by-cyclic in general, but in the setting of the following lemma one can prove something in this direction. This lemma is useful when one wants to prove that a manifold is a punctured-torus bundle by studying finite-sheeted coverings of the manifold.

Lemma 3.13. *Let Λ be a torsion-free group and let $\Gamma < \Lambda$ be a subgroup of index d. Suppose that $b_1(\Lambda) = b_1(\Gamma) = 1$. If $\Gamma \cong F_2 \rtimes_\phi \mathbb{Z}$, then $\Lambda \cong F_2 \rtimes_\psi \mathbb{Z}$, with $\phi = \psi^d$.*

Proof. As $b_1(\Gamma) = b_1(\Lambda) = 1$, there are unique normal subgroups $I \triangleleft \Gamma$ and $J \triangleleft \Lambda$ such that $\Gamma/I \cong \Lambda/J \cong \mathbb{Z}$. The image of Γ has finite index in Λ/J, so $I = J \cap \Gamma$. But $I \cong F_2$, and F_2 is not a proper subgroup of finite index in any torsion-free group. To see this, note that such a group would be torsion-free and virtually free of rank 2, and so it must be free of rank 2 (and index 1). Thus $I = J = F_2$, and the image of Γ/I in $\Lambda/J \cong \mathbb{Z}$ has index d. □

4 PROFINITE COMPLETIONS OF 3-MANIFOLD GROUPS

We shall make use of several results about the profinite completions of 3-manifold groups. We summarize these in the following theorem, and quote their origins in the proof. Note that in what follows, by the statement *a compact 3-manifold M contains an incompressible Klein bottle* $\mathbf{K}$, we shall mean that the induced map $\pi_1(\mathbf{K}) \hookrightarrow \pi_1(M)$ is injective.

Theorem 4.1. *Let X be a compact connected 3-manifold with infinite fundamental group.*

1. *If $\widehat{\pi_1(X)} \cong \widehat{\pi_1(M)}$, then $H_1(X, \mathbb{Z}) \cong H_1(M, \mathbb{Z})$.*
2. *$\pi_1(X)$ is good.*
3. *$b_1^{(2)}(X) = 0$ if and only if $\pi_1(X)$ is virtually infinite cyclic or X is aspherical (hence irreducible) and ∂X consists of a (possibly empty) disjoint union of tori and Klein bottles.*
4. *If X is closed and Γ is either free-by-cyclic or else the fundamental group of a non-compact finite volume hyperbolic 3-manifold, then $\widehat{\pi_1(X)}$ and $\widehat{\Gamma}$ are not isomorphic.*
5. *Suppose that X is a Seifert fibered space and that Γ is either the fundamental group of a non-compact finite volume hyperbolic 3-manifold or else of the form $F_r \rtimes_\phi \mathbb{Z}$ where $r \geq 2$ and $[\phi]$ has infinite order in $\mathrm{Out}(F_r)$. Then $\widehat{\pi_1(X)}$ and $\widehat{\Gamma}$ are not isomorphic.*

Proof. (1) is a special case of Lemma 2.1. Using [1] and [37], goodness of compact 3-manifold groups follows from [8] (see also [27] Theorem 7.3 and [4] §6G.24).

Part (3) follows from the calculations of Lott and Lück [21, Theorem 0.1]. Their theorem is stated only for orientable manifolds but this is not a serious problem because, by Lück approximation (which we used to define $b_1^{(2)}(X)$), if X is a non-orientable compact 3-manifold with infinite fundamental group and $Y \to X$ is its orientable double cover, then $b_1^{(2)}(Y) = 2\,b_1^{(2)}(X)$.

To prove (4) we follow the proof of [27, Theorem 8.3]. Item (3) and Proposition 2.8 tell us that $b_1^{(2)}(\Gamma) = 0$, so by Proposition 2.7 we have $b_1^{(2)}(X) = 0$. Moreover, every subgroup of finite index in Γ has non-cyclic finite quotients, so the opposite implication in (3) tells us that X is aspherical. We know that $\pi_1(X)$ is good (item (2)), so since X is closed, $\mathrm{cd}(\widehat{\pi_1 N}) = 3$ by Lemma 2.3. But Γ is also good (by (2) or Lemma 2.6), and $\mathrm{cd}(\widehat{\Gamma}) = 2$, by Lemma 2.3.

To prove the last part, we argue as follows. First, since every subgroup of finite index in Γ has infinitely many finite quotients that are not solvable, we can quickly eliminate all possibilities for X apart from those with geometric structure modelled on $\mathbb{H}^2 \times \mathbb{R}$ and $\widetilde{\mathrm{SL}}_2$. In these two remaining cases, the projection of X onto its base orbifold gives a short exact sequence of profinite groups

$$1 \to \overline{Z} \to \widehat{\pi_1 X} \to \widehat{\Lambda} \to 1,$$

where Z is the infinite cyclic center of $\pi_1 X$ and Λ is a non-elementary discrete group of isometries of $\mathbb{H}^2$. (In fact, following the discussion in §2, we know that $\overline{Z} = \widehat{Z}$, because $\pi_1(X)$ is LERF, but we do not need this.) Note that $\overline{Z}$ is central in $\widehat{\pi_1(X)}$, because if $[z, g] \neq 1$ for some $z \in \overline{Z}$ and $g \in \widehat{\pi_1 X}$, then for some finite quotient $q : \widehat{\pi_1(X)} \to Q$ we would have $[q(z), q(g)] \neq 1$, which would contradict the centrality of Z, since $q(\overline{Z}) = q(Z)$ and $q(\widehat{\pi_1(X)}) = q(\pi_1(X))$.

It is easy to see that the fundamental group of a finite volume hyperbolic 3-manifold has trivial center. A trivial calculation shows that $F_r \rtimes_\phi \mathbb{Z}$ also has trivial center, except when a power of ϕ is an inner automorphism. (In the exceptional case, if t is the generator of the $\mathbb{Z}$ factor and ϕ^r is conjugation by $u \in F_r$, then $t^{-r}u$ is central.)

Since Γ has trivial center, its image under any isomorphism $\widehat{\Gamma} \to \widehat{\pi_1(X)}$ would intersect $\overline{Z}$ trivially, and hence project to a dense subgroup of $\widehat{\Lambda}$. But using Lück approximation, it is easy to see that $b_1^{(2)}(\Lambda) \neq 0$ since Λ is either virtually free of rank at least 2 or virtually a surface group of genus at least 2. And in the light of Theorem 4.1(3) and Proposition 2.8, this would contradict Proposition 2.7.   □

Remark 4.2. In the setting of Theorem 4.1(5), when Γ is the fundamental group of a non-compact finite volume hyperbolic 3-manifold, one can use Proposition 6.6 of [35] together with [37] to prove the stronger statement that $\widehat{\Gamma}$ has trivial center.

Corollary 4.3. *Let X be a compact orientable 3-manifold. If $\widehat{\pi_1(X)} \cong \widehat{\Gamma}$, where $\Gamma = F_r \rtimes \mathbb{Z}$, then X is irreducible and its boundary is a union of t incompressible tori where $1 \leq t \leq b_1(\Gamma)$.*

Proof. Theorem 4.1 (3), (4), (5) tells us that X is irreducible and has non-empty toral boundary. If one of the boundary tori were compressible, then by irreducibilty

X would be a solid torus, which it is not since Γ has non-abelian finite quotients and $\widehat{\Gamma} = \widehat{\pi_1(X)}$.

The upper bound on the number of tori comes from the well-known "half lives, half dies" phenomenon described in the following standard consequence of Poincaré-Lefschetz duality. $\square$

Proposition 4.4. *Let M be a compact orientable 3-manifold with non-empty boundary. The rank of the image of*

$$H_1(\partial M, \mathbb{Z}) \to H_1(M, \mathbb{Z})$$

is $b_1(\partial M)/2$.

It is more awkward to state the analogue of Corollary 4.3 for non-orientable manifolds M. One way around this is to note that an index-2 subgroup of $F_r \rtimes_\phi \mathbb{Z}$ is either $F_r \rtimes_{\phi^2} \mathbb{Z}$ or else is of the form $F_{2r-1} \rtimes \mathbb{Z}$, so the orientable double cover X is of the form described in Corollary 4.3 and each boundary torus in X either covers a Klein bottle in ∂X or a torus (1-to-1 or 2-to-1).

5 THE PROOF OF THEOREMS B AND C

The results in this section form the technical heart of the paper. In the statement of Theorem B, we assumed that $b_1(M) = 1$, so the following theorem applies in that setting. Indeed, in the light of Corollary 4.3 and the comment that follows Proposition 4.4, Theorem 5.1 completes the proof of Theorem B.

Theorem 5.1. *Let M be a compact, irreducible 3-manifold whose boundary is a non-empty union of incompressible tori and Klein bottles. Suppose that there is an isomorphism $\pi_1(M) \to \widehat{F \rtimes \mathbb{Z}}$ such that $\pi_1 M$ has cyclic image under the composition*

$$\pi_1 M \to \widehat{\pi_1 M} \to \widehat{F \rtimes \mathbb{Z}} \to \widehat{\mathbb{Z}},$$

where F is finitely generated and free, and $\widehat{F \rtimes \mathbb{Z}} \to \widehat{\mathbb{Z}}$ is induced by the obvious surjection $F \rtimes \mathbb{Z} \to \mathbb{Z}$. Then, M fibers over the circle with compact fiber (equivalently, $\pi_1(M)$ is of the form $F_r \rtimes \mathbb{Z}$).

Remark 5.2. Given an arbitrary isomorphism $\widehat{\pi_1(M)} \to \widehat{F \rtimes \mathbb{Z}}$, the image of $\pi_1 M$ in $\widehat{\mathbb{Z}}$ will be a finitely generated, dense, free abelian group, but one has to contend with the fact that it might not be cyclic. In Theorem B we overcame this by requiring $b_1(M) = 1$. Boileau and Friedl [5] avoid the same problem by restricting attention to isomorphisms $\widehat{\Gamma}_1 \to \widehat{\Gamma}_2$ that induce an isomorphism on $H^1(\Gamma_i, \mathbb{Z})$. In Theorem 5.1 we remove the difficulty directly with the hypothesis on $\pi_1(M) \to \widehat{\mathbb{Z}}$.

Theorem 5.1 is proved by applying the following result to the map from $\pi_1 M$ to $\widehat{\mathbb{Z}}$. In more detail, in Theorem 5.1 we assume that we have an isomorphism $\widehat{\pi_1(M)} \to \widehat{F \rtimes \mathbb{Z}}$, which we compose with $\widehat{F \rtimes \mathbb{Z}} \to \widehat{\mathbb{Z}}$ to obtain an epimorphism of profinite groups $\Phi : \widehat{\pi_1 M} \to \widehat{\mathbb{Z}}$. Let f denote the restriction of Φ to $\pi_1 M$. The kernel of Φ is isomorphic to the free profinite group $\widehat{F}$; in particular it has cohomological dimension 1. From Proposition 2.4 it follows that the closure of ker f also has cohomological dimension 1. And we have assumed that the image of f is isomorphic to $\mathbb{Z}$, so Theorem 5.3 implies that ker f is finitely generated and free.

Theorem 5.3. *Let M be a compact, irreducible 3-manifold with non-empty boundary which is a non-empty union of incompressible tori and Klein bottles. Let f: $\pi_1(M) \to \mathbb{Z}$ be an epimorphism. Then, either ker f is finitely generated and free (and M is fibered) or else the closure of ker f in $\widehat{\pi_1 M}$ has cohomological dimension at least 2.*

Proof. In Theorem 3 of [12], Freedman and Freedman prove that if M is a compact 3-manifold with each boundary component either a torus or a Klein bottle and $f : \pi_1(M) \to \mathbb{Z}$ has a non-trivial restriction to each boundary component, then either (1) the infinite cyclic covering $M^f \to M$ corresponding to f is homeomorphic to a compact surface times $\mathbb{R}$, or (2) the cover M^f contains a closed 2-sided embedded incompressible surface S.

In Theorem 5.3, we are assuming that the boundary components are incompressible, so if the restriction of f to some component T were trivial, then $\pi_1(T)$ would lie in the kernel of f. Let $A \cong \mathbb{Z}^2$ be a subgroup of finite index in $\pi_1(T)$. From [16], we know that $\pi_1(M)$ induces the full profinite topology on A. Therefore, by Lemma 2.3, the closure of ker f in $\widehat{\pi_1(M)}$ has cohomological dimension at least 2, as required. Thus we may assume that f satisfies the hypotheses of Theorem 3 of [12]. If alternative (1) of that theorem holds, then the kernel of f is a finitely generated free group and we are done. So we assume that alternative (2) holds and consider the closed embedded 2-sided incompressible surface S.

Since S is compact, it does not intersect its translates by suitably high powers of the basic deck transformation of M^f, so we can factor out by such a power to obtain a finite-cyclic covering $N \to M$ in which S is embedded. The proof will be complete if we can prove that $\pi_1(N)$ (equivalently $\pi_1(M)$) induces the full profinite topology on $\pi_1(S)$ (since, by construction, $\pi_1(S)$ is contained in ker f).

It will be convenient in what follows to pass to a finite cover of N and S (which we continue to call N and S) so that both are orientable and S continues to be embedded. Standard separability properties ensure that if $\pi_1(M)$ induces the full profinite topology on $\pi_1(S)$, then this will get promoted to the original surface. Henceforth we assume N and S are orientable.

In more topological language, what we must prove is the following.

Proposition 5.4. *If $p\colon S_1 \to S$ is the finite-sheeted covering corresponding to an arbitrary finite-index subgroup of $\pi_1(S)$, then there is a finite-sheeted covering $q\colon N_1 \to N$ so that p is the restriction of q to a connected component of $q^{-1}(S)$.*

This will be proved in the next section, thus completing the proof of Theorem 5.3. Note that in the hyperbolic case, Proposition 5.4 follows immediately from the separability results of Wise [37]. $\square$

5.1 Surface separability, controlled Dehn filling and aspiral surfaces

What we need in Proposition 5.4 is close to a recent theorem of Przytycki and Wise in [26] (see also [25] and [24]). They prove that for every closed incompressible surface S embedded in a compact 3-manifold M, the image of $\pi_1(S)$ is closed in the profinite topology of $\pi_1(M)$. But we need more (recall §2.1): all finite-index subgroups of $\pi_1(S)$ have to be closed in $\pi_1 M$ as well. In order to prove their results, Przytycki and Wise developed a technology for merging finite covers of the blocks in the JSJ decomposition of the manifold [24], [26], [25]. Liu [19] and his coauthors [10] refined this technology to establish further results, and Liu's refinement in [19] will serve us well here. But since his criterion applies only to closed manifolds, we have to perform Dehn filling on N to get us into this situation.

We introduce some notation. Assume that N has t boundary components all of which are incompressible tori. For a choice of slopes $r_1, \ldots, r_t$, denote by $N(r_1, \ldots, r_t)$ the manifold obtained by performing r_i-Dehn filling on the torus T_i for $i = 1, \ldots, t$.

Lemma 5.5. *Let N and $p: S_1 \to S$ be as above. Then:*

1. *There exist infinitely many collections of slopes $(r_1, \ldots, r_t)$ so that the Dehn fillings $N(r_1, \ldots, r_t)$ of N are irreducible and the image of S in $N(r_1, \ldots, r_t)$ remains incompressible (and embedded).*
2. *There exists a finite cover $q: N' \to N(r_1, \ldots, r_t)$ and an embedding $\iota: S_1 \to N'$ such that $q \circ \iota = p$.*

Proof of Proposition 5.4. Since $S \subset N(r_1, \ldots, r_t)$ lies in the complement of the filling core curves of the filled tori, $S_1 \subset N'$ lies in the complement of the preimage of these curves, so by deleting them we get the desired finite cover $N_1 \to N$. □

Proof of Lemma 5.5. For part (1), incompressibility can be deduced from [9, Theorems 2.4.2 and 2.4.3], as we shall now explain.

One possibility in [9] is that N is homeomorphic to $T^2 \times I$. But we can exclude this possibility because our N contains a closed incompressible surface that is not boundary parallel whereas $T^2 \times I$ contains no such surface.

We now proceed to apply [9] one torus at a time. To ensure that S remains incompressible upon r_1-Dehn filling on T_1, we simply arrange by [9] to choose (from infinitely many possibilities) a slope r_1 that has large distance (i.e., > 2) from any slope for which filling compresses S, as well as from any slope that co-bounds an annulus with some essential simple closed curve on S. Repeating this for each torus T_i in turn ensures that S remains incompressible in $N(r_1, \ldots, r_t)$.

To ensure that $N(r_1, \ldots, r_t)$ is irreducible, we make the additional stipulation (possible by [15]) that for $i = 1, \ldots, t-1$ each r_i-Dehn filling avoids the three possible slopes that result in a reducible manifold upon Dehn filling the torus T_i. (Note that [15] is stated only for manifolds with a single torus boundary component but the proof and result still apply to our context.) Now, for the final torus T_t, we simply choose r_t-Dehn filling so as to avoid the finite number of boundary slopes, and hence we avoid any essential embedded planar surface that could give rise to a reducing sphere in $N(r_1, \ldots, r_t)$.

For the proof of (2), we need to recall some of Liu's terminology [19]. Theorem 1.1 of [19] includes the statement:

A surface Σ is aspiral in the almost fiber part if and only if S is virtually essentially embedded.

Since S is embedded and incompressible in $N(r)$, it is aspiral. Assertion (2) of the lemma is equivalent to the statement that S_1 is virtually essentially embedded, so what we must argue is that S_1 is aspiral.

Let M be a closed orientable 3-manifold and Σ a closed essential surface (of genus ≥ 1) *immersed* into M. The JSJ decomposition of M induces a canonical decomposition of Σ. Following [19], the *almost fiber part* of Σ, denoted $\Phi(\Sigma)$, is the union of all its horizontal subsurfaces in the Seifert manifold pieces, together with the geometrically infinite (i.e., virtual fiber) subsurfaces in the hyperbolic pieces. Note that, by definition, for any finite covering $\Sigma_1 \to \Sigma$, the preimage of the almost fiber part $\Phi(\Sigma)$ is $\Phi(\Sigma_1)$.

To define aspiral we recall Liu's construction of the *spirality character* [19]. Given a principal $\mathbb{Q}^*$-bundle P over $\Phi(\Sigma)$, the spirality character of P, denoted $s(P)$, is the element of $H^1(\Phi(S), \mathbb{Q}^*)$ constructed as follows. For any closed loop $\alpha : [0,1] \to \Phi(\Sigma)$ based at $x_0 \in \Phi(\Sigma)$, each choice of lift for x_0 determines a lift $\tilde{\alpha} : [0,1] \to P$, and a ratio $\tilde{\alpha}(1)/\tilde{\alpha}(0) \in \mathbb{Q}^*$. This ratio depends only on $[\alpha] \in H_1(\Phi(\Sigma), \mathbb{Z})$ and so induces a homomorphism $s(P) : H_1(\Phi(\Sigma), \mathbb{Z}) \to \mathbb{Q}^*$; i.e., an element of $H^1(\Phi(\Sigma), \mathbb{Q}^*)$. The principal bundle P is called *aspiral* if $s(P)$ takes only the values ± 1 for all $[\alpha] \in H_1(\Phi(\Sigma), \mathbb{Z})$. The proof of Theorem 1.1 of [19] involves the construction of a particular principal $\mathbb{Q}^*$-bundle $\mathcal{H}$ (see Proposition 4.1 of [19]). Then, Σ is defined to be *aspiral in the almost fiber part* if the bundle $\mathcal{H}$ is aspiral.

For brevity, set $s = s(\mathcal{H})$. To see that aspirality passes to finite covers of Σ, one argues as follows. First, as above, we note that for a finite covering $\Sigma_1 \to \Sigma$, the almost fibre part $\Phi(\Sigma_1)$ is the inverse image of $\Phi(\Sigma)$. Next we observe that the naturality established in the proof of [19, Proposition 4.1] (more specifically Formula 4.5 and the paragraph following it), shows that s pulls back to the spirality character s_1 of $\Phi(\Sigma_1)$. Indeed the bundle $\mathcal{H}$ in the definition of s, which is independent of the data chosen in the construction, builds in the data for all finite covers of Σ; see §4.2 of [19].

Thus, in our setting, S_1 is aspiral in $\Phi(S_1)$ as required. □

For emphasis, we repeat the key fact that we have extracted from [26] and [19].

Theorem 5.6. *If M is a closed orientable 3-manifold and $S \subset M$ is a closed embedded incompressible surface, then $\pi_1(M)$ induces the full profinite topology on $\pi_1(S)$.*

6   PROFINITE RIGIDITY FOR THE FIGURE-EIGHT KNOT COMPLEMENT

As before, let $\mathcal{K}$ be the figure-eight knot and let $\Pi = \pi_1(\mathbb{S}^3 \smallsetminus \mathcal{K})$.

Theorem 6.1. *Let M be a compact connected 3-manifold. If $\widehat{\pi_1(M)} = \widehat{\Pi}$, then the interior of M is homeomorphic to $\mathbb{S}^3 \smallsetminus \mathcal{K}$.*

Proof. Theorem B tells us that M is a fibered manifold with fundamental group of the form $F_r \rtimes \mathbb{Z}$, and Proposition 3.9 then tells us that $r = 2$ and $\pi_1(M) \cong \Pi =$

$F_2 \rtimes_\phi \mathbb{Z}$. The fundamental group of the fiber in M is the kernel of the unique map $\pi_1(M) \to \mathbb{Z}$, so it is free of rank 2. The only compact surface with fundamental group F_2 that supports a hyperbolic automorphism is the punctured torus, so M is the once-punctured torus bundle with monodromy ϕ, i.e., the complement of the figure-eight knot. $\qquad\square$

Similarly, using Proposition 3.10 one can show that the complement of the trefoil knot and the Gieseking manifold are determined up to homeomorphism by the profinite completions of their fundamental groups.

7   RELATED RESULTS

We have built our narrative around the complement of the figure-eight knot, but our arguments also establish results for larger classes of manifolds. In this section we record some of these results. We write M_ϕ to denote the punctured-torus bundle with monodromy $\phi \in \mathrm{GL}(2, \mathbb{Z})$.

Theorem 7.1. *Let M_ϕ be a once-punctured torus bundle that is hyperbolic. Then there are at most finitely many compact orientable 3-manifolds $M_1, M_2, \ldots, M_n$ so that $\widehat{\pi_1(M_i)} \cong \widehat{\pi_1(M_\phi)}$ and these are all hyperbolic once-punctured torus bundles M_{ϕ_i} with $tr(\phi) = tr(\phi_i)$ and $\det \phi = \det \phi_i$.*

Proof. This follows the proof of Theorem 6.1, except that now we have the finite ambiguity provided by Proposition 3.2 rather than uniqueness. $\qquad\square$

Theorem 7.2. *Let $\mathcal{K}_1, \mathcal{K}_2 \subset S^3$ be knots and let $\Gamma_i = \pi_1(S^3 \smallsetminus K_i)$. Assume that the complement of $\mathcal{K}_1$ fibers over the circle with fiber a surface of genus g. If $\widehat{\Gamma}_1 \cong \widehat{\Gamma}_2$, then the complement of $\mathcal{K}_2$ fibers over the circle with fiber a surface of genus g.*

Proof. From Theorem B we know that $S^3 \smallsetminus K_2$ is fibered and from Lemma 3.1 we know that the fibers have the same euler characteristic. That the genus is the same now follows, since in both cases the fiber is a Seifert surface and so the surface has a single boundary component, hence the same genus. $\qquad\square$

Theorem 7.3. *Let M and N be compact orientable 3-manifolds with $\widehat{\pi_1(N)} \cong \widehat{\pi_1(M)}$. Suppose that ∂M is an incompressible torus and that $b_1(M) = 1$. If M fibers over the circle with fiber a surface of euler characteristic χ, then N fibers over the circle with fiber a surface of euler characteristic χ, and ∂N is a torus.*

Proof. This follows from Theorem B, Lemma 3.1 and Corollary 4.3. $\qquad\square$

Theorem 4.1(5) shows that the fundamental group of a torus knot complement cannot have the same profinite completion as that of a Kleinian group of finite co-volume.

Theorem 7.4. *Let $K_1, K_2 \subset S^3$ be knots whose complements are geometric. Let $\Gamma_i = \pi_1(S^3 \smallsetminus K_i)$ and assume that $\widehat{\Gamma}_1 \cong \widehat{\Gamma}_2$. Then:*

- K_1 is hyperbolic if and only if K_2 is hyperbolic.
- K_1 is fibered if and only if K_2 is fibered.

The following is simply a summary of earlier results.

Theorem D. *Let N and M be compact connected 3-manifolds. Assume $b_1(N) = 1$ and $\pi_1(N) \cong F_r \rtimes_\phi \mathbb{Z}$. If $\widehat{\pi_1(N)} \cong \widehat{\pi_1(M)}$, then $b_1(M) = 1$ and*

1. *$\pi_1 M \cong F_r \rtimes_\psi \mathbb{Z}$, for some $\psi \in Out(F_r)$,*
2. *M is fibered, and*
3. *∂M is either a torus or a Klein bottle.*

Moreover, when $r = 2$,

4. *for each N there are only finitely many possibilities for M,*
5. *M is hyperbolic if and only if N is hyperbolic, and*
6. *if M is hyperbolic, it is a once-punctured torus bundle.*

REFERENCES

[1] I. Agol, *The virtual Haken conjecture*, with an appendix by I. Agol, D. Groves, and J. Manning, Documenta Math. **18** (2013), 1045–1087.

[2] I. Agol, *Virtual properties of 3-manifolds*, Proceedings of the International Congress of Mathematicians, Seoul 2014, **Vol 1**, 141–170, Kyung Moon Sa, Seoul, 2014.

[3] M. Aka, *Arithmetic groups with isomorphic finite quotients*, J. Algebra **352**, (2012), 322–340.

[4] M. Aschenbrenner, S. Friedl and H. Wilton, *3-Manifold groups*, EMS Series of Lectures in Mathematics. European Mathematical Society, Zürich, 2015.

[5] M. Boileau and S. Friedl, *The profinite completion of 3-manifold groups, fiberedness and the Thurston norm*, arXiv:1505.07799, and elsewhere in this volume.

[6] M. R. Bridson, M. Conder and A. W. Reid, *Determining Fuchsian groups by their finite quotients*, Israel J. Math. **214** (2016), 1–41.

[7] M. R. Bridson, A. W. Reid and H. Wilton, *Profinite rigidity and surface bundles over the circle*, Bull. Lond. Math. Soc. **49** (2017), 831–841.

[8] W. Cavendish, *Finite-sheeted covering spaces and solenoids over 3-manifolds*, PhD Thesis, Princeton University (2012).

[9] M. Culler, C. M. Gordon, J. Luecke and P. B. Shalen, *Dehn surgery on knots*, Annals of Math. **125** (1987), 237–300.

[10] P. Derbez, Y. Liu and S.-C. Wang, *Chern–Simons theory, surface separability, and volumes of 3-manifolds*, J. Topol. **8** (2015), 933–974.

[11] J. D. Dixon, E. W. Formanek, J. C. Poland and L. Ribes, *Profinite completions and isomorphic finite quotients*, J. Pure Appl. Algebra **23** (1982), 227–231.

[12] B. Freedman and M. H. Freedman, *Kneser-Haken finiteness for bounded 3-manifolds, locally free groups and cyclic covers*, Topology **37** (1998), 133–147.

[13] L. Funar, *Torus bundles not distinguished by TQFT invariants*, Geometry and Topology **17** (2013), 2289–2344.

[14] D. Gaboriau, *Invariants ℓ^2 de relations d'équivalence et de groupes*, Publ. Math. I.H.E.S. **95** (2002), 93–150.

[15] C. McA. Gordon and J. Luecke, *Reducible manifolds and Dehn surgery*, Topology **35** (1996), 385–409.

[16] E. Hamilton, *Abelian subgroup separability of Haken 3-manifolds and closed hyperbolic n-orbifolds*, Proc. London Math. Soc. **83** (2001), 626–646.

[17] J. Hempel, *Some 3-manifold groups with the same finite quotients*, arXiv:1409.3509.

[18] A. Jaikin-Zapirain, *Recognition of being fibred for compact 3-manifolds*, 2018 preprint, to appear in Geometry and Topology.

[19] Y. Liu, *A characterization of virtually embedded subsurfaces in 3-manifolds*, Trans. Amer. Math. Soc. **369** (2017), 1237–1264.

[20] Y. Liu, *Erratum to A characterization of virtually embedded subsurfaces in 3-manifolds*, Trans. Amer. Math. Soc. **369** (2017), 1513–1515.

[21] J. Lott and W. Lück, *L^2-topological invariants of 3-manifolds*, Invent. Math. **120** (1995), 15–60.

[22] W. Lück, *L^2-invariants of regular coverings of compact manifolds and CW-complexes*, Handbook of Geometric Topology, North-Holland, Amsterdam (2002), 735–817.

[23] N. Nikolov and D. Segal, *On finitely generated profinite groups. I. Strong completeness and uniform bounds*, Annals of Math. **165** (2007), 171–238.

[24] P. Przytycki and D. T. Wise, *Graph manifolds with boundary are virtually special*, J. Topology **7** (2014), 419–435.

[25] P. Przytycki and D. T. Wise, *Mixed 3-manifolds are virtually special*, J. Amer. Math. Soc. **31** (2018), 319–347.

[26] P. Przytycki and D. T. Wise, *Separability of embedded surfaces in 3-manifolds*, Compositio Math. **150** (2014), 1623–1630.

[27] A. W. Reid, *Profinite properties of discrete groups*, in "Groups St Andrews 2013," 73–104, London Math. Soc. Lecture Note Ser. 422, Cambridge Univ. Press, Cambridge, 2015.

[28] L. Ribes and P. A. Zalesskii, *Profinite groups*, Ergeb. der Math. **40**, Springer-Verlag (2000).

[29] R. Riley, *A quadratic parabolic group*, Math. Proc. Camb. Phil. Soc. **77** (1975), 281–288.

[30] J-P. Serre, *Galois cohomology*, Springer-Verlag (1997).

[31] J. R. Stallings, *On fibering certain 3-manifolds*, Topology of 3-manifolds and related topics (Proc. Univ. of Georgia Institute, 1961), Prentice-Hall, Englewood Cliffs, NJ, 1962, 95–100.

[32] P. Stebe, *Conjugacy separability of groups of integer matrices*, Proc. Amer. Math. Soc. **32** (1972), 1–7.

[33] O. Taussky-Todd, *Introduction into connections between algebraic number theory and integral matrices*, Appendix to: H.Cohn, A classical invitation to algebraic numbers and class fields, Universitext, Springer-Verlag, 1978.

[34] W. P. Thurston, *The geometry and topology of 3-manifolds*, Princeton University mimeographed notes, (1979).

[35] H. Wilton and P. A. Zalesskii, *Distinguishing geometries using finite quotients*, Geom. Topol. **21** (2017), 345–384.

[36] H. Wilton and P. A. Zalesskii, *Profinite detection of 3-manifold decompositions*, Compositio Math. **155** (2019), 246–259.

[37] D. T. Wise, *The structure of groups with a quasi-convex hierarchy*, preprint 2012.

Coxeter Groups and Random Groups

DANNY CALEGARI

1 INTRODUCTION

The recent combined work of Wise [7], Ollivier-Wise [6] and Agol [1] shows that random groups at density $<1/6$ virtually embed in right-angled Artin groups. Our main theorem is complementary to this (at least in flavor), and is concerned with virtually embedding certain classes of Coxeter groups in random groups (at any density).

A *superideal* hyperbolic polyhedron is obtained by taking a polyhedron P in real projective space (of some dimension) and intersecting it with the region H bounded by a quadric (which is identified with hyperbolic space in the Klein model), so that one obtains an infinite hyperbolic polyhedron $P \cap H$. If P is a regular polyhedron and H is centered at the center of P, one obtains a *regular superideal polyhedron* $P \cap H$, for which the symmetries of P are realized as hyperbolic isometries of $P \cap H$. For judiciously chosen P, the dihedral angles of $P \cap H$ might be of the form $2\pi/m$ for some integer m, and hyperbolic space can then be tiled by copies of $P \cap H$. The symmetry group of this tiling is a *superideal reflection group*.

In every dimension d, there are two interesting infinite families of convex cocompact hyperbolic reflection groups—the *superideal simplex* reflection group, and the *superideal cube* reflection group. We denote the simplex groups by $\Delta(m, d)$ and the cube groups by $\square(m, d)$, where the dihedral angles of the superideal regular polyhedra in each case are $2\pi/m$.

These reflection groups are Coxeter groups; for $\Delta(m, d)$ the Coxeter diagram is

whereas for $\square(m, d)$ the Coxeter diagram is

where in each case there are $d + 1$ nodes.

The main result we prove in this paper is that for any d, a random finitely presented group at any density less than a half (or in the few relators model) contains quasiconvex subgroups which are commensurable with some $\Delta(m, d)$ for some large m, and with some $\square(m, d)$ for some large (possibly different) m, with overwhelming probability.

The two cases are treated in a similar way, but the combinatorial details of the argument are different in each case. The precise statements of our main theorems are the Superideal Simplex Theorem and the Superideal Cube Theorem, stated in § 2. The proof of these theorems is carried out in § 3 and § 4.

These theorems generalize previous joint work of the author with Alden Walker [2] and with Henry Wilton [3], and the architecture of the proof greatly resembles the proofs of the main theorems in those papers. The new ideas in this paper are mainly combinatorial in nature.

2 STATEMENTS OF THE MAIN THEOREMS

We first discuss the superideal simplex groups $\Delta(m, d)$.

Example 2.1. Some low-dimensional examples are familiar:

- In any dimension d, taking $m = 3$ gives the finite Coxeter group A_{d+1}, which is just the symmetric group S_{d+2}. Similarly, in any dimension d, taking $m = 4$ gives the finite Coxeter group BC_{d+1}, the symmetry group of the $(d+1)$-dimensional cube (or equivalently of the $(d+1)$-dimensional cross-polytope), known as the *hyperoctahedral group*.
- In dimension $d = 2$ there is an (ordinary) hyperbolic simplex with angles $2\pi/m$ whenever $m \geq 7$. The groups $\Delta(m, 2)$ are all commensurable, and are commensurable with the fundamental groups of all closed surfaces of genus at least 2.
- In dimension $d = 3$ the ideal regular simplex has $m = 6$; the group $\Delta(6, 3)$ is commensurable with the fundamental group of the complement of the figure 8 knot. For $m \geq 7$ the superideal "simplices" have infinite volume, but the groups they generate are convex cocompact, with limit set a Sierpinski carpet. As $m \to \infty$ the limit sets converge to the Apollonian gasket, with Hausdorff dimension approximately 1.3057.
- In dimension $d \geq 4$ the simplices are genuinely superideal whenever $m \geq 5$.

Our first main theorem says that for every fixed dimension d, a "generic" finitely presented group will contain some subgroup isomorphic to a finite index subgroup of some element of the $\Delta(m, d)$ family. Explicitly, we show:

Superideal Simplex Theorem. *For any fixed d, a random group at any density $< 1/2$ or in the few relators model contains (with overwhelming probability) a subgroup commensurable with the Coxeter group $\Delta(m, d)$ for some $m \geq 7$, where $\Delta(m, d)$ is the superideal simplex group with Coxeter diagram*

where there are $d + 1$ nodes.

Example 2.2. Some special cases of this theorem were already known:

- Taking $d = 2$, this theorem implies that a random group contains surface subgroups (with overwhelming probability). This is the main theorem of Calegari-Walker in the paper [2].

- Taking $d = 3$, this theorem implies that a random group contains a subgroup isomorphic to the fundamental group of a hyperbolic 3-manifold with totally geodesic boundary. This is the main theorem of Calegari-Wilton in the paper [3]. In fact, the Commensurability Theorem proved in that paper is *exactly* the statement of our main theorem in the case $d = 3$.

We now discuss the superideal cube groups $\square(m, d)$.

Example 2.3. The low-dimensonal examples are again familiar:

1. In any dimension d, taking $m = 3$ gives the finite Coxeter group BC_{d+1}, the hyperoctahedral group. Taking $m = 4$ gives the *hypercubic honeycomb* $\tilde{C}_{d+1}$, which is virtually abelian, and acts cocompactly on Euclidean space.
2. In dimension $d = 2$ the group $\square(m, 2)$ is the symmetry group of the tiling by regular right-angled m-gons, whenever $m \geq 5$. These groups are all commensurable with each other, and with the fundamental groups of all closed surfaces of genus at least 2 (and with all $\Delta(m', 2)$).
3. In dimension $d = 3$ the group $\square(5, 3)$ is the symmetry group of the tiling by regular right-angled dodecahedra; this is commensurable with the (orbifold) fundamental group of the orbifold with underlying space the 3-sphere, and cone angle π singularities along the components of the Borromean rings, as made famous by Thurston. For $m \geq 6$ the groups $\square(m, 3)$ are convex cocompact but not cocompact. Similarly, in dimension $d = 4$ the group $\square(5, 4)$ is the symmetry group of the tiling by regular right-angled 120-cells, whereas $\square(m, 4)$ is convex cocompact but not cocompact for $m \geq 6$.
4. In dimension $d \geq 5$ the groups $\square(m, d)$ are never cocompact when $m \geq 5$.

The analog of the Superideal Simplex Theorem for such groups is the following:

Superideal Cube Theorem. *For any fixed d, a random group at any density $< 1/2$ or in the few relators model contains (with overwhelming probability) a subgroup commensurable with the Coxeter group $\square(m, d)$ for some $m \geq 5$, where $\square(m, d)$ is the superideal cube group with Coxeter diagram*

where there are $d + 1$ nodes.

The case $d = 2$ reduces to the existence of closed surface subgroups, proved by Calegari-Walker in [2], but all other cases are new.

We prove these theorems under the following two simplifying assumptions:

1. that the length n of the relations is divisible by a finite list of specific integers (implicitly depending on the density D); and
2. that the number k of free generators is sufficiently large depending on d in either case; for the Superideal Simplex Theorem we assume $2k - 1 \geq d + 1$, whereas for the the Superideal Cube Theorem we assume $2k - 1 \geq 2d + 1$.

These assumptions simplify the combinatorics, but they are superfluous and at the end of § 4.8 we indicate why they can be dispensed with, and the theorems are true in full generality.

3   SPINES

The proof of our main theorems is entirely direct and constructive; it depends on building certain 2-dimensional complexes subject to local and global combinatorial constraints from pieces which correspond to relators in the random group (in a precise way). In this section we describe these combinatorial constraints.

3.1   Sets as modules

We frequently need to discuss finite sets together with an action of a finite group. If a finite group H acts on a set A we say that A is an H-*module*. The cases of interest in this paper are:

- the symmetric group S_d acting on a d-element set by the standard permutation action; and
- the hyperoctahedral group BC_d acting on a $2d$-element set of d pairs by the standard action, which permutes the pairs and acts on each pair as $\mathbb{Z}/2\mathbb{Z}$.

By abuse of notation, we refer to a set together with such structure as an S_d-module or BC_d-module respectively. If A is a $2d$ element set with a BC_d-module structure, and a is an element of A, the other element of the pair containing a is said to be *antipodal* to a.

3.2   Graphs and 2-complexes

Let G be a random group with k generators at density $D < 1/2$ and length n. That means G is the group defined by a presentation

$$G := \langle x_1, \cdots, x_k \mid r_1, r_2, \cdots, r_\ell \rangle$$

where $\ell = (2k-1)^{nD}$, and where each r_i is chosen randomly and independently from the set of all (cyclically) reduced words in the free group $F_k := \langle x_1, \cdots, x_k \rangle$ of length n, with the uniform distribution. For an introduction to random groups see [4] or [5].

We let $r = r_1$, and let G_r denote the random 1-relator group with presentation

$$G_r := \langle x_1, \cdots, x_k \mid r \rangle.$$

The group G is the fundamental group of a 2-complex K, with one vertex, with one edge for each generator, and with one 2-cell for each relator. We denote the 1-skeleton of K by X; thus X is a k-fold rose. The group G_r is the fundamental group of a subcomplex K_r, whose 1-skeleton is equal to X, and which contains only one 2-cell, the cell attached by the relation $r = r_1$.

Definition 3.1. *A graph over X is a graph Σ together with a simplicial immersion from Σ into X.*

A morphism over X *between oriented graphs* $\Sigma \to X$, $\Sigma' \to X$ *over* X *is a simplicial immersion* $\Sigma \to \Sigma'$ *such that the composition* $\Sigma \to \Sigma' \to X$ *is the given immersion from* $\Sigma \to X$.

If Σ is a graph over X, the oriented edges of Σ can be labeled by the x_i and their inverses, by pulling back the labels from the edges of X. Conversely, a graph with oriented edges labeled by the x_i and their inverses, has the structure of a graph over X provided no adjacent oriented edges have inverse labels.

Let L be a finite union of simplicial circles (i.e., circles subdivided into edges) with oriented edges labeled by generators or their inverses, in such a way that the cyclic word on each component of L is r. Then L is a graph over X.

We will build a simplicial graph Σ over X together with a morphism $L \to \Sigma$ over X satisfying certain conditions. These conditions are different for the Superideal Simplex Theorem and for the Superideal Cube Theorem, but they have common features, which we now describe.

Definition 3.2. *An* (m, d)*-regular simplicial spine or* (m, d)*-regular cubical spine is a graph* Σ *over* X, *together with a morphism* $L \to \Sigma$ *over* X *with the following properties:*

1. *The graph* Σ *is obtained by subdividing a regular graph. That is, all the vertices are either 2-valent (we call these* internal*) or of the same fixed valence* > 2 *(we call these* genuine*). In the simplex case, each genuine vertex is* $d+1$*-valent; in the cube case, each genuine vertex is* $2d$*-valent and further admits the structure of a* BC_d*-module. By abuse of notation, we omit the adjective "genuine" when discussing higher valence vertices unless the meaning would be ambiguous.*
2. *The map* $L \to \Sigma$ *is* d *to 1 in the simplex case, and* $2(d-1)$ *to 1 in the cube case; i.e., each edge of* Σ *is the preimage of exactly* d *(resp.* $2(d-1)$*) edges of* L. *Furthermore, the set of preimages of each edge has the structure of an* S_d *(resp.* BC_{d-1}*)-module.*
3. *For each vertex* v *of* Σ *and each pair of distinct incident edges* e, e' *which are not antipodal in the cube case, there is exactly one segment of* L *of length 2 whose midpoint maps to* v *and whose adjacent edges map to* e *and* e'. *In the cubical case, if* $L(e)$ *is the set of* $2(d-1)$ *edges of* L *mapping to* e *with its* BC_{d-1}*-module structure, then the* $2(d-1)$ *edges in* L *adjacent to* $L(e)$ *map bijectively to the* $2(d-1)$ *edges of* Σ *not equal to* e *or its antipode; this latter set also has a natural* BC_{d-1}*-module structure by restriction, and we require that this map of* BC_{d-1}*-modules respect the module structure; see Figure 1 for the local picture in the simplicial case and Figure 2 for the cubical case in dimensions 2 and 3.*
4. *Each component of* L *has exactly* m *vertices which map to genuine vertices of* Σ.
5. Σ *satisfies the cocycle condition (to be defined shortly).*

Let $L \to \Sigma$ be a spine satisfying the first three conditions to be (m, d)-regular. The mapping cylinder M is a 2-complex which admits a canonical local embedding into $\mathbb{R}^d$ in such a way that the (combinatorial) symmetries of the link of each vertex or edge are realized by isometries of the ambient embedding (taking module structure into account). Taking a tubular neighborhood of this canonical thickening gives rise to a D^{d-2}-bundle over M with a flat connection with holonomy in the symmetric group S_{d-1} in the simplicial case, or BC_{d-2} in the cubical case, acting by the standard representation. The restriction of this bundle to each component

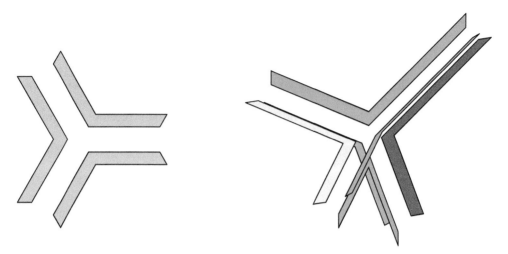

Figure 1: Local structure of simplicial spine near vertex in dimensions 2 and 3.

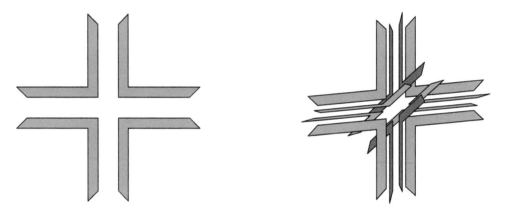

Figure 2: Local structure of cubical spine near vertex in dimensions 2 and 3.

of L therefore determines a conjugacy class in S_{d-1} or BC_{d-2} respectively. The *cocycle condition* is the condition that this conjugacy class is trivial, for each component of L.

Let $L \to \Sigma$ be an (m, d)-regular spine. Form the mapping cylinder M and then let $\overline{M}$ be the 2-complex obtained from M by attaching a disk to each component of L. Since each component of L has the label r, the tautological immersion $L \to X$ extends to a cellular immersion $\overline{M} \to K_r$.

The defining properties of an (m, d)-regular simplicial (resp. cubical) spine ensure that the 2-complex $\overline{M}$ is locally isomorphic to the 2-dimensional spine $\mathcal{S}$ which is dual to the tiling of d-dimensional hyperbolic space by regular superideal simplices (resp. cubes) with dihedral angles π/m. To see this, observe that the vertex stabilizers in the symmetry group of $\mathcal{S}$ are isomorphic to S_d (resp. BC_{d-1}) and their action on vertex links of $\mathcal{S}$ agrees with the module structure on the vertex links of $\overline{M}$. There is therefore a developing map from the universal cover of $\overline{M}$ to $\mathcal{S}$ which is an immersion, and therefore an isomorphism. This realizes the fundamental group of $\overline{M}$ as a subgroup of $\Delta(m, d)$ (resp. $\square(m, d)$) which is of finite index, since $\overline{M}$ is compact.

We will show in the next section that for r a sufficiently long random word, we can construct an (m, d)-regular spine (of either kind) in which all the maximal paths in Σ containing only internal vertices in their interior (informally, the "topological edges" of Σ) are bigger than any specified constant λ. This is the main ingredient necessary to show that the immersion $\overline{M} \to K_r \to K$ is π_1-injective, and stays injective when a further $(2k-1)^{nD} - 1$ random relators are added.

4   CONSTRUCTING SPINES

We start with L, a finite union of simplicial circles each labeled with the relator r. The spine Σ is obtained from L by identifying oriented edges of L in groups of d or $2(d-1)$, respecting labels, in such a way as to satisfy the conditions of a regular spine. We thus obtain Σ as the limit of a (finite) sequence of quotients $L \to \Sigma_i$ limiting to Σ, where each $L \to \Sigma_i$ is a morphism over X. Thus, each intermediate Σ_i should immerse in X. We call an identification of subgraphs of Σ_{i-1} with the same labels, giving rise to a quotient $\Sigma_{i-1} \to \Sigma_i$ over X a *legal quotient* if Σ_i is immersed in X. We assume in the sequel that our quotients are always legal.

In our intermediate quotients Σ_i, some edges are in the image of d (resp. $2(d-1)$) distinct edges of L, and some are in the image of a unique edge of L. We call edges of the first kind *glued*, and edges of the second kind *free*.

In order to simplify notation, in the sequel we let $\eth$ denote either d in the simplicial case, or $2(d-1)$ in the cubical case.

We choose a big number λ divisible by $\eth$, and an even number N with $N \gg \lambda$, and assume for the sake of convenience that $\lambda \cdot N$ divides n, the length of the relator r (the size of λ that we need will ultimately depend only on the density D, but not on the length n of the relators). Then we subdivide each component of L into consecutive *segments* of length λ.

4.1   Creating beachballs

A *block* in L is a sequence of N consecutive segments. Since we assume that $\lambda \cdot N$ divides n, each component of L can be subdivided into $n/(\lambda \cdot N)$ consecutive blocks. Each block is made up of alternating *odd segments* and *even segments*; by abuse of notation we refer to the segment immediately before the first segment in a block (which is contained in the previous block) as the "first even segment." We say that a $\eth$-tuple of blocks is *compatible* if their odd segments can all be identified in groups of $\eth$ (in the order in which they appear in the blocks) in such a way that the resulting quotient is legal. This means that each $\eth$-tuple of even segments at the same location in the $\eth$ blocks must start and end with *different* letters, and the same must be true for the last $\eth$-tuple of even segments in the blocks immediately preceding the given collection of blocks. The existence of a compatible collection of blocks requires $2k - 1 \geq \eth$.

When $\eth$ blocks are glued together along their odd segments, each collection of even segments *except* for the first and last one is identified at their vertices, producing a *beachball*—a graph with two vertices and $\eth$ edges, each joining one vertex to the other. See Figure 3.

We say that a $(\eth + 1)$-tuple of blocks is *supercompatible* if each sub $\eth$-tuple is compatible. Evidently, the existence of supercompatible tuples requires $2k - 1 \geq \eth + 1$, which is our second simplifying assumption.

Figure 3: A beachball for $\mathfrak{d} = 9$.

Recall that $\mathfrak{d} = d$ in the simplicial case, and $\mathfrak{d} = 2(d-1)$ in the cubical case. Suppose r is a random reduced word of length n. Since r is random, if n is sufficiently long it is possible to partition a proportion $(1 - \epsilon)$ of the blocks into compatible $\mathfrak{d}$-tuples, and glue their odd segments, for ϵ depending only on n, and therefore $1/\epsilon \gg N, \lambda$, with probability $1 - O(e^{-n^c})$. This leaves a proportion ϵ of the blocks left unglued. For each of these unglued blocks, find a collection of $\mathfrak{d}$ glued blocks so that the $\mathfrak{d} + 1$ tuple is supercompatible. If we unglue the collection of $\mathfrak{d}$ glued blocks, and then take $\mathfrak{d} + 1$ copies of our graph, we can glue the resulting $\mathfrak{d}(\mathfrak{d} + 1)$ blocks in compatible groups of $\mathfrak{d}$.

In the cubical case, in order to give each collection of edges the natural structure of a BC_{d-1}-module, it is convenient to first take some finite number of disjoint copies of L and partition them into $2(d-1)$ subsets of equal size. The group BC_{d-1} acts on each set of $2(d-1)$ copies of L as a standard representation, giving these components the structure of a BC_{d-1}-module. When we partition blocks into $\mathfrak{d}$-tuples (where $\mathfrak{d} = 2(d-1)$ in the cubical case), each tuple should contain exactly one block from each subset, so that the elements of the block inherit a natural BC_{d-1} action.

Thus at the end of this step, we have constructed Σ_1 which consists of collections of glued edges of length λ (each the image of $\mathfrak{d}$ edges of L), a *reservoir* of beachballs with every edge of length λ and all possible edge labels in almost uniform distribution of possible edge labels, and a *remainder* consisting of a $\mathfrak{d}$-valent graph in which every edge has length λ. Note that the remainder is obtained as the union of the first and last (even) segments from each block, so the mass of the remainder is of order $O(1/N)$, where by assumption we have chosen $N \gg \lambda$. Furthermore in the cubical case, for each component X of the reservoir or remainder, the edges of X incident to each vertex inherit a natural BC_{d-1}-module structure.

4.2   Covering move

The next move is called the *covering move*, and its effect is to *undo* some of the gluing of some previous step, and then to *reglue* in such a way that the net effect is to transform some collection of beachballs into more complicated graphs, each of which is a finite cover of a beachball, with prescribed topology and edge labels. This

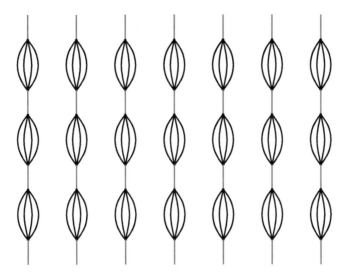

Figure 4: A matrix of 7 columns of 3 consecutive beachballs. In this and subsequent figures, glued edges are in red and free edges are in black.

move does *not* necessarily preserve the total mass of beachballs, and can be used to adjust the total mass of beachballs of any specific kind, at the cost of transforming some other (prescribed) set of beachballs into covers.

To describe this move, we first take some consecutive strings of beachballs and arrange them in a "matrix" form, where each string of beachballs is a column of the matrix, and the different strings make up the rows of the matrix. See Figure 4 for a matrix consisting of 7 columns of 3 consecutive beachballs.

The covering move takes as input a matrix of s columns of r consecutive beach-balls with compatible labels on corresponding glued segments. This means the fol-lowing. In each column there are $r + 1$ consecutive glued segments interspersed with the r beachballs. Thus between each pair of consecutive rows of beachballs there is a row of glued segments, each consisting of ∂ glued edges with some label. We require that all the labels on glued edges in the same row are equal.

In the cubical case, the edges in each column have a natural structure of a BC_{d-1}-module. Thus it is possible to "trivialize" the module structure on the union of edges in each row, giving them the structure of a product of a standard BC_{d-1}-module with an s-element set.

The covering move pulls apart each interior row of glued segments (i.e., the rows between a pair of consecutive rows of beachballs) into ∂s edges, all with the same label, then permutes them into s new sets of ∂ elements and reglues them. More specifically, if we identify the set of ∂s edges with the product $\{1, \cdots, s\} \times \{1, \cdots, \partial\}$ where BC_{d-1} acts on the second factor, we permute each $\{1, \cdots, s\} \times i$ factor by some element of S_s. Thus the move is described on each of the rows by an element $p_j \in S_s^{\partial}$, the product of ∂ copies of the symmetric group S_s, for $0 \leq j \leq r$ where p_0 and p_r are the identity permutation.

For each $1 \leq j \leq r$ the s beachballs in the jth row are rearranged by this move into a graph which is a (possibly disconnected) covering space of a beachball of degree s; which covering space is determined by the permutations p_{j-1} and p_j. If $p_{j-1} = p_j$ it is a trivial covering space, in the sense that it consists just of s disjoint beachballs. However, even if it is trivial as a covering space, if p_{j-1} and p_j are not

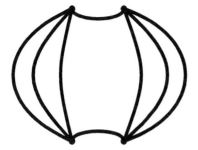 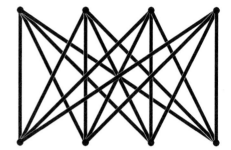

Figure 5: A barrel and a bipart for $\mathfrak{d} = 4$.

the identity, the edges making up each beachball might be different from the edges making up the beachballs before the move. We allow two possibilities for the labels on the new beachball covers in each row:

1. the beachball cover in the given row can have arbitrary topology, and the labels on the edges are legal; or
2. the beachball cover in the given row is trivial (i.e., $p_{j-1} = p_j$) and on each beachball the edges all have the *same* label.

In the former case we do nothing else to the row. In the latter case, since all the edge labels are equal (and the gluing is not legal as it stands), we collapse the beachball to the interior of a glued segment of length 3λ. So the net effect of the covering move is to transform some set of beachballs into covering spaces of the same mass, while possibly eliminating some set of beachballs of comparable mass.

Two kinds of beachball covering spaces are especially important in what follows; these are

- a 2-fold cover of a beachball called a *barrel*; and
- a $\mathfrak{d}$-fold cover of a beachball called a *bipart*, whose underlying graph is the complete bipartite graph $K_{\mathfrak{d},\mathfrak{d}}$.

See Figure 5.

We now describe two especially useful examples of covering moves which will be used in the sequel.

Example 4.1. **Elimination** The elimination move trades $2\mathfrak{d}$ beachballs for 2 biparts, and eliminates a collection of $\mathfrak{d}$ beachballs with the same labels. It is convenient to perform this move in such a way that the biparts created have labels of *covering type*. This means that the edge labels on the bipart are pulled back from some legal edge labeling on a beachball under the covering projection. See Figure 6.

Example 4.2. **Rolling barrels** The rolling move trades 4 beachballs for 2 barrels. This move is self-explanatory. See Figure 7.

4.3   Tearing the remainder

We now describe a modification of the gluing which replaces the remainder with a new remainder of similar mass, but composed entirely of barrels. In the process, a set of beachballs of similar mass are transformed into biparts.

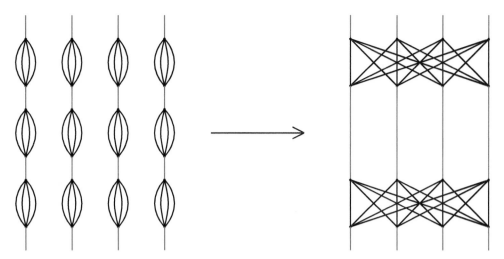

Figure 6: The elimination move for $\mathfrak{d} = 4$.

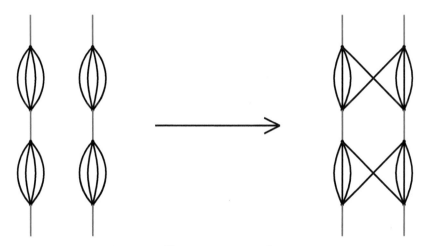

Figure 7: The rolling move for $\mathfrak{d} = 4$.

The tear move is applied to two vertices v, v' in a component Y of the remainder. It uses up the two beachballs in the reservoir which are adjacent to v and v', and a further $3(\mathfrak{d} - 1)$ beachballs drawn from the middle of the reservoir in consecutive groups of three. The result of the move "tears" Y apart at v and v', producing $\mathfrak{d}$ copies of each vertex v_i, v'_i for $1 \le i \le \mathfrak{d}$, each v_i joined to v'_i by $\mathfrak{d} - 1$ new edges (these new $\mathfrak{d}(\mathfrak{d} - 1)$ edges coming from $(\mathfrak{d} - 1)$ of the beachballs), and transforms the remaining $2\mathfrak{d}$ beachballs into two biparts. In the process, $2\mathfrak{d}$ glued edges are unglued, and then reglued in a different configuration; thus it is necessary for the labels on each set of $\mathfrak{d}$ edges to agree. The tear move is illustrated in Figure 8; as in § 4.2, the segments before and after the move have the same vertical coordinate, and the gluing respects BC_{d-1}-module structure on edges in the cubical case.

Applying a tear move at a pair of vertices v, v' in Y creates a collection of *half-barrels*, i.e., a beachball with one edge replaced by a pair of *free edges* joined to nearby *free vertices*; see Figure 9.

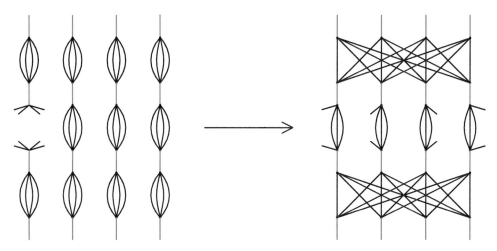

Figure 8: The tear move for $\mathfrak{d} = 4$.

Figure 9: A half-barrel for $\mathfrak{d} = 4$.

We refer to the non-free edges in a half-barrel as *barrel edges*, and observe that the number of non-barrel edges in Y is unchanged by a tear move. Thus after applying the tear move once to create several half-barrels, we can apply the tear move again to the pair of free vertices at the end of one of the half-barrels, in the process creating an honest barrel which may be transferred to the reservoir, reducing the total number of non-barrel edges in Y. We can use total number of non-barrel edges of Y as a measure of complexity, and observe that after applying $O(1/N)$ tear moves, we can tranform the entire remainder to a mass of $O(1/N)$ barrels and biparts, while the reservoir still contains almost exactly the same number of beachballs in almost exactly the uniform distribution.

4.4   Hypercube gluing

At this stage of the construction we have a reservoir consisting of an almost equidistributed collection of beachballs, and a relatively small mass of pieces consisting of biparts and barrels. In this section and the next we explain how to glue up beachballs. There are two different (but related) constructions depending on whether we are in the simplicial or cubical case. In the simplicial case we use the *hypercube gluing*.

A finite cover of the hypercube gluing can be used to glue up all the barrels and biparts produced by the tear moves, together with some other complementary

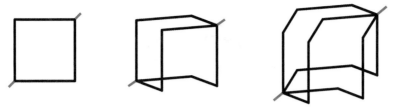

Figure 10: Beachball immersed in the 1-skeleton of a cube in dimensions 2, 3 and 4.

collection of pieces produced by the covering move. The existence of such a combinatorial cover in which the pieces from the remainder can be included follows by
LERF for free groups, and it is straightforward to use the covering move to produce
complementary pieces of the correct topology and labels. An explicit description of
the combinatorics of the necessary covers in dimension 3 is given in [3], § 3.10, and
the general case is very similar.

After dealing with the remainder, almost all the mass of the reservoir can be
completely glued up using hypercube gluing, and the error can be transformed away
using the elimination move, while simultaneously transforming some prescribed
collection of beachballs into biparts of covering type, which can themselves be glued
up by a covering of the hypercube gluing.

It remains to describe the hypercube gluing (recall that we are in the simplicial
case). Let C denote the 1-skeleton of the d-dimensional cube, and with each edge of
length λ/d. Let $e_1, \cdots, e_d$ denote vectors of length λ/d aligned positively along the d
coordinate axes in order. There is a path of length λ from the vertex $(0, 0, \cdots, 0)$ to
the vertex $(\lambda/d, \lambda/d, \cdots, \lambda/d)$ of C, obtained by concatenating straight segments of
length λ/d in the order $e_1, e_2, \cdots, e_d$. Taking d cyclic permutations of this sequence
of paths gives d paths of length λ between extremal vertices of C, whose union is a
beachball. This is illustrated in Figure 10 in dimensions 2, 3 and 4.

There are 2^{d-1} pairs of extremal vertices of C, and the union of 2^{d-1} suitably
labeled beachballs can be arranged along the 1-skeleton of C as above. Gluing the
beachballs in this pattern is legal, and the resulting graph locally satisfies the first
three conditions to be an (m, d)-regular simplicial spine.

4.5   Lens gluing

The analog of the hypercube gluing for cubical spines is the *lens gluing*. This is the
most complicated move in our construction, and we first describe this move in low
dimensions before giving the definition in generality.

In dimension 2 a beachball has degree 2 (i.e., it has 2 edges); $BC_1 = \mathbb{Z}/2\mathbb{Z}$ acts
as the full permutation group. Two beachballs with suitable labels can be identified
along their boundaries, laid out along a circle; see Figure 11.

In dimension 3 a beachball has degree 4; $BC_2 = D_4$ acts by the dihedral group,
preserving or reversing a circular order on the 4 edges. Eight beachballs with suitable
labels can be glued up along a $K_{4,4}$ graph in S^3, where each edge of each beachball
runs along a segment of length 2 in the $K_{4,4}$ graph, as in Figure 12 (where one of
the vertices is at "infinity" in the figure).

The lens gluing in dimension d is closely related to the hypercube gluing in
dimension $d - 1$. Recall the definition of the hypercube gluing move from § 4.4.
In this move, 2^{d-2} beachballs of degree $d - 1$ are draped along the 1-skeleton of a

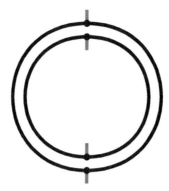

Figure 11: Lens gluing $d=2$.

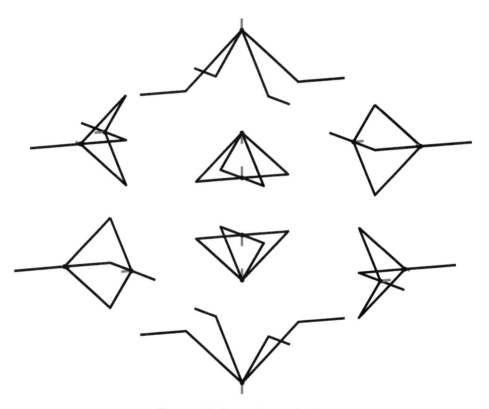

Figure 12: Lens gluing $d=3$.

$d-1$-dimensional cube; each edge of each beachball is subdivided into $d-1$ segments which run between antipodal vertices of the cube. We let C denote the 1-skeleton of the cube, and denote the immersion of a beachball b into C by $b \to C$.

For any graph L, the *spherical graph* associated to L is the graph with one S^0 (i.e., two vertices) for each vertex in L, and one $S^0 * S^0$ for each edge in L. If v is a vertex of L, we denote the corresponding pair of vertices in $S(L)$ by $v0$ and $v1$. Similarly, if e is an edge of L, we denote the corresponding edges in $S(L)$ by

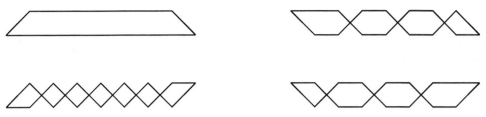

Figure 13: Height graphs for $d = 9$.

$e00$, $e01$, $e10$ and $e11$ (in fact this construction generalizes in an obvious way to simplicial complexes of arbitrary dimension). There is a canonical simplicial map from $S(L) \to L$ which forgets the 01 labels.

Now we apply this construction to C; the lens gluing will map $4 \times 2^{d-2}$ beach-balls of degree $2(d-1)$ into $S(C)$ in such a way that the projection to C will map each 4 beachballs onto each degree $(d-1)$ beachball mapping into C in the hyper-cube gluing, with each edge of the beachball downstairs in the image of two edges upstairs.

If $b \to C$ is a beachball, and $b' \to S(C)$ is one of the four beachballs upstairs mapping to it, each edge of b maps to a path in C of length $d - 1$, which is covered by two paths in $S(C)$ of length $d - 1$. Each such path is determined (given its image in C) by a word of length d in the alphabet $0, 1$. Our beachballs upstairs have the property that the pair of lifts of each edge of b are described by the *same* pair of 01 words. Moreover, the collection of 4 pairs of 01 words describing the lifts of b are the same for all $b \to C$. So to describe the lens move we just need to give 4 sets of pairs of 01 words of length d. We call these *height pairs*. An explicit formula for $d > 2$ is given by

$$
\text{height pairs are} \quad
\begin{cases}
(00^{d-2}0, & 01^{d-2}0) \\[4pt]
(0(01)^{\frac{d-2}{2}}1, & 0(10)^{\frac{d-2}{2}}1) \\[4pt]
(1(1100)^{\frac{d-2}{4}}x, & 1(0011)^{\frac{d-2}{4}}x) \\[4pt]
(1(1001)^{\frac{d-2}{4}}(1-x), & 1(0110)^{\frac{d-2}{4}}(1-x))
\end{cases}
$$

where x is 0 or 1 depending on the parity of d, and where an expression like $w^{p/q}$ for w a word of length q means the initial string of length p of the word $w^{\infty} := www \cdots$. The meaning of this formula is best explained by a picture; see Figure 13.

The case $d = 2$ is degenerate; the case $d = 3$ gives rise to the height pairs

$$
(000, 010), (001, 011), (111, 101), (110, 100)
$$

which reproduces four of the beachballs in the lens gluing for $d = 3$ described above (the other four beachballs are in the preimage of the other beachball in the hyper-cube gluing).

Note that there is an ambiguity in our choice of labels of each pair of vertices of $S(C)$ over v by $v0$ and $v1$; thus the group $\mathbb{Z}/2\mathbb{Z}^{\mathrm{vert}(C)}$ acts by automorphisms of $S(C)$ over C. Thus although this is not evident in the formulae, under this symmetry group all four height pairs are in the same orbit.

Figure 14: Four beachballs in the $d=4$ lens gluing which project to one beachball in the $d=3$ hypercube gluing.

The case $d=4$ is hard to draw without the diagram becoming cluttered; Figure 14 shows the 4 beachballs of degree 6 in the $d=4$ lens gluing which project to one beachball of degree 3 in the $d=3$ hypercube gluing.

4.6   Regularity

The third condition for (m,d)-regularity is that each component of L has exactly m vertices which map to high-valent vertices of Σ. The only move which adjusts the number of such vertices on each component of L is the elimination move. This move takes pieces which may be drawn from any component L; so we simply take enough disjoint copies of L, and symmetrize the components from which the moves are drawn, so that we apply this move the same number of times to every component. This will ensure the third condition for (m,d)-regularity.

4.7   Cocycle condition

The cocycle is sensitive to more combinatorial data than we have been using so far; it depends not only on the set of labels appearing as the edges in a beachball, but the way in which these labels come from consecutive even segments on each of $\eth$ blocks in a compatible $\eth$-tuple. Gluing different compatible $\eth$-tuples gives rise to the same beachball collections but with different cocycles (relative to a trivialization along the odd segments of the blocks). If the original gluing was done randomly, there are pairs of compatible segments whose gluings give rise to the same set of beachballs, but with cocycles differing by any element in the symmetric group S_{d-1} in the simplicial case, or any element of BC_{d-1} in the cubical case; adjusting the gluing by interchanging elements of these pairs gives the same collection of beachballs and the same remainder (and therefore the same gluing problem) but with the cocycle adjusted in the desired away. Boundedly many moves of this kind for each component produces an (m,d)-regular simplicial or cubical spine with all topological edges of length at least λ.

4.8   Few generators

We now indicate how to modify our arguments and constructions so that they hold without our simplifying assumptions.

The first assumption—that the length n of the relators is divisible by some big fixed constant $\lambda \cdot N$ (where also $\eth$ divides λ)—is easy to dispense with. We let ρ be the remainder when dividing n by $\lambda \cdot N$, then find $\eth$ copies of a subword of r of length $\lambda \cdot N + \rho$ so that it is legal to glue these $\eth$ words in $\eth$ distinct copies of r; what is left after this step is a graph with edges whose lengths are divisible by $\lambda \cdot N$, and the rest of the construction can go through as before.

The second assumption—that the number k of free generators satisfies $2k - 1 \geq 2d + 1$—can be finessed by looking more carefully at the construction. It can be verified that we do not really use the hypothesis that the word r is random with respect to the uniform measure on reduced words of length n; instead we can make do with a much weaker hypothesis, namely that we generate random words by *any* stationary ergodic Markov process, whose subword distribution of any fixed length has enough symmetry, and such that there are at least $\eth + 1$ letters that may follow any legal substring.

The relevance of this is as follows: a random reduced word of length n in a free group of rank k can be thought of as a random reduced word of length n/s in a free group of rank $k(2k - 1)^{s-1}$ (whose symmetric generating set is the set of reduced words in F_k of length s) generated by a certain stationary ergodic Markov process. We can perform matching of subwords and build spines in this new "alphabet"; the result will not immerse in X, since there might be folding of subtrees of diameter at most $2s$. But the spine will still have the crucial properties that the fundamental group of the 2-complex $\overline{M}$ is commensurable with $\Delta(m, d)$, and that the number of immersed paths of length ν grows like $(2k - 1)^{s\nu/\lambda}$ where λ is as big as we like. Taking s big enough to "simulate" a group of rank at least $d + 1$, we can then take λ much bigger so that s/λ is arbitrarily small. This small exponential growth rate of subpaths is the key to proving that $\pi_1(\overline{M})$ maps injectively with quasiconvex image.

4.9   Conclusion of the theorem

The remainder of the argument is almost identical to the arguments in [2] and [3], and makes use in the same way of the mesoscopic small cancellation theory for random groups developed by Ollivier [5]. It depends on two ingredients:

1. a *bead decomposition*, exactly analogous to that in [2], § 5 or [3], § 4; and
2. a *mesoscopic small cancellation argument*, which applies techniques of Ollivier as in [2], § 6 or [3], § 6.

We very briefly summarize the argument below just to indicate that no new ideas are needed, but refer the reader to the cited references for details.

A bead decomposition is an (m, d)-regular simplicial or cubical spine Σ which is made up of $O(n^\delta)$ pieces of size $O(n^{1-\delta})$ arranged in a circle, with each piece separated from the next by a *neck* (an unusually long glued segment) of length $C \log n$. We can construct a spine as in [3], § 4 as follows. Fix some small positive constant δ and write r as a product

$$r = r_1 s_1 r_2 s_2 \cdots r_m s_m$$

where each r_i has length approximately $n^{1-\delta}$ and each s_i has length approximately n^δ, and we choose lengths so that m is divisible by $\eth$. We then fix a small positive constant $C < \delta / \log(2k - 1)$ and look for a common subword x of length $C \log n$ in $s_i, s_{i+m/\eth}, \cdots, s_{i+(\eth-1)m/\eth}$ with indices taken mod m, and then glue these distinct copies of x into unusually long segments (called *lips*) of what will become the spine Σ.

The lips partition the remainder of L into subsets b_i each with mass $O(n^{1-\delta})$, and we can extend the partial gluing along the lips into an (m, d)-regular simplicial or cubical spine by gluing intervals in each b_i independently. As in [2], § 5 and [3], § 4 the resulting spine Σ has the property that for any positive β, any immersed segment $\gamma \to \Sigma$ with length βn whose image in X lifts to r or r^{-1} already lifts to L (i.e., it appears in the boundary of a disk of $\overline{M}$). From this and the fact that a random 1-relator group is $C'(\mu)$ for any positive μ it follows that $\overline{M} \to K_r$ is π_1-injective and its image is quasiconvex.

Finally, the fact that the valence of Σ is uniformly bounded and the length of every segment is at least λ where we may choose λ as big as we like implies that $\overline{M} \to K_r$ stays π_1-injective and quasiconvex when we attach the disks corresponding to the remaining $(2k-1)^{nD} - 1$ random relations. This depends on Ollivier's mesoscopic small cancellation theory [5]; the argument is exactly the same as that in [2], § 6 and [3], § 6.

This completes the proof of the Superideal Simplex Theorem and Superideal Cube Theorem.

5 ACKNOWLEDGMENTS

Danny Calegari was supported by NSF grant DMS 1405466. The LaTeX macro for the symbol $\square$ was copied from pmtmacros20.sty by F. J. Yndeeráin. I would like to thank the anonymous referee for their helpful comments.

REFERENCES

[1] I. Agol, *The virtual Haken conjecture*, With an appendix by Agol, Daniel Groves, and Jason Manning. Doc. Math. **18** (2013), 1045–1087.

[2] D. Calegari and A. Walker, *Random groups contain surface subgroups*, Jour. AMS **28** (2015), 383–419.

[3] D. Calegari and H. Wilton, *3-manifolds everywhere*, preprint; arXiv:1404.7043.

[4] M. Gromov, *Asymptotic invariants of infinite groups*, Geometric group theory, vol. 2; LMS Lecture Notes **182** (Niblo and Roller eds.), Cambridge University Press, Cambridge, 1993.

[5] Y. Ollivier, *Some small cancellation properties of random groups*, Internat. J. Algebra Comput. **17** (2007), no. 1, 37–51.

[6] Y. Ollivier and D. Wise, *Cubulating random groups at density less than 1/6*, Trans. AMS **363** (2011), no. 9, 4701–4733.

[7] D. Wise, *The structure of groups with a quasiconvex hierarchy*, preprint.

Cries and Whispers in Wind-Tree Forests

Vincent Delecroix and Anton Zorich

To the memory of Bill Thurston with admiration for his fantastic imagination.

Ты, мой лес и вода! кто объедет, а кто, как сквозняк,
проникает в тебя, кто глаголет, а кто обиняк...

И. Бродский

"You, my forest and water! One swerves, while the other shall spout
Through your body like draught; one declares, while the first has a
doubt."
J. Brodsky

"Mein Vater, mein Vater, und hoerest du nicht,
Was Erlenkoenig mir leise verspricht?"
"Sei ruhig, bleib ruhig, mein Kind!
In duerren Blaettern saeuselt der Wind."

J. W. Goethe

1 INTRODUCTION

The classical wind-tree model corresponds to a billiard in the plane endowed with $\mathbb{Z}^2$-periodic obstacles of rectangular shape; the sides of the rectangles are aligned along the lattice, see Figure 1.

The wind-tree model (in a slightly different version) was introduced by P. Ehrenfest and T. Ehrenfest [Eh] about a century ago and studied, in particular, by J. Hardy and J. Weber [HaWe]. All these studies had physical motivations.

Several considerable advances were obtained recently using the powerful technology of deviation spectrum of measured foliations on surfaces and the underlying dynamics in the moduli space. For all parameters of the obstacle and for almost all directions the trajectories are known to be recurrent, see [HLT] and [AH]; there are examples of divergent trajectories constructed in [D]; the non-ergodicity is proved in [FrU]; the growth rate of the number of periodic orbits of bounded length is computed in [Par1, Par2]. It was proved in [DHL] that the diffusion rate is $\frac{2}{3}$ in almost every direction. It does not depend either on the concrete values of

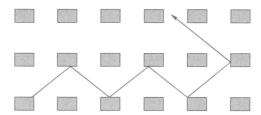

Figure 1: Original wind-tree model.

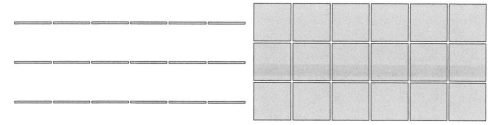

Figure 2: The diffusion rate $\frac{2}{3}$ does not depend on particular values of the parameters of the rectangular scatterer: it is the same for the plane with horizontal walls having tiny periodic holes, for narrow periodic corridors between "chocolate plates" and for any other periodic billiard as in Figure 1.

parameters of the obstacle or on the position of the starting point, see Figure 2. In [CrLanH] it is proven that when considering all possible directions and not only generic ones, one can achieve any diffusion rate between 0 and 1. See also papers [P] and [JSch] on related subjects.

In other words, the maximal deviation of the trajectory from the starting point during the time t has the order of $t^{\frac{2}{3}}$ for large t in the following sense:

$$\lim_{t \to \infty} \frac{\log \operatorname{diam}(\text{trajectory for time interval}[0,t])}{\log t} = \frac{2}{3}.$$

Thus, this behavior is quite different from the brownian motion, random walk in the plane, or billiards in the plane with periodic dispersing scatterers: for all of them the diffusion has the order $\sqrt{t}$ (and, thus, the diffusion rate is $\frac{1}{2}$).

We address the natural question, "What happens if we change the shape of the obstacle?" We do not have ambition to solve this problem in the current paper in the most general setting. We just plant the wind-tree forest with several interesting families of obstacles and study the diffusion rate as the combinatorics of the obstacle inside the family becomes more complicated. We show, in particular, that if the obstacle is a connected symmetric right-angled polygon as in Figure 3, then the diffusion rate in the corresponding wind-tree model tends to zero as the number of corners of the obstacle grows; see Theorem 1 for a more precise statement. This result gives an explicit affirmative answer to a question addressed by J.-C. Yoccoz.

Now, when we have showed that for certain species of wind-trees the sound in the wind-tree forest propagates "as a whisper" (in the sense that the diffusion rate tends to zero), the challenge is to prove that for certain other species it propagates "as a cry" with the diffusion rate approaching 1.

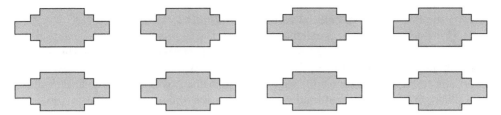

Figure 3: The diffusion rate in this wind-tree forest tends to zero when the number of corners of the obstacle grows.

Question 1. *Are there periodic wind-tree billiards with diffusion rate arbitrary close to 1? Are there continuous families of wind-tree billiards like this? What are the shapes of the obstacles which provide diffusion rate arbitrary close to 1?*

There are certain restrictions on $\mathrm{GL}(2,\mathbb{R})$-invariant suborbifolds which can serve as orbit closures of flat surfaces associated to such hypothetical wind-tree billiards. For example, by Theorem 8.2 of Chen and Möller in [CMö] combined with the density argument in [BnEWl] the Lyapunov exponent λ_1^+ of any invariant suborbifold defined over $\mathbb{Q}$ in any stratum of meromorphic quadratic differentials in genus 1 is bounded above by $\frac{2}{3}$. This might be an indication that if examples as in the Question above exist, they might correspond to large genera.

1.1 Strategy of the proof

We develop the approach originated in the pioneering work of V. Delecroix, P. Hubert, and S. Lelièvre [DHL] who applied the results from dynamics in the moduli space to the wind-tree model.

Following the techniques of [Fo] and [Z] on deviation spectrum, [DHL] relates diffusion in the wind-tree model to Lyapunov exponents of the Kontsevich-Zorich cocycle. This result is quite general and applies to our situation. The first step is to associate for each wind-tree billiard a compact translation surface S and integer cocycles on S that describe the $\mathbb{Z}^2$-covering associated to the initial infinite billiard. In our situation, we show that the diffusion rate of the wind-tree billiard as in Figure 3 is equal to the Lyapunov exponent $\lambda^+(h^*) = \lambda^+(v^*)$ of certain very specific integer cocycles $h^*, v^* \in H^1(S,\mathbb{Z})$, where the flat surface S is considered as a point of the orbit closure $\mathcal{L}(S)$ in the ambient stratum of meromorphic quadratic differentials; the vector space $H^1(S,\mathbb{C})$ containing h^* and v^* is considered as a fiber over S of the complex Hodge bundle over $\mathcal{L}(S)$, and the Lyapunov exponents are the Lyapunov exponents of the complex Hodge bundle $H^1_{\mathbb{C}}$ over $\mathcal{L}(S)$ with respect to the Teichmüller geodesic flow.

In [DHL], the fact that the diffusion rate is independent of the size of the obstacles follows from the fact that the Lyapunov exponent does not depend on the $\mathrm{SL}(2,\mathbb{R})$-orbit closure and a deep result of J. Chaika and A. Eskin [CkE] (based on fundamental advances of A. Eskin, M. Mirzakhani, A. Mohammadi [EMi], [EMiMo]). The latter ensures that for *any* translation surface, almost every direction is Birkhoff-generic and Lyapunov-generic. In the more general wind-tree billiards we consider in this paper, the Lyapunov exponents do depend on the orbit closure and we exhibit examples with exotic exponents.

In the current paper we intentionally focus on the family of billiards as in Figure 3 since for this family the rest of the computation is particularly transparent.

Namely, the flat surface S belongs to the *hyperelliptic* locus $\mathcal{L} := \mathcal{Q}^{hyp}(1^{2m}, -1^{2m})$ over $\mathcal{Q}(1^m, -1^{m+4})$. (In our particular situation, the *hyper*elliptic locus is, actually, "elliptic": the genus of the covering surface is 1.) Moreover, applying the arguments analogous to those in [AtEZ], one immediately verifies that the family $\mathcal{B}$ of billiards is so large (in dimension) that it is transversal to the unstable foliation in the ambient invariant submanifold $\mathcal{L}$, and, thus, for almost every billiard table Π in $\mathcal{B}$ the orbit closure $\mathcal{L}(S)$ of the associated flat surface $S(\Pi)$ coincides with the entire locus $\mathcal{L}$.

The flat surface S has genus one, so the complex Hodge bundle $H_{\mathbb{C}}^1 = H_+^1$ has single positive Lyapunov exponent λ_1^+. The fact that h^*, v^* are *integer* immediately implies that $\lambda^+(h^*) = \lambda^+(v^*) = \lambda_1^+$.

Formula (2.3) from [EKZ2] expresses the sum $\sum_{i=1}^g \lambda_i^+$ of all positive Lyapunov exponents of H_+^1 in terms of the degrees of zeroes (and poles) in the ambient locus and in terms of the Siegel–Veech constant $c_{area}(\mathcal{L})$. Since in our particular case the genus of the surface is equal to one, we get a formula for the individual Lyapunov exponent λ_1^+ in which we are interested.

Developing Lemma 1.1 from [EKZ2] we relate the Siegel–Veech constant $c_{area}(\mathcal{L})$ of the hyperelliptic locus $\mathcal{L} := \mathcal{Q}^{hyp}(1^{2m}, -1^{2m})$ over $\mathcal{Q}(1^m, -1^{m+4})$ to the Siegel–Veech constants $c_{\mathbb{C}}(\mathcal{Q}(1^m, -1^{m+4}))$ of the underlying stratum in genus zero. Plugging in the resulting expression the explicit values of $c_{\mathbb{C}}(\mathcal{Q}(1^m, -1^{m+4}))$ obtained in the recent paper [AtEZ] and proving certain combinatorial identity for the resulting hypergeometric sum we obtain the desired explicit value of $\lambda_1^+(\mathcal{Q}^{hyp}(1^{2m}, -1^{2m}))$, which represents the diffusion rate in almost every billiard under consideration.

Remark 1. Our results provide certain evidence that when the genus is fixed and the number of simple poles grows, the Lyapunov exponents of the Hodge bundle tend to zero (see [GrH] for the original conjecture).

1.2   Structure of the paper

In section 2 we state the main result in two different forms. In section 3 we show how to reduce the problem of the diffusion rate in a generalized wind-tree billiard to the problem of evaluation of the top Lyapunov exponent λ_1^+ of the complex Hodge bundle over an appropriate hyperelliptic locus of quadratic differentials. In section 3.1 we revisit the original paper [DHL] where this question is treated in all details for the original wind-tree model with periodic rectangular scatterers. We suggest, however, several simplifications. Namely, in section 3.2 we describe the hyperelliptic locus over certain stratum of meromorphic quadratic differentials in genus zero where lives the flat surface S corresponding to the wind-tree billiard and we show that the diffusion rate corresponds to the top Lyapunov exponent λ_1^+ of the complex Hodge bundle over the $\mathrm{PSL}(2, \mathbb{R})$-orbit closure of S. Following an analogous statement in [AtEZ] we prove in section 3.3 that for almost any initial billiard table Π the $\mathrm{PSL}(2, \mathbb{R})$-orbit closure of the associated flat surface $S(\Pi)$ coincides with the entire hyperelliptic locus. At this stage we reduce the problem of evaluation of the diffusion rate for almost any billiard table in the family to evaluation of the single positive Lyapunov exponent of the complex Hodge bundle $H_+^1 = H_{\mathbb{C}}^1$ over certain specific hyperelliptic locus.

In section 4 we evaluate this Lyapunov exponent. We start by recalling in section 4.2 the technique from [EKZ2]; we also relate the Siegel–Veech constant of

Figure 4: The diffusion rate depends only on the number of corners of the obstacle and not on the particular values of (almost all) length parameters nor on the particular shape of the obstacle.

the hyperelliptic locus with the Siegel–Veech constant of the corresponding stratum in genus 0. In section 4.3 we summarize the necessary material on cylinder configurations in genus zero and on the related Siegel–Veech constants from [Bo] and [AtEZ]. Finally, in section 4.4 we prove the key Theorem 2 evaluating the desired Lyapunov exponent λ_1^+. The proof uses a combinatorial identity for certain hypergeometric sum; this identity is proved separately in section 5.

Following the title "*What's next?*" of the conference, we discuss in appendix B directions of further research in the area relevant to the context of this paper.

2   MAIN RESULTS

Denote by $\mathcal{B}(m)$ the family of billiards such that the obstacle has $4m$ corners with (internal) angle $\pi/2$, $4(m-1)$ corners with angle $3\pi/2$ and admits horizontal and vertical axis of symmetry. Say, all billiards from the original wind-tree family as in Figures 1 and 2 live in $\mathcal{B}(1)$; the billiard in Figure 3 belongs to $\mathcal{B}(3)$; the billiard in Figure 4 belongs to $\mathcal{B}(17)$.

Theorem 1. *For almost all billiard tables in the family $\mathcal{B}(m)$ and for almost all directions the diffusion rate $\delta(m)$ is the same and equals*

$$\delta(m) = \frac{(2m)!!}{(2m+1)!!} \, .$$

When $m \to +\infty$ the diffusion rate $\delta(m)$ has asymptotics

$$\delta(m) = \frac{\sqrt{\pi}}{2\sqrt{m}} \left(1 + O\left(\frac{1}{m} \right) \right) \, .$$

Here the double factorial means the product of all even (correspondingly odd) natural numbers from 2 to $2m$ (correspondingly from 1 to $2m+1$). For the original wind-tree, when the obstacle is a rectangle, we have $m=1$ and we get the value $\delta(1) = \frac{2}{3}$ found in [DHL].

Following the strategy described in section 1.1 we derive Theorem 1 from the following result.

Theorem 2. *The locus $\mathcal{Q}^{hyp}(1^{2m}, -1^{2m})$ over $\mathcal{Q}(1^m, -1^{m+4})$ is connected and invariant under the action of $\mathrm{PGL}(2, \mathbb{R})$. The measure induced on $\mathcal{Q}_1^{hyp}(1^{2m}, -1^{2m})$ from the Masur–Veech measure on $\mathcal{Q}_1(1^m, -1^{m+4})$ is $\mathrm{PSL}(2, \mathbb{R})$-invariant and ergodic under the action of the Teichmüller geodesic flow. The Lyapunov exponent $\lambda_1^+(m)$ of the Hodge bundle H_+^1 over the locus $\mathcal{Q}^{hyp}(1^{2m}, -1^{2m})$ under consideration has the following value:*

$$\lambda_1^+(m) = \frac{(2m)!!}{(2m+1)!!}. \tag{2.1}$$

Remark 2. Note that there is a very important difference between the case of $m = 1$ (corresponding to the classical wind-tree) and the cases $m \geq 2$. Namely, the locus $\mathcal{Q}^{hyp}(1^2, -1^2)$ over $\mathcal{Q}(1, -1^5)$ is *nonvarying*: the Lyapunov exponent λ_1^+ has the same value $\lambda_1^+ = \frac{2}{3}$ for *all* flat surfaces in the locus $\mathcal{Q}^{hyp}(1^2, -1^2)$. For $m \geq 2$ it is not true anymore, which can be justified by the following argument. For each integer $m \geq 2$, taking appropriate unramified cover of degree m of a flat surface in the stratum $\mathcal{Q}(1^2, -1^2)$ we get a flat surface in the hyperelliptic locus of $\mathcal{Q}^{hyp}(1^{2m}, -1^{2m})$ over $\mathcal{Q}(1^m, -1^{m+4})$. By construction, the Lyapunov exponent λ_1^+ of the resulting Teichmüller curve in $\mathcal{Q}^{hyp}(1^{2m}, -1^{2m})$, and, hence, the diffusion rate δ for the corresponding wind-tree billiard does not change: $\delta = \lambda_1^+ = 2/3$. Furthermore, for $m = 2$ we were able to find examples of square-tiled surfaces for which the value is neither the generic value $\delta(2) = 8/15 = 0.5333\ldots$ nor $2/3 = 0.6666\ldots$.

permutations r and u	λ_1^+
$r = (1, 2, 3, 4, 5, 6, 7)(8, 9, 10, 11, 12, 13, 14)$ $u = (1, 3, 13, 8, 2, 14)(4, 6, 11, 5, 10, 12)(7, 9)$	$\frac{20}{33} = 0.6060\ldots$
$r = (1, 2, 3, 4, 5, 6, 7, 8)(9, 10, 11, 12, 13, 14)$ $u = (1, 2, 3, 14, 9)(4, 13)(5, 6, 7, 11, 12)(8, 10)$	$\frac{6}{11} = 0.5454\ldots$

See also the example in appendix A with $m = 3$.

Question 2. *What are the extremal values of λ_1^+ over all closed $\mathrm{PSL}(2, \mathbb{R})$-invariant suborbifolds in a given stratum (given locus)?*

Same question for the hyperelliptic locus $\mathcal{Q}^{hyp}(1^{2m}, -1^{2m})$ over $\mathcal{Q}(1^m, -1^{m+4})$? By Theorem 8.2 of Chen and Möller in [CMö] combined with the density argument in [BnEWl], the Lyapunov exponent λ_1^+ of any invariant suborbifold defined over $\mathbb{Q}$ in this locus is bounded above by $\frac{2}{3}$. Is this bound attained? If so, for which invariant suborbifolds?

What are the shapes of billiards for which these extremal values are achieved (if there are any wind-tree billiards corresponding to these invariant suborbifolds)?

3 FROM BILLIARDS TO FLAT SURFACES

3.1 Original wind-tree revisited

Recall that in the classical case of a billiard in a rectangle we can glue a flat torus out of four copies of the billiard table and unwind billiard trajectories to flat geodesics on the resulting flat torus.

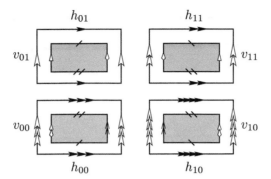

Figure 5: The flat surface X obtained as a quotient over $\mathbb{Z} \oplus \mathbb{Z}$ of an unfolded wind-tree billiard table.

In the case of the wind-tree model we also start from gluing a flat surface out of four copies of the billiard table. The resulting surface is $\mathbb{Z}^2$-periodic with respect to translations by vectors of the original lattice. We pass to the quotient over $\mathbb{Z}^2$ to get a compact flat surface without boundary. For the case of the original wind-tree billiard the resulting flat surface X is represented in Figure 5. It has genus 5; it belongs to the stratum $\mathcal{H}(2^4)$ (see section 3 of [DHL] for details).

One of the key statements of [DHL] can be stated as follows.

Let Π be the original rectangular obstacle; define the corresponding wind-tree billiard by the same symbol Π. Let $X = X(\Pi)$ be the flat surface as in Figure 5 constructed by the wind-tree billiard defined by the obstacle Π.

Consider the $\mathrm{SL}(2, \mathbb{R})$-orbit closure $\mathcal{L}(X) \subset \mathcal{H}(2^4)$ of the flat surface X. Consider the cohomology classes $h^*, v^* \in H^1(X, \mathbb{Z})$ Poincaré-dual to cycles

$$h = h_{00} - h_{01} + h_{10} - h_{11}$$

$$v = v_{00} - v_{10} + v_{01} - v_{11}$$

(see Figure 5) as elements of the fiber over the point $X \in \mathcal{L}(X)$ of the complex Hodge bundle $H_{\mathbb{C}}^1$ over $\mathcal{L}(X)$.

Theorem (V. Delecroix, P. Hubert, S. Lelièvre [DHL]). *The diffusion rate in the original wind-tree billiard Π coincides with the Lyapunov exponent $\lambda(h^*) = \lambda(v^*)$ of the complex Hodge bundle $H_{\mathbb{C}}^1$ with respect to the Teichmüller geodesic flow on $\mathcal{L}(X)$.*

Note that any resulting flat surface X as in Figure 5 has (at least) the group $(\mathbb{Z}/2\mathbb{Z})^3$ as a group of isometries. As three generators we can choose the isometries τ_h and τ_v interchanging the pairs of flat tori with holes in the same rows (correspondingly columns) by parallel translations and the isometry ι acting on each of the four tori with holes as the central symmetry with the center in the center of the hole.

Consider the quotient $\tilde{S}$ of the flat surface $\tilde{S}$ over the subgroup $(\mathbb{Z}/2\mathbb{Z})^2$ of isometries spanned by τ_h and $\iota \circ \tau_v$. The resulting surface $\tilde{S}$ as on the left side of the Figure 6 belongs to the stratum $\mathcal{Q}(1^2, -1^2)$; in particular, it has genus 1. The surface S obtained as the quotient of the original flat surface X over the entire group $(\mathbb{Z}/2\mathbb{Z})^3$ as on the right side of the Figure 6 belongs to the stratum $\mathcal{Q}(1, -1^5)$; in particular, it has genus 0. Clearly, $\tilde{S}$ is a ramified double cover over S with ramification points at four (out of five) simple poles of the flat surface S.

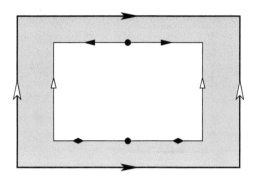 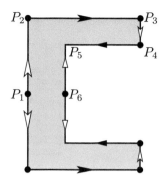

Figure 6: A surface $\tilde{S}$ in the hyperelliptic locus $\mathcal{Q}^{hyp}(1^2, -1^2)$ (on the left) is a double cover over the underlying surface S in $\mathcal{Q}(1, -1^5)$ (on the right) ramified at four simple poles represented by bold dots.

Lemma 3.1. *Consider the natural projection* $p\colon X \to \tilde{S}$. *The following inclusion is valid:*
$$h^* \in p^*(H^1(\tilde{S}; \mathbb{Z})) .$$

Proof. Recall that
$$h = h_{00} - h_{01} + h_{10} - h_{11} ,$$

where the cycles h_{ij}, $i, j = 0, 1$ are indicated in Figure 5. Now note that

$$\tau_h(h_{00}) = h_{10} \qquad \iota \circ \tau_v(h_{00}) = -h_{01} \qquad \iota \circ \tau_v \circ \tau_h(h_{00}) = -h_{11} \quad .$$

Thus, the cycle h is invariant under the action of the commutative group Γ spanned by τ_h and $\iota \circ \tau_v$. This implies that the Poincaré-dual cocycle h^* is also invariant under the action of Γ and, hence, is induced from some cocycle in cohomology of $\tilde{S} = X/\Gamma$. $\qquad\square$

Corollary 1. *Consider the orbit closure* $\mathcal{M}(\tilde{S})$ *of the flat surface* $\tilde{S}(\Pi) \in \mathcal{Q}(1^2, -1^2)$. *The diffusion rate in the original wind-tree billiard* Π *in vertical direction coincides with the positive Lyapunov exponent* λ_1^+ *of the complex Hodge bundle* $H_{\mathbb{C}}^1$ *with respect to the Teichmüller geodesic flow on* $\mathcal{M}(\tilde{S})$.

Proof. From the theorem of Delecroix–Hubert–Lelièvre cited above we know that the diffusion rate coincides with the Lyapunov exponent $\lambda(h^*)$. By the previous Lemma $h^* = p^*(\alpha)$ where $\alpha \in H^1(\tilde{S}; \mathbb{Z})$. It is immediate to see that $\lambda(h^*) = \lambda(\alpha)$, where $\lambda(\alpha)$ is the Lyapunov exponent of the complex Hodge bundle $H_{\mathbb{C}}^1$ with respect to the Teichmüller geodesic flow on $\mathcal{M}(\tilde{S})$. Since $g(\tilde{S}) = 1$ the corresponding cocycle has Lyapunov exponents $\pm\lambda_1^+$. Since α is an integer covector, $\lambda(\alpha)$ cannot be strictly negative. Hence $\lambda(\alpha) = \lambda_1^+$. $\qquad\square$

3.2 Hyperelliptic locus and diffusion in the generalized wind-tree billiard

The hyperelliptic (to be more precise—the *elliptic*) locus $\mathcal{Q}^{hyp}(1^{2m}, -1^{2m})$ in $\mathcal{Q}(1^{2m}, -1^{2m})$ is obtained by the following construction. For any flat surface S in $\mathcal{Q}(1^m, -1^{m+4})$ consider all possible quadruples of unordered simple poles. For each

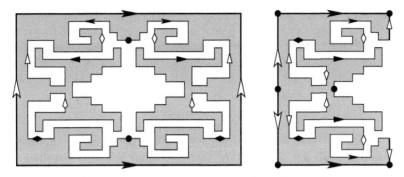

Figure 7: A surface $\tilde{S}$ in the hyperelliptic locus $\mathcal{Q}^{hyp}(1^{2m}, -1^{2m})$ is a double cover over the underlying surface S in $\mathcal{Q}(1^m, -1^{m+4})$ branched at the four simple poles represented by bold dots.

quadruple construct a ramified double cover with four ramification points exactly at the chosen quadruple of points. By construction the induced flat surface belongs to the stratum $\mathcal{Q}(1^{2m}, -1^{2m})$ in genus 1. Considering all possible flat surfaces S in $\mathcal{Q}(1^m, -1^{m+4})$ and all covers over all quadruples of simple poles we get the locus $\mathcal{Q}^{hyp}(1^{2m}, -1^{2m})$ over $\mathcal{Q}(1^m, -1^{m+4})$.

It is immediate to see that when the obstacle is a rectangle, as in the original wind-tree billiard, the surface $\tilde{S}$ constructed in section 3.1 belongs to the hyperelliptic locus $\mathcal{Q}^{hyp}(1^2, -1^2)$ over the stratum $\mathcal{Q}(1, -1^5)$. Similarly, when the obstacle has $4m$ corners with the angle $\pi/2$, the analogous surface $\tilde{S}$ (as on the left of Figure 7) belongs to the hyperelliptic locus $\mathcal{Q}^{hyp}(1^{2m}, -1^{2m})$ over $\mathcal{Q}(1^m, -1^{m+4})$.

The arguments of Delecroix–Hubert–Lelièvre [DHL] extend to the more general case of obstacles Π symmetric with respect to vertical and horizontal axes with arbitrary numbers of angles $\frac{\pi}{2}$, basically, line by line. Applying exactly the same consideration as above we prove the following statement generalizing Corollary 1 to arbitrary $m \in \mathbb{N}$.

Let Π be a connected obstacle having $4m$ corners with angle $\frac{\pi}{2}$ and $4(m-1)$ corners with angle $\frac{3\pi}{2}$ aligned in such way that all its sides are vertical or horizontal (see the white domain in the left picture in Figure 7). Suppose that Π is symmetric under reflections over some vertical and some horizontal lines.

Proposition 1. *Consider the orbit closure* $\mathcal{M}(\tilde{S}) \subset \mathcal{Q}^{hyp}(1^{2m}, -1^{2m})$ *of the flat surface* $\tilde{S}(\Pi)$ *in the ambient hyperelliptic locus* $\mathcal{Q}^{hyp}(1^{2m}, -1^{2m})$. *The diffusion rate in the wind-tree billiard with periodic obstacles* Π *aligned with the lattice coincides with the positive Lyapunov exponent* λ_1^+ *of the complex Hodge bundle* $H_{\mathbb{C}}^1$ *over* $\mathcal{M}(\tilde{S})$ *with respect to the Teichmüller geodesic flow on* $\mathcal{M}(\tilde{S})$.

3.3   Orbit closures

Consider the original wind-tree billiard with rectangular obstacles. Such a billiard is described by five real parameters: by two lengths of the sides of the external rectangle defining the lattice; by two lengths of the sides of the inner rectangle represented by the obstacle; and by the angle defining the direction of trajectory, see Figure 6. Continuously varying these parameters we obtain a continuous family of billiards.

Starting with a more general obstacle as in Figure 7 having $4m$ corners of angle $\frac{\pi}{2}$, $4(m-1)$ corners of angle $\frac{3\pi}{2}$ and symmetric with respect to vertical and horizontal axes of symmetry we get an analogous continuous family of billiards which we denote by $\mathcal{B}(m)$. We denote by $\mathbb{S}^1/\mathrm{Sym}$ the quotient of the circle under the action of the group spanned by symmetries with respect to vertical and horizontal axes. It is immediate to check that

$$\dim_{\mathbb{R}} \mathcal{B}(m) = 2m + 2. \tag{3.1}$$

Here the direction of the billiard flow is not considered as a parameter of the family $\mathcal{B}$. For the original wind-tree billiard with rectangular obstacles this gives $\dim_{\mathbb{R}} \left(\mathcal{B}(1) \times (\mathbb{S}^1/\mathrm{Sym}) \right) = 4 + 1 = 5$, as we have already seen.

Proposition 2. *For Lebesgue-almost every directional billiard in $\Pi \in \mathcal{B}(m)$ the $GL(2,\mathbb{R})$-orbit closure of $S(\Pi)$ in $\mathcal{Q}(1^m, -1^{m+4})$ coincides with the entire ambient stratum $\mathcal{Q}(1^m, -1^{m+4})$.*

Proof. The Proposition is a straightforward corollary of Lemma 3.2 below.   □

The following Lemma is completely analogous to Proposition 3.2 in [AtEZ].

Lemma 3.2. *Consider the canonical local embedding*

$$\mathcal{B}(m) \times (\mathbb{S}^1/\mathrm{Sym}) \hookrightarrow \mathcal{Q}(1^m, -1^{m+4}).$$

For almost all pairs (Π, θ) in $\mathcal{B}(m) \times (\mathbb{S}^1/\mathrm{Sym})$ the projection of the tangent space $T_*\left(\mathcal{B}(m) \times (\mathbb{S}^1/\mathrm{Sym}) \right)$ to the unstable subspace of the Teichmüller geodesic flow is a surjective map.

Proof. Let us first prove the statement for $m = 1$ and then make the necessary adjustments for the most general case. Consider a broken line following the upper part of the polygon as on the right side of Figure 6 starting at the point P_0 and finishing at the corner P_4. Turn the figure by the angle θ. Consider the associated flat surface $S \in \mathcal{Q}(1, -1^5)$. Consider the canonical orienting double cover $\hat{S} \in \mathcal{H}(2)$. The six Weierstrass points of the cover correspond to the corners of our polygonal pattern and to the two bold points on the horizontal axis of symmetry. Thus the vectors of the resulting broken line considered as complex numbers are exactly the basic half-periods of the holomorphic 1-form ω corresponding to a standard basis of cycles on the hyperelliptic surface $\hat{S}$ (see sections 3.1 and 3.3 in [AtEZ] for more details). The first cohomology $H^1(\hat{S}; \mathbb{C})$ serve as local coordinates in $\mathcal{Q}(1, -1^5)$. The components of the projection of the vector of periods to the space $H^1(\hat{S}; \mathbb{R})$ are of the form

$$\pm 2\sin(\phi)|P_i P_{i+1}| \quad \text{or} \quad \pm 2\cos(\phi)|P_i P_{i+1}| \quad \text{where } i = 1, \dots, 4.$$

Thus, for ϕ different from an integer multiple of $\pi/2$ the composition map

$$T_*\left(\mathcal{B}(m) \times (\mathbb{S}^1/\mathrm{Sym}) \right) \to H^1(\hat{S}; \mathbb{R})$$

is a surjective map.

It is clear from the proof that to extend it from $m = 1$ to arbitrary $m \in \mathbb{N}$ it is sufficient to show that the real dimension of $\mathcal{B}(m)$ coincides with the complex dimension of $\mathcal{Q}(1^m, -1^{m+4})$. Recalling (3.1) and the classical formula for the dimension of a stratum $\mathcal{Q}(d_1, \ldots, d_n)$ of quadratic differentials

$$\dim_{\mathbb{C}} \mathcal{Q}(d_1, \ldots, d_n) = 2g + n - 2$$

we conclude that

$$\dim_{\mathbb{R}} \mathcal{B}(m) = 2m + 2 = \dim_{\mathbb{C}} \mathcal{Q}(1^m, -1^{m+4})$$

which completes the proof of the Lemma. □

Combining Propositions 1 and 2 we obtain the following Corollary, which proves the first part of Theorem 1.

Corollary 2. *For Lebesgue-almost every directional billiard in $\Pi \in \mathcal{B}(m)$ the diffusion rate $\delta(m)$ in almost all directions $\theta \in \mathbb{S}^1$ in the wind-tree billiard Π coincides with the positive Lyapunov exponent $\lambda_1^+(m)$ of the complex Hodge bundle $H_{\mathbb{C}}^1$ over the hyperelliptic locus $\mathcal{Q}^{hyp}(1^{2m}, -1^{2m})$ with respect to the Teichmüller geodesic flow on this locus, $\delta(m) = \lambda_1^+(m)$.*

It remains to evaluate the Lyapunov exponent $\lambda_1^+(m)$ of the complex Hodge bundle $H_{\mathbb{C}}^1$ over the hyperelliptic locus $\mathcal{Q}^{hyp}(1^{2m}, -1^{2m})$, which we do in the next section.

4   SIEGEL–VEECH CONSTANTS AND SUM OF THE LYAPUNOV EXPONENTS OF THE HODGE BUNDLE OVER HYPERELLIPTIC LOCI

In this section we relate the Siegel–Veech constants of the hyperelliptic locus $\mathcal{Q}^{hyp}(1^{2m}, -1^{2m})$ and of the underlying stratum $\mathcal{Q}(1^m, -1^{m+4})$ of meromorphic quadratic differentials with at most simple poles on $\mathbb{C}\mathrm{P}^1$. Applying the technique from [EKZ2] and developing the results from [AtEZ] this allows us to get an explicit value for the desired Lyapunov exponent $\lambda_1^+(m)$ of the hyperelliptic locus and, thus, to prove Theorem 2.

4.1   Siegel–Veech constants

We start this section with a brief reminder of the notion of *Siegel–Veech constant*.

Let S be a flat surface in some stratum of Abelian or quadratic differentials. Together with every closed regular geodesic γ on S we have a bunch of parallel closed regular geodesics filling a maximal cylinder *cyl* having a conical singularity at each of the two boundary components. By the *width* w of a cylinder we call the flat length of each of the two boundary components.

The number of maximal cylinders on a flat surface S filled with regular closed geodesics of bounded length $w(cyl) \le L$ is finite. Thus, for any $L > 0$ the following quantity is well-defined:

$$N_{area}(S, L) := \frac{1}{\text{Area}(S)} \sum_{\substack{cyl \subset S \\ w(cyl) < L}} \text{Area}(cyl).$$ (4.1)

The following theorem is a special case of a fundamental result of W. Veech, [Ve2] considered by Y. Vorobets in [Vo]:

Theorem (W. Veech; Y. Vorobets). *Let ν_1 be an ergodic $SL(2, \mathbb{R})$-invariant probability measure supported on an $SL(2, \mathbb{R})$-invariant suborbifold $\mathcal{M}_1$ in some stratum $\mathcal{H}_1(m_1, \dots, m_n)$ of Abelian differentials of area one. Then, the following ratio is constant (i.e., does not depend on the value of a positive parameter L):*

$$\frac{1}{\pi L^2} \int_{\mathcal{M}_1} N_{area}(S, L)\, d\nu_1 = c_{area}(\nu_1) = c_{area}(\mathcal{M}_1).$$ (4.2)

This formula is called a *Siegel–Veech formula*, and the corresponding constant $c_{area}(\nu_1) = c_{area}(\mathcal{M}_1)$ is called the *Siegel–Veech constant*.

Analogous formula is valid for $PSL(2, \mathbb{R})$-invariant suborbifolds in the strata of meromorphic quadratic differentials with at most simple poles; see [AtEZ] and [Gj2] for normalization conventions.

The function $N_{area}(S, L)$ counts maximal cylinders filled with periodic geodesic, where each cylinder is counted with weight $\text{Area}(cyl)/\text{Area}(S)$. One can consider analogous counting function, where each cylinder is counted with weight 1. One can consider further analogous functions counting saddle connections, or even certain specific *configurations* of homologous saddle connections. All these counting functions define their own Siegel–Veech constants; see [EMs] for details. Some of them can be expressed in terms of the others.

The Siegel–Veech constants corresponding to the connected components of the strata of Abelian differentials were expressed in terms of the Masur–Veech volumes of the *principal boundary* strata in [EMsZ].

For the connected components of the strata of quadratic differentials analogous formulae were obtained in [Gj2] using the description of the principal boundary strata obtained in [MsZ].

Simple and very explicit expression for the Siegel–Veech constants corresponding to strata of quadratic differentials in genus 0 (which is particularly relevant for this paper) was obtained in [EKZ2].

By the fundamental result of Eskin and Masur [EMs] for ν_1-almost all surfaces S in $\mathcal{M}_1$ one has

$$\lim_{L \to +\infty} \frac{N_{area}(S, L)}{\pi L^2} = c_{area}(\nu_1) = c_{area}(\mathcal{M}_1).$$

4.2 Sum of the Lyapunov exponents of the complex Hodge bundle over hyperelliptic loci of quadratic differentials

We need the following result.

Theorem (Theorem 2 in [EKZ2]). *Consider a stratum $\mathcal{Q}_1(d_1, \dots, d_n)$ in the moduli space of quadratic differentials with at most simple poles, where $d_1 + \cdots + d_n = 4g - 4$. Let $\mathcal{M}_1$ be any regular $PSL(2, \mathbb{R})$-invariant suborbifold of $\mathcal{Q}_1(d_1, \dots, d_n)$.*

The Lyapunov exponents $\lambda_1^+ \geq \cdots \geq \lambda_g^+$ of the complex Hodge bundle $H_+^1 = H_{\mathbb{C}}^1$ over $\mathcal{M}_1$ along the Teichmüller flow satisfy the following relation:

$$\lambda_1^+ + \cdots + \lambda_g^+ = \kappa + \frac{\pi^2}{3} \cdot c_{area}(\mathcal{M}_1),\tag{4.3}$$

where

$$\kappa = \frac{1}{24} \sum_{j=1}^{n} \frac{d_j(d_j+4)}{d_j+2}\tag{4.4}$$

and $c_{area}(\mathcal{M}_1)$ is the Siegel–Veech constant corresponding to the suborbifold $\mathcal{M}_1$. By convention the sum in the left-hand side of equation (4.3) is defined to be equal to zero for $g = 0$.

In the context of this paper we are particularly interested in the case when $\mathcal{M}$ is a hyperelliptic locus over some stratum of meromorphic quadratic differentials with at most simple poles in genus zero.

Let $\mathcal{M}_1$ be a closed $SL(2, \mathbb{R})$-invariant (correspondingly $PSL(2, \mathbb{R})$-invariant) suborbifold in a stratum of Abelian (correspondingly quadratic) differentials; let ν be the associated $SL(2, \mathbb{R})$-ergodic (correspondingly $PSL(2, \mathbb{R})$-ergodic) measure on $\mathcal{M}_1$.

Consider a locus $\tilde{\mathcal{M}}_1$ of all possible double covers of fixed profile over flat surfaces from $\mathcal{M}_1$. Suppose that it is closed, connected and $SL(2, \mathbb{R})$- (correspondingly $PSL(2, \mathbb{R})$-invariant), and that $\tilde{\nu}$ is the ergodic measure on $\tilde{\mathcal{M}}$ such that ν is the direct image of $\tilde{\nu}$ with respect to the natural projection $\tilde{\mathcal{M}}_1 \to \mathcal{M}_1$.

Let $c_{\mathcal{C}}$ be an area Siegel–Veech constant associated to the counting of *multiplicity one configuration $\mathcal{C}$ of cylinders* weighted by the area of the cylinder. (Here the notion "configuration" is understood in the sense of [MsZ] and [W2]; "multiplicity one" means that $\mathcal{C}$ contains a single cylinder.) The reader might think of $\mathcal{M}$ as of some stratum of quadratic differentials in genus 0; then there are only two types of configurations (see section 4.3 below or the original paper [Bo]), and both configurations contain a single cylinder.

We assume that the profile of the double cover does not admit branching points inside the cylinders of the configuration $\mathcal{C}$. Then the configuration $\mathcal{C}$ induces a cylinder configuration $\tilde{\mathcal{C}}$ on the double cover. Let $\tilde{c}$ be the associated area Siegel–Veech constant. The Lemma below relates c and $\tilde{c}$; it is a slight generalization of Lemma 1.1 in [EKZ2].

Lemma 4.1. *If the circumference of the cylinder in the configuration has nontrivial monodromy (that is, if the lift of the cylinder in the double cover is a unique cylinder twice wider), then $\tilde{c} = c/2$.*

If the circumference of the cylinder has trivial monodromy (that is, if the preimage of the cylinder consists in two cylinders isometric to the one on the base), then $\tilde{c} = 2c$.

Proof. The second case corresponds to Lemma 1.1 in [EKZ2]. The first case is proved analogously. □

In the same setting, let κ be the expression (4.4) in degrees of singularities of the flat surfaces in the invariant manifold $\mathcal{M}$ and $\tilde{\kappa}$ be analogous expression in degrees of singularities of the double covers of fixed profile in the invariant manifold $\tilde{\mathcal{M}}$.

Define

$$\Delta \tilde{\kappa} := \tilde{\kappa} - 2\kappa \,,$$

$$\Delta \tilde{c}_{area} := c_{area}(\tilde{\mathcal{M}}_1) - 2c_{area}(\mathcal{M}_1) \,.$$

Lemma 4.2. *For any ramified double cover the degrees of zeroes of the quadratic differential on the underlying surface and of the induced quadratic differential on the double cover satisfy the following relation:*

$$\Delta \tilde{\kappa} = \tilde{\kappa} - 2\kappa = \frac{1}{4} \sum_{\substack{ramification \\ points}} \frac{1}{d_j + 2} \,. \tag{4.5}$$

Proof. The nonramified zeroes cancel out. For the other zeroes (and poles) a singularity of degree d_j at the ramification point gives rise to a zero of degree $2d_j + 2$. Hence,

$$\tilde{\kappa} - 2\kappa = \frac{1}{24} \sum_{\substack{ramification \\ points}} \left(\frac{(2d_j + 2)(2d_j + 6)}{2d_j + 4} - 2\frac{d_j(d_j + 4)}{d_j + 2} \right)$$

$$= \frac{1}{12} \sum_{\substack{ramification \\ points}} \frac{(d_j + 1)(d_j + 3) - d_j(d_j + 4)}{d_j + 2}$$

$$= \frac{1}{12} \sum_{\substack{ramification \\ points}} \frac{3}{d_j + 2} \,. \qquad \square$$

The following notational Lemma would help to simplify certain bulky computations.

Lemma 4.3. *For any locus $\tilde{\mathcal{M}}_1$ of double covers of fixed profile as above over a $\mathrm{PSL}(2, \mathbb{R})$-invariant orbifold $\mathcal{M}_1$ in some stratum of meromorphic quadratic differentials with at most simple poles on $\mathbb{CP}^1$ the sum of Lyapunov exponents*

$$\Lambda^+ = \lambda_1^+ + \cdots + \lambda_g^+$$

satisfies the following relation

$$\Lambda^+ = \Delta \tilde{\kappa} + \frac{\pi^2}{3} \cdot \Delta \tilde{c}_{area} \,. \tag{4.6}$$

Proof. For any invariant submanifold $\mathcal{M}_1$ in a stratum of meromorphic quadratic differentials with at most simple poles on $\mathbb{CP}^1$ the sum of Lyapunov exponents is null, so formula (4.3) gives

$$0 = \kappa + \frac{\pi^2}{3} \cdot c_{area}(\mathcal{M}_1) \,.$$

By the same formula (4.3) gives we have

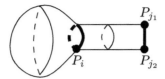

Figure 8: A "pocket" configuration with a cylinder bounded on one side by a saddle connection joining two simple poles, and by a saddle connection joining a zero to itself on the other side.

$$\Lambda^+ = \tilde{\kappa} + \frac{\pi^2}{3} \cdot c_{area}(\tilde{\mathcal{M}}_1).$$

Extracting from the latter relation twice the previous one we obtain the desired relation (4.6).   □

4.3 Configurations for the strata in genus zero and corresponding Siegel–Veech constants (after Boissy and Athreya–Eskin–Zorich)

In this section we recall briefly the results from [Bo] describing configurations of periodic geodesics for flat surfaces in genus zero, and the results from [AtEZ] providing the values of the corresponding Siegel–Veech constants. By $\mathcal{Q}(d_1, \ldots, d_k)$ we denote a stratum of meromorphic quadratic differentials with at most simple poles, where $d_i \in \{-1, 1, 2, \ldots\}$ denote all zeroes and poles, and $\sum_{i=1}^k d_i = -4$. By convention, considering the strata we assume that all the zeroes and poles are labeled (numbered).

A "pocket." In this configuration we have a single cylinder filled with closed regular geodesics, such that the cylinder is bounded by a saddle connection joining a fixed pair of simple poles P_{j_1}, P_{j_2} on one side and by a separatrix loop emitted from a fixed zero P_i of order $d_i \geq 1$ on the other side.

By convention, the affine holonomy associated to this configuration corresponds to the closed geodesic and *not* to the saddle connection joining the two simple poles. (Such a saddle connection is twice as short as the closed geodesic.)

By Theorem 4.5 and formula (4.28) in [AtEZ], the Siegel–Veech constant $c_{j_1,j_2;i}^{pocket}$ corresponding to this configuration has the form

$$c_{j_1,j_2;i}^{pocket} = \frac{d_i + 1}{(k-4)} \cdot \frac{1}{2\pi^2}, \tag{4.7}$$

where k is the total number of zeroes and poles in the stratum $\mathcal{Q}(d_1, \ldots, d_k)$.

One can consider the union of several configurations as above fixing the pair of simple poles P_{j_1}, P_{j_2} but considering *any* zero P_i on the boundary of the cylinder. By Corollary 4.7 and formula (4.36) in [AtEZ], the resulting Siegel–Veech constant c_{j_1,j_2}^{pocket} corresponding to this configuration has the form

$$c_{j_1,j_2}^{pocket} = \frac{1}{2\pi^2}. \tag{4.8}$$

A "dumbbell." For the second configuration we still have a single cylinder filled with closed regular geodesics. But this time the cylinder is bounded by a separatrix loop on each side. We assume that the separatrix loop bounding the

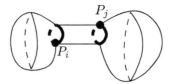

Figure 9: A "dumbbell" composed of two flat spheres joined by a cylinder. Each boundary component of the cylinder is a saddle connection joining a zero to itself.

cylinder on one side is emitted from a fixed zero P_i of order $d_i \geq 1$ and that the separatrix loop bounding the cylinder on the other side is emitted from a fixed zero P_j of order $d_j \geq 1$.

Such a cylinder separates the original surface S in two parts; let $P_{i_1}, \ldots, P_{i_{k_1}}$ be the list of singularities (zeroes and simple poles) which get to the first part and $P_{j_1}, \ldots, P_{j_{k_2}}$ be the list of singularities (zeroes and simple poles) which get to the second part. In particular, we have $i \in \{i_1, \ldots, i_{k_1}\}$ and $j \in \{j_1, \ldots, j_{k_2}\}$. We assume that S does not have any marked points. Denoting as usual by d_k the order of the singularity P_k we can represent the sets with multiplicities $\alpha := \{d_1, \ldots, d_k\}$ as a disjoint union of the two subsets

$$\{d_1, \ldots, d_k\} = \{d_{i_1}, \ldots, d_{i_{k_1}}\} \sqcup \{d_{j_1}, \ldots, d_{j_{k_2}}\}.$$

(Recall that $\{d_1, \ldots, d_k\}$ denotes the degrees of all zeroes and poles.) This information is considered to be part of the configuration.

By Theorem 4.8 and equation (4.38) in [AtEZ] the corresponding Siegel–Veech constant $c_{i,j}^{dumbell}$ is expressed as follows:

$$c_{i,j}^{dumbell} = \frac{(d_i + 1)(d_j + 1)}{2} \cdot \frac{(k_1 - 3)!\,(k_2 - 3)!}{(k - 4)!} \cdot \frac{1}{\pi^2}. \tag{4.9}$$

According to [Bo] and [MsZ], almost any flat surface S in any stratum $\mathcal{Q}_1(d_1, \ldots, d_k)$ of meromorphic quadratic differentials with at most simple poles different from the pillowcase stratum $\mathcal{Q}_1(-1^4)$ does not have a single regular closed geodesic not contained in one of the two families described above.

Finally, by Corollary 4.10 from [AtEZ] (generalizing the theorem of Vorobets from [Vo]) the Siegel–Veech constant c_{area} is expressed in terms of the above Siegel–Veech constants as follows. For any stratum $\mathcal{Q}_1(d_1, \ldots, d_k)$ of meromorphic quadratic differentials with simple poles on $\mathbb{CP}^1$ the Siegel–Veech constant c_{area} is expressed in terms of the Siegel–Veech constants of configurations as follows:

$$c_{area} = \frac{1}{k-3} \cdot \sum_{\substack{Configurations\ \mathcal{C} \\ containing\ a\ cylinder}} c_{\mathcal{C}}. \tag{4.10}$$

4.4   Proof of Theorem 2

Now everything is ready for the proof of Theorem 2.

Proof. The connectedness of our hyperelliptic locus (with *unlabeled* zeroes and poles) follows from the fact that it can be seen as a $\mathbb{C}^*$-bundle over the space

$\mathcal{C}(m, m, 4)$ of configurations of points on $\mathbb{CP}^1$ considered up to a modular transformation. More precisely, $\mathcal{C}(m, m, 4)$ can be seen as a set of configurations of $2m + 4$ distinct points arranged into three groups ("colored into three colors") of cardinalities $\{m, m, 4\}$, where the points inside each group are named. The first group represents the simple zeroes, the second one—unramified simple poles, and the last one—the ramified simple poles. Clearly, such space is connected.

By the results of H. Masur [Ms] and of W. Veech [Ve1], the Teichmüller geodesic flow on the underlying stratum is ergodic, and, moreover, "sufficiently hyperbolic." Thus, the induced flow on any connected finite cover is also ergodic. The hyperelliptic locus $\mathcal{Q}^{hyp}(1^{2m}, -1^{2m})$ over $\mathcal{Q}(1^m, -1^{m+4})$ is a finite cover and, as we have just proved, it is connected. This proves the ergodicity of the Teichmüller flow on the hyperelliptic locus.

Let us compute $\Delta\tilde{c}^{pocket}_{area}$, see section 4.2. To follow notations of section 4.2 it would be convenient to label zeroes and poles of the base S of the cover $\tilde{S} \to S$ from now on. Without loss of generality we may assume that the four branching points of the cover are projected to the poles P_1, P_2, P_3, P_4.

When both simple poles P_{j_1}, P_{j_2} involved in the "pocket" configuration are unramified points or when they are both ramified, the holonomy of the hyperelliptic cover $\tilde{S} \to S$ along the perimeter of the cylinder is trivial, so by Lemma 4.1 such configurations do not contribute to $\Delta\tilde{c}^{pocket}_{area}$. By Lemma 4.1, a configuration when one of P_{j_1}, P_{j_2} is ramified and the other one is nonramified contributes to $\Delta\tilde{c}^{pocket}_{area}$ with a weight $-\frac{3}{2}$. Since we have 4 ramified poles and m nonramified, there are $4m$ such configurations. Applying (4.8) and (4.10) with $k = m + (m + 4)$ we get

$$\frac{\pi^2}{3} \cdot \Delta\tilde{c}^{pocket}_{area} = \frac{\pi^2}{3} \cdot 4m \cdot \left(-\frac{3}{2}\right) \cdot \left(\frac{1}{2m+1} \cdot \frac{1}{2\pi^2}\right) = -\frac{m}{2m+1}. \qquad (4.11)$$

Let us proceed to computation of $\Delta\tilde{c}^{dumbbell}_{area}$. When the number of ramified simple poles on two parts of the "dumbbell" is even, the holonomy of the cover along the perimeter of the cylinder is trivial, so by Lemma 4.1 such configurations do not contribute to $\Delta\tilde{c}^{dumbbell}_{area}$. By Lemma 4.1, a configuration when the number of ramified simple poles on each side of the dumbbell is odd contributes to $\Delta\tilde{c}^{dumbbell}_{area}$ with a weight $-\frac{3}{2}$. To compute the number of such configurations we remark that we have to split m named simple zeroes into two groups of m_1 and $m - m_1$ ones; we also have to split 4 ramified simple poles into 1 and 3; finally we have to split m unramified simple poles into $m_1 + 1$ and $m - m_1 - 1$ ones to have in total $m_1 + 2$ simple poles on one side of the "dumbbell" and $m - m_1 + 2$ on the other side. Note that the fact that there is a single ramified simple pole on one part and 3 ramified simple poles on the other makes our count of configurations asymmetric. Finally note that we have to choose one of m_1 zeroes to be located at the boundary of the cylinder on one side and one of $m - m_1$ zeroes to be located at the boundary of the cylinder on the other side.

For any given m_1, where $1 \leq m_1 \leq m - 1$ our count gives

$$\binom{m}{m_1} \cdot \binom{4}{1} \cdot \binom{m}{m_1 - 1} \cdot (m_1 \cdot (m - m_1)).$$

Applying the general formulae (4.9) and (4.10) to $c^{dumbbell}_{i,j;area}$; taking into consideration that in our particular case we have $d_i = d_j = 1$; $k_1 = m_1 + (m_1 + 2)$; $k_2 = (m - m_1) + (m - m_1 + 2)$; and $k = m + (m + 4)$; and applying (4.10) we get

$$\frac{\pi^2}{3}\Delta\tilde{c}_{area}^{dumbell}$$

$$= \frac{\pi^2}{3}\sum_{m_1=1}^{m-1}\left(\binom{m}{m_1}\cdot\binom{4}{1}\cdot\binom{m}{m_1-1}\cdot(m_1\cdot(m-m_1))\right)\cdot\left(-\frac{3}{2}\right)$$

$$\cdot\frac{1}{(2m+4)-3}\cdot\frac{(1+1)(1+1)}{2}\cdot\frac{((2m_1+2)-3)!\cdot((2m-2m_1+2)-3)!}{(2m+4)-4)!}\cdot\frac{1}{\pi^2}$$

$$= -\frac{1}{2m+1}\cdot\sum_{m_1=1}^{m-1}\binom{m}{m_1}\binom{m}{m_1-1}\cdot\frac{(2m_1)!\cdot(2m-2m_1)!}{(2m)!}$$

$$= -\frac{1}{2m+1}\cdot\sum_{m_1=1}^{m-1}\frac{\binom{m}{m_1}\binom{m}{m_1-1}}{\binom{2m}{2m_1}} = -\frac{1}{2m+1}\cdot\sum_{m_1=1}^{m-1}\frac{\binom{m}{m_1}\binom{m}{m_1+1}}{\binom{2m}{2m_1}}, \quad (4.12)$$

where the last equality in this chain combines the invariance of binomial coefficients under the symmetry $m_1\mapsto m-m_1$ with inversion of the order of summation.

Summing up (4.11) and (4.12) and applying the standard convention

$$\binom{m}{m+1}:=0 \qquad \text{and} \qquad \binom{m}{0}:=1$$

we obtain

$$\frac{\pi^2}{3}\Delta\tilde{c}_{area} = -\frac{1}{2m+1}\sum_{m_1=0}^{m}\frac{\binom{m}{m_1}\binom{m}{m_1+1}}{\binom{2m}{2m_1}} = -1+\frac{(2m)!!}{(2m+1)!!}, \quad (4.13)$$

where the second equality is the combinatorial identity (5.2) proved in Proposition 3.

By formula (4.5) we have

$$\Delta\tilde{\kappa} = \frac{1}{4}\cdot\sum_{i=1}^{4}\frac{1}{(-1+2)} = 1. \quad (4.14)$$

Plugging the results (4.13) and (4.14) of our calculation in the general formula (4.6) for the sum Λ^+ we get

$$\Lambda^+ = 1 + \left(-1+\frac{(2m)!!}{(2m+1)!!}\right).$$

It remains to note that the stratum $\mathcal{Q}(1^{2m}, -1^{2m})$ corresponds to genus one, so the spectrum of Lyapunov exponents of H_+^1 contains a single entry and $\Lambda^+ = \lambda_1^+$, which proves Theorem 2.   □

Corollary 3. *The Lyapunov exponent $\lambda_1^+(m)$ tends to zero as m tends to infinity. More precisely,*

$$\lambda_1^+(m) = \frac{\sqrt{\pi}}{2\sqrt{m}}\left(1+O\left(\frac{1}{m}\right)\right). \quad (4.15)$$

Proof. Rewriting the double factorials in terms of usual factorials as

$$\frac{(2m)!!}{(2m+1)!!} = \frac{((2m)!!)^2}{(2m+1)!} = \frac{1}{2m+1} \cdot 2^{2m} \frac{(m!)^2}{(2m)!}$$

and applying Stirling's formula

$$n! = \sqrt{2\pi n} \left(\frac{n}{e}\right)^n \left(1 + O\left(\frac{1}{n}\right)\right)$$

to both factorials and simplifying the resulting expression, we get

$$\frac{(2m)!!}{(2m+1)!!} = \frac{\sqrt{\pi m}}{2m+1} \left(1 + O\left(\frac{1}{m}\right)\right).$$

Clearly, the latter expression tends to zero as m tends to infinity.   □

5   COMBINATORIAL IDENTITIES

Proposition 3. *For any $m \in \mathbb{N}$ the following identities hold:*

$$\sum_{k=0}^{m} \frac{\binom{m}{k}\binom{m}{k}}{\binom{2m}{2k}} = \frac{(2m)!!}{(2m-1)!!} = 4^m \frac{(m!)^2}{(2m)!}, \tag{5.1}$$

$$\sum_{k=0}^{m} \frac{\binom{m}{k}\binom{m}{k+1}}{\binom{2m}{2k}} = 2m+1 - \frac{(2m)!!}{(2m-1)!!}, \tag{5.2}$$

$$\sum_{k=0}^{m} \frac{\binom{m}{k}\binom{m+1}{k+1}}{\binom{2m}{2k}} = 2m+1. \tag{5.3}$$

Proof of Proposition 3. First note that

$$\binom{m+1}{k+1} = \binom{m}{k+1} + \binom{m}{k}.$$

Thus, the expression in the left-hand side of (5.3) is the sum of the corresponding expressions in the left-hand sides of (5.1) and (5.2). Hence, any two out of three identities (5.1)–(5.3) imply the remaining one.

Proof of Identity (5.3). Developing binomial coefficients into factorials in (5.3) and moving the common factorials to the right-hand side, we can rewrite this identity as

$$\sum_{k=0}^{m} \frac{(2k)!}{k!\,(k+1)!} \cdot \frac{(2m-2k)!}{(m-k)!\,(m-k)!} \stackrel{?}{=} \binom{2m+1}{m}. \tag{5.4}$$

Denote the sum in the left-hand side of (5.4) by $s_3(m)$. We prove the latter identity by induction. For $m=0$ it clearly holds. Assuming that (5.4) holds for some integer m we are going to prove that

$$(m+2) \cdot s_3(m+1) - 2(2m+3) \cdot s_3(m) = 0. \tag{5.5}$$

Since the right-hand side of (5.4) satisfies the same relation (which is an exercise), this completes the step of induction. It remains to prove (5.5) under assumption (5.4).

$$(m+2) \cdot s_3(m+1) - 2(2m+3) \cdot s_3(m)$$

$$= (m+2) \sum_{k=0}^{m+1} \frac{(2k)!}{k! \, (k+1)!} \cdot \frac{(2m+2-2k)!}{(m+1-k)! \, (m+1-k)!}$$

$$- 2(2m+3) \sum_{k=0}^{m} \frac{(2k)!}{k! \, (k+1)!} \cdot \frac{(2m-2k)!}{(m-k)! \, (m-k)!}$$

$$= (m+2) \frac{(2m+2)!}{(m+1)!(m+2)!} + \sum_{k=0}^{m} \frac{(2k)!}{k! \, (k+1)!} \cdot \frac{(2m-2k)!}{(m-k)! \, (m-k)!}$$

$$\cdot \left((m+2) \frac{(2m+2-2k)(2m+1-2k)}{(m+1-k)(m+1-k)} - 2(2m+3) \right)$$

$$= 2\frac{(2m+1)!}{(m+1)! \, m!} + \sum_{k=0}^{m} \frac{(2k)!}{k! \, (k+1)!} \cdot \frac{(2m-2k)!}{(m-k)! \, (m-k)!} \cdot \left(-2\frac{k+1}{m+1-k} \right)$$

$$= 2\binom{2m+1}{m} - 2\sum_{k=0}^{m} \frac{(2k)!}{k! \, k!} \cdot \frac{(2m-2k)!}{(m-k)! \, (m-k+1)!}$$

$$= 2\binom{2m+1}{m} - 2\sum_{j=0}^{m} \frac{(2(m-j))!}{(m-j)! \, (m-j)!} \cdot \frac{(2j)!}{j! \, (j+1)!}$$

$$= 2\binom{2m+1}{m} - 2s_3(m) = 0 \,,$$

where the last equality is the induction assumption. Identity (5.4) and hence identity (5.3) is proved.

Proof of identity (5.1). Developing binomial coefficients into factorials in (5.1) and simplifying common factorials in the right- and left- hand side of (5.1), we can rewrite this identity as

$$\sum_{j=0}^{m} \binom{2j}{j} \binom{2m-2j}{m-j} = 4^m \,,$$

which is identity (3.90) in [Go].
   Proposition 3 is proved.   □

APPENDIX A.   REMOVING SOME SQUARES IN THE WIND-TREE MODEL

Let us consider the original wind-tree model as in Figure 1 with particular parameters. We assume that the obstacles are squares and that the fundamental domain of the lattice is also a square with the side twice longer than the side of the obstacle.

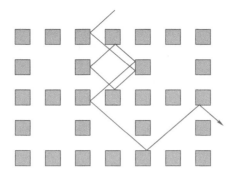

Figure 10: A square-tiled wind-tree from which we regularly remove one obstacle in every repetitive pattern of four.

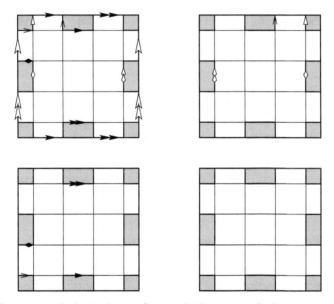

Figure 11: A square-tiled wind-tree from which we regularly remove one obstacle of four in every repetitive pattern.

If we remove periodically one obstacle out of four, we can still perform the same construction as before and end up with a surface in $\mathcal{Q}^{hyp}(1^6, -1^6)$ over $\mathcal{Q}(1^3, -1^7)$.

Namely, unfolding the wind-tree in Figure 10 and taking the quotient over $\mathbb{Z} \oplus \mathbb{Z}$ we get a compact translation surface represented in Figure 11. This translation surface corresponds to the wind-tree in Figure 10 exactly in the same way as the compact flat surface in Figure 5 corresponds to the original wind-tree in Figure 1.

As in the previous examples, the flat surface X in Figure 11 has the group $(\mathbb{Z}/2\mathbb{Z})^3$ as a group of isometries (we have already seen exactly the same group of symmetries for the surface in Figure 5). As before, we can choose as generators the isometries τ_h and τ_v interchanging the pairs of flat tori with holes in the same rows (correspondingly columns) by parallel translations and the isometry ι acting on each of the four tori with holes as the central symmetry with the center in the center of the square. Passing to the quotient over $(\mathbb{Z}/2\mathbb{Z})$ spanned by τ_v we get the square-tiled surface $\hat{S} \in \mathcal{H}(2^6)$ as in Figure 12.

	8		9	
6	7	4	5	6
	3	1	2	
12	13	10	11	12
	8		9	

	21		22	
19	20	17	18	19
	16	14	15	
25	26	23	24	25
	21		22	

Figure 12: The flat surface $\hat{S} = X/\tau_v$.

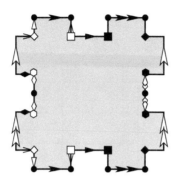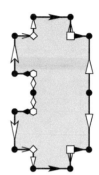

Figure 13: The flat surfaces $S = X/\langle \tau_h, \tau_v \rangle \in \mathcal{Q}^{hyp}(1^6, -1^6)$ (on the left) and $S = X/\langle \tau_h, \tau_v, \iota \rangle \in \mathcal{Q}(1^3, -1^{3+4})$ (on the right). We invert the shading in this picture with respect to Figures 12 and 11: now we shade the surface, and not the obstacles. Tiny black discs at the vertices represent the simple poles; the other vertices correspond to simple zeroes.

It is encoded by the following two permutations:

$r = (1, 2, 16, 14, 15, 3)(4, 5, 6, 7)(8, 22)(9, 21)(10, 11, 12, 13)(14, 15, 16)(17, 18, 19, 20)(23, 24, 25, 26)$

$u = (1, 4, 10)(2, 5, 9, 11)(3, 7, 8, 13)(6, 12)(14, 17, 23)(15, 18, 22, 24)(16, 20, 21, 26)(19, 25).$

We construct its $\mathrm{SL}(2, \mathbb{R})$-orbit, analyze its elements, and apply formula (2.12) from [EKZ2] to find

$$\lambda_1(\hat{S}) + \cdots + \lambda_7(\hat{S}) = \frac{3088}{1053} \tag{A.1}$$

as the sum of the Lyapunov exponents of the resulting arithmetic Teichmüller curve $\mathcal{L}(\hat{S})$.

Now, following the strategy of section 3.1 we pass to the second quotient $\tilde{S} = X/\langle \tau_h, \tau_v \rangle \in \mathcal{Q}^{hyp}(1^6, -1^6)$ (on the left of Figure 13) followed by the quotient $S = X/\langle \tau_h, \tau_v, \iota \rangle \in \mathcal{Q}(1^3, -1^{3+4})$ (on the right of Figure 13).

The surface $\hat{S} \in \mathcal{H}(2^6)$ is the orienting double cover of the surface $\tilde{S}$. Hence, the Lyapunov spectrum (A.1) of the arithmetic Teichmüller curve $\mathcal{L}(\hat{S})$ is the union of the spectra of Lyapunov exponents of the arithmetic Teichmüller curve $\mathcal{L}(\tilde{S})$:

$$\{\lambda_1(\hat{S}), \ldots, \lambda_7(\hat{S})\} = \{\lambda_1^-(\tilde{S}), \ldots, \lambda_6^-(\tilde{S})\} \cup \{\lambda_1^+(\tilde{S})\}.$$

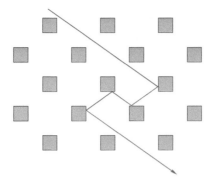

Figure 14: A square-tiled wind-tree where we regularly remove every second square obstacle has the original diffusion rate $\delta = 2/3$.

By formula (2.4) from [EKZ2] we have

$$\left(\lambda_1^-(\tilde{S}) + \cdots + \lambda_6^-(\tilde{S})\right) - \lambda_1^+ = \frac{1}{4} \cdot \sum_{\substack{j \text{ such that} \\ d_j \text{ is odd}}} \frac{1}{d_j + 2} = \frac{1}{4} \cdot 6 \cdot \left(\frac{1}{3} + 1\right) = 2 \,.$$

Together with (A.1) this implies that

$$\lambda_1^+(\tilde{S}) = \frac{1}{2} \left(\frac{3088}{1053} - 2\right) = \frac{491}{1053}$$

and, hence, the diffusion rate for the square-tiled wind-tree as in Figure 10 is given by

$$\delta = \lambda_1^+(\tilde{S}) = \frac{491}{1053} \,.$$

We shall see that in all the other cases, removing some of the obstacles out of every repetitive block of 2×2 obstacles in the wind-tree model with any parameters a and b we keep the diffusion rate $\delta = \lambda_1^+ = 2/3$. There are three cases to consider:

(1) removing two obstacles which are in the same row or in the same column;
(2) removing two obstacles which are on the same diagonal;
(3) removing three obstacles.

In the first and in the last case we can just choose a new fundamental domain of the rectangular lattice (duplicating it in the first case and choosing it 2×2 bigger in the third case) to reduce the situation to the original wind-tree with different parameters. We have seen that the diffusion rate in the original wind-tree as in Figure 1 does not depend on the parameters of the lattice and it does not depend on the parameters of the obstacle. Hence in both cases $\delta = \lambda_1^+ = 2/3$.

Let us consider the case where we remove two obstacles on the same diagonal as in Figure 14. In this case, we can modify the construction a little bit. We unfold the billiard as before, but then we quotient the resulting periodic surface by the integer sublattice spanned by the integer vectors $(1,1)$ and $(1,-1)$ in $\mathbb{Z} \oplus \mathbb{Z}$ and not by the entire lattice $\mathbb{Z} \oplus \mathbb{Z}$ as before. The quotient belongs to the same locus as the original wind-tree. We deduce that we again have $\delta = \lambda_1^+ = 2/3$.

Let us emphasize that the above construction is very specific to the sublattice $\mathbb{Z}(1,1) \oplus \mathbb{Z}(1,-1)$ which is invariant under reflexions by the horizontal and vertical axes. In such a situation the quotient keeps a $(\mathbb{Z}/2\mathbb{Z})^3$ group of symmetry.

APPENDIX B.   WHAT IS NEXT?

The name of the conference and the personality of Bill Thurston suggested we discuss the current directions of research in the area. We focus on those aspects which are somehow related to the current paper. The selection represents our particular taste and might be subjective. These problems are, certainly, known in our community; some of them are already subjects of intensive investigation, so we do not claim novelty in this discussion.

The Magic Wand Theorem of Eskin–Mirzakhani–Mohammadi enormously amplified the importance of the classification of the orbit closures. Ideally, any problem about an individual flat surface (say, diffusion rate for a wind-tree model with obstacles of some specific "rational shape") should be solved as follows: touch the corresponding surface with the Magic Wand and find the corresponding orbit closure $\mathcal{L}$. Now touch the complex Hodge bundle over $\mathcal{L}$ with the Magic Wand and find its irreducible component containing the two integer cocycles responsible for the diffusion. Find or estimate the corresponding Lyapunov exponents and the problem is solved. This strategy evokes three problems: How to find an orbit closure? How to find irreducible components of the Hodge bundle? How to estimate the top Lyapunov exponent of a given invariant flat subbundle?

B.1   Classification of invariant suborbifolds

Problem 1. *Classify all* $\mathrm{GL}(2,\mathbb{R})$*-invariant suborbifolds in* $\mathcal{H}(d_1,\ldots,d_n)$.

The invariant suborbifolds in the strata of quadratic differentials are not mentioned since replacing any flat surface corresponding to a quadratic differential by its canonical ramified double cover on which the induced quadratic differential becomes a global square of a holomorphic 1-form we get an associated invariant suborbifold in the corresponding stratum of Abelian differentials.

In genus 2 the problem is solved by C. McMullen [Mc1]. There are serious advances in this problem for the stratum $\mathcal{H}(4)$, due to Aulicino, Nguyen and Wright [ANW] for $\mathcal{H}^{odd}(4)$ and Nguyen and Wright [NW] for $\mathcal{H}^{hyp}(4)$. Certain important finiteness results for the number of primitive Teichmüller curves are obtained by Bainbridge and Möller [BM], by Matheus and Wright [MaWr] and by Lanneau, Nguyen and Wright [LNW].

The situation in the Prym loci is very well understood due to the work of Lanneau and Nguyen [LN1], [LN2] and [LN3]. The situation for hyperelliptic components of the strata is described in [Ap]. The case of full rank is solved by Mirzakhani and Wright [MiW1], [MiW2].

The theorem of Eskin–Mirzakhani–Mohammadi states that such suborbifolds correspond to complex linear subspaces in period coordinates. Due to the result of Avila–Eskin–Möller [AEMö] the projection of such affine subspace to absolute cohomology should be a symplectic subspace: the restriction of the intersection form to this subspace is nondegenerate. Results of Wright [W1] impose conditions on the field of definition. At some point all these constraints started to seem so restrictive that there was a doubt whether any really new invariant suborbifolds are left. Namely, there is a separate question of classification of Teichmüller curves (both arithmetic and non-arithmetic), and there are plenty of invariant suborbifolds which can be obtained from Teichmüller curves or from connected components of the strata by various ramified covering constructions. The question was whether there is anything else (with exception of some sporadic series of invariant suborbifolds specific for very small genera).

Examples of invariant submanifolds which do not fit any of the known schemes mentioned above were recently found in [McMkW], [EMcMkW] by Eskin, McMullen, Mukamel and Wright. They indicate that potential classification might be highly nontrivial.

B.2 Classification of invariant subbundles of the complex Hodge bundle

The complex Hodge bundle $H^1_{\mathbb{C}}$ over the moduli space $\mathcal{M}_g$ of curves has the homology space $H^1(C; \mathbb{C})$ as a fiber over the point represented by the curve C. This bundle can be pulled back to any stratum $\mathcal{H}(m_1, \ldots, m_n)$.

Problem 2. *Find the decomposition of the complex Hodge bundle $H^1_{\mathbb{C}}$ over any given $GL(2, \mathbb{R})$-invariant suborbifold into irreducible $GL(2, \mathbb{R})$-equivariant subbundles.*

The fact that such decomposition exists is proved by S. Filip [Fi1]. Here "irreducible" should be understood in the sense that there is no further splitting even when we pass to any finite ramified cover of the $GL(2, \mathbb{R})$-invariant suborbifold.

Major advance in this direction was recently obtained by Eskin, Filip and Wright in the fundamental paper [EFiW]. The general arithmetic Teichmüller curves were studied by Matheus, Möller and Yoccoz in [MMöY]; certain special arithmetic Teichmüller curves were studied by Forni, Matheus and Zorich [FoMZ] using the method of Eskin.

A closely related question is simplicity of the spectrum of Lyapunov exponents, referred to as *Kontsevich–Zorich conjecture*. For connected components of strata of Abelian differentials the Kontsevich–Zorich conjecture was proved by Avila and Viana [AV]. A proof of this conjecture for most of the strata of quadratic differentials was recently obtained by Gutierrez [Gu2].

One more related question follows.

Problem 3. *What groups are realizable as Zariski closures of leafwise monodromy groups of equivariant irreducible blocks of the complex Hodge bundle $H^1_{\mathbb{C}}$ over $GL(2, \mathbb{R})$-invariant suborbifolds?*

The original guess of Forni–Matheus–Zorich [FoMZ] states that this group is always $SU(p, q)$ for appropriate p and q. The paper of S. Filip [Fi2] shows that general Hodge-theoretical arguments admit a priori larger list (including some more sophisticated representations of $SU(p, q)$). However, it is not clear yet which of these groups (representations) are realizable as Zariski closures of leafwise monodromy groups of equivariant subbundles of the complex Hodge bundle over $GL(2, \mathbb{R})$-invariant suborbifolds. Note that both "leafwise" and "$GL(2, \mathbb{R})$-invariant" impose restrictions with respect to most general flat subbundles of the complex Hodge bundle over abstract submanifolds of the moduli space. Progress in this direction was recently obtained by Filip, Forni and Matheus [FiFoM]; by Gutierrez [Gu1], [Gu2]; and by Hubert and Matheus [HM].

B.3   Estimates for individual Lyapunov exponents

The article [EKZ2] provides a formula for the sum of the positive Lyapunov exponents of the complex Hodge bundle along the Teichmüller geodesic flow. However, there is no reason to hope for exact values of individual Lyapunov exponents. The Fei Yu conjecture (stated in [Yu] and proved by Eskin, Kontsevich, Möller, and Zorich in [EKMöZ]) shows that partial sums of Lyapunov exponents bound normalized degrees of holomorphic vector subbundles of the associated local system over a Teichmüller curve. Combined with results of Bonatti, Eskin and Wilkinson [BnEWl] this allowed Fei Yu [Yu] and Fougeron [Fg1] to obtain information on large genus asymptotics of all leading Lyapunov exponents for certain strata of Abelian and quadratic differentials.

More generally (though less precisely):

Problem 4. *Study extremal properties of the "curvature" of the Lyapunov subbundles compared to holomorphic subbundles of the Hodge bundle. Estimate the individual Lyapunov exponents.*

An estimate for the top Lyapunov exponent λ_1^+ of the strata $\mathcal{Q}(1^n, -1^m)$, where $n + m = 4k$, and $k = 0, 1, \ldots$, represents particular interest in this context.

Consider a wind-tree billiard where the periodic obstacle is constructed by superposing r different rectangles aligned with the lattice. The rectangles might overlap or not. Depending on the positioning of the rectangles, the obstacle gets from one to r connected components. By the recent result of Fougeron [Fg3], for generic parameters of the rectangles the diffusion rate of such a wind-tree billiard is given by the Lyapunov exponent λ_1^+ of one of the strata $\mathcal{Q}(1^n, -1^m)$, where n and m are determined in a simple way by geometry of the resulting obstacle. Conjecturally, $\lambda_1^+(\mathcal{Q}(1^{4g-4}))$ tends to $\frac{1}{2}$ as $g \to +\infty$ while conjecturally $\lambda_1^+(\mathcal{Q}(1^n, -1^m))$ tends to 0 if n remains bounded and $m \to +\infty$; see [Fg3] and [GrH] respectively. It would be very interesting to prove any of these conjectures. One can speculate, that letting both n and m tend to $+\infty$ in an appropriate way, one might make λ_1^+ tend to any value in the interval $[0, \frac{1}{2}]$; see [Fg3].

B.4   Siegel–Veech constants in terms of an adequate
        intersection theory

The sum of positive Lyapunov exponents of the complex Hodge bundle over an $\mathrm{SL}(2, \mathbb{R})$-invariant suborbifold $\mathcal{L}$ in a stratum $\mathcal{H}_1(m_1, \ldots, m_n)$ is expressed in [EKZ2] as

$$\lambda_1 + \cdots + \lambda_g = \frac{1}{12} \sum_{i=1}^{n} \frac{m_i(m_i + 2)}{m_i + 1} + \frac{\pi^2}{3} \cdot c_{area}(\mathcal{L}),$$

where $c_{area}(\mathcal{L})$ is the Siegel–Veech constant of $\mathcal{L}$. Currently there are two formulae for $c_{area}(\mathcal{L})$ for two extremal cases of $\mathcal{L}$. When $\mathcal{L}$ is a connected component of a stratum, the Siegel–Veech constant $c_{area}(\mathcal{L})$ is expressed as a polynomial in volumes of simpler *principal boundary strata* (normalized by the volume of the initial stratum); see [EMsZ] for the strata of Abelian differentials and [Gj1] for the strata of quadratic differentials. When $\mathcal{L}$ is a Teichmüller curve, $c_{area}(\mathcal{L})$ is expressed as the integral of the Chern class of the determinant bundle over $\mathcal{L}$ normalized by the Euler characteristic of $\mathcal{L}$, see [K1], [BwMo], [EKZ1]. The challenge is to construct a bridge between these two cases:

Problem 5. *Express $c_{area}(\mathcal{L})$ in terms of an appropriate intersection theory.*

Here we do not mean some kind of asymptotic limit formulae, but something in the spirit of ELSV-formula for Hurwitz numbers, see [ELSV].

There is certain resemblance between the "hyperbolic regime" studied in [Mi1]–[Mi3] by M. Mirzakhani and the "flat regime" studied in [EMsZ]. M. Mirzakhani used dynamics on moduli space to relate the length functions of simple geodesics on hyperbolic surfaces to the Weil–Peterson volumes of the moduli spaces $\mathcal{M}_{g,n}$ of punctured Riemann surfaces, and also to relate the Weil–Peterson volumes to the intersection numbers of tautological line bundles over $\mathcal{M}_{g,n}$.

Morally, we have a somehow similar situation in a parallel flat world. The step of relating the counting functions for simple *flat* geodesics (for the flat metrics in the same conformal class as the original hyperbolic metric) to polynomial in volumes of the strata with respect to Masur–Veech volume form is already performed in [EMsZ] and [Gj1]. The challenge is to accomplish the second step and to relate these volumes to an adequate intersection theory. A recent paper [DGZZ] by Delecroix, Goujard, Zograf and Zorich provides a solution of this problem for the principal stratum of meromorphic quadratic differentials. Recent papers by Sauvaget [S] and by Chen, Möller, Sauvaget, Zagier [CMöSZ] solve the problem for all strata of Abelian differentials. A very recent paper [CMöS] provides an alternative solution for the principal stratum of meromorphic quadratic differentials and suggests a conjectural formula for all other strata.

Certain parallels between hyperbolic and flat word manifest in further aspects. For example, Mirzakhani and Zograf obtained a simple asymptotic formula for large genera for the Weil–Peterson volumes [MiZ]. Simple asymptotic formula for Masur–Veech volumes conjectured in [EZ] was recently proved by Chen, Möller and D. Zagier [CMöZ] for the principal stratum $\mathcal{H}(1^{2g-2})$; by Sauvaget [S] for the minimal stratum $\mathcal{H}(2g-2)$; and by Aggarwal [Ag] for general strata.

B.5 Dynamics on other families of complex varieties.

One more challenging direction of study is dynamics of the complex Hodge bundle over geodesic flows over moduli spaces of higher-dimensional complex manifolds.

Problem 6. *Study dynamics of the Hodge bundle over geodesic flows on other families of compact varieties. Are there other dynamical systems (compared to billiards in rational polygons) which admit renormalization leading to dynamics on families of complex varieties?*

Some results (partly experimental) for families of Calabi–Yau varieties and similar local systems were recently obtained by Kontsevich [K2]; Eskin, Kontsevich, Möller and Zorich [EKMöZ]; and Ch. Fougeron [Fg2].

S. Filip studied in [Fi3] families of K3-surfaces.

Acknowledgments

The first author is deeply indebted to the organizers of the conference "What's Next? The Mathematical Legacy of Bill Thurston" for the extraordinary job which they have done. We thank MPIM, where part of this paper was written, for an extremely stimulating research atmosphere.

The numerical computations with square-tiled surfaces have been done with the computer software SageMath [S+09] and the surface dynamics package [SD0.3].

REFERENCES

[Ag]　A. Aggarwal, *Large genus asymptotics for volumes of strata of Abelian differentials*, arXiv:1804.05431 (2018), to appear in J. Amer. Math. Soc.

[Ap]　P. Apisa, $GL(2,\mathbb{R})$ *orbit closures in hyperelliptic components of strata*, Duke Math. J. **167** (2018), no. 4, 679–742.

[AtEZ]　J. Athreya, A. Eskin, A. Zorich, *Right-angled billiards and volumes of moduli spaces of quadratic differentials on* $\mathbb{CP}^1$, Ann. Scient. ENS, 4ème série **49** (2016), 1307–1381.

[ANW]　D. Aulicino, D.-M. Nguyen, A. Wright, *Classification of higher rank orbit closures in* $\mathcal{H}^{odd}(4)$, J. Eur. Math. Soc. (JEMS) **18** (2016), no. 8, 1855–1872.

[AH]　A. Avila, P. Hubert, *Recurrence for the wind-tree model*, Annales de l'Institut Henri Poincaré. Anal. Non Linéaire **37** (2020), no. 1, 1–11.

[AEMö]　A. Avila, A. Eskin, M. Möller, *Symplectic and isometric* $SL(2,\mathbb{R})$-*invariant subbundles of the Hodge bundle*, J. Reine Angew. Math. **732** (2017), 1–20.

[AV]　A. Avila, M. Viana, *Simplicity of Lyapunov spectra: proof of the Zorich-Kontsevich conjecture*, Acta Math. **198** (2007), no. 1, 1–56.

[BM]　M. Bainbridge, M. Möller, *Deligne–Mumford compactification of the real multiplication locus and Teichmüller curves in genus three*, Acta Math. **208** (2012), 1–92.

[Bo]　C. Boissy. *Configurations of saddle connections of quadratic differentials on* $\mathbb{CP}^1$ *and on hyperelliptic Riemann surfaces*, Comment. Math. Helv. **84:4** (2009), 757–791.

[BnEWl]　Ch. Bonatti, A. Eskin, A. Wilkinson, *Projective cocycles over* $SL(2,\mathbb{R})$-*actions: measures invariant under the upper triangular group*, arXiv: 1709.02521, Astérisque. **412** (2020), 155–178.

[BwMo]　I. Bouw, M. Möller, *Teichmüller curves, triangle groups and Lyapunov exponents*, Ann. of Math. **172** (2010), 139–185.

[CkE]　J. Chaika, A. Eskin, *Every flat surface is Birkhoff and Oseledets generic in almost every direction*, J. Mod. Dyn. **9** (2015), 1–23.

[CMö] D. Chen, M. Möller, *Quadratic differentials in low genus: exceptional and non–varying strata*, Ann. Sci. Éc. Norm. Supér. (4) **47** (2014), no. 2, 309–369.

[CMöS] D. Chen, M. Möller, A. Sauvaget, with an appendix by G. Borot, A. Giacchetto, D. Lewanski, *Masur–Veech volumes and intersection theory: the principal strata of quadratic differentials*, arXiv:1912.02267 (2019).

[CMöSZ] D. Chen, M. Möller, A. Sauvaget, D. Zagier, *Masur–Veech volumes and intersection theory on moduli spaces of Abelian differentials*, arXiv: 1901.01785.

[CMöZ] D. Chen, M. Möller, D. Zagier, *Quasimodularity and large genus limits of Siegel–Veech constants*, J. Amer. Math. Soc. **31** (2018), no. 4, 1059–1163.

[CrLanH] S. Crovisier, E. Lanneau, P. Hubert, *Diffusion rate in non generic directions in the wind-tree model*, preprint, to appear.

[D] V. Delecroix, *Divergent directions in some periodic wind-tree models*, Journal of Modern Dynamics **7:1** (2013), 1–29.

[DGZZ] V. Delecroix, E. Goujard, P. Zograf, A. Zorich, Masur–Veech volumes, *frequencies of simple closed geodesics and intersection numbers of moduli spaces of curves*, arXiv:1908.08611.

[DHL] V. Delecroix, P. Hubert, S. Lelièvre, *Diffusion for the periodic wind-tree model*, Annales scientifiques de l'ENS **47:6** (2014), 1085–1110.

[SD0.3] V. Delecroix et al., *Surface_dynamics SageMath package* version 0.3 (2017), https://pypi.org/project/surface_dynamics/.

[Eh] P. and T. Ehrenfest, Begriffliche Grundlagen der statistischen Auffassung in der Mechanik Encykl. d. Math. Wissensch. IV 2 II, Heft 6, 90 S (1912) (in German); translated in: The conceptual foundations of the statistical approach in mechanics (trans. Moravicsik, M. J.), 10–13, Cornell University Press, Itacha NY (1959).

[ELSV] T. Ekedahl, S. Lando, M. Shapiro, A. Vainshtein, *Hurwitz numbers and intersections on moduli spaces of curves*, Invent. Math. **146:2** (2001), 297–327.

[EFiW] A. Eskin, S. Filip, A. Wright, *The algebraic hull of the Kontsevich–Zorich cocycle*, Ann. of Math. (2) **188** (2018), no. 1, 281–313.

[EKMöZ] A. Eskin, M. Kontsevich, M. Möller, A. Zorich, *Lower bounds for Lyapunov exponents of flat bundles on curves*, Geom. Topol. **22** (2018), no. 4, 2299–2338.

[EKZ1] A. Eskin, M. Kontsevich, A. Zorich, *Lyapunov spectrum of square-tiled cyclic covers*, Journal of Modern Dynamics **5**, no. 2 (2011), 319–353.

[EKZ2]   A. Eskin, M. Kontsevich, A. Zorich, *Sum of Lyapunov exponents of the Hodge bundle with respect to the Teichmüller geodesic flow*, Publications de l'IHES **120:1** (2014), 207–333.

[EMs]   A. Eskin, H. Masur, Asymptotic formulas on flat surfaces, *Ergodic theory and dynamical systems* **21** (2) (2001), 443–478.

[EMsZ]   A. Eskin, H. Masur, A. Zorich, *Moduli spaces of Abelian differentials: the principal boundary, counting problems and the Siegel–Veech constants*, Publications Mathématiques de l'IHÉS **97** (1) (2003), 61–179.

[EMcMkW] A. Eskin, C. McMullen, R. Mukamel, A. Wright, *Billiards, quadrilaterals and moduli spaces*, to appear in J. Amer. Math. Soc.

[EMi]   A. Eskin, M. Mirzakhani, *Invariant and stationary measures for the* $SL(2, \mathbb{R})$ *action on moduli space*, Publ. Math. Inst. Hautes Études Sci. **127** (2018), 95–324.

[EMiMo]   A. Eskin, M. Mirzakhani, A. Mohammadi, *Isolation, equidistribution, and orbit closures for the* $SL(2, \mathbb{R})$-*action on moduli space*, Ann. of Math. (2) **182** (2015), no. 2, 673–721.

[EZ]   A. Eskin, A. Zorich, *Volumes of strata of Abelian differentials and Siegel–Veech constants in large genera*, Arnold Mathematical Journal, **1:4** (2015), 481–488.

[Fi1]   S. Filip, *Semisimplicity and rigidity of the Kontsevich–Zorich cocycle*, Invent. Math. **205** (2016), no. 3, 617–670.

[Fi2]   S. Filip, *Zero Lyapunov exponents and monodromy of the Kontsevich-Zorich cocycle*, Duke Math. J. **166** (2017), no. 4, 657–706.

[Fi3]   S. Filip, *Families of K3 surfaces and Lyapunov exponents*, Israel J. Math. **226** (2018), no. 1, 29–69.

[FiFoM]   S. Filip, G. Forni, C. Matheus, *Quaternionic covers and monodromy of the Kontsevich–Zorich cocycle in orthogonal groups*, J. Eur. Math. Soc. (JEMS) **20** (2018), no. 1, 165–198.

[Fo]   G. Forni, *Deviation of ergodic averages for area-preserving flows on surfaces of higher genus*, Annals of Math. **155** no. 1 (2002), 1–103.

[FoMZ]   G. Forni, C. Matheus, A. Zorich, *Zero Lyapunov exponents of the Hodge bundle*, Commentari Math. Helvetici **89:2** (2014), 489–535.

[Fg1]   Ch. Fougeron, *Lyapunov exponents of the Hodge bundle over strata of quadratic differentials with large number of poles*, Math. Res. Letters. **25:4** (2018), 1213–1225.

[Fg2]   Ch. Fougeron, *Parabolic degrees and Lyapunov exponents for hypergeometric local systems*, arXiv:1701.08387, to appear in Experimental Mathematics. DOI: 10.1080/10586458.2019.1580632.

[Fg3]   Ch. Fougeron, *Diffusion rate of windtree models and Lyapunov exponents*, arXiv:1803.10717, to appear in Bulletin de la SMF.

[FrU]   K. Fraczek, C. Ulcigrai, *Non-ergodic $\mathbb{Z}$-periodic billiards and infinite translation surfaces*, Invent. Math. **197** (2014), no. 2, 241–298.

[Gj1]   E. Goujard, *Siegel–Veech constants for strata of moduli spaces of quadratic differentials*, GAFA **25** (2015), no. 5, 1440–1492.

[Gj2]   E. Goujard, *Volumes of strata of moduli spaces of quadratic differentials: getting explicit values*, Ann. Inst. Fourier (Grenoble) **66** (2016), no. 6, 2203–2251.

[Go]   H. W. Gould, *Combinatorial identities. A standardized set of tables listing 500 binomial coefficient summations*, Rev. ed. (English) Morgantown (1972).

[GrH]   J. Grivaux, P. Hubert, *Loci in strata of meromorphic differentials with fully degenerate Lyapunov spectrum*, Journal of Modern Dynamics **8**, no. 1 (2014), 61–73.

[Gu1]   R. Gutiérrez-Romo, *Classification of Rauzy–Veech groups: proof of the Zorich conjecture*, Invent. Math. **215** (2019), no. 3, 741–778.

[Gu2]   R. Gutiérrez-Romo, *Simplicity of the Lyapunov spectra of certain quadratic differentials*, arXiv:1711.02006.

[HaWe] J. Hardy, J. Weber, *Diffusion in a periodic wind-tree model*, J. Math. Phys. **21** (7) (1980), 1802–1808.

[HLT]   P. Hubert, S. Lelièvre, S. Troubetzkoy, *The Ehrenfest wind-tree model: periodic directions, recurrence, diffusion*, J. Reine Angew. Math. **656** (2011), 223–244.

[HM]   P. Hubert, C. Matheus, *An origami of genus 3 with arithmetic Kontsevich–Zorich monodromy*, to appear in Mathematical Proceedings of the Cambridge Philosophical Society, arXiv:1806.09855.

[JSch] Ch. Johnson, M. Schmoll, *Pseudo–Anosov eigenfoliations on Panov planes*, Electron. Res. Announc. Math. Sci. **21** (2014), 89–108.

[K1]   M. Kontsevich, *Lyapunov exponents and Hodge theory*, "The mathematical beauty of physics" (Saclay, 1996), (in Honor of C. Itzykson), 318–332, Adv. Ser. Math. Phys. 24, World Sci. Publishing, River Edge, NJ, 1997.

[K2]   M. Kontsevich, *Kähler random walks and Lyapunov exponents*, talk at the conference "Control, index, traces and determinants" related to the work of Jean-Michel Bismut, http://www.math.u-psud.fr/~rep surf/ERC/Bismutfest/Bismutfest.html.

[LN1]   E. Lanneau, D.-M. Nguyen, *Teichmueller curves generated by Weierstrass Prym eigenforms in genus three and genus four*, J. Topol. **7**, no. 2 (2014), 475–522.

[LN2]   E. Lanneau, D.-M. Nguyen, $GL^{+}(2, \mathbb{R})$-*orbits in Prym eigenform loci*, Geom. Topol. **20** (2016), no. 3, 1359–1426.

[LN3]   E. Lanneau, D.-M. Nguyen, *Connected components of Prym loci in genus three*, Math. Ann. **371** (2018), no. 1–2, 753–793.

[LNW]   E. Lanneau, D.-M. Nguyen, A. Wright, *Finiteness of Teichmüller curves in non-arithmetic rank 1 orbit closures*, Amer. J. Math. **139** (2017), no. 6, 1449–1463.

[Ms]   H. Masur, *Interval exchange transformations and measured foliations*, Ann. of Math. **115** (1982), 169–200.

[MsZ]   H. Masur, A. Zorich, *Multiple saddle connections on flat surfaces and principal boundary of the moduli spaces of quadratic differentials*, Geom. Funct. Anal. **18** (2008), no. 3, 919–987.

[MMöY]   C. Matheus, M. Möller, J.-C. Yoccoz, *A criterion for the simplicity of the Lyapunov spectrum of square-tiled surfaces* **202** (2015), no. 1, 333–425.

[MaWr]   C. Matheus, A. Wright, *Hodge–Teichmüller planes and finiteness results for Teichmüller curves*, Duke Math. J. **164** (2015), no. 6, 1041–1077.

[Mc1]   C. McMullen, *Dynamics of* $\mathrm{SL}(2,\mathbb{R})$ *over moduli space in genus two*, Ann. of Math. (2) **165** (2007), no. 2, 397–456.

[McMkW]   C. McMullen, R. Mukamel, A. Wright, *Cubic curves and totally geodesic subvarieties of moduli space*, Ann. of Math. (2) **185** (2017), no. 3, 957–990.

[Mi1]   M. Mirzakhani, *Weil–Petersson volumes and intersection theory on the moduli space of curves*, Journal of Amer. Math. Soc. **20** (2007), no. 1, 1–23.

[Mi2]   M. Mirzakhani, *Simple geodesics and Weil-Petersson volumes of moduli spaces of bordered Riemann surfaces*, Invent. Math. **167** (2007), no. 1, 179–222.

[Mi3]   M. Mirzakhani, *Growth of the number of simple closed geodesics on hyperbolic surfaces*, Annals of Math. (2) **168** (2008), no. 1, 97–125.

[MiZ]   M. Mirzakhani, P. Zograf, *Towards large genus asymtotics of intersection numbers on moduli spaces of curves*, GAFA **25** (2015), no. 4, 1258–1289.

[MiW1]   M. Mirzakhani, A. Wright, *The boundary of an affine invariant submanifold*, Invent. Math. **209** (2017), no. 3, 927–984.

[MiW2]   M. Mirzakhani, A. Wright, *Full-rank affine invariant submanifolds*, Duke Math. J. **167** (2018), no. 1, 1–40.

[NW]   D.-M. Nguyen, A. Wright, *Non–Veech surfaces in* $\mathcal{H}^{hyp}(4)$ *are generic*, GAFA **24** (2014), no. 4, 1316–1335.

[P]   D. Panov, *Foliations with unbounded deviation on* $\mathbb{T}^2$, J. Mod. Dyn. **3** (2009), no. 4, 589–594.

[Par1]   A. Pardo, *Counting problem on wind-tree models*, Geom. Topol. **22** (2018), no. 3, 1483–1536.

[Par2]   A. Pardo, *Quantitative error term in the counting problem on Veech wind-tree models*, arXiv:1704.07428, to appear in Annali della Scuola normale superiore di Pisa, Classe discienze.

  [S]   A. Sauvaget, *Volumes and Siegel–Veech constants of $\mathcal{H}(2g-2)$ and Hodge integrals*, GAFA, **28:6** (2018), 1756–1779.

[S+09]   W. A. Stein et al., *Sage Mathematics Software (Version 6.4)*, Sage Development Team, 2014, http://www.sagemath.org.

 [Ve1]   W. Veech, *Gauss measures for transformations on the space of interval exchange maps*, Annals of Math. **115** (1982), 201–242.

 [Ve2]   W. A. Veech, Siegel measures, *Annals of Math.* **148** (1998), 895–944.

  [Vo]   Y. Vorobets, *Periodic geodesics on generic translation surfaces*, Algebraic and topological dynamics, 205–258, Contemp. Math. 385, Amer. Math. Soc., Providence, RI, 2005.

 [W1]   A. Wright, *The field of definition of affine invariant submanifolds of the moduli space of Abelian differentials*, Geometry and Topology **18**, no. 3 (2014), 1323–1341.

 [W2]   A. Wright, *Cylinder deformations in orbit closures of translation surfaces*, Geometry and Topology **19** (2015), no. 1, 413–438.

 [Yu]   Fei Yu, *Eigenvalues of curvature, Lyapunov exponents and Harder–Narasimhan filtrations*, Geom. Topol. **22** (2018), no. 4, 2253–2298.

  [Z]   A. Zorich, *How do the leaves of a closed 1-form wind around a surface*, In the collection: "Pseudoperiodic Topology," *AMS Translations*, Ser. 2, **197**, AMS, Providence, RI (1999), 135–178.

Groups of Homeomorphisms of One-Manifolds, I: Actions of Nonlinear Groups

BENSON FARB AND JOHN FRANKS

In memory of Bill

1 INTRODUCTION

In this paper we consider finitely presented infinite groups acting by diffeomorphisms on one-dimensional manifolds. For lattices in higher rank semisimple Lie groups such actions are essentially completely understood:

Theorem 1.1 (Ghys [Gh], Burger-Monod [BM]). *Let Γ be a lattice in a simple Lie group of $\mathbf{R}$-rank at least two. Then any C^0-action of Γ on S^1 has a finite orbit, and any C^1-action of Γ on S^1 must factor through a finite group.*

Theorem 1.1 is the solution in dimension one of Zimmer's program of classifying actions of higher rank lattices in simple Lie groups on compact manifolds. In [La] (see in particular §8), Labourie describes possible extensions of Zimmer's program to other "big" groups. A. Navas [Na1] proved that any $C^{1+\alpha}, \alpha > 1/2$ action of a group with Kazhdan's property T factors through a finite group. In this paper we consider three basic examples of nonlinear groups: mapping class groups of surfaces, (outer) automorphism groups of free groups, and Baumslag-Solitar groups.

Mapping class groups and automorphism groups of free groups

Let $\mathrm{Mod}(g, k)$ denote the group of isotopy classes of diffeomorphisms of the genus g surface with k punctures. Unlike the case of lattices (Theorem 1.1), the group $\mathrm{Mod}(g, 1), g \geq 1$ does have a faithful, C^0-action on S^1 without a global fixed point. This is a classical result of Nielsen. However, imposing a small amount of regularity changes the situation dramatically.

Theorem 1.2 (Mapping class groups). *For $g \geq 3$ and $k = 0, 1$, any C^2 action of $\mathrm{Mod}(g, k)$ on S^1 or on $I = [0, 1]$ is trivial.*

Both authors are supported in part by the NSF. This paper was first released in June 2001. We have kept the paper close to the original. However, we have included pointers to related work that appeared since the first version of this paper was released.

The S^1 case of Theorem 1.2 was announced by E. Ghys several years ago (see §8 of [La]), and for real-analytic actions was proved by Farb-Shalen [FS].

Another class of basic examples of "big groups" are automorphism groups of free groups. Let $\mathrm{Aut}(F_n)$ (resp. $\mathrm{Out}(F_n)$) denote the group of automorphisms (resp. outer automorphisms) of the free group F_n of rank n. It is known that $\mathrm{Aut}(F_n), n > 2$ is not linear [FP].

The techniques we develop to prove the results above allow us to prove the following.

Theorem 1.3 (Automorphism groups of free groups). *For $n \geq 6$, any homomorphism from $\mathrm{Aut}(F_n)$ to $\mathrm{Diff}^2_+(S^1)$ factors through $\mathbf{Z}/2\mathbf{Z}$. Any homomorphism from $\mathrm{Aut}(F_n)$ to $\mathrm{Diff}^2_+(I)$ is trivial. The analagous results hold for $\mathrm{Out}(F_n)$.*

M. Bridson and K. Vogtmann [BV] have proved a stronger result: any homomorphism from $\mathrm{Aut}(F_n)$ to a group which does not contain a symmetric group must have an image which is at most $Z/2Z$. See also [Pa]. However, their proof uses torsion elements in an essential way, hence does not extend to finite index subgroups. On the other hand our techniques have implications for finite index subgroups of $\mathrm{Aut}(F_n)$ and $\mathrm{Mod}(g,k)$; such subgroups are typically torsion free.

Theorem 1.4 (Finite index and other subgroups). *Let H be any group of C^2 diffeomorphisms of I or S^1 with the property that no nontrivial element of H has an interval of fixed points (e.g., H is a group of real-analytic diffeomorphisms). Then H does not contain any finite index subgroup of:*

1. *$\mathrm{Mod}(g,k)$ for $g \geq 3, k \geq 0$;*
2. *$\mathrm{Aut}(F_n)$ or $\mathrm{Out}(F_n)$ for $n \geq 6$;*
3. *the Torelli group $T_{g,k}$ for $g \geq 3, k \geq 0$.*

In [FF3] we construct a C^1 action of $T_{g,k}$ on I and on S^1. Note also that $\mathrm{Out}(F_2)$ has a free subgroup of finite index, which admits a faithful, C^ω action on I and on S^1.

Baumslag-Solitar groups

The *Baumslag-Solitar groups* $\mathrm{BS}(m,n)$ are defined by the presentation

$$\mathrm{BS}(m,n) = <a, b : ab^m a^{-1} = b^n>.$$

When $n > m > 1$ the group $\mathrm{BS}(m,n)$ is not residually finite; in particular it is not a subgroup of any linear group (see, e.g., [LS]). We will give a construction which shows that $\mathrm{BS}(m,n)$ is a subgroup of one of the "smallest" infinite-dimensional Lie groups.

Theorem 1.5 (Baumslag-Solitar groups: existence). *The group $\mathrm{Diff}^\omega_+(\mathbf{R})$ of real-analytic diffeomorphisms of $\mathbf{R}$ contains a subgroup isomorphic to $\mathrm{BS}(m,n)$ for any $n > m \geq 1$. The analogous result holds for $\mathrm{Homeo}_+(S^1)$ and $\mathrm{Homeo}_+(I)$.*

It is not difficult to construct pairs of diffeomorphisms $a, b \in \mathrm{Diff}^\omega_+(\mathbf{R})$ which satisfy the relation $ab^m a^{-1} = b^n$; the difficulty is to prove that (in certain situations) this is the only relation. To do this we use a Schottky type argument.

While Ghys-Sergiescu [GS] showed that $\text{Diff}^\infty(S^1)$ contains Thompson's infinite simple (hence non-residually finite) group T, they also showed that T admits no real-analytic action on S^1 (see also [FS]); indeed we do not know of any subgroups of $\text{Diff}^\omega(S^1)$ which are not residually finite.

The construction in the proof of Theorem 1.5 gives an abundance of analytic actions of $\text{BS}(m,n)$ on $\mathbf{R}$, and C^0 actions of $\text{BS}(m,n)$ on S^1 and on I. The loss of regularity in moving from $\mathbf{R}$ to S^1 is no accident; we will show in contrast to Theorem 1.5 that (for typical m, n) there are no C^2 actions of $\text{BS}(m,n)$ on S^1 or I.

Theorem 1.6 (Baumslag-Solitar groups: non-existence). *No subgroup of* $\text{Diff}^2_+(I)$ *is isomorphic to* $\text{BS}(m,n), n > m > 1$. *If further* m *does not divide* n, *then the same holds for* $\text{Diff}^2_+(S^1)$.

The hypothesis $m > 1$ in Theorem 1.6 is also necessary since $\text{BS}(1,n), n \geq 1$ is a subgroup of $\text{PSL}(2, \mathbf{R})$, hence of $\text{Diff}^\omega_+(S^1)$.

Remarks on more recent work

Since this paper was originally posted in June 2001, there has been a great deal of progress on the types of problems discussed in this paper. We mention only a few papers here, and we refer the reader to the references contained in these papers.

Rigidity of Baumslag-Solitar group actions on one-manifolds was addressed in [BW]. Motivation for a general study of this phenomenon in affine groups is given in [BMNR]. For $g \geq 6$, Parwani [Pa] generalized Theorem 1.2 from C^2 actions to C^1 actions. Other theorems on the rigidity of mapping class group actions on S^1 include work of Baik-Kim-Koberda [BKK] and Mann-Wolff [MW]. There have also appeared a number of papers on the study of other types of groups acting on one-manifolds, for example, Navas's work [Na1, Na2] on groups with property T. For an introduction to this topic, and more results, we refer the reader to Navas's book [Na3].

Acknowledgments

We would like to thank Katie Mann as well as an anonymous referee for a number of useful comments that helped to improve this paper.

2   TOOLS

In this section we recall some properties of diffeomorphisms of one-manifolds which will be used throughout the paper.

2.1   Kopell's Lemma and Hölder's Theorem

Our primary tool is the following remarkable result of Nancy Kopell which is Lemma 1 of [K].

Theorem 2.1 (Kopell's Lemma). *Suppose* f *and* g *are* C^2, *orientation-preserving diffeomorphisms of an interval* $[a, b)$ *such that* $fg = gf$. *If* f *has no fixed point in* (a, b) *and* g *has a fixed point in* (a, b) *then* $g = id$.

Another useful result is the following theorem, which is classical.

Theorem 2.2 (Hölder's Theorem). *Suppose a group G of homeomorphisms of R acts freely and effectively on a closed subset of R. Then G is abelian.*

2.2   The translation number and mean translation number

If $f \in \mathrm{Homeo}_+(S^1)$, i.e, f is an orientation preserving homeomorphism of S^1, then there is a countable collection of lifts of f to orientation-preserving homeomorphisms of the line. If F is one such lift then it satisfies $FT = TF$ where $T(x) = x + 1$, and all others are of the form FT^n, $n \in \mathbf{Z}$. Any orientation-preserving homeomorphism of R which commutes with T is a lift of an element of $\mathrm{Homeo}_+(S^1)$. We will denote by $\mathrm{Homeo}_{\mathbf{Z}}(R)$ the group of homeomorphisms of R which commute with T, or equivalently the group of all lifts of elements of $\mathrm{Homeo}_+(S^1)$.

There are two important and closely related functions from $\mathrm{Homeo}_+(S^1)$ and $\mathrm{Homeo}_{\mathbf{Z}}(R)$ to S^1 and R respectively, which we now recall.

Definition 2.3. *If $F \in \mathrm{Homeo}_{\mathbf{Z}}(R)$ its translation number $\tau(F) \in R$ by*

$$\tau(F) = \lim_{n \to \infty} \frac{F^n(x) - x}{n}.$$

If $f \in \mathrm{Homeo}_+(S^1)$ its rotation number, $\rho(f) \in S^1$, by

$$\rho(f) = \tau(F) \mod (1),$$

where F is any lift of f.

We summarize some basic properties of the rotation and translation numbers. Proofs of these and additional properties can be found, for example, in [dMvS].

Proposition 2.4 (Properties of rotation number). *If $F \in \mathrm{Homeo}_{\mathbf{Z}}(R)$ the number $\tau(F)$ always exists and is independent of the $x \in R$ used to define it. It satisfies*

$$\tau(F^n) = n\tau(F)$$
$$\tau(FT^n) = \tau(F) + n$$
$$\tau(F_0 F F_0^{-1}) = \tau(F) \text{ for any } F_0 \in \mathrm{Homeo}_{\mathbf{Z}}(R).$$

F has a fixed point if and only if $\tau(F) = 0$.

If $f \in \mathrm{Homeo}_+(S^1)$ then $\rho(f)$ is independent of the lift F used to define it. It satisfies

$$\rho(f^n) = n\rho(f)$$
$$\rho(f_0 f f_0^{-1}) = \rho(f) \text{ for any } f_0 \in \mathrm{Homeo}_+(S^1).$$

f has a fixed point if and only if $\rho(f) = 0$.

Suppose G is a subgroup of $\mathrm{Homeo}_+(S^1)$ and $\bar{G}$ is the group of all lifts of elements of G to R. If G preserves a Borel probability measure μ_0 on S^1 then this measure may be lifted to a $\bar{G}$ invariant measure μ on $\mathbf{R}$ which is finite on compact sets and which is preserved by the covering translation $T(x) = x + 1$. This permits us to define the *mean translation number* with respect to μ.

Definition 2.5. *If* $F \in \mathrm{Homeo}_{\mathbf{Z}}(\mathbf{R})$ *define its* mean translation number $\tau_\mu(F)$ $\in \mathbf{R}$ *by*

$$\tau_\mu(F) = \begin{cases} \mu([x, F(x))) & \text{if } F(x) > x, \\ 0 & \text{if } F(x) = x, \\ -\mu([F(x), x)) & \text{if } F(x) < x, \end{cases}$$

where $x \in \mathbf{R}$.

We enumerate some of the well-known basic properties of the mean translation number which we will use later.

Proposition 2.6. *Suppose* G *is a subgroup of* $\mathrm{Homeo}_+(S^1)$ *which preserves the Borel probability measure and* $\bar{G}$ *is the group of all lifts of elements of* G. *If* $F \in \bar{G}$ *the mean translation number* $\tau_\mu(F)$ *is independent of the point* x *used to define it. Indeed* $\tau_\mu(F) = \tau(F)$. *Moreover the function* $\tau_\mu : \bar{G} \to \mathbf{R}$ *is a homomorphism (and hence so is* $\tau : \bar{G} \to \mathbf{R}$).

Proof. Consider the function

$$\nu(x, y) = \begin{cases} \mu([x, y)) & \text{if } y > x, \\ 0 & \text{if } y = x, \\ -\mu([y, x)) & \text{if } y < x. \end{cases}$$

It has the property that $\nu(x, y) + \nu(y, z) = \nu(x, z)$.

We also note that $\tau_\mu(F) = \nu(x, F(x))$. To see this is independent of x note for any $y \in \mathbf{R}$, $\nu(x, F(x)) = \nu(x, y) + \nu(y, F(y)) + \nu(F(y), F(x)) = \nu(y, F(y)) + \nu(x, y) - \nu(F(x), F(y))$. But F preserves the measure μ and the orientation of $\mathbf{R}$ so $\nu(x, y) = \nu(F(x), F(y))$. Hence $\nu(x, F(x)) = \nu(y, F(y))$.

Since μ is the lift of a probability measure on S^1 we know that $\mu([x, x + k)) = k$ for any $k \in \mathbf{Z}$. So if $F^n(x) \in [x + k, x + k + 1)$ we see that

$$F^n(x) - x - 1 \le k \le \nu(x, F^n(x)) \le k + 1 \le F^n(x) - x + 1.$$

To see that $\tau_\mu(F) = \tau(F)$ we note

$$\begin{aligned} \tau(F) &= \lim_{n \to \infty} \frac{F^n(x) - x}{n} \\ &= \lim_{n \to \infty} \frac{\nu(x, F^n(x))}{n} \\ &= \lim_{n \to \infty} \frac{1}{n} \sum_{i=0}^{n-1} \nu(F^i(x), F^{i+1}(x)) \\ &= \nu(x, F(x)) \\ &= \tau_\mu(F). \end{aligned}$$

$\diamond$

There are two well-known corollaries of this result which we record here for later use.

Corollary 2.7. *Suppose G is a subgroup of* $\mathrm{Homeo}_+(S^1)$ *which preserves the Borel probability measure μ_0 and $\bar{G}$ is the group of all lifts of elements of G. Then each of the functions $\tau_\mu : \bar{G} \to \mathbf{R}$, $\tau : \bar{G} \to \mathbf{R}$ and $\rho : G \to S^1$ is a homomorphism.*

Proof. To see that $\tau_\mu : \bar{G} \to \mathbf{R}$ is a homomorphism we suppose $f, g \in \bar{G}$ and consider ν as defined above. Then

$$\tau_\mu(fg) = \nu(x, f(g(x))) = \nu(x, g(x)) + \nu(g(x), f(g(x))) = \tau_\mu(g) + \tau_\mu(f).$$

Since $\tau_\mu = \tau$ we know that τ is also a homomorphism. There is a natural homomorphism $\pi : \bar{G} \to G$ which assigns to g its projection on S^1. If $p : \mathbf{R} \to S^1 = \mathbf{R}/\mathbf{Z}$ is the natural projection then for any $f, g \in \bar{G}$,

$$\rho(\pi(fg)) = p(\tau(fg)) = p(\tau(f)) + p(\tau(g)) = \rho(\pi(f)) + \rho(\pi(g)).$$

Hence $\rho : G \to S^1$ is a homomorphism. $\diamond$

Corollary 2.8. *If G is an abelian subgroup of* $\mathrm{Homeo}_+(S^1)$ *and $\bar{G}$ is the group of all lifts to $\mathbf{R}$ of elements of G, then $\bar{G}$ is abelian and both $\tau : \bar{G} \to \mathbf{R}$ and $\rho : G \to S^1$ are homomorphisms.*

Proof. Since G is abelian it is amenable and there is a Borel probability measure μ_0 on S^1 invariant under G. Let μ be the lift of this measure to $\mathbf{R}$. Then Corollary 2.7 implies that $\tau : \bar{G} \to \mathbf{R}$ and $\rho : G \to S^1$ are homomorphisms.

As above let $\pi : \bar{G} \to G$ be the natural projection. The kernel of π is $\{T^n\}_{n \in \mathbf{Z}}$, where $T(x) = x + 1$, i.e., the lifts of the identity. If $f, g \in \bar{G}$ then $[f, g]$ is in the kernel of π so $[f, g] = T^k$, for some $k \in \mathbf{Z}$. But since $\bar{\rho}$ is a homomorphism $\rho([f, g]) = 0$ which implies $k = 0$. Hence $\bar{G}$ is abelian. $\diamond$

The following easy lemma turns out to be very useful.

Lemma 2.9. *Suppose G is a subgroup of* $\mathrm{Homeo}_+(S^1)$ *which preserves the Borel probability measure μ_0 and $f_0, g_0 \in G$ satisfy $\rho(f_0) = \rho(g_0)$. Then $f_0(x) = g_0(x)$ for all x in the support of μ_0.*

Proof. Let μ be the lift of the measure μ_0 to $\mathbf{R}$. Pick lifts f and g of f_0 and g_0 respectively which satisfy $\tau_\mu(f) = \tau_\mu(g)$. Then for any $x \in \mathbf{R}$ we have $\mu([x, f(x))) = \tau_\mu(f) = \tau_\mu(g) = \mu([x, g(x)))$.

Suppose that $g(x) > f(x)$. It then follows that $\mu([f(x), g(x))) = 0$, so for a sufficiently small $\epsilon > 0$, $\mu([f(x), f(x) + \epsilon]) = 0$ and $\mu([g(x) - \epsilon, g(x))) = 0$. Applying gf^{-1} to $[f(x), f(x) + \epsilon)$ we see that $\mu([g(x), g(x) + \epsilon')) = 0$. Hence $\mu([g(x) - \epsilon, g(x) + \epsilon')) = 0$ which is not possible if x and hence $g(x)$ is in the support of μ.

We have shown that $x \in \mathrm{supp}(\mu)$ implies $f(x) \leq g(x)$. The inequality $g(x) \leq f(x)$ is similar. Projecting back to S^1 we conclude that $f_0 = g_0$ on $\mathrm{supp}(\mu_0)$. $\diamond$

One additional well-known property we will use is the following.

Proposition 2.10. *If $g \in \mathrm{Homeo}_+(S^1)$ is an irrational rotation then its centralizer $Z(g)$ in $\mathrm{Homeo}_+(S^1)$ is the group of rigid rotations of S^1.*

Proof. Let f be an element of $Z(g)$. Then the group generated by f and g is abelian and hence amenable so there it has an invariant Borel probability measure. But Lebesgue measure is the unique Borel probability measure invariant by the irrational rotation g. Since f preserves orientation and Lebesgue measure, it is a rotation. ◇

3 MAPPING CLASS GROUPS

In this section we prove Theorem 1.2. Several of the results, in particular those in §3.1, will also be used to prove Theorem 1.3.

3.1 Fully supported diffeomorphisms

One of the main techniques of this paper is to use Kopell's Lemma (Theorem 2.1) to understand actions of commuting diffeomorphisms. The key dichotomy that arises is the behavior of diffeomorphisms with an interval of fixed points versus those which we call *fully supported*.

If f is a homeomorphism of a manifold M, we will denote by $\partial \mathrm{Fix}(f)$ the frontier of $\mathrm{Fix}(f)$, i.e., the set $\partial \mathrm{Fix}(f) = \mathrm{Fix}(f) \setminus \mathrm{Int}(\mathrm{Fix}(f))$.

Definition 3.1 (Fully supported homeomorphism). *A homeomorphism f of a manifold M is* fully supported *provided that $\mathrm{Int}(\mathrm{Fix}(f)) = \emptyset$, or equivalently $\partial \mathrm{Fix}(f) = \mathrm{Fix}(f)$. A subgroup G of $\mathrm{Homeo}(M)$ will be called* fully supported *provided that every nontrivial element is fully supported.*

It is a trivial but useful observation that if M is connected then for every nontrivial homeomorphism f of M, the set $\mathrm{Fix}(f) \neq \emptyset$ if and only if $\partial \mathrm{Fix}(f) \neq \emptyset$.

The following lemma is a consequence of Kopell's Lemma (Theorem 2.1).

Lemma 3.2 (Commuting diffeomorphisms on I). *Suppose f and g are commuting orientation-preserving C^2 diffeomorphisms of I. Then f preserves every component of $\mathrm{Fix}(g)$ and vice versa. Moreover, $\partial \mathrm{Fix}(f) \subset \mathrm{Fix}(g)$ and vice versa. In particular if f and g are fully supported then $\mathrm{Fix}(f) = \mathrm{Fix}(g)$.*

Proof. The proof is by contradiction. Assume X is a component of $\mathrm{Fix}(g)$ and $f(X) \neq X$. Since f and g commute $f(\mathrm{Fix}(g)) = \mathrm{Fix}(g)$ so $f(X) \neq X$ implies $f(X) \cap X = \emptyset$. Let x be an element of X and without loss of generality assume $f(x) < x$. Define

$$a = \lim_{n \to \infty} f^n(x) \text{ and } b = \lim_{n \to -\infty} f^n(x).$$

Then a and b are fixed under both f and g and f has no fixed points in (a, b). Then Kopell's Lemma (Theorem 2.1) implies $g(y) = y$ for all $y \in [a, b]$, contradicting the hypothesis that X is a component of $\mathrm{Fix}(g)$. The observation that $\partial \mathrm{Fix}(f) \subset \mathrm{Fix}(g)$ follows from the fact that $x \in \partial \mathrm{Fix}(f)$ implies that either $\{x\}$ is a component of

$\mathrm{Fix}(f)$ or x is the endpoint of an interval which is a component of $\mathrm{Fix}(f)$ so, in either case, $x \in \mathrm{Fix}(g)$.   ◇

There is also a version of Lemma 3.2 for the circle. For $g \in \mathrm{Homeo}_+(S^1)$ let $\mathrm{Per}(g)$ denote the set of periodic points of g.

Lemma 3.3 (Commuting diffeomorphisms on S^1). *Suppose f and g are commuting orientation-preserving C^2 diffeomorphisms of S^1. Then f preserves every component of $\mathrm{Per}(g)$ and vice versa. Moreover, $\partial \mathrm{Per}(f) \subset \mathrm{Per}(g)$ and vice versa. In particular if neither $\mathrm{Per}(f)$ or $\mathrm{Per}(g)$ has interior then $\mathrm{Per}(f) = \mathrm{Per}(g)$.*

Proof. First consider the case that $\mathrm{Per}(g) = \emptyset$. The assertion that f preserves components of $\mathrm{Per}(g)$ is then trivial. Also, g is topologically conjugate to an irrational rotation which by Proposition 2.10 implies that f is topologically conjugate to a rotation and hence $\mathrm{Per}(f) = \emptyset$ or $\mathrm{Per}(f) = S^1$. In either case we have the desired result.

Thus we may assume both f and g have periodic points. Since for any circle homeomorphism all periodic points must have the same period, we can let p be the least common multiple of the periods and observe that $\mathrm{Per}(f) = \mathrm{Fix}(f^p)$ and $\mathrm{Per}(g) = \mathrm{Fix}(g^p)$. Since f and g commute and both f^p and g^p have fixed points, if $x \in \mathrm{Fix}(f^p)$ then $y = \lim_{n \to \infty} g^{np}(x)$ will exist and be a common fixed point for f^p and g^p. If we split the circle at y we obtain two commuting diffeomorphisms, f^p and g^p, of an interval to which we may apply Lemma 3.2 and obtain the desired result.   ◇

Lemma 3.4. *Let $g_0 \in \mathrm{Diff}_+^2(S^1)$ and let $Z(g_0)$ denote its centralizer. Then the rotation number $\rho : Z(g_0) \to S^1$ is a homomorphism.*

Proof. We first observe the result is easy if g_0 has no periodic points. In this case g_0 is conjugate in $\mathrm{Homeo}_+(S^1)$ to an irrational rotation by Denjoy's Theorem, and the centralizer of an irrational rotation is abelian by Proposition 2.10. So the centralizer of g in $\mathrm{Diff}_+^1(S^1)$ is abelian. But then ρ restricted to an abelian subgroup of $\mathrm{Homeo}_+(S^1)$ is a homomorphism by Proposition 2.8.

Thus we may assume g_0 has a periodic point, say of period p. Then $h_0 = g_0^p$ has a fixed point. Let h be a lift of h_0 to $\mathbf{R}$ which has a fixed point. Let G denote the group of all lifts to $\mathbf{R}$ of all elements of $Z(g_0)$. Note that every element of this group commutes with h because by Corollary 2.8 any lifts to $\mathbf{R}$ of commuting homeomorphisms of S^1 commute.

Let $X = \partial \mathrm{Fix}(h)$ and observe that $g(X) = X$ for all $g \in G$, so G acts on the unbounded closed set X and G acts on $X_0 = \partial \mathrm{Fix}(h_0)$ which is the image of X under the covering projection to S^1. We will show that if $\mathrm{Fix}(g) \cap X \neq \emptyset$ then $X \subset \mathrm{Fix}(g)$. Indeed applying Lemma 3.3 to h_0 and map g_1 of S^1 which g covers, we observe that $\partial \mathrm{Fix}(h) \subset \mathrm{Fix}(g_1)$ which implies $X \subset \partial \mathrm{Fix}(g)$ if g has has a fixed point.

It follows that if $H = \{g \in G \mid g(x) = x \text{ for all } x \in X\}$ is the stabilizer of X then G/H acts freely on X and hence is abelian by Hölder's Theorem. Also G/H acts on X_0 and hence there is a measure μ_0 supported on X_0 and invariant under G/H. Clearly this measure is also invariant under the action of $Z(g_0)$ on S^1. The measure μ_0 lifts to a G-invariant measure μ.

It follows from Corollary 2.7 that the translation number $\tau: G \to \mathbf{R}$ and the rotation number $\rho: Z(g_0) \to S^1$ are homomorphisms. ⬦

The *commutativity graph* of a set S of generators of a group is the graph consisting of one vertex for each $s \in S$ and an edge connecting elements of S which commute.

Theorem 3.5 (Abelian criterion for I). *Let $\{g_1, g_2, \ldots, g_k\}$ be a set of fully supported elements of $\mathrm{Diff}_+^2(I)$ and let G be the group they generate. Suppose that the commutativity graph of this generating set is connected. Then G is abelian.*

Proof. By Lemma 3.2 we may conclude that for each j_1, j_2 we have $\mathrm{Fix}(g_{j_1}) = \mathrm{Fix}(g_{j_2})$. Call this set of fixed points F. Clearly F is the set of global fixed points of G.

Fix a value of i and consider $Z(g_i)$. Restricting to any component U of the complement of F we consider the possibility that there is an $h \in Z(g_i)$ with a fixed point in U. Kopell's Lemma (Theorem 2.1), applied to the closure of the open interval U, tells us that such an h is the identity on U. Thus the restriction of $Z(g_i)$ to U is free and hence abelian by Hölder's Theorem. But U was an arbitrary component of the complement of F and obviously elements of $Z(g_i)$ commute on their common fixed set F. So we conclude that $Z(g_i)$ is abelian.

We have hence shown that if g_j and g_k are joined by a path of length two in the commutativity graph, they are joined by a path of length one. A straightforward induction shows that any two generators are joined by a path of length one, i.e., any two commute. ⬦

There is also a version of Theorem 3.5 for the circle.

Corollary 3.6 (Abelian criterion for S^1). *Let $\{g_1, g_2, \ldots, g_k\}$ be a set of fully supported elements of $\mathrm{Diff}_+^2(S^1)$, each of which has a fixed point, and let G be the group they generate. Suppose that the commutativity graph of these generators is connected. Then G is abelian.*

Proof. By Lemma 3.3 we may conclude that for each j, k we have $\mathrm{Per}(g_j) = \mathrm{Per}(g_k)$. But since each g_i has a fixed point $\mathrm{Per}(g_i) = \mathrm{Fix}(g_i)$. Hence there is a common fixed point for all of the generators. Splitting S^1 at a common fixed point we see G is isomorphic to a subgroup of $\mathrm{Diff}_+^2(I)$ satisfying the hypothesis of Theorem 3.5. It follows that G is abelian. ⬦

We will need to understand relations between fixed sets not just of commuting diffeomorphisms, but also of diffeomorphisms with another basic relation which occurs in mapping class groups as well as in automorphism groups of free groups.

Lemma 3.7 (*aba* lemma). *Suppose a and b are elements of $\mathrm{Homeo}_+(I)$ or elements of $\mathrm{Homeo}_+(S^1)$ which have fixed points. Suppose also that a and b satisfy the relation $a^{n_1} b^{m_3} a^{n_2} = b^{m_1} a^{n_3} b^{m_2}$ with $m_1 + m_2 \neq m_3$. If J is a nontrivial interval in $\mathrm{Fix}(a)$ then either $J \subset \mathrm{Fix}(b)$ or $J \cap \mathrm{Fix}(b) = \emptyset$.*

Proof. Suppose $J \subset \mathrm{Fix}(a)$ and suppose $z \in J \cap \mathrm{Fix}(b)$. We will show that this implies that $J \subset \mathrm{Fix}(b)$.

Let $x_0 \in J$. Observe that $a^{n_1} b^{m_3} a^{n_2} = b^{m_1} a^{n_3} b^{m_2}$ implies $a^{-n_2} b^{-m_3} a^{-n_1} = b^{-m_2} a^{-n_3} b^{-m_1}$. Hence we may assume without loss of generality that $b^{m_3}(x_0)$ is in the subinterval of J with endpoints x_0 and z (otherwise replace a and b by their inverses). In particular $b^{m_3}(x_0) \in J$. Note that

$$b^{m_3}(x_0) = a^{n_1} b^{m_3}(x_0) = a^{n_1} b^{m_3} a^{n_2}(x_0) = b^{m_1} a^{n_3} b^{m_2}(x_0)$$

$$= b^{m_1} b^{m_2}(x_0) = b^{m_1 + m_2}(x_0).$$

This implies $b^{m_1 + m_2 - m_3}(x_0) = x_0$ and since $m_1 + m_2 - m_3 \neq 0$ we can conclude that $b(x_0) = x_0$. Since $x_0 \in J$ was arbitrary we see $J \subset \mathrm{Fix}(b)$.   ◇

3.2   Groups of type $MC(n)$

We will need to abstract some of the properties of the standard generating set for the mapping class group of a surface in order to apply them in other circumstances.

Definition 3.8 (Groups of type $MC(n)$). *We will say that a group G is of type $MC(n)$ provided it is non-abelian and has a set of generators $\{a_i, b_i\}_{i=1}^n \cup \{c_j\}_{j=1}^{n-1}$ with the properties that*

1. *$a_i a_j = a_j a_i$, $b_i b_j = b_j b_i$, $c_i c_j = c_j c_i$, $a_i c_j = c_j a_i$, for all i, j,*
2. *$a_i b_j = b_j a_i$, if $i \neq j$,*
3. *$b_i c_j = c_j b_i$, if $j \neq i, i - 1$,*
4. *$a_i b_i a_i = b_i a_i b_i$ for $1 \leq i \leq n$, and $b_i c_j b_i = c_j b_i c_j$ whenever $j = i, i - 1$.*

A useful consequence of the $MC(n)$ condition is the following.

Lemma 3.9. *If G is a group of type $MC(n)$ and $\{a_i, b_i\}_{i=1}^n \cup \{c_j\}_{j=1}^{n-1}$ are the generators guaranteed by the definition then any two of these generators are conjugate in G.*

Proof. The relation $aba = bab$ implies $a = (ab)^{-1} b(ab)$, so a and b are conjugate. Therefore property (4) of the definition implies a_i is conjugate to b_i and b_i is conjugate to c_i and c_{i-1}. This proves the result.   ◇

The paradigm of a group of type $MC(n)$ is the mapping class group $\mathrm{Mod}(n, 0)$.

Proposition 3.10 ($\mathrm{Mod}(n,0)$ and $\mathrm{Mod}(n,1)$). *For $n > 2$ and $k = 0, 1$ the group $\mathrm{Mod}(n, k)$ contains a set of elements*

$$S = \{a_i, b_i\}_{i=1}^n \cup \{c_j\}_{j=1}^{n-1}$$

with the following properties:

1. *For $k = 0$ the set S generates all of $\mathrm{Mod}(n, 0)$.*
2. *The group generated by S is of type $MC(n)$.*

3. *There is an element of $g \in \mathrm{Mod}(n, k)$ such that $g^{-1}a_1 g = a_n$, $g^{-1}b_1 g = b_n$, $g^{-1}b_2 g = b_{n-1}$, and $g^{-1}c_1 g = c_{n-1}$.*
4. *If G_0 is the subgroup of $\mathrm{Mod}(n, k)$ generated by the subset*

$$\{a_i, b_i\}_{i=1}^{n-1} \cup \{c_j\}_{j=1}^{n-2}$$

then G_0 is of type $MC(n-1)$.

Proof. Let Σ denote the surface of genus n with k punctures. Let $i(\alpha, \beta)$ denote the geometric intersection number of the (isotopy classes of) closed curves α and β on Σ. Let a_i, b_i, c_j with $1 \leq i \leq n, 1 \leq j \leq n$ be Dehn twists about essential, simple closed curves $\alpha_i, \beta_i, \gamma_j$ with the following properties:

- $i(\alpha_i, \alpha_j) = i(\beta_i, \beta_j) = i(\gamma_i, \gamma_j) = 0$ for $i \neq j$.
- $i(\alpha_i, \beta_i) = 1$ for each $1 \leq i \leq n$.
- $i(\gamma_j, \beta_j) = 1 = i(\gamma_j, \beta_{j+1})$ for $1 \leq j \leq n-1$.

It is known (see, e.g., [Iv]) that there exist such a_i, b_i, c_j which generate $\mathrm{Mod}(n, 0)$; we choose such elements. These are also nontrivial in $\mathrm{Mod}(n, 1)$. Since Dehn twists about simple closed curves with intersection number zero commute, and since Dehn twists a, b about essential, simple closed curves with intersection number one satisy $aba = bab$ (see, e.g., Lemma 4.1.F of [Iv]), properties (1)–(4) in Definition 3.8 follow. We note that the proofs of these relations do not depend on the location of the puncture, as long as it is chosen off the curves $\alpha_i, \beta_i, \gamma_j$.

To prove item (2) of the lemma, consider the surface of genus $g - 2$ and with 2 boundary components obtained by cutting Σ along α_1 and γ_1. The loop β_1 becomes a pair of arcs connecting 2 pairs of boundary components, and β_2 becomes an arc connecting 2 boundary components. The genus, boundary components, and combinatorics of arcs is the exact same when cutting Σ along α_n and γ_{n-1}. Hence by the classification of surfaces it follows that there is a homeomorphism between resulting surfaces, inducing a homeomorphism $h : \Sigma \to \Sigma$ with $h(\alpha_1) = \alpha_n, h(\gamma_1) = \gamma_{n-1}$ and $h(\beta_i) = \beta_{n-i+1}, i = 1, 2$. The homotopy class of h gives the required element g.

To prove item (3), let τ be an essential, separating, simple closed curve on Σ such that one of the components Σ' of $\Sigma - \tau$ has genus one and contains a_n and b_n. Then there is a homeomorphism $h \in \mathrm{Homeo}_+(\Sigma)$ taking any element of $\{a_i, b_i\}_{i=1}^{n-1} \cup \{c_j\}_{j=1}^{n-2}$ to any other element, and which is the identity on Σ'. Then the isotopy class of h, as an element of $\mathrm{Mod}(n, 0)$, is the required conjugate, since it lies in the subgroup of $\mathrm{Mod}(0, n)$ of diffeomorphisms supported on $\Sigma - \Sigma'$, which is isomorphic to $\mathrm{Mod}(n - 1, 1)$ and equals G_0. ◇

3.3   Actions on the interval

In this subsection we consider actions on the interval $I = [0, 1]$.

Theorem 3.11 ($MC(n)$ actions on I). *Any C^2 action of a group G of type $MC(n)$ for $n \geq 2$ on an interval I is abelian.*

Proof. We may choose a set of generators $\{a_i, b_i\}_{i=1}^n \cup \{c_j\}_{j=1}^{n-1}$ for G with the properties listed in Definition 3.8.

If G has global fixed points other than the endpoints of I we wish to consider the restriction of the action to the closure of a component of the complement of these global fixed points. If all of these restricted actions are abelian then the original action was abelian. Hence it suffices to prove that these restrictions are abelian. Thus we may consider an action of G on a closed interval I_0 with no interior global fixed points. None of the generators above can act trivially on I_0 since the fact that they are all conjugate would mean they all act trivially.

We first consider the case that one generator has a nontrivial interval of fixed points in I_0 and show this leads to a contradiction. Since they are all conjugate, each of the generators has a nontrivial interval of fixed points.

Choose an interval J of fixed points which is maximal among all intervals of fixed points for all of the a_i. That is, J is a nontrivial interval of fixed points for one of the a_i, which we assume (without loss of generality) is a_2, and there is no J' which properly contains J and which is pointwise fixed by some a_j.

Suppose $J = [x_0, x_1]$. At least one of x_0 and x_1 is not an endpoint of I_0 since a_2 is not the identity. Suppose it is x_0 which is in $\text{Int}(I_0)$. Then $x_0 \in \partial \text{Fix}(a_2)$, so by Lemma 3.2 we know that x_0 is fixed by all of the generators except possibly b_2, since b_2 is the only generator with which a_2 does not commute. The point x_0 cannot be fixed by b_2 since otherwise it would be an interior point of I_0 fixed by all the generators, but there are no global fixed points in I_0 other than the endpoints. The identity $a_2 b_2 a_2 = b_2 a_2 b_2$ together with Lemma 3.7 then tells us that $b_2(J) \cap J = \emptyset$.

Now assume without loss of generality that $b_2(x_0) > x_0$ (otherwise replace all generators by their inverses). Define

$$y_0 = \lim_{n \to -\infty} b_2^n(x_0) \text{ and } y_1 = \lim_{n \to +\infty} b_2^n(x_0).$$

Since $b_2(J) \cap J = \emptyset$ we have $b_2^{-1}(J) \cap J = \emptyset$ and hence $y_0 < x_0 < x_1 < y_1$. But x_0 is a fixed point of a_1 (as well as the other generators which commute with a_2) and a_1 commutes with b_2. Hence y_0 and y_1 are fixed by a_1 (since b_2 preserves $\text{Fix}(a_1)$). We can now apply Kopell's Lemma (Theorem 2.1) to conclude that a_1 is the identity on $[y_0, y_1]$ which contradicts the fact that $J = [x_0, x_1]$ is a maximal interval of fixed points among all the a_i.

We have thus contradicted the supposition that one of the generators has a nontrivial interval of fixed points in I_0, so we may assume that each of the generators, when restricted to I_0, has fixed point set with empty interior. That is, each generator is fully supported on I_0.

We now note that given any two of the generators above, there is another generator h with which they commute. Thus we may conclude from Lemma 3.5 that the action of G on I_0 is abelian. ◇

Proof of Theorem 1.2 for I: By Proposition 3.10, the mapping class group $\text{Mod}(g, k), g \geq 2, k = 0, 1$ is a group of type $MC(g)$. Since every abelian quotient of $\text{Mod}(g, k), g \geq 2$ is finite (see, e.g., [Iv]), in fact trivial for $g \geq 3$, and since finite groups must act trivially on I, the statement of Theorem 1.2 for I follows.

3.4   Actions on the circle

We are now prepared to prove Theorem 1.2 in the case of the circle S^1.

Theorem 1.28 (Circle case). *Any C^2 action of the mapping class group $M(g, k)$ of a surface of genus $g \geq 3$ with $k = 0, 1$ punctures on S^1 must be trivial.*

As mentioned in the introduction, the C^2 hypothesis is necessary.

Proof. **Closed case.** We first consider the group $G = \text{Mod}(g, 0)$, and choose generators $\{a_i, b_i\}_{i=1}^{n} \cup \{c_j\}_{j=1}^{n-1}$ for G with the properties listed in Definition 3.8 and Proposition 3.10. All of these elements are conjugate by Lemma 3.9, so they all have the same rotation number.

We wish to consider a subgroup G_0 of type $MC(n-1)$. We let $u = a_n^{-1}$ and for $1 \leq i \leq n-1$ we set $A_i = a_i u$ and $B_i = b_i u$. For $1 \leq i \leq n-2$ we let $C_i = c_i u$. Since u commutes with any element of $\{a_i, b_i\}_{i=1}^{n-1} \cup \{c_j\}_{j=1}^{n-2}$ we have the relations

1. $A_i A_j = A_j A_i$, $B_i B_j = B_j B_i$, $C_i C_j = C_j C_i$, $A_i C_j = C_j A_i$, for all i, j,
2. $A_i B_j = B_j A_i$, if $i \neq j$,
3. $B_i C_j = C_j B_i$, if $j \neq i, i+1$,
4. $A_i B_i A_i = B_i A_i B_i$ and $B_i C_j B_i = C_j B_i C_j$ whenever $j = i, i-1$.

We now define G_0 to be the subgroup of G generated by $\{A_i, B_i\}_{i=1}^{n-1} \cup \{C_j\}_{j=1}^{n-2}$. The fact that u commutes with each of $\{a_i, b_i\}_{i=1}^{n-1} \cup \{c_j\}_{j=1}^{n-2}$ and has the opposite rotation number implies that the rotation number of every element of $\{A_i, B_i\}_{i=1}^{n-1} \cup \{C_j\}_{j=1}^{n-2}$ is 0. Thus each of these elements has a fixed point.

We note that given any two of these generators of G_0 there is another generator with which they commute. If all of these generators have fixed point sets with empty interior then we may conclude from Lemma 3.2 that any two of them have equal fixed point sets, i.e., that $\text{Fix}(A_i) = \text{Fix}(B_j) = \text{Fix}(C_k)$ for all i, j, k. So in this case we have found a common fixed point for all generators of G_0. If we split S^1 at this point we get an action of G_0 on an interval I which must be abelian by Theorem 3.11. It follows that all of the original generators $\{a_i, b_i\}_{i=1}^{n} \cup \{c_j\}_{j=1}^{n-1}$ commute with each other except possibly c_{n-1} may not commute with b_{n-1} and b_n, and a_n and b_n may not commute.

Let $\psi : G \to \text{Diff}_+^2(S^1)$ be the putative action, and consider the element g guaranteed by Proposition 3.10. We have that

$$[\psi(a_n), \psi(b_n)] = \psi(g)^{-1}[\psi(a_1), \psi(b_1)]\psi(g) = id$$

and

$$[\psi(c_{n-1}), \psi(b_n)] = \psi(g)^{-1}[\psi(c_1), \psi(b_1)]\psi(g) = id$$

and

$$[\psi(c_{n-1}), \psi(b_{n-1})] = \psi(g)^{-1}[\psi(c_1), \psi(b_2)]\psi(g) = id.$$

Hence G is abelian, hence trivial since (see [Iv]) abelian quotients of $\text{Mod}(g, k), g \geq 3$ are trivial.

Thus we are left with the case that one generator of G_0 has a nontrivial interval of fixed points in S^1. Since they are all conjugate, each of the generators of G_0 has a nontrivial interval of fixed points. Also we may assume no generator fixes every point of S^1 since if one did the fact that they are all conjugate would imply they all act trivially.

Choose a maximal interval of fixed points J for any of the subset of generators $\{A_i\}$. That is, J is a nontrivial interval of fixed points for one of the $\{A_i\}$, which we assume (without loss of generality) is A_2, and there is no J' which properly contains J and which is pointwise fixed by some A_j.

Suppose x_0 and x_1 are the endpoints of J. Then $x_0, x_1 \in \partial \operatorname{Fix}(A_2)$ so by Lemma 3.3, x_0 and x_1 are fixed points for all of the generators except B_2, since B_2 is the only one with which A_2 does not commute. If the point $x_0 \in \operatorname{Fix}(B_2)$ we have found a common fixed point for all the generators of G_0 and we can split S^1 at this point obtaining an action of G_0 on an interval which implies G_0 is abelian by Theorem 3.11. So suppose $x_0 \notin \operatorname{Fix}(B_2)$.

Recall that the diffeomorphism B_2 has rotation number 0, so it fixes some point. Thus the identity $A_2 B_2 A_2 = B_2 A_2 B_2$, together with the fact that B_2 cannot fix x_0, implies by Lemma 3.7 that $B_2(J) \cap J = \emptyset$ and $J \cap B_2^{-1}(J) = \emptyset$. Define

$$y_0 = \lim_{n \to -\infty} B_2^n(x_0) \text{ and } y_1 = \lim_{n \to \infty} B_2^n(x_0).$$

We will denote by K the interval in S^1 with endpoints y_0 and y_1 which contains x_0. Note that K properly contains J and that $B_2(K) = K$.

Since $x_0 \in \operatorname{Fix}(A_1)$ and A_1 commutes with B_2 we conclude y_0 and y_1 are fixed by A_1. We can now apply Kopell's Lemma (Theorem 2.1) to the interval K with diffeomorphisms B_2 and A_1 to conclude that A_1 is the identity on K. But J is a proper subinterval of K which contradicts the fact that J was a maximal interval of fixed points for any one of the A_i. Thus in this case too we have arrived at a contradiction.

Punctured case

We now consider the group $\operatorname{Mod}(g, 1)$, and choose elements $\{a_i, b_i\}_{i=1}^n \cup \{c_j\}_{j=1}^{n-1}$ as above. The argument above shows that the subgroup G of $\operatorname{Mod}(g, 1)$ generated by these elements[1] acts trivially on S^1. We claim that the normal closure of G in $\operatorname{Mod}(g, 1)$ is all of $\operatorname{Mod}(g, 1)$, from which it follows that $\operatorname{Mod}(g, 1)$ acts trivially, finishing the proof.

To prove the claim, let S_g denote the closed surface of genus g and recall the exact sequence (see, e.g., [Iv]):

$$1 \to \pi_1(S_g) \to \operatorname{Mod}(g, 1) \to \operatorname{Mod}(g, 0) \to 1$$

where $\pi_1(\Sigma_g)$ is generated by finitely many "pushing the point" homeomorphisms p_τ around generating loops τ in $\pi_1(S_g)$ with basepoint the puncture. Each generating loop τ has intersection number one with exactly one of the loops, say β_i, corresponding to one of the twist generators of G. Conjugating b_i by p_τ gives a twist about a loop β_i' which together with β_i bounds an annulus containing the puncture. The twist about β_i composed with a negative twist about β_i' gives the

---

[1] We do not know whether or not G actually equals $\operatorname{Mod}(g, 1)$; there seems to be some confusion about this point in the literature.

isotopy class of p_τ. In this way we see that the normal closure of G in $\text{Mod}(g, 1)$ contains G together with each of the generators of the kernel of the above exact sequence, proving the claim.   ◇

4 $\text{Aut}(F_N)$, $\text{Out}(F_N)$ AND OTHER SUBGROUPS

In §4.1 we prove Theorem 1.3. Note that the result of Bridson-Vogtmann mentioned in the introduction implies this theorem since it is easy to see that any finite subgroup of $\text{Homeo}_+(S^1)$ is abelian so their result implies ours. We give our proof here because it is short, straightforward and provides a good illustration of the use of the techniques developed above. While several aspects of the proof are similar to the case of $\text{Mod}(g, 0)$, we have not been able to find a single theorem from which both results follow.

In §4.2 we prove theorem 1.4, extending the application to finite index subgroups of $\text{Mod}(g, k)$ and $\text{Aut}(F_n)$.

4.1   Actions of $\text{Aut}(F_n)$ and $\text{Out}(F_n)$

We begin with a statement of a few standard facts about generators and relations of $\text{Aut}(F_n)$.

Lemma 4.1 (Generators for $\text{Aut}(F_n)$). *The group $\text{Aut}(F_n)$ has a subgroup of index two which has a set of generators $\{A_{ij}, B_{ij}\}$ with $i \neq j$, $1 \leq i \leq n$ and $1 \leq j \leq n$. These generators satisfy the relations*

$$A_{ij}A_{kl} = A_{kl}A_{ij} \text{ and } B_{ij}B_{kl} = B_{kl}B_{ij} \text{ if } \{i, j\} \cap \{k, l\} = \emptyset \tag{i}$$

$$[A_{ij}, A_{jk}] = A_{ik}^{-1} \text{ and } [A_{ij}, A_{jk}^{-1}] = A_{ik} \tag{ii}$$

$$[B_{ij}, B_{jk}] = B_{ik}^{-1} \text{ and } [B_{ij}, B_{jk}^{-1}] = B_{ik} \tag{iii}$$

$$A_{ij}A_{ji}^{-1}A_{ij} = A_{ji}^{-1}A_{ij}A_{ji}^{-1} \text{ and } B_{ij}B_{ji}^{-1}B_{ij} = B_{ji}^{-1}B_{ij}B_{ji}^{-1}. \tag{iv}$$

Proof. Let $\{e_i\}_{i=1}^n$ be the generators of F_n and define automorphisms A_{ij} and B_{ij} by

$$A_{ij}(e_k) = \begin{cases} e_i e_j, & \text{if } i = k, \\ e_k, & \text{otherwise, and} \end{cases}$$

$$B_{ij}(e_k) = \begin{cases} e_j e_i, & \text{if } i = k, \\ e_k, & \text{otherwise.} \end{cases}$$

Then $\{A_{ij}, B_{ij}\}$ with $i \neq j$, generate the index two subgroup of $\text{Aut}(F_n)$ given by those automorphisms which induce on the abelianization $\mathbf{Z}^n$ of F_n an automorphism of determinant one (see, e.g., [LS]). A straightforward but tedious computation shows that relations (i)–(iv) are satisified.   ◇

We next find some fixed points.

Lemma 4.2. *Suppose $n > 4$ and $\phi \colon \mathrm{Aut}(F_n) \to \mathrm{Diff}^2_+(S^1)$ is a homomorphism with $a_{ij} = \phi(A_{ij})$ and $b_{ij} = \phi(B_{ij})$ where A_{ij} and B_{ij} are the generators of Lemma 4.1. Then each of the diffeomorphisms a_{ij} and b_{ij} has a fixed point.*

Proof. We fix i, j and show a_{ij} has a fixed point. Since $n > 4$ there is an a_{kl} with $\{i, j\} \cap \{k, l\} = \emptyset$. Let $Z(a_{kl})$ denote the centralizer of a_{kl}, so $a_{ij} \in Z(a_{kl})$. Also, since $n > 4$ there is $1 \leq q \leq n$ which is distinct from i, j, k, l, so $a_{iq}, a_{qj} \in Z(a_{kl})$.

By Lemma 3.4 we know that the rotation number $\rho \colon Z(a_{kl}) \to \mathbf{R}$ is a homomorphism so $\rho(a_{ij}) = \rho([a_{iq}, a_{qj}]) = 0$. This implies that a_{ij} has a fixed point. $\diamond$

We can now prove the main result of this section.

Theorem 1.3. *For $n \geq 6$, any homomorphisms from $\mathrm{Aut}(F_n)$ or $\mathrm{Out}(F_n)$ to $\mathrm{Diff}^2_+(I)$ or $\mathrm{Diff}^2_+(S^1)$ factors through $\mathbf{Z}/2\mathbf{Z}$.*

Proof. Let H be the index two subgroup of $\mathrm{Aut}(F_n)$ from Lemma 4.1. Let ϕ be a homomorphism H to $\mathrm{Diff}^2_+(I)$ or $\mathrm{Diff}^2_+(S^1)$ and let $a_{ij} = \phi(A_{ij})$ and $b_{ij} = \phi(B_{ij})$ where A_{ij} and B_{ij} are the generators from Lemma 4.1. Then $\{a_{ij}, b_{ij}\}$ are generators for $Image(\phi)$. We will show in fact that the subgroup G_a generated by $\{a_{ij}\}$ is trivial, as is the subgroup G_b generated by $\{b_{ij}\}$. The arguments are identical so we consider only the set $\{a_{ij}\}$. Since $n \geq 6$, we have by Lemma 4.1 that the commutativity graph for the generators $\{a_{ij}\}$ of G_a is connected. Hence by Corollary 3.6, if these generators are fully supported then G_a is abelian. But the relations $[a_{ij}, a_{jk}^{-1}] = a_{ik}$ then imply that G_a is trivial.

Hence we may assume at least one of these generators is not fully supported. Let J be an interval which is a maximal component of fixed point sets for any of the a_i. More precisely, we choose J so that there is a_{pq} with J a component of $\mathrm{Fix}(a_{pq})$ and so that there is no a_{kl} such that $\mathrm{Fix}(a_{kl})$ properly contains J.

We wish to show that J is fixed pointwise by each a_{ij}. In case $\{p, q\} \cap \{i, j\} = \emptyset$ we note that at least the endpoints of J are fixed by a_{ij} because a_{pq} and a_{ij} commute and the endpoints of J are in $\partial \mathrm{Fix}(a_{pq})$. So Lemma 3.2 or Lemma 3.3 implies these endpoints are fixed by a_{ij}.

For the general case we first show $a_{ij}(J) \cap J \neq \emptyset$. In fact there is no a_{ij} with the property that $a_{ij}(J) \cap J = \emptyset$ because if there were then the interval J' defined to be the smallest interval containing $\{a_{ij}^n(J)\}_{n \in \mathbf{Z}}$ is an a_{ij}-invariant interval with no interior fixed points for a_{ij}. (In case $G_a \subset \mathrm{Diff}^2_+(S^1)$ we need the fact that a_{ij} has a fixed point, which follows from Lemma 4.2, in order to know this interval exists.) But since $n \geq 6$ there is some a_{kl} which commutes with both a_{ij} and a_{pq}, and it leaves both J and J' invariant. Applying Kopell's Lemma (Theorem 2.1) to a_{kl} and a_{ij} we conclude that $J' \subset \mathrm{Fix}(a_{kl})$, which contradicts the maximality of J. Hence we have shown that $a_{ij}(J) \cap J \neq \emptyset$ for all i, j.

Now by Lemma 4.1 the relation $a_{pq} a_{qp}^{-1} a_{pq} = a_{qp}^{-1} a_{pq} a_{qp}^{-1}$ holds so we may apply Lemma 3.7 to conclude that J is fixed pointwise by a_{qp}.

We next consider the generator a_{pk}. If x_0 is an endpoint of J then since $a_{qk}(J) \cap J \neq \emptyset$ at least one of $a_{qk}(x_0)$ and $a_{qk}^{-1}(x_0)$ must be in J. Hence the relations $[a_{pq}, a_{qk}] = a_{pk}^{-1}$ and $[a_{pq}, a_{qk}^{-1}] = a_{pk}$ imply that $a_{pk}(x_0) = x_0$. This holds for the other endpoint of J as well, so J is invariant under a_{pk} for all k. The same argument shows J is invariant under a_{qk} for all k. But since a_{pq} is the identity on J we conclude from $[a_{pq}, a_{qk}] = a_{pk}^{-1}$ that a_{pk} is the identity on J. Similarly a_{qk} is the identity on J.

Next the relation $a_{qk}a_{kq}^{-1}a_{qk} = a_{kq}^{-1}a_{qk}a_{kq}^{-1}$, together with Lemma 3.7, implies that a_{kq} is the identity on J. A similar argument gives the same result for a_{kp}. Finally, the relation $[a_{ip}, a_{pj}^{-1}] = a_{ij}$ implies that a_{ij} is the identity on J.

Thus we have shown that any subgroup of $\mathrm{Diff}_+^2(I)$ or $\mathrm{Diff}_+^2(S^1)$ which is a homomorphic image of H, the index two subgroup of $\mathrm{Aut}(F_n)$, has an interval of global fixed points. In the case of $\mathrm{Diff}_+^2(S^1)$ we can split at a global fixed point to get a subgroup of $\mathrm{Diff}_+^2(I)$ which is a homomorphic image of H. In the I case we can restrict the action to a subinterval on which the action has no global fixed point. But our result then says that for the restricted action there is an interval of global fixed points, which is a contradiction. We conclude that the subgroup of G_a generated by $\{a_{ij}\}$ is trivial and the same argument applies to the subgroup G_b generated by $\{b_{ij}\}$. So $\phi(H)$ is trivial. Since $\mathrm{Aut}(F_n)/H$ has order two any homomorphism ϕ from $\mathrm{Aut}(F_n)$ to $\mathrm{Diff}_+^2(I)$ or $\mathrm{Diff}_+^2(S^1)$ has an image whose order is at most two. The fact that $\mathrm{Diff}_+^2(I)$ has no elements of finite order implies that in this case ϕ is trivial.

Since there is a natural homomorphism from $\mathrm{Aut}(F_n)$ onto the group $\mathrm{Out}(F_n)$, the statement of Theorem 1.3 for $\mathrm{Out}(F_n)$ holds. $\diamond$

4.2   Finite index and other subgroups

In this section we prove Theorem 1.4. Recall that the Torelli group $T_{g,k}$ is the subgroup of $\mathrm{Mod}(g,k)$ consisting of diffeomorphisms of the k-punctured, genus g surface $\Sigma_{g,k}$ which act trivially on $H^1(\Sigma_{g,k}; \mathbf{Z})$. Of course any finite index subgroup of $\mathrm{Mod}(g,k)$ contains a finite index subgroup of $T_{g,k}$, so we need only prove the theorem for the latter group; that is, (3) implies (1).

So, let H be any finite index subgroup of $T_{g,k}, g \geq 3, k \geq 0$ or of $\mathrm{Aut}(F_n)$ or $\mathrm{Out}(F_n)$ for $n \geq 6$. Suppose that H acts faithfully as a group of C^2 diffeomorphisms of I or S^1, so that no nontrivial element of H has an interval of fixed points (i.e., the action is fully supported, in the terminology of §3.1). In each case we claim that H contains infinite order elements a, b, c with the property that, for each $r \geq 2$, the elements a^r, b^r generate an infinite, non-abelian group which commutes with c^s for all s.

To prove the claim for $T_{g,k}$ one takes a, b, c to be Dehn twists about homologically trivial, simple closed curves with a and b having positive intersection number and c disjoint from both. For r sufficiently large the elements a^r, b^r, c^r clearly lie in H. As powers of Dehn twists commute if and only if the curves have zero intersection number, the claim is proved for T_g, k. A direct calculation using appropriate generators A_{ij} as in Lemma 4.1 gives the same claim for $\mathrm{Aut}(F_n)$ and $\mathrm{Out}(F_n)$ for $n > 2$.

Theorem 1.4 in the case of I now follows immediately from Theorem 3.5 and the fact that the group $<a^r, b^r, c^r>$ has connected commuting graph but is non-abelian. For the case of S^1, note that if c^r has a fixed point for any $r > 0$ so do a^r, b^r by Lemma 3.2, so after taking powers we are done by the same argument, this time applying Corollary 3.6.

If no positive power of c has a fixed point on S^1, then c must have irrational rotation number, in which case the centralizer of c in $\mathrm{Homeo}(S^1)$ is abelian by 2.10. But $<a, b>$ lie in this centralizer and do not commute, a contradiction.

The proof of Theorem 1.4 for $\mathrm{Aut}(F_n)$ and $\mathrm{Out}(F_n)$ is similar to the above, so we leave it to the reader.

5   BAUMSLAG-SOLITAR GROUPS

In this section we study actions of the groups

$$\mathrm{BS}(m,n) = < a, b : ab^m a^{-1} = b^n >.$$

For $n > m > 1$ these groups are not residually finite, hence are not linear (see, e.g., [LS]).

5.1   An analytic action of $BS(m, n)$ on R

In this subsection we construct an example of an analytic action of $BS(m, n)$, $1 < m < n$ on **R**. It is straightforward to find diffeomorphisms of **R** which satisfy the Baumslag-Solitar relation.

Proposition 5.1 (Diffeomorphisms satisfying the Baumslag-Solitar relation). *Let $f_n : S^1 \to S^1$ be any degree n covering and let $f_m : S^1 \to S^1$ be any degree m covering. Let g_n and g_m be lifts of f_n and f_m respectively to the universal cover $\boldsymbol{R}$. Then the group of homeomorphisms of $\boldsymbol{R}$ generated by $g = g_m g_n^{-1}$ and the covering translation $h(x) = x + 1$ satisfies the relation $gh^n g^{-1} = h^m$.*

Proof. We have

$$g_n^{-1}(h^n(x)) = g_n^{-1}(x + n)$$
$$= g_n^{-1}(x) + 1$$
$$= h(g_n^{-1}(x)),$$

and

$$g_m(h(x)) = g_m(x + 1)$$
$$= g_m(x) + m$$
$$= h^m(g_m(x)).$$

So

$$g(h^n(x)) = g_m(g_n^{-1}(h^n(x)))$$
$$= g_m(h(g_n^{-1}(x)))$$
$$= h^m(g_m(g_n^{-1}(x)))$$
$$= h^m(g(x)). \qquad \diamond$$

We will now show how to choose the diffeomorphisms constructed in Proposition 5.1 more carefully so that we will be able to use a Schottky type argument to show that the diffeomorphisms will satisfy no other relations. Part of the difficulty will be that the "Schottky sets" will have to have infinitely many components.

We first construct, for each $j \geq 1$, a C^∞ diffeomorphism Θ_j of **R** which is the lift of a map of degree j on S^1. Let $a = 1/10$. We choose a C^∞ function

$\Theta_j : (-a/2, 1+a/2) \to \mathbf{R}$ with the following properties:

$$\Theta_j(x) = \frac{a}{2}x \text{ for } x \in (-a/2, a/2),$$

$$\Theta_j(x) = j + \frac{a}{2}(x-1) \text{ for } \in (a, 1+a/2),$$

$$\Theta_j'(x) > 0.$$

Note that for $x \in (-a/2, a/2)$ we have $\Theta_j(x+1) = \Theta_j(x) + j$. It follows that we can extend Θ_j to all of $\mathbf{R}$ by the rule $\Theta_j(x+k) = \Theta_j(x) + jk$. Thus the extended Θ_j is the lift of a map of degree j on S^1. This map is in fact a covering map since Θ_j is a diffeomorphism.

We shall be particularly interested in the intervals $A = [-a, a]$, $A^s = [-a, 0]$, $A^u = [0, a]$, and $C = [a, 1-a]$. Note that $[0, 1] = A^u \cup C \cup (A^s + 1)$. Also by construction $\Theta_j(A^s) \subset A^s$ and $\Theta_j(C) \subset (A^s + j)$. More importantly we have the following inclusions concerning the union of integer translates of these sets. For a set $X \subset \mathbf{R}$ and a fixed i_0, we will denote by $X + i_0$ the set $\{x + i_0 \mid x \in X\}$, by $X + \mathbf{Z}$ the set $\{x + i \mid x \in X, i \in \mathbf{Z}\}$, and by $X + m\mathbf{Z}$ the set $\{x + mi \mid x \in X, i \in m\mathbf{Z}\}$.

Suppose k is not congruent to 0 modulo j. Then we have

$$\Theta_j(A^s + \mathbf{Z}) \subset A^s + j\mathbf{Z},$$

$$\Theta_j(C + \mathbf{Z}) \subset A^s + j\mathbf{Z},$$

$$\Theta_j(A^u + j\mathbf{Z} + k) \subset A^s + j\mathbf{Z},$$

$$\Theta_j^{-1}(C + \mathbf{Z}) \subset A^u + \mathbf{Z}.$$

We now define two diffeomorphisms of $\mathbf{R}$. Let $g_n(x) = \Theta_n(x)$ and let $g_m(x) = \Theta_m(x - 1/2) + 1/2$. We define $B = A + 1/2$, $B^s = A^s + 1/2$, $B^u = A^u + 1/2$, and $C_2 = C + 1/2$. Then from the equations above we have

$$g_n(C + \mathbf{Z}) \subset A^s + n\mathbf{Z}$$

$$g_n^{-1}(C + \mathbf{Z}) \subset A^u + \mathbf{Z}$$

$$g_n^{-1}(A + k) \subset A^u + \mathbf{Z} \text{ if } k \notin n\mathbf{Z} \tag{1}$$

$$g_m(C_2 + \mathbf{Z}) \subset B^s + m\mathbf{Z}$$

$$g_m^{-1}(C_2 + \mathbf{Z}) \subset B^u + \mathbf{Z} \tag{2}$$

$$g_m^{-1}(B + k) \subset B^u + \mathbf{Z} \text{ if } k \notin m\mathbf{Z}. \tag{3}$$

The diffeomorphisms g_m and g_n can be approximated by analytic diffeomorphisms which still satisfy the equations above and are still lifts of covering maps on S^1. More precisely, the function $\phi(x) = g_n(x) - nx$ is a periodic function on $\mathbf{R}$ and may be C^1 approximated by a periodic analytic function. If we replace g_n by $\phi(x) + nx$ then the new analytic g_n will have positive derivative and be a lift of a degree n map on S^1. If the approximation is sufficiently close, it will also still satisfy the equations above. In a similar fashion we may perturb g_m to be analytic while retaining its properties.

Note that $B + \mathbf{Z} \subset C + \mathbf{Z}$ and $A + \mathbf{Z} \subset C_2 + \mathbf{Z}$, so we have the following key properties:

$$g_n(B + \mathbf{Z}) \subset A^s + n\mathbf{Z}, \tag{4}$$

$$g_n^{-1}(B + \mathbf{Z}) \subset A^u + \mathbf{Z}, \tag{5}$$

$$g_m(A + \mathbf{Z}) \subset B^s + m\mathbf{Z}, \tag{6}$$

$$g_m^{-1}(A + \mathbf{Z}) \subset B^u + \mathbf{Z}. \tag{7}$$

Define $g(x) = g_n(g_m^{-1}(x))$ and $h(x) = x + 1$, and let G be the group generated by g and h. Propostion 5.1 implies that $gh^m g^{-1} = h^n$. Hence there is a surjective homomorphism $\Phi : BS(m, n) \to G$ sending b to g and a to h. We need only show that Φ is injective.

Lemma 5.2 (Normal forms). *In the group $BS(m, n)$ with generators b and a satisfying the relation $ba^m b^{-1} = a^n$ every nontrivial element can be written in the form*

$$a^{r_n} b^{e_n} a^{r_{n-1}} b^{e_{n-1}} \ldots a^{r_1} b^{e_1} a^{r_0}$$

where $e_i = \pm 1$ and $r_i \in \mathbf{Z}$ have the property that whenever $e_i = 1$ and $e_{i-1} = -1$ we have $r_i \notin m\mathbf{Z}$, and whenever $e_i = -1$ and $e_{i-1} = 1$ we have $r_i \notin n\mathbf{Z}$.

While Lemma 5.2 follows immediately from the normal form theorem for HNN extensions, we include a proof here since it is so simple in this case.

Proof. If we put no restrictions on the integers r_i and, in particular, allow them to be 0, then it is trivially true that any element can be put in the form above. If for any i, $e_i = 1$, $e_{i-1} = -1$ and $r_i = mk \in m\mathbf{Z}$ then the fact that $ba^{mk}b^{-1} = a^{nk}$ allows us to substitute and obtain another expression in the form above, representing the same element of $BS(m, n)$, but with fewer occurrences of the terms b and b^{-1}. Similarly if $e_i = -1$, $e_{i-1} = 1$ and $r_i = nk \in n\mathbf{Z}$ then $b^{-1} a^{nk} b$ can be replaced with a^{mk} further reducing the occurrences of the terms b and b^{-1}.

These substitutions can be repeated at most a finite number of times after which we have the desired form. ◇

We now show that $BS(m, n)$ is a subgroup of $\mathrm{Diff}_+^\omega(\mathbf{R})$. Note that $\mathrm{Diff}_+^\omega(\mathbf{R}) \subset \mathrm{Homeo}_+(I) \subset \mathrm{Homeo}_+(S^1)$, the first inclusion being induced by the one-point compactification of $\mathbf{R}$. Hence the following implies Theorem 1.5.

Proposition 5.3. *The homomorphism $\Phi : BS(m, n) \to G$ defined by $\Phi(b) = g$ and $\Phi(a) = h$ is an isomorphism.*

Proof. By construction Φ is surjective; we now prove injectivity. To this end, consider an arbitrary nontrivial element written in the form of Lemma 5.2. After conjugating we may assume that $r_n = 0$ (replacing r_0 by $r_0 - r_n$). So we must show that (Φ applied to) the element

$$\alpha = g^{e_n} h^{r_{n-1}} g^{e_{n-1}} \ldots h^{r_1} g^{e_1} h^{r_0}$$

acts nontrivially on $\mathbf{R}$. If $n = 0$, i.e., if $\alpha = h^{r_0}$, then α clearly acts nontrivially if $r_0 \neq 0$, so we may assume $n \geq 1$.

Let x be an element of the interior of $C \cap C_2$. We will prove by induction on n that $\alpha(x) \in (A^s + n\mathbf{Z}) \cup (B^s + m\mathbf{Z})$. This clearly implies $\alpha(x) \neq x$.

Let $n = 1$. Then assuming $e_1 = 1$, we have $\alpha(x) = g(h^{r_0}(x)) = g_n(g_m^{-1}(h^{r_0}(x)))$. But $x \in C_2 + \mathbf{Z}$ implies $h^{r_0}(x) = x + r_0 \in C_2 + \mathbf{Z}$. So equation (2) implies $g_m^{-1}(x + r_0)$

$\in B^u + \mathbf{Z}$, and then equation (4) implies $g_n(g_m^{-1}(x + r_0)) \in A^s + n\mathbf{Z}$. Thus $e_1 = 1$ implies $\alpha(x) \in A^s + n\mathbf{Z}$. One shows similarly if $e_1 = -1$ then $\alpha(x) = g(h^{r_0}(x)) \in B^s + m\mathbf{Z}$.

Now as induction hypothesis assume that

$$y = g^{e_k} h^{r_{k-1}} g^{e_{k-1}} \dots h^{r_1} g^{e_1} h^{r_0}(x)$$

and that either

$$e_k = 1 \text{ and } y \in A^s + n\mathbf{Z}, \text{ or}$$

$$e_k = -1 \text{ and } y \in B^s + m\mathbf{Z}.$$

We wish to establish the induction hypothesis for $k + 1$. There are four cases corresponding to the values of ± 1 for each of e_k and e_{k+1}. If $e_k = 1$ and $e_{k+1} = 1$ then $y \in A^s + n\mathbf{Z}$ so $h^{r_i}(y) = y + r_i \in A^s + \mathbf{Z}$ and equation (7) implies that $g_m^{-1}(y + r_i) \in B^u + \mathbf{Z}$. Hence by equation (4), $g(h^{r_i}(y)) = g_n(g_m^{-1}(y + r_i)) \in A^s + n\mathbf{Z}$ as desired. On the other hand if $e_k = 1$ and $e_{k+1} = -1$ then $r_k \notin n\mathbf{Z}$. Hence $y \in A^s + n\mathbf{Z}$ implies $h^{r_i}(y) = y + r_i \in A^s + p$ with $p \notin n\mathbf{Z}$. Consequently $y' = g_n^{-1}(y + r_i) \in A^u + \mathbf{Z}$ by equation (1). So equation (6) implies $g^{-1}(h^{r_i}(y)) = g_m(y') \in B^s + m\mathbf{Z}$. Thus we have verified the induction hypothesis for $k + 1$ in the case $e_k = 1$.

If $e_k = -1$ and $e_{k+1} = -1$ then $y \in B^s + m\mathbf{Z}$ so $h^{r_i}(y) = y + r_i \in B^s + \mathbf{Z}$ and equation (5) implies that $g_n^{-1}(y + r_i) \in A^u + \mathbf{Z}$. Hence by equation (6) $g(h^{r_i}(y)) = g_m(g_n^{-1}(y + r_i)) \in B^s + m\mathbf{Z}$ as desired. Finally for the case $e_k = -1$ and $e_{k+1} = 1$ we have $r_k \notin m\mathbf{Z}$. Hence $y \in B^s + m\mathbf{Z}$ implies $h^{r_i}(y) = y + r_i \in B^s + p$ with $p \notin m\mathbf{Z}$. Consequently $y' = g_m^{-1}(y + r_i) \in B^u + \mathbf{Z}$ by equation (3). So equation (4) implies $g(h^{r_i}(y)) = g_n(y') \in A^s + n\mathbf{Z}$. Thus we have verified the induction hypothesis for $k + 1$ in the case $e_k = -1$ also. $\diamond$

Remark. We showed in the proof of Proposition 5.3 that any element of the form

$$g^{e_n} h^{r_{n-1}} g^{e_{n-1}} \dots h^{r_1} g^{e_1} h^{r_0}$$

acts nontrivially on $\mathbf{R}$ provided $e_i = \pm 1$, $r_i \in \mathbf{Z}$, whenever $e_i = 1$ and $e_{i-1} = -1$ we have $r_i \notin m\mathbf{Z}$, and whenever $e_i = -1$ and $e_{i-1} = 1$ we have $r_i \notin n\mathbf{Z}$. As a corollary we obtain the following lemma which we will need later.

Lemma 5.4. *In the group $BS(m, n)$ with generators b and a satisfying the relation $ba^m b^{-1} = a^n$ every element of the form*

$$b^{e_n} a^{r_{n-1}} b^{e_{n-1}} \dots a^{r_1} b^{e_1} a^{r_0}$$

is nontrivial provided $e_i = \pm 1$, $r_i \in \mathbf{Z}$, and whenever $e_i = 1$, $e_{i-1} = -1$ we have $r_i \notin m\mathbf{Z}$ and whenever $e_i = -1$, $e_{i-1} = 1$ we have $r_i \notin n\mathbf{Z}$.

Of course this lemma can also be obtained from the normal form theorem for HNN extensions.

5.2   General properties of $BS(m, n)$ actions

In this short subsection we note two simple implications of the Baumslag-Solitar relation.

Lemma 5.5. *If $BS(m,n)$ acts by homeomorphisms on S^1 with generators g, h satisfying $gh^m g^{-1} = h^n$ then h has a periodic point whose period is a divisor of $|n-m|$.*

Proof. The basic properties of ρ, together with the hypothesis, imply that, mod 1 :

$$n\rho(h) = \rho(h^n) = \rho(gh^m g^{-1}) = \rho(h^m) = m\rho(h)$$

and so $(m-n)\rho(h) = 0$ mod 1. Since h has a rational rotation number, h has a point whose period is a divisor of $|n-m|$.   ◇

Lemma 5.6. *Suppose g and h are homeomorphisms of $\mathbf{R}$ and satisfy $gh^m g^{-1} = h^n$. Then if h is fixed point free, g has a fixed point.*

Proof. Assume, without loss of generality, that $h(x) = x+1$ and $n > m$ and $g(0) \in [p, p+1]$. Then

$$h^{nk}(g(0)) = (g(0) + nk) \in [p+nk, p+nk+1].$$

But

$$h^n g = gh^m, \text{ so}$$

$$h^{nk} g = gh^{mk}, \text{ and}$$

$$h^{nk}(g(0)) = g(h^{mk}(0)) = g(mk), \text{ so}$$

$$g(mk) \in [p+nk, p+nk+1].$$

For k sufficiently large this implies $g(mk) > mk$ and for k sufficiently negative that $g(mk) < mk$. The intermediate value theorem implies g has a fixed point.   ◇

5.3 C^2 actions of $BS(m,n)$

In contrast to the analytic actions of $BS(m,n)$ on $\mathbf{R}$, we will show that there are no such actions, even C^2 actions, on either I or S^1.

Lemma 5.7. *Suppose g and h are orientation-preserving C^2 diffeomorphisms of I satisfying $gh^m g^{-1} = h^n$. Then h and ghg^{-1} commute.*

Proof. If $x \in \text{Fix}(h)$ then $g(x) = g(h^m(x)) = h^n(g(x))$ so $g(x)$ is a periodic point for h. But the only periodic points of h are fixed. We conclude that $g(\text{Fix}(h)) = \text{Fix}(h)$.
   Let (a, b) be any component of the complement of $\text{Fix}(h)$ in $[0,1]$. Since $g(\text{Fix}(h)) = \text{Fix}(h)$ we have that $ghg^{-1}([a,b]) = [a,b]$.
   Let H be the centralizer of $h^n = gh^m g^{-1}$. Note that h and ghg^{-1} are both in H. Let f be any element of H. Then if f has a fixed point in (a,b), since it commutes with h^n we know by Kopell's Lemma (Theorem 2.1) that $f = id$. In other words H acts freely (though perhaps not effectively) on (a,b). Hence the restrictions of any elements of H to $[a,b]$ commute by Hölder's Theorem. In particular the restrictions

of h and ghg^{-1} to $[a, b]$ commute. But (a, b) was an arbitrary component of the complement of $\mathrm{Fix}(h)$. Since the restrictions of h and ghg^{-1} to $\mathrm{Fix}(h)$ are both the identity we conclude that h and ghg^{-1} commute on all of I.   ◇

As a corollary we have:

Theorem 1.6. *for I:* *If m and n are greater than 1 then no subgroup of $\mathrm{Diff}^2_+(I)$ is isomorphic to $BS(m, n)$.*

Proof. If there is such a subgroup, it has generators a and b satisfying $a^n = ba^m b^{-1}$. Lemma 5.7 asserts that a and $c = bab^{-1}$ commute. But the commutator $[c, a]$ is $ba^{-1}b^{-1}abab^{-1}a$ and this element is nontrivial by Lemma 5.4. This contradicts the existence of such a subgroup.   ◇

Showing that $BS(m, n)$ has no C^2 action on the circle is a bit harder.

Theorem 1.6. *for S^1:* *If m is not a divisor of n then no subgroup of $\mathrm{Diff}^2_+(S^1)$ is isomorphic to $BS(m, n)$.*

Proof. Suppose there is such a subgroup, and we have generators g, h satisfying $gh^m g^{-1} = h^n$. By Lemma 5.5, h has a periodic point. Let $Per(h)$ be the closed set of periodic points of h (all of which have the same period, say p). If $x \in Per(h)$ then $g(x) = gh^{mp} = h^{np}g(x)$ which shows $g(Per(h)) = Per(h)$.

The omega limit set $w(x, g)$ of a point x under g is equal to a subset of the set of periodic points of g if g has periodic points, and is independent of x if g has no periodic points. Since g is C^2 we know by Denjoy's theorem that $w(g, x)$ is either all of S^1 or is a subset of $Per(g)$ (see [dMvS]). In the case at hand $w(g, x)$ cannot be all of S^1 since $x \in Per(h)$ implies $w(x, g) \subset Per(h)$ and this would imply $h^p = id$. Hence there exists $x_0 \in Per(g) \cap Per(h)$. Then $x_0 \in \mathrm{Fix}(h^p) \cap \mathrm{Fix}(g^q)$ where q is the period of points in $Per(g)$.

The fact that x_0 has period p for h implies that its rotation number under h (well defined as an element of $\mathbf{R}/\mathbf{Z}$) is $k/p + \mathbf{Z}$ for some k which is relatively prime to p. Using the fact that $\rho(h^m) = \rho(gh^m g^{-1})$, we conclude that

$$m(k/p) + \mathbf{Z} = \rho(h^m) = \rho(gh^m g^{-1}) = \rho(h^n) = n(k/p) + \mathbf{Z}.$$

Consequently $mk \equiv nk \mod p$ and since k and p are relatively prime $m \equiv n \mod p$. The fact that $1 < m < n$ and $n - m = rp$ for some $r \in \mathbf{Z}$ tells us that $p \notin m\mathbf{Z}$ because if it were then n would be a multiple of m. Similarly $p \notin n\mathbf{Z}$.

Let G be the group generated by $h_0 = h^p$ and $g_0 = g^q$. Now G has a global fixed point x_0, so we can split S^1 at x_0 to obtain a C^2 action of G on I. Note that $g_0 h_0^{m^q} g_0^{-1} = g^q h^{pm^q} g^{-q} = h^{pm^q} = h_0^{n^q}$.

We can apply Lemma 5.7 to conclude that h_0 and $g_0 h_0 g_0^{-1}$ commute. Thus h^p and $c = g^q h^p g^{-q}$ commute. But their commutator

$$[c, h^p] = g^q h^{-p} g^{-q} h^{-p} g^q h^p g^{-q} h^p.$$

Since $p \notin m\mathbf{Z}$ and $p \notin n\mathbf{Z}$ this element is nontrivial by Lemma 5.4. This contradicts the existence of such a subgroup.   ◇

REFERENCES

[BKK] H. Baik, S. Kim and T. Koberda, Unsmoothable group actions on compact one-manifolds. preprint. arXiv:1601.05490 [math.GT].

[BMNR] C. Bonatti, I. Monteverde, A. Navas and C. Rivas, Rigidity for C^1 actions on the interval arising from hyperbolicity I: solvable groups, *Math. Zeit.* 286(3–4): 919–949, 2017.

[BV] M. Bridson and K. Vogtmann, Homomorphisms from automorphism groups of free groups, *Bull. London Math. Soc.* 35 (2003), no. 6, 785–792.

[BM] M. Burger and N. Monod, Bounded cohomology of lattices in higher rank Lie groups, *J. Eur. Math. Soc.* 1 (1999), no. 2, 199–235.

[BW] L. Burslem and A. Wilkinson, Global rigidity of solvable group actions on S^1, *Geometry & Topology*, Vol. 8 (2004), 877–924.

[dMvS] W. de Melo and S. van Strien, *One-dimensional Dynamics*, Springer Verlag, Berlin (1991).

[FF3] B. Farb and J. Franks, Groups of homeomorphisms of one-manifolds, III: nilpotent subgroups, *Ergodic Theory Dynam. Systems* 23 (2003), no. 5, 1467–1484.

[FS] B. Farb and P. Shalen, Groups of real-analytic diffeomorphisms of the circle, *Ergodic Theory and Dynam. Syst.*, Vol. 22 (2002), 835–844.

[FP] E. Formanek and C. Procesi, The automorphism group of a free group is not linear, *J. Algebra* 149 (1992), no. 2, 494–499.

[Gh] E. Ghys, Actions de réseaux sur le cercle, *Inventiones Math.* 137 (1999), no. 1, 199–231.

[GS] E. Ghys and V. Sergiescu, Sur un groupe remarquable de difféomorphismes du cercle, *Comm. Math. Helv.*, Vol. 62 (1987), 185–239.

[Iv] N. Ivanov, *Mapping Class Groups*, Handbook of geometric topology, 523–633, North-Holland, Amsterdam, 2002.

[K] N. Kopell, Commuting Diffeomorphisms in *Global Analysis, Proceedings of the Symposium on Pure Mathematics* **XIV**, Amer. Math. Soc., Providence, RI (1970), 165–184.

[La] F. Labourie, Large groups actions on manifolds, Proc. I.C.M. (Berlin, 1998), *Doc. Math.* 1998, Extra Vol. II, 371–380.

[LS] R. C. Lyndon and P. Schupp, *Combinatorial Group Theory*, Springer-Verlag, 1977.

[MW] K. Mann and M. Wolff, Rigidity of mapping class group actions on S^1, preprint. arXiv:1808.02979 [math.GT].

[Na1] A. Navas, Actions de groupes de Kazhdan sure le cercle, *Ann. Sci. École Norm. Sup.* (4) 35 (2002), no. 5, 749–758.

[Na2] A. Navas, Quelques nouveaux phénomènes de rang 1 pour les groupes de difféomorphismes du cercle, *Comment. Math. Helv.*, 80(2): 355–375, 2005.

[Na3] A. Navas, Groups of circle diffeomorphisms, Chicago Lectures in Mathematics, Univ. of Chicago Press, 2011.

[Pa] K. Parwani, C^1 actions on the mapping class on the circle, *Alg. and Geom. Top.* 8 (2008), 935–944.

Morse Spectra, Homology Measures, Spaces of Cycles and Parametric Packing Problems

MISHA GROMOV

An "ensemble" $\Psi = \Psi(X)$ of (finitely or infinitely many) particles in a space X, e.g., in the Euclidean 3-space, is customarily characterized by the set function

$$U \mapsto ent_U(\Psi) = ent(\Psi_{|U}), \ U \subset X,$$

that assigns the *entropies* of *the U-reductions* $\Psi_{|U}$ *of* Ψ, to all bounded open subsets $U \subset X$. In the physicists' parlance, this entropy is

the logarithm of the number of the states of Ψ
that are effectively observable from U.

This "definition," in the context of mathematical statistical mechanics, is translated to the language of the measure/probability theory.[1]

But what happens if the *"effectively observable number of states"* is replaced by

the number of effective/persistent degrees of freedom
of ensembles of moving particles?

We suggest in this paper several mathematical counterparts to the idea of *"persistent degrees of freedom"* and formulate specific questions, many of which are inspired by Larry Guth's results and ideas on the Hermann Weyl kind of asymptotics of *the Morse (co)homology spectra* of the *volume energy function* on the spaces of *cycles* in balls.[2] And often we present variable aspects of the same idea in different sections of this paper.

Hardly anything that can be called a "new theorem" can be found in our paper but we reshuffle many known results and expose them from a particular angle. This article is meant as an introductory chapter to something yet to be written with much of what we present here extracted from my yet unfinished manuscript *Number of Questions.*

---

[1] See: Lanford's *Entropy and equilibrium states in classical statistical mechanics, Lecture Notes in Physics, Volume 20, pp. 1–113, 1973* and Ruelle's *Thermodynamic formalism : the mathematical structures of classical equilibrium statistical mechanics*, 2nd Edition, Cambridge Mathematical Library 2004, where the emphasis is laid upon (discrete) *lattice* systems. Also a *categorical rendition* of Boltzmann-Shannon entropy is suggested in *"In a Search for a Structure, Part 1: On Entropy,"* www.ihes.fr/~gromov/PDF/structre-serch-entropy-july5-2012.pdf.

[2] *Minimax problems related to cup powers and Steenrod squares*, by Guth [33] and *Weyl law for the volume spectrum* by Liokumovich, Margues, and Neves [42].

1   OVERVIEW OF CONCEPTS AND EXAMPLES

We introduce below the idea of *"parametric packings"* and of related concepts which are expanded in detail in the rest of the paper.

A. Let X be a topological space, e.g., a manifold, and I be a countable index set that may be finite, especially if X is compact.

A collection of I-tuples of non-empty open (sometimes closed) subsets $U_i \subset X$, $i \in I$, is called *a packing* or an *I-packing* of X if these subsets *do not intersect*.

Denote by $\Psi(X; I)$ the space of these packings with some natural topology, where, observe, there are several candidates for such a topology if X is *non-compact*.

B. *Homotopies constrained by inradii and waists.* We are interested in the homotopy, especially (co)homology, properties of subspaces $\mathcal{P} \subset \Psi(X; I)$ defined by imposing *lower bounds* on the sizes of U_i, where the two such invariants of $U \subset X$ we shall often (albeit sometimes implicitly) use in this paper are the *inradius of U* and the *k-waist of U*, $k = 1, 2, \ldots, dim(X) - 1$, defined later on in the metric and the symplectic categories.

C. *Metric category and packings by balls.* Let X be a metric space, let $r_i \geq 0$ be non-negative numbers and take metric balls in X of radii $R_i \geq r_i$ for U_i. Packings by such balls are traditionally called *sphere packings* where one is especially concerned with packing homogeneous spaces (e.g., spheres and Euclidean spaces) by *equal* balls.

The corresponding space $\mathcal{P} = \mathcal{P}(X; \{\geq r_i\}_{i \in I})$ naturally embeds into the *Cartesian power space*

$$X^I = \underbrace{X \times X \times \cdots \times X}_{I}$$

of I-tuples of points in X where it is distinguished by the inequalities

$$dist(x_i, x_j) \geq d_{ij} = r_i + r_j, \ i, j \in I, i \neq j,$$

and where, observe, all these spaces with $d_{ij} > 0$ lie in the Cartesian power space X^I minus diagonals,

$$X^I \setminus \bigcup_{i,j \in I} Diag_{ij} \text{ for } Diag_{ij} \subset X^I \text{ defined by the equations } dist(x_i, x_j) = 0.$$

And if X is Riemannian manifold with *the convexity (injectivity?) radius $\geq R$*, then, clearly, the inclusion

$$\mathcal{P}(X; \{\geq r_i\}_{i \in I}) \hookrightarrow X^I \setminus \bigcup_{i,j \in I} Diag_{ij}$$

is *a homotopy equivalence* for $\sum_{i \in I} r_i < R/2$; moreover, if all r_i are mutually equal, this homotopy equivalence is *equivariant* for the permutation group that acts on I and thus on X^I and on $X^I \setminus \bigcup_{i,j \in I} Diag_{ij} \subset X^I$ and $\mathcal{P}(X; \{\geq r_i = r\}) \subset X^I \setminus \bigcup_{i,j \in I} Diag_{ij}$.

The packing spaces behave *covariantly functorially* under *expanding* maps between metric spaces $X \to Y$.

But *contravariant* functoriality under *contracting*, i.e., *distance decreasing*, maps $f : X \to Y$ needs the following extension of the concept of ball packings to I-tuples of *subsets* (rather than points) $V_i \subset X$ instead of points $x_i \in X$.

D. *Packings by tubes.* These are I-tuples of closed subsets $V_i \in X$, such that mutual distances[3] between them satisfy $dist(V_i, V_j) \geq d_{ij}$.

E. *Packings by cycles.* The above becomes interesting if all $V_i \subset X$ support given *nonzero* homology classes h_i of dimension k, $k = 0, 1, \ldots$, in X, or if they support k-cycles some of which are linked in X, either individually or "parametrically" (compare **P** below and section 18).

F. *Packings by maps.* This is yet another variation of the concept of "packing." Here *subsets* $U_i \subset X$ are replaced by *maps* $\psi_i : U \to X$, where, in general, the domain U of ψ_i may depend on $i \in I$.

Now, "packing by ψ_i" means packing by the images of these maps, i.e., these images should not intersect with specific "packing conditions" expressed in terms of geometry of U and of these maps.

For instance, if X and U are equidimensional Riemannian manifolds one may require the maps ψ_i to be expanding. Or ψ_i may belong to a particular category (pseudogroup) of maps.

G. *Symplectic packings by balls.* Here $X = (X, \omega)$ is a $2m$-dimensional symplectic manifold, U_i are balls of radii R_i in the standard symplectic space $\mathbb{R}^{2n} = (\mathbb{R}^2)^n$ and symplectic packings are given by I-tuples of symplectic embeddings $U_i \to X$ with disjoint images.

Another attractive class of symplectic packings is that by polydiscs and by R-tubes around Lagrangian submanifolds in X.

H. *Holomorphic packings.* These make sense and look interesting for packing by holomorphic maps $U_i \to \mathbb{C}^n$, $n \geq 2$, with *Jacobian one*, in particular, by *symplectic holomorphic* maps for n even, but I have not thought about these.

I. *Essential homotopy.* This is the part of the homotopy, e.g., cohomology, structure of a *geometrically* defined subspace $\mathcal{P} \subset \Psi$ that comes from the ambient space Ψ, where Ψ itself is defined in purely *topological* terms.

One thinks of Ψ as the background that supports the *geometric information* on $\mathcal{P}$ written in the *homotopy theoretic* language of Ψ.

This information concerns the *relative homotopy size* of $\mathcal{P}$ in Ψ, often expressed by particular (quasi)numerical invariants, such as, for instance, *homotopy height, cell numbers, and cohomology valued measures.*

J. *Example: Packings by two balls.* The space $\mathcal{P}(R) \subset X \times X$, $R \geq 0$, of packings of a metric space X by two R-balls is defined by the inequality

$$X \times X \supset \mathcal{P}(R) =_{def} \{x_1, x_2\}_{dist(x_1, x_2) \geq 2R},$$

---

[3] Recall that $dist(V_1, V_2)$ between two subsets in a metric space X is defined as the *infimum* of the distances between points $x_1 \in V_1$ and $x_2 \in V_2$.

where, observe, the distance function d on $X \times X$ is related to the distance in $X \times X$ to the diagonal $X_{dia} = Diag_{12} \subset X \times X$ by

$$d = dist_X(x_1, x_2) = \sqrt{2} \cdot dist_{X \times X}((x_1, x_2), X_{dia}).$$

The *algebraic topology* of the distance function d on $X \times X$, more specifically the (co)homologies of the *inter-levels*

$$d^{-1}[R_1, R_2] \subset X \times X$$

(that carries, in general, more *geometric* information about X than what we call "essential homotopy" of these subsets) can be thought of (be it essential or non-essential) as *a cohomology valued "measure-like" set function* on the real line, namely,

$$[a, b] \mapsto H^*(d^{-1}[a, b]) \text{ for all segments } [a, b] \subset \mathbb{R}.$$

Exceptionally, e.g., if X is a symmetric space, the distance function is "*Morse perfect*"[4]: all the homotopy topology (e.g., homology) of f is "essential," quite transparent instances of which are *projective spaces over* $\mathbb{R}, \mathbb{C}$ *and* $\mathbb{H}$ as well as *flat tori*, where these functions are obviously induced from (Morse perfect distance) functions $d_0 : X \to \mathbb{R}$ for $d_0(x) = dist(x, x_0)$.

On the other hand, much of the geometrically significant *geometric information* carried by *the topology* of the distance functions on manifolds of *negative curvature* does not quite conform to such a concept of "essential." (The simplest part of the information encoded by d, that is represented by the set of the lengths of *undistorted*[5] closed geodesics in X, is homotopically, but not necessarily homologically, essential.)

K. *Permutation symmetries.* The Cartesian power space X^I is acted upon by the symmetric group $\mathbb{S}_N = Sym(I)$, $N = card(I)$ and this action is *free* in the complement to the diagonals.

Let G be a subgroup in $\mathbb{S}_N = Sym(I)$, $N = card(I)$, e.g., $G = \mathbb{S}_N$, and let $\mathcal{P} \subset X^I$ be a subspace invariant under this action. We are especially interested in those homotopy characteristics of $\mathcal{P}$ that are encoded by the kernel of the cohomology homomorphism $\kappa^* : H^*(BG) \to H^*(\mathcal{P}/\!/G)$ for the classifying map κ from the *homotopy quotient space* $\mathcal{P}/\!/G$ to the classifying space BG of the group G,

$$\kappa : \mathcal{P}/\!/G \to BG.$$

(Recall that the homotopy quotient $\mathcal{P}/\!/G$ is defined as the quotient space $(\widetilde{BG} \times \mathcal{P})/G$ for the diagonal action of G, where $\widetilde{BG}$ is the universal covering of BG with the Galois action of G and observe that $\mathcal{P}/\!/G$ can be replaced by the ordinary quotient space $\mathcal{P}/G$ if the action of G on $\mathcal{P}$ is free, e.g., if $\mathcal{P} \subset X^I$ is contained in the complement to the diagonals.)

---

[4] "Morse perfect" functions do not have to be "textbook Morse": they may be *non-smooth* and have *positive dimensional sets of critical points.*

[5] A submanifold Y in a Riemannian manifold X is called *undistorted* if the distance in Y associated with the *induced Riemannian* structure coincide with the *restriction of the distance function* from $X \supset Y$. For instance the shortest non-contractible curve in X is undistorted.

Some of the questions we want to know the answer to are as follows:[6]

L. Let F_0 be a G-equivariant map from a topological space S with an action of a group $G = G(I) \subset \mathbb{S}_N$, $N = card(I)$ on it, to the Cartesian power space X^I,

$$F_0 : S \to X^I,$$

and let $[F_0]_G$ denote the equivariant homotopy class of this map.

What is the maximal radius $R = R(S, [F_0])$, such that F_0 admits an equivariant homotopy to an equivariant map F from S to the space $\mathcal{P}(X; I, R) \subset X^I$, of I-packings of X by balls of radii R,

$$F : S \to \mathcal{P}(X; I, R) \subset X^I, \ F \in [F_0]_{\mathbb{S}_N}?$$

What is the supremum $R_{max}(k)$ of these $R = R(S, [F_0])$, over all $(S, [F_0])$ with $dim(S) = k$?

In other words, what is the maximal $R = R_{max}(k) = R(X, N, k)$, such that every k-dimensional G-invariant subset in the complement of the diagonals in X^I (where the action of $\mathbb{S}_N \supset G$ is free) admits an equivariant homotopy to $\mathcal{P}(X; I, R)$ $\subset X^I$?

M. Let

$$K^* = K^*(N, \varepsilon) \subset H^*(BG; \mathbb{F}_p), \ \mathbb{F}_p = Z/pZ, \ \text{for } G = G(N), N = card(I)$$

be the kernels of the homomorphisms

$$\kappa* = \kappa^*_{\mathbb{F}_p} : H^*(BG; \mathbb{F}_p) \to H^*(\mathcal{P}/G; \mathbb{F}_p),$$

where $\mathcal{P} = \mathcal{P}(X; I, \varepsilon)$ is the space of packings of a given Riemannian manifold X, e.g., of the unit sphere S^l or the torus $\mathbb{T}^l$, by N balls of radius ε.

(M₁) What is the behavior of the (graded) ranks of these kernels $K^*(N, \varepsilon)$ as functions of ε?

(M₂) What is the asymptotic behavior of these rank $(K^*(N, \varepsilon))$ for $N \to \infty$ and $\varepsilon \to 0$, where a particular case of interest is that of

$$\varepsilon = \varepsilon_N = const \cdot N^\alpha \text{ for some } \alpha < 0?$$

N. *Packing energies and Morse packing spectra.* The space $\mathcal{P} = \mathcal{P}_\varepsilon = \mathcal{P}(X; I, \varepsilon)$ can be seen as *the sublevel* of a suitable *"energy function"* E on the ambient space $\Psi = X^I \supset \mathcal{P}_\varepsilon$, where any monotone decreasing function in

$$\rho(\psi) \text{ for } \psi = \{x_i\} \text{ and } \rho(\psi) = \min_{x_i \neq x_j} dist(x_i, x_j)$$

---

[6]The cohomology of $\mathbb{S}_N$ are well understood, (e.g., see [1]) but I shamefully failed to extract a rough estimate of the ranks $rank(H^i(\mathbb{S}_N, \mathbb{F}_p))$ from what I read. But a referee to this paper pointed out that *if $n > 2i$, then $H^i(\mathbb{S}_N)$ is independent of N by Nakaoka stability, which shows, by looking at the bar resolution, that the ranks of $(H^i(\mathbb{S}_N; \mathbb{F})) \leq 4^i$ for all N and all fields F.*

will do[7] and where simple candidates for such functions are

$$E(\psi) = \frac{1}{\rho(\psi)} \quad \text{or} \quad E(\psi) = -\rho(\psi),$$

or, as seems most appropriate from a certain perspective,

$$E(\psi) = -\log \rho(\psi).$$

Notice that $\rho(\psi)$ equals $\sqrt{2}$ times the distance from $\{x_i\} \in X^I$ to the union of the diagonals $Diag_{ij} \subset X^I$ that are defined by the equations $x_i = x_j$,

$$\rho(\psi) = \sqrt{2} \cdot \min_{ij} dist_{X^I}(\psi, Diag_{ij}), \quad \psi = \{x_i\}.$$

N*. We are predominantly interested in *the homotopy significant (Morse) spectra* of such energy functions $E : \Psi \to \mathbb{R}$ on topological spaces Ψ, where such a spectrum is the set of those values $e \in \mathbb{R}$ where the homotopy type of the sublevel $E^{-1}(-\infty, e]$ undergoes an *irreversible change* (precise definitions are given in section 4) and the above $(\mathbf{M}_1)$ concerns such changes that are recorded by the variation of the kernels $K^*(N, \varepsilon)$.

O. *"Duality" between homology spectra of packings and of cycles.* Evaluation of the homotopy (or homology) spectrum of packings, in terms of the above **A**, needs establishing of two opposite geometric inequalities, similar to how it goes for the spectra of *Laplacians* associated to *Dirichlet's energies*.

O_I *Upper bounds on packing spectra.* Such a bound for packings of a manifold X by R-balls means an inequality $R \leq \rho_{up}$ for some $\rho_{up} = \rho_{up}(S, F_0)$ (or several such inequalities for various S and F_0) that would guarantee that a map $F_0 : S \to X^I$ is homotopic (or at least homologous) to a map with image in the packing space $\mathcal{P}(X; I, R) \subset X^I$.

This, in all examples that I know of, is achieved by *explicit constructions* of specific "homotopically (or homologically) significant" packing families $P_s \in \mathcal{P}(X; I, R)$, $s \in S$, for $R \geq \rho_{up}$.

O_{II} *Lower spectral bounds.*] Such a bound $R \geq \rho_{low} = \rho_{low}(S, F_0)$ is supposed to signify that $F_0 : S \to X^I$ is *not homotopic (or not even homologous)* to a map $S \to \mathcal{P}(X; I, R) \subset X^I$.

All bounds of this kind that I know of are obtained by *(parametric) homological localization*, that is, by confronting such maps F_0, think of them as *families of $N = card(I)$ moving balls in X parametrized by S*, with *families of cycles* in X where the two families have a nontrivial (co)homology pairing between them.

A simple, yet instructive, instance of this is where there is *a (necessarily nonzero) homology class $h_\circ \in H_k(X)$ for some $k = 1, 2, \ldots, n = dim(X)$*, such that the image

---

[7]The role of real numbers $\mathbb{R}$ here reduces to indexing the subsets $\Psi_r \subset \Psi$, $r \in \mathbb{R}$, according to their order by inclusion: $\Psi_{r_1} \subset \Psi_{r_2}$ for $r_1 \leq r_2$.

In fact, our "spectra" make sense for functions with values in an arbitrary *lattice* (that is, a partially ordered set that admits *inf* and *sup*), while *additivity*, the most essential feature of the physical energy, becomes visible only for spaces Ψ that split as $\Psi = \Psi_1 \times \Psi_2$.

$h_S \in H_*(X^I)$ of some homology class from $H_*(S)$ under the induced homomorphism

$$(F_0)_* : H_*(S) \to H_*(X^I)$$

has *nonzero homology intersection with the power class* $h_\circ^{\otimes I} \in H_*(X^I)$,

$$h_S \frown h_\circ^{\otimes I} \neq 0.$$

Obviously, in this case, *if a map* $F_0 : S \to X^I$ *is homotopic to a map into the packing space* $\mathcal{P}(X; I, R) \subset X^I$, *then every closed subset* $Y \subset X$ *that supports the class* $h_\circ$ *admits a packing by* N*-balls* $U_i \subset Y$, $i \in I$, $N = card(I)$, *of radii* R *for the restriction of the distance function from* X *to* Y. *Consequently,*

$$N \cdot R^k \leq const_X < \infty.$$

Example. Let X be a Riemannian product of two closed connected Riemannian manifolds, $X = Y \times Z$, let $S = Z^I$ and let $F_0 : Z^I = (Z \times y_0)^I \to X^I$, $y_0 \in Y$, be the tautological embedding. If this $S \subset X^I$ can be moved by a homotopy in X^I to $\mathcal{P}(X; I, R) \subset X^I$, then Y can be packed by N-balls of radius R.

Notice that the converse is also true in this case. In fact, if balls $U_{y_i}(R) \subset Y$ pack Y, then the Cartesian product

$$S = \underset{i \in I}{\times} (Z \times y_i) \subset X^I$$

is contained in $\mathcal{P}(X; I, R)$.

In general, when a cycle moves, this kind of argument, besides suitable nontrivial (co)homology pairing, needs lower bounds on spectra of the *volume-energies* in spaces of k-cycles, in particular lower bounds on k-*waists* of our manifolds that correspond to the bottoms of such spectra.

In particular, such bounds on *symplectic waists* are used in symplectic geometry for proving *non-existence of individual packings, as well as of multi-parametric families of certain symplectic packings.*

P. *Packings by tubes around cycles.* The concepts of spaces of packings and of spaces of cycles can be brought to a common ground by introducing the space of I-tuples of *disjoint* k-*cycles* V_i in X and of (the homotopy spectrum of) the function $E\{V_i\}$ that would somehow incorporate $vol_k(V_i)$ along with $\log dist_X(V_i, V_j)$.[8]

(One may replace the distances $dist_X(V_i, V_j)$ in X by distances in *the flat metric* in the space of cycles, where $dist_{flat}(V_i, V_j)$ is defined as the $(k+1)$-volume of *the minimal* $(k+1)$-*chain* between V_i and V_j in X, but this would lead to a quite different picture.)

Q. *Spaces of infinite packings.* If X is a non-compact manifold, e.g. the Euclidean space $\mathbb{R}^n$, then there are many candidates for THE SPACE OF PACKINGS, all of which are infinite dimensional spaces with infinite dimensional (co)homologies, where the infinities may be (partly) offset by actions of infinite groups on these spaces.

---

[8]One may think of these V_i as images of k-manifolds mapped to X that faithfully correspond to cycles with $\mathbb{Z}_2$-coefficients.

For instance, spaces $\mathcal{P}$ of packings of $\mathbb{R}^n$ by countable sets of R-balls $U_i(R) \subset \mathbb{R}^n$ are acted upon by the isometry group $iso(\mathbb{R}^n)$ that, observe, commutes with the action of the group $Sym(I)$ of bijective transformations of the (infinite countable) set I.

Perhaps the simplest instance of an interesting infinite packing space $\mathcal{P} = \mathcal{P}(\mathbb{R}^n; I, R)$ is where $I = \mathbb{Z}^n \subset \mathbb{R}^n$ is the integer lattice, where the ambient space Ψ equals the space of *bounded displacements* of $\mathbb{Z}^n \subset \mathbb{R}^n$: maps $\psi : i \mapsto x_i \in \mathbb{R}^n$, such that

$$dist(i, x_i) \leq C = C(\psi) < \infty \text{ for all } i \in \mathbb{Z}^n$$

and where $\mathcal{P} \subset \Psi$ is distinguished by the inequalities $dist(x_i, x_j) \geq 2R$.

The essential part of the infinite dimensionality of this Ψ comes from the infinite group $\Upsilon = \mathbb{Z}^n \subset iso(\mathbb{R}^n)$ that acts on it. In fact, Ψ naturally (and Υ-equivariantly) embeds into the union of compact infinite product spaces

$$\Psi \subset \bigcup_{C>0} B(C)^{\Upsilon},$$

where $B(C) \subset \mathbb{R}^n$ is the Euclidean ball of radius C with center at the origin, and where several entropy-like topological invariants, such as the *mean dimension* $dim(\mathcal{P} : \Upsilon)$ and *polynomial Morse entropy* $H^*(\mathcal{P}) : \Upsilon$, are available as in [22] and [8].

The quotient space of the above space Ψ, or rather of this Ψ minus the diagonals, by the infinite "permutation group"[9] $Sym(I)$ consists of the set of certain discrete subsets $\Delta \subset \mathbb{R}^n$. Each Δ has the property that the intersections of it with all bounded open subsets $V \subset \mathbb{R}^n$ satisfy *the uniform density condition*,

❋ $$card(V \cap \Delta) - card(V \cap \mathbb{Z}^n) \leq const_\Delta \cdot vol(U_1(\partial V))$$

where $U_1(\partial V) \subset \mathbb{R}^n$ denotes the union of the unit balls with their centers in the boundary of V.

But there are by far more uniformly dense subsets (i.e., subsets that satisfy ❋) than the images of $\mathbb{Z}^n$ under bounded displacement.

In fact, it is far from clear what topology (or rather homotopy) structure should be used on the space of uniformly dense subsets, that, for instance, would render this space *(path)connected*.

R. *On stochasticity.* Traditional probability may be brought back to this picture, if, for instance, packing spaces are defined by the inequalities

$$dist(x_i, x_j) \geq \rho(i, j),$$

where $\rho(i, j)$ assumes two values, $R_1 > 0$ and $R_2 > R_1$, taken independently with given probabilities $p(i, j) = p(i - j)$, $i, j \in I = \Upsilon = \mathbb{Z}^n$.

S. *From packings to partitions and back.* An I-packing P of a metric space X by subsets U_i can be canonically extended to the corresponding *Dirichlet-Voronoi*

---

[9]One only needs here the subgroup of $Sym(I = \mathbb{Z}^n)$ that consists of *bounded* bijective displacements $\mathbb{Z}^n \to \mathbb{Z}^n$, i.e., where these "displacements" $i \mapsto j$ satisfy $dist(i, j) \leq C < \infty$.

partition[10] by subsets $U_i^+ \supset U_i$, where each $U_i^+ \subset X$, $i \in I$, consists of the points $x \in X$ nearest to U_i, i.e., such that

$$dist(x, U_i) \leq dist(X, U_j), \quad j \neq i.$$

If, for instance, X is a convex[11] Riemannian space with constant curvature and if all U_i are convex then U_i^+ are *convex polyhedral sets*.

Conversely, convex subsets $U \subset X$ can be often be *canonically shrunk* to single points $u \in U$ by families of *convex* subsets, $U_t \subset U$, $0 \leq t \leq 1$,

$$\text{where } U_0 = U, \ U_1 = u \text{ and } U_{t_2} \subset U_{t_1} \text{ for } t_2 \geq t_1 \ .$$

For instance, if X has *positive* curvature, such a shrinking can be accomplished with the *inward equidistant deformation* of the boundary ∂U.

This shows, in particular, that the space of convex I-partitions (as well as of convex I-packings) of a convex space X of constant curvature is $\mathbb{S}_N$-*equivariantly*, $N = card(I)$, *homotopy equivalent* to the space of I-tuples of distinct points $x_i \in X$. (The case of non-positive curvature reduces to that of the positive one via projective isomorphisms between bounded convex spaces of constant curvatures.)

Partitions of metric spaces, especially convex ones whenever these are available, reflects finer aspects the geometry of X than sphere packings. For instance, families of convex partitions obtained by consecutive division of convex sets by hyperplanes are used for sharp evaluation of *waists* of spheres as we shall explain later on.

On the other hand, a typical Riemannian manifold X of dimension $n \geq 3$ admits only approximately convex partitions (along with convex packings), where the geometric significance of these remains problematic.

T. *Composition of packings, (multi)categories and operads.* If U_i pack X then packings of U_i, $i \in I$, by U_{ij}, $j \in J_i$, define a packing of X by all these U_{ij}.

Thus, for instance, in the case of X and all U being Euclidean balls, this composition defines a *topological/homotopy operad* structure in the space of packings of balls by balls.

The significance of such a structure is questionable for round ball packings, especially for those of *high density*, since composed packings constitute only a small part of the space of all packings.

But packings *of cubes by smaller cubes* and *symplectic* packings of balls by smaller balls seem more promising in this respect, since even quite dense packings in these cases, even partitions, may have a significant amount of (persistent) degrees of freedom.[12]

U. *On faithfulness of the "infinitesimal packings" functor.* The (multi)category structures in spaces of packings define, in the limit, similar structures in spaces of packings of spaces X by "infinitely many infinitesimally small" subsets $U_i \subset X$.

---

[10]Here, "*I-Partition*" means a covering of X by closed subsets $V_i \subset X$, $i \in I$, with non-empty non-intersecting interiors, where we often tacitly assume certain regularity of the boundaries of these U_i.

[11]"*Convex*" means that every two points are joint by a unique geodesic.

[12]Besides composition, there are other operations on (nested) packings. For instance, (nearby) large balls (cubes) may "exchange" small balls (cubes) in them.

*Question **U1***. How much of the geometry of a (compact) space X, say with a metric or symplectic geometry, can be seen in the homotopies of spaces of packings of X by such U_i?

*Question **U2***. Is there a good category of "abstract packing-like objects" that are not, a priori, associated to actual packings of geometric spaces?

Concerning Question U1, notice that the above-mentioned pairing between "cycles" and packings shows that the volumes of certain *minimal subvarieties* in a Riemannian manifold X can be reconstructed from the homotopies of packings of X by arbitrarily small balls.

For instance, if X is a complete Riemannian manifold with *non-positive sectional curvatures*, then *the lengths of its closed geodesics* are (easily) seen in the homotopy spectra of these packing spaces (see section 9).

And if X is an *orientable surface*, then this remains true with *no assumption on the curvatures* for the geodesics that are *length minimizing in their respective homotopy classes*.

Similarly, much of the geometry of *waists* of a convex set X, say in the sphere S^n, may be seen in the homotopies of spaces of partition of X into convex subsets, see [25].

*Question **U3***. Would it be useful to enhance the homotopy structure of a packing space of an X, say by (infinitesimally) small balls, by keeping track of (infinitesimal) geometric sizes of the homotopies in such a space?

V. *Limited intersections.* A similar (but perhaps less interesting) to a space of packings is that of N-tuples of balls with *no k-multiple intersections* between them (this space contains the space of $(k-1)$-tuples of packings) and can be seen with the distance function to the union of the *k-diagonals*—there are $\binom{N}{k}$ of them— that are the pullbacks of the principal diagonal $\{x_1 = x_2 = \cdots = x_k\}$ in X^J, where $card(J) = k$, under maps $X^I \to X^J$ corresponding to $N!/(N-k)!$ embedding $J \to I$.

W. *Spaces of coverings.* Individual packings often go together with coverings, say, with minimal covering of metric spaces by r-balls. Possibly, this companionship extends to that between *spaces* of packings and *spaces* of coverings.

2   A FEW WORDS ON NON-PARAMETRIC PACKINGS

Classically, one is concerned with *maximally dense* packings of spaces X by *disjoint balls*, rather than with the homotopy properties of families of moving balls in X.

Recall that *a sphere packing* or, more precisely, *a packing of a metric space X by balls of radii r_i, $i \in I$, $r_i > 0$*, for a given *indexing set I* of finite or countable cardinality $N = card(I)$ is, by definition, a collection of (closed or open) balls $U_{x_i}(r_i) \subset X$, $x_i \in X$, with mutually non-intersecting interiors.

Obviously, points $x_i \in X$ serve as centers of such balls if and only if

$$dist(x_i, x_j) \geq d_{ij} = r_i + r_j.$$

Basic problem.   What is the *maximal radius* $r = r_{max}(X; N)$ such that X admits a packing by N balls of radius r?

In particular, *what is the asymptotics of $r_{max}(X; N)$ for $N \to \infty$?*

If X is a compact n-dimensional Riemannian manifold (possibly with boundary), then the principal term of this asymptotics depends only on the volume of X, namely, one has the following (nearly obvious) *asymptotic packing equality*

$$\lim_{N \to \infty} \frac{N \cdot r_{max}(X; N)^n}{vol_n(X)} = \odot_n,$$

where $\odot_n > 0$ is a universal (i.e., independent of X) *Euclidean packing constant* that corresponds in an obvious way to the *optimal density* of the sphere packings of the Euclidean space $\mathbb{R}^n$.

(Probably, the full asymptotic expansion of $r_{max}(X; N)_{N \to \infty}$ is expressible in terms of the derivatives of the curvature of X and similarly to the Minakshisundaram-Pleijel formulae for spectral asymptotics.)

The explicit value of $\odot_n$ is known only for $n = 1, 2, 3, 8, 24$. In fact, the optimal, i.e., maximal, packing density of $\mathbb{R}^n$ for $n \leq 3$ can be implemented by a $\mathbb{Z}^m$-*periodic* (i.e., invariant under some discrete action of $\mathbb{Z}^n$ on $\mathbb{R}^n$) packing, where the case $n = 1$ is obvious, the case $n = 2$ is due to Lagrange (who proved that the optimal packing is the hexagonal one) and the case of $n = 3$, conjectured by Kepler and resolved by Thomas Hales.

More recently (2016) Maryna Viazovska proved that the E_8 *root lattice* is the densest sphere packing for $n = 8$ and optimality of the *Leech lattice* in dimension 24 was proven in a paper by Cohn, Kumar, Miller, Radchenko, and Viazovska (see [15] and references therein).

Notice that $\mathbb{R}^3$, unlike $\mathbb{R}^2$ where *the only* densest packing is the hexagonal one, admits *infinitely many* different packings; most of these are *not* $\mathbb{Z}^3$-*periodic*, albeit they are $\mathbb{Z}_2$-periodic.

Probably, none of the densest packings of $\mathbb{R}^n$ is $\mathbb{Z}^n$-periodic for sufficiently large n, possibly for $n > 24$. Moreover, *the topological entropy* of the action of $\mathbb{R}^n$ on the space of optimal packings may be nonzero.

Also, there may be infinitely many algebraically independent numbers among $\odot_1, \odot_2, \ldots$; moreover, the number of algebraically independent numbers among $\odot_1, \odot_2, \ldots, \odot_n$ may grow as $const \cdot n$, $const > 0$.

3   HOMOLOGICAL INTERPRETATION OF THE DIRICHLET-LAPLACE SPECTRUM

Let Ψ be a topological space and $E : \Psi \to \mathbb{R}$ be a continuous real valued function, thought of as an energy $E(\psi)$ of states $\psi \in \Psi$ or as a Morse-like function on Ψ.

The subsets

$$\Psi_e = \Psi_{\leq e} = E^{-1}(\infty, e] \subset \Psi, \ r \in \mathbb{R},$$

are called the (closed) *e-sublevels of E*.

A number $e_\circ \in \mathbb{R}$ is said *to lie in the homotopy significant spectrum of E* if the homotopy type of Ψ_r undergoes a *significant*, that is, *irreversible*, change as e passes through the value $e = e_\circ$. This may be understood as *non-existence of a homotopy* of the subset $\Psi_{e_\circ}$ in Ψ that would bring it to the sublevel $\Psi_{e < e_\circ} \subset \Psi_{e_\circ}$.

Basic quadratic example. Let Ψ be an infinite dimensional projective space and E equal the ratio of two quadratic functionals. More specifically, let E_{Dir} be the Dirichlet function(al) on differentiable functions $\psi = a(x)$ normalized by the L_2-norm on a compact Riemannian manifold X,

$$E_{Dir}(\psi) = \frac{||d\psi||_{L_2}^2}{||\psi||_{L_2}^2} = \frac{\int_X ||d\psi(x)||^2 dx}{\int_X \psi^2(x) dx}.$$

The eigenvalues $e_0, e_1, e_2, \ldots, e_N, \ldots$ of E_{Dir} (i.e., of the corresponding Laplace operator) are *homotopy significant* since the rank of the inclusion homology homomorphism $H_*(\Psi_r; \mathbb{Z}_2) \to H_*(\Psi; \mathbb{Z}_2)$ *strictly* increases (for $* = N$) as e passes through e_N.

An essential feature of Dirichlet energy that, as we shall see, is shared by many other examples is *homological localization*.

Let X be partitioned by closed subsets U_i, $i \in I$, $card(I) = n$, with piecewise smooth boundaries. Then *the N-th eigenvalue $e_N = e_N(X)$ is bounded from below by the minimum of the first eigenvalues of U_i,*

$$e_N(X) \geq \min_{i \in I} e_1(U_i).$$

Indeed, by linear algebra, every N-dimensional projective space of functions on X, say $S = P^N \subset \Psi = P^\infty$, contains a (necessarily nonzero) function $\psi_\star(x)$ such that

$$\star_i \qquad\qquad \int_{U_i} \psi_\star(x) dx = 0 \text{ for all } i \in I.$$

Therefore,

$$\sup_{a \in P^n} E_{Dir}(\psi) \geq \max_{i \in I} E_{Dir}(\psi_{\star | U_i}) \geq e_i,$$

where the key feature of this argument, *simultaneous solvability of the equations $\star_i$,* does not truly need the linear structure in P^n. but only the fact that $S \subset \Psi$ *supports a cohomology class $h \in H^1(\Psi; \mathbb{Z}_2) = \mathbb{Z}_2 = \mathbb{Z}/2\mathbb{Z}$, with non-vanishing $\smile$power $h^{\smile N} \in H^N(\Psi; \mathbb{Z}_2)$.*

If U_i are small approximately round subsets, then

$$e_1(U_i) \geq \varepsilon_n \left(\frac{1}{vol(U_i)} \right)^{\frac{2}{n}}, n = dim X,$$

and, with suitable partitions into such subsets, one bounds $e_N(X)$ from below[13] by

$$e_N(X) \geq \varepsilon_X \cdot \left(\frac{N}{vol(X)} \right)^{\frac{2}{n}}.$$

---

[13]It is obvious that $e_N(X) \leq C_X \cdot \left(\frac{N}{vol(X)} \right)^{\frac{2}{n}}$. In fact, the numbers $N_{sp \leq e}(U)$ of the eigenvalues $e_i(U) \leq e$ of open subsets $U \subset X$ satisfy *Hermann Weyl's asymptotic formula* $N_{sp \leq e}(U) \asymp D_n vol(U)^{\frac{n}{2}}$, $n = dim(X)$, where the existence of the limits $N_\asymp(U)$ of $N_{sp \leq e}(U) e^{\frac{2}{n}}$ for $e \to \infty$ and additivity of the set function $U \to N_\asymp(U)$ follows from the locality of the $\smile$product, while the evaluation of D_n, that happens to be equal to $2\pi^{-n} vol(B^n(1))$ for $B^n(1)$ being the unit Euclidean ball, depends on the (Riemannian) geometry of the Dirichlet energy.

This can be quantified in terms of the geometry of X.

For instance, if $Ricci(X) \geq -n\delta$, $\delta \geq 0$, and $diam(X) \leq D$, then the *homology localization* for $(n-1)$-*volume energy*

$$E_{vol}(\psi) = vol_{n-1}(\psi^{-1}(0))$$

in conjunction with Cheeger's spectral inequality implies that

$$e_N(X) \geq \varepsilon_n^{1+D\sqrt{\delta}} D^{-2} N^{\frac{2}{n+1}}.$$

(See section 7 and "Paul Levy Appendix" in [26].)

4 INDUCED ENERGIES ON (CO)HOMOTOPIES AND ON (CO)HOMOLOGIES

On stable and unstable critical points. If E is a Morse function on a smooth manifold Ψ, then the homotopy type of the energy sublevels $\Psi_e = E^{-1}(-\infty, e] \subset \Psi$ does change at all critical values e_{cri} of E. However, only exceptionally rarely, for the so-called *perfect Morse functions*, such as for the above quadratic energies, these changes are irreversible. In fact, every value $r_0 \in \mathbb{R}$ can be made critical by an *arbitrarily small C^0-perturbation*[14] E' of a smooth function $E(\varphi)$, such that E' equals E outside the subset $E^{-1}[r_0 - \varepsilon, r_0 + \varepsilon] \subset \Psi$; thus, the topology change of the sublevels of E' at r_0 is insignificant.

But the spectra of Morse-like functions introduced below have such homotopy significance built into their very definitions.

$\mathcal{H}_\circ(\Psi)$, $E_\circ$ *and the homotopy spectrum.* Let $\mathcal{S}$ be a class of topological spaces S and let $\mathcal{H}_\circ(\Psi) = \mathcal{H}_\circ(\Psi; \mathcal{S})$ be the category where the objects are homotopy classes of continuous maps $\phi : S \to \Psi$ and morphisms are homotopy classes of maps $\varphi_{12} : S_1 \to S_2$, such that the corresponding triangular diagrams are (homotopy) commutative, i.e., the composed maps $\phi_2 \circ \varphi_{12} : S_1 \to \Psi$ are homotopic to ϕ_1.

Extend functions $E : \Psi \to \mathbb{R}$ from Ψ to $\mathcal{H}_\circ(\Psi)$ as follows. Given a continuous map $\phi : S \to \Psi$ let

$$E(\phi) = E_{max}(\phi) = \sup_{s \in S} E \circ \phi(s),$$

denote by $[\phi] = [\phi]_{hmt}$ the homotopy class of ϕ. and set

$$E_\circ[\phi] = E_{mnmx}[\phi] = \inf_{\phi \in [\phi]} E(\phi).$$

In other words, $E_\circ[\phi] \leq e \in \mathbb{R}$ *if and only if the map* $\phi = \phi_0$ *admits a homotopy of maps* $\phi_t : S \to A$, $0 \leq t \leq 1$, *such that* ϕ_1 *sends* S *to the sublevel* $\Psi_e = E^{-1}(-\infty, e] \subset \Psi$.

The covariant (homotopy) $\mathcal{S}$-spectrum of E *is the set of values* $E_\circ[\phi]$ *for some class* $\mathcal{S}$ *of (homotopy types of) topological spaces* S *and (all) continuous maps* $\phi : S \to \Psi$.

---

[14] "C^0" refers to the uniform topology in the space of continuous functions.

For instance, one may take for S the set of homeomorphism classes of countable (or just finite) cellular spaces. In fact, the set of sublevels Ψ_r, $r \in \mathbb{R}$, themselves is sufficient for most purposes.

Lower and upper bounds on spectra. Lower bounds on homotopy spectra say, in effect, that "homotopically *complicated/large*" maps ϕ (that may need complicated parameter spaces S supporting them) necessarily have *large energies* $E_\circ[\phi]$.

Conversely, upper bounds depend on construction of *complicated* $\phi : S \to \Psi$ with *small energies.*

On topology, homotopy and on semisimplicial spaces. The topology of a space Ψ per se is not required for the definition of homotopy (and cohomotopy below) spectra. What is needed is a *"homotopy structure"* in Ψ defined by distinguishing a class of maps from "simple spaces" S into Ψ.

If such a structure in Ψ is associated with polyhedra taken for "simple S," then Ψ is called a *semisimplicial (homotopy) space* with its "homotopy structure" defined via the contravariant functor $S \rightsquigarrow maps(S \to \Psi)$ from the category of simplicial complexes and simplicial maps to the category of sets.

$\mathcal{H}^\circ(\Psi)$, E° *and the cohomotopy S-spectra.* Now, instead of $\mathcal{H}_\circ(\Psi)$ we extend E to the category $\mathcal{H}^\circ(\Psi)$ of homotopy classes of maps $\varphi : \Psi \to T$, $T \in \mathcal{S}$, by defining $E^\circ(\varphi)$ as the supremum of those $e \in \mathbb{R}$ for which the restriction map of φ to the energy sublevel $\Psi_e = E^{-1}(-\infty, e] \subset \Psi$,

$$\varphi_{|\Psi_e} : \Psi_e \to T,$$

is *contractible.*[15] This $E^\circ(\varphi)$ depends only on the homotopy class $[\varphi]$ of φ and the set of the values $E^\circ[\varphi]$, is called *the contravariant homotopy (or cohomotopy) S-spectrum of E.*

For instance, if $\mathcal{S}$ is comprised of the *Eilenberg-MacLane $K(\Pi, n)$-spaces*, $n = 1, 2, 3, \ldots$, then this is called *the Π-cohomology spectrum of E.*

Relaxing contractibility via cohomotopy operations. Let us express "contractible" in writing as $[\varphi] = 0$, let $\sigma : T \to T'$ be a continuous map and let us regard the (homotopy classes of the) compositions of σ with $\varphi : \Psi \to T$ as an *operation* $[\varphi] \overset{\sigma}{\mapsto} [\sigma \circ \varphi]$.

Then define $E^\circ[\varphi]_\sigma \geq E^\circ[\varphi]$ by maximizing over those e where $[\sigma \circ \varphi_{|\Psi_e}] = 0$ rather than $[\varphi_{|\Psi_e}] = 0$.

Pairing between homotopy and cohomotopy. Given a pair of maps (ϕ, φ), where $\phi : S \to \Psi$ and $\varphi : \Psi \to T$, write

$$[\varphi \circ \phi] = 0 \ \textit{if the composed map } S \to T \textit{ is contractible,}$$
$$[\varphi \circ \phi] \neq 0 \ \textit{otherwise.}$$

Think of this as a function with value "0" and "$\neq 0$" on these pairs.[16]

---

[15]In some cases, e.g., for maps φ into discrete spaces T such as Eilenberg-MacLane's $K(\Pi; 0)$, *"contractible"* must be replaced by *"contractible to a marked point serving as zero"* in T that is expressed in writing as $[\varphi] = 0$.

[16]If the space T is *disconnected*, it should be better endowed with a marking $t_0 \in T$ with "contractible" understood as "contractible to t_0."

E_*, E^* and the (co)homology spectra.   If h is a homology class in the space Ψ then $E_*(h)$ denotes *the infimum* of $E_\circ[\phi]$ over all (homotopy classes) of maps $\phi : S \to \Psi$ such that h is *contained in the image* of the homology homomorphism induced by ϕ.

Dually, the energy $E^*(h)$ on a cohomology class $h \in H^*(\Psi; \Pi)$ for an Abelian group Π is defined as $E^\circ[\varphi_h]$ for the h-inducing map from Ψ to the product of Eilenberg-MacLane spaces:

$$\varphi_h : \Psi \to \underset{n}{\times} K(\Pi, n), \; n = 0, 1, 2, \ldots \; .$$

In simple words, $E^*(h)$ equals the supremum of those e for which h vanishes on $\Psi_e = E^{-1}(\infty, e] \subset \Psi$.[17]

Then one defines the (co)homology spectra as the sets of values of these energies E_* and E^* on homology and on cohomology.

Homotopy dimension (height) growth.   The roughest invariant one wishes to extract from the (co)homotopy spectra of an energy $E : \Psi \to \mathbb{R}$ is *the rate of the growth of the homotopy dimension* of the sublevels $\Psi_e = E^{-1}(-\infty, e] \subset \Psi$, where the homotopy dimension a subset $B \subset A$ is the minimal d such that B, or at least every polyhedral space P mapped to B, is contractible in A to a subset $Q \subset A$ of dimension d.[18]

In many cases, this dimension is known to satisfy a polynomial bound *homdim* $(\Psi_e) \leq c e^\delta$ for some constants $c = c(E)$ and $\delta = \delta(E)$, where such an inequality amounts to *a lower* bound on the spectrum of E.

In the simplest case of Ψ homotopy equivalent to P^∞, this dimension as function of e carries *all* spectral information about E.

For instance if E is the Dirichlet energy of function on a Riemannian n-manifold X where the eigenvalues are bounded from below by $e_N(X) \geq \varepsilon_X N^{\frac{2}{n}}$, one has $homdim(\Psi_e) \leq c \cdot e^{\frac{n}{2}}$ for $c = \varepsilon_X^{-\frac{n}{2}}$.

Multidimensional spectra.   Let $\mathcal{E} = \{E_j\}_{j \in J} : \Psi \to \mathbb{R}^J$ be a continuous map. Let h be a cohomology class of Ψ and define the *spectral hypersurface* $\Sigma_h \subset \mathbb{R}^J$ in the Euclidean space $\mathbb{R}^J = \mathbb{R}^{l = card(J)}$ as the boundary of the subset $\Omega_h \subset \mathbb{R}^J$ of the J-tuples of numbers (e_j) such that the class h vanishes on the subset $\Psi_{<e_j} \subset \Psi$ defined by the inequalities

$$E_j(\psi) < e_j, \; j \in J,$$

$$\Sigma_h = \partial \Omega_h, \; \Omega_h = \{e_j\}_{h | \times_{j \in J} \Psi_{e_j} = 0}.$$

(This also makes sense for general cohomotopy classes h on Ψ with $h = 0$ understood as contractibility of the map $\psi : \Psi \to T$ that represent h to a marked "zero" point in T where marking is unnecessary for connected spaces T.)

More generally, given a continuous map $\mathcal{E} : \Psi \to Z$, one "measures" open subsets $U \subset Z$ according to the sizes of "the parts" of the homology of Ψ that are

---

[17]The definitions of energy on homology and on cohomology obviously extend to generalized homology and and cohomology theories.

[18]This is called *essential dimension* in [21] and it equals the homotopy height of (the homotopy class of) the inclusion $B \hookrightarrow A$.

(A referee to this paper suggested that *Moore spaces* may provide examples where the homotopy dimension $hd(X) = hd(X \underset{id}{\subset} X)$ would satisfy $hd(X \times Y) < hd(X) + hd(Y)$.)

"contained" in $\mathcal{E}^{-1}(U) \subset \Psi$, that are the images of the homology homomorphism $H_*(U) \to H_*(\Psi)$; similarly, kernels of the cohomology homomorphisms $H^*(\Psi \setminus U) \to H^*(\Psi)$ serve as a measure-like function $U \mapsto \mu^*(U)$ on Z (see section 11).

If Ψ and Z are smooth manifolds, and $\mathcal{E}$ is a proper smooth map, then the set $\Sigma_{\mathcal{E}} \subset Z$ of critical values of $\mathcal{E}$ "cuts" Z into subsets where the measure μ^* is (nearly) constant. (It is truly constant on the connected components of $Z \setminus \Sigma_{\mathcal{E}}$ but may vary at the boundaries of these subsets, since these boundaries are contained in $\Sigma_{\mathcal{E}}$.) Here, "the cohomology spectrum of $\mathcal{E}$" should be somehow defined via "coarse-graining(s)" of the "partition" of Z into these subsets according to the values of μ^*.

Packing example.   Take Ψ equal the I-Cartesian power of a Riemannian manifold, $\Psi = X^I$, let J consist of unordered pairs (i_1, i_2) $i_1 \neq i_2$, thus $card(J) = l = N(N-1)/2$, $N = card(I)$, and let $\mathcal{E}$ be given by the reciprocals of the l functions $E_j = dist_X(x_{i_1}, x_{i_2})$. (This map is equivariant for the natural actions of the permutation group $Sym_N = aut(I)$ on Ψ and on $\mathbb{R}^J$ and the most interesting aspects of the topology of this $\mathcal{E}$ that are indicated below become visible only in the equivariant setting of section 12.)

Spectral families.   A similar (or perhaps dual) picture arises when one has a family of functions $E_z : \Psi_z \to \mathbb{R}$ parametrized by a topological space $Z \ni z$, where the family $homospec_z \subset \mathbb{R}$ of homotopy spectra of E_z is seen as the *spectral hypersurface* Σ of $\{E_z\}_{z \in Z}$,

$$\Sigma = \bigcup_{z \in Z} homospec_z \subset Z \times \mathbb{R}.$$

On positive and negative spectra.   Our definitions of homotopy and homology spectra are best adapted to functions $E(\psi)$ bounded from below but they can be adjusted to more general functions E such as $E(x) = \sum_k a_k x_k^2$ where there may be infinitely many negative as well as positive numbers among the ψ_k.

For instance, one may define the spectrum of an E unbounded from below as the limit of the homotopy spectra of $E_\sigma = E_\sigma(\psi) = \max(E(\psi), -\sigma)$ for $\sigma \to +\infty$.

But often, e.g., for the action-like functions in the symplectic geometry, one needs something more sophisticated than a simple-minded cut-off of "undesirable infinities."[19]

On continuous homotopy spectra.   There also is a homotopy theoretic rendition/generalization of *continuous spectra* with some Fredholm-like notion of homotopy,[20] such that, for instance, the natural inclusion of the projectivized Hilbert subspace $PL_2[0, t] \subset PL_2[0, 1]$, $0 < t < 1$, would not contract to any $PL_2[0, t - \varepsilon]$.

5   FAMILIES OF K-DIMENSIONAL ENTITIES, SPACES OF CYCLES AND SPECTRA OF K-VOLUME ENERGIES

The *"k-dimensional size"* of a metric space X, e.g., of a Riemannian manifold, may be defined in terms of the (co)homotopy or the (co)homology spectrum of the k-*volume* or a similar energy function on a *space of "virtually k-dimensional entities"*

---

[19]It seems, however, that neither a general theory nor a comprehensive list of examples exist for the moment.

[20]See P. Benevieri and M. Furi [7].

Y in X. These spectra and related invariants of X can be defined by means of families of such "entities" as in the following examples.

(A) If X and S are topological spaces of dimensions $n = dim(X)$ and $m = n - k = dim(S)$, then continuous maps $\varsigma : X \to S$ define S-families of the fibers $Y_s = \varsigma^{-1}(s) \subset X$.

(B) Given a pair of spaces T and $T_0 \subset T$, where $dim(T_0) = dim(T) - m$, and an S-family of maps $\phi_s : X \to T$, $s \in S$, one arrives at an S-family of "virtually m-codimensional" subsets in X by taking the pullbacks $Y_s = \phi_s^{-1} \subset X$.

(C) In the case of *smooth manifolds* X and S and *generic smooth maps* $\varsigma : X \to S$, the families from (A) and similarly for (B) can be seen geometrically as maps from S to the *space of k-manifolds* with the natural (homotopy) semisimplicial structure defined by bordisms.
And for general continuous maps ς we think of $Y_s = \varsigma^{-1}(s) \subset X$ as S-families of *virtual k-(sub)manifolds* in X.

(D) More generally, if X and S are *pseudomanifolds* of dimensions n and $m = n - k$ and $\varsigma : X \to S$ is a *simplicial* map, then the fibers $Y_s \subset X$ are k-dimensional pseudomanifolds for all s in S away from the $(m-1)$-skeleton of S with the map $s \mapsto Y_s$ being semisimplicial for the natural semisimplicial structure in the space of pseudomanifolds.

(E) In order to admit families of *mutually intersecting* subsets $Y_s \subset X$ we need an auxiliary space Σ mapped to S by $\varsigma : \Sigma \to S$. Then we let $\tilde{Y}_s = \varsigma^{-1}(s) \subset \Sigma$ and define S-families $Y_s \subset X$ via maps $\chi : \Sigma \to X$ by taking images $Y_s = \chi(\tilde{Y}_s) \subset X$.

(Ẽ) Such a Σ may be constructed starting from a family of subsets $Y_s \subset X$ as the subset $\Sigma = \tilde{X} \subset X \times S$ that consists of the pairs (x, s) such that $x \in Y_s$. However, this $\tilde{X} \to X$, unlike more general $\Sigma \to X$, does not account for possible self-intersections of $\tilde{Y}_s$ mapped to X.

(F) If we want to keep track of multiplicities of maps $\tilde{Y}_s \to Y_s \subset X$ it is worth-while to regard the maps

$$\chi_{|\tilde{Y}_s} : \tilde{Y}_s \to X$$

themselves, rather than their images, as our (virtual) "k-dimensional entities in X."

(G = C + F) The space of smooth k-submanifolds in an n-dimensional manifold X can be represented by the space of continuous maps σ from X to the Thom space T of the universal m-dimensional vector bundle for $m = n - k$, where "virtual k-submanifolds" in X come as the σ-pullbacks of the zero section of this bundle.
Then *the space of bordisms* associated to arbitrary topological space X may be defined with n-manifolds Σ, maps $\sigma : \Sigma \to T$ and maps $\chi : \Sigma \to X$.

Space $\mathcal{C}_k(X; \Pi)$ of k-cycles in X

There are several homotopy equivalent candidates for *the space of k-cycles with Π-coefficients* of a topological space X.
For instance, one may apply the construction of a *semisimplicial space of k-cycles* associated to a chain complex of Abelian groups (see section 2.2 of [24])

to the complex $\{C_i \overset{\partial_i}{\to} C_{i-1}\}_{i=0,1,\ldots,k,\ldots}$ of *singular* Π-*chains* that are $\sum_\nu \pi_\nu \sigma_\nu$ for $\pi_\nu \in \Pi$, where Π is an Abelian group and where $\sigma_\nu : \triangle^i \to X$ are continuous maps of the i-simplex to X.

For instance, k-dimensional pseudomanifolds mapped to X define singular $\mathbb{Z}_2$-cycles in X ($\mathbb{Z}$-cycles if Σ and S are oriented) where the above S-families agree with the semisimplicial structure in $\mathcal{C}_k(X; \mathbb{Z}_2)$.

More generally, l-chains in the space of k-cycles in X can be represented by l-dimensional families of k-cycles in X that are $(k+l)$-chains in X. This defines a map between the corresponding spaces of cycles

$$\top : \mathcal{C}_l(\mathcal{C}_k(X; \Pi); \Pi) \to \mathcal{C}_{k+l}(X; \Pi \otimes_{\mathbb{Z}} \Pi)$$

for all Abelian (coefficient) groups Π, as well as a natural homomorphism of degree $-k$ from the cohomology of a space X to the cohomology of the space of k-cycles in X with Π-coefficients, denoted

$$\top^{[-k]} : H^n(X; \Pi) \to H^{n-k}(\mathcal{C}_k(X; \Pi); \Pi), \text{ for all } n \geq k,$$

provided $\Pi = \mathbb{Z}$ or $\Pi = \mathbb{Z}_p = \mathbb{Z}/p\mathbb{Z}$.

Almgren-Dold-Thom Theorem for the spaces of cycles. Let X be a triangulated space and $f : X \to \mathbb{R}^m$ a generic piecewise linear map. Then the "slicings" of generic $(m+k)$-cycles $V \subset X$ by the fibers of f, define homomorphisms from the homology groups $H_{m+k}(X; \Pi)$ to the homotopy groups $\pi_m(\mathcal{C}_k(X; \Pi))$ of the spaces $\mathcal{C}_k(X; \Pi)$ of k-dimensional Π-cycles in X for all Abelian groups Π.

In fact, the map $r \mapsto f^{-1}(r) \cap V$, $r \in \mathbb{R}^m$, sends $\mathbb{R}^m$ to $\mathcal{C}_k(X; \Pi)$, where the complement to the (compact!) image $f(V) \subset \mathbb{R}^m$ goes to the the zero cycle.

The Almgren-Dold-Thom Theorem claims that these homomorphisms

$$H_{m+k}(X; \Pi) \to \pi_m(\mathcal{C}_k(X; \Pi))$$

are isomorphisms. This is easy for the above semisimplicial spaces (see [24]) and, as it was shown by Almgren, this remains true for spaces of *rectifiable cycles with flat topology.*

Recall, that the *flat/filling distance* between two homologous k-cycles C_1, C_2 in X with $\mathbb{Z}$- or $\mathbb{Z}_p$-coefficients is defined as the infimum of $(k+1)$-volumes (see below) of $(k+1)$-chains D such that $\partial D = C_1 - C_2$.

Also notice, that if Π is a field, then $\mathcal{C}_k(X; \Pi)$ *splits into the product of the Eilenberg-MacLane spaces* corresponding to the homology groups of X,

$$\mathcal{C}_k(X; \Pi) = \underset{n=0,1,2,\ldots}{\times} K(\Pi_n, n), \text{ where } \Pi_n = H_{k+n}(X; \Pi),$$

where, observe, $H^n(K(\Pi, n); \Pi)$ is *canonically* isomorphic to Π for the cyclic groups Π.

k-Volume and Volume-like Energies

- *Hausdorff measures of spaces and maps.* The k-dimensional Hausdorff measure of a *semimetric*[21] space Y is defined for all positive real numbers $k \geq 0$ as

---

[21] "Semi" allows a possibility of $dist(y_1, y_2) = 0$ for $y_1 \neq y_2$.

$$Haumes_k(Y) = \beta_k \cdot \inf_{\{r_i\}} \sum_{i \in I} r_i^{-k},$$

where I is a countable set, $\beta_k = \frac{\pi^{k/2}}{\Gamma(\frac{k}{2}+1)}$ is the normalizing constant that, for integer k, equals the volume of the unit ball B^k and where the infimum is taken over all I-tuples $\{r_i\}$ of positive numbers, such that Y admits an I-covering by balls of radii r_i.

The corresponding *Hausdorff measure of a map* $f : Y \to X$, where X is a metric space, is, by definition, the Hausdorff measure of Y with the semimetric induced by f from X.

If f is one-to-one then $Haumes(f) = Haumes(f(Y))$ but if f has a "significant multiplicity" then $Haumes(f) > Haumes f(Y)$.

δ-neighborhoods $U_\delta(Y) \subset X$ and Minkowski volume. Minkowski k-volume of a subset Y in an n-dimensional Riemannian manifold X and/or in a similar space with a distinguished measure regarded as the n-volume is defined as

$$Mink_k(Y) = \liminf_{\delta \to 0} \frac{vol_n(U_\delta(Y))}{\delta^{n-k} vol_{n-k}(B^{n-k})}.$$

In general, the Minkowski k-volume may be much greater than the Hausdorff k-measure but the two are equal for "regular" subsets $Y \subset X$ where "regular" includes

- compact smooth and piecewise smooth submanifolds in smooth manifolds;
- compact real analytic and semianalytic subspaces in real analytic spaces;
- compact minimal subvarieties in Riemannian manifolds.

Besides Minkowski volumes themselves, *the normalized n-volumes of the δ-neighbourhoods of subsets $Y \subset X$,*

$$\delta\text{-}Mink_k(Y) =_{def} \frac{vol_n(U_\delta(Y))}{\delta^{n-k} vol_{n-k}(B^{n-k})},$$

also can be used as "volume-like energies" that have interesting homotopy spectra.

The pleasant, albeit obvious, feature of the volume $\delta\text{-}Mink_k(Y)$ for $\delta > 0$ is its *continuity* with respect to *the Hausdorff metric* in the space of subsets $Y \subset X$.

On δ-covers and δ-packings of Y. The *minimal* number of δ-balls *needed to cover* Y provides an alternative to $\delta\text{-}Mink_k(Y)$ and the *maximal* number of *disjoint* δ-balls in X with *centers in* Y plays a similar role.

The definitions of these numbers *does not depend* on vol_n; yet, they are closely (and obviously) related to $\delta\text{-}Mink_k(Y)$.

6   MINMAX VOLUMES AND WAISTS

Granted a space Ψ of *"virtually k-dimensional entities"* in X and a volume-like energy function $E = E_{vol_k} : \Psi \to \mathbb{R}_+$, "the first eigenvalue"—the bottom of the homotopy/(co)homology spectrum of this E—is called the k-*waist of X.*

To be concrete, we define waist(s) below via the two basic operations of producing S-parametric families of subsets—taking pullbacks and images of maps represented by the following diagrams

$$\mathcal{D}_X = \{X \xleftarrow{\chi} \Sigma \xrightarrow{\varsigma} S\}$$

where S and Σ are simplicial (i.e., triangulated topological) spaces of dimensions $m = \dim(S)$ and $m + k = dim(\Sigma)$ and where our "entities" are the pullbacks

$$\tilde{Y}_s = \varsigma^{-1}(s) \subset \Sigma, \ s \in S,$$

that are mapped to X by χ.

Definitions

[A] *The maximal k-volume*—be it Hausdorff, Minkowski, δ-Minkowski, etc.,—of such a family is defined as the supremum of the corresponding volumes of the image restrictions of the map χ to Y_s, that is,

$$\sup_{s \in S} vol_k(\chi(\tilde{Y}_s)).$$

(It is more logical to use the *volumes of the maps* $\chi_{|Y_s} : \tilde{Y}_s \to X$ rather than of their images but this is not essential at this point.)

[B] *The minmax k-volume* of the pair of the *homotopy classes* of maps ς and χ denoted $vol_k[\varsigma, \chi]$ is defined as

$$\inf_{\varsigma, \chi} \sup_{s \in S} Haumes_k(\chi(Y_s)),$$

where the infimum is taken over all pairs of maps (ς, χ) in a given homotopy class $[\varsigma, \chi]$ of (pairs of) maps.

[C] *The k-waist* of a Riemannian manifold X, possibly with a boundary, is

$$waist_k(X) = \inf_{\mathcal{D}_X} \sup_{s \in S} vol_k(\chi(Y_s)),$$

where the infimum (that, probably, leads to the same outcome as taking *minimum*) is taken over all diagrams $\mathcal{D}_X = \{X \xleftarrow{\chi} \Sigma \xrightarrow{\varsigma} S\}$ that represent "*homologically substantial*" families of subsets $Y_s = \chi(\tilde{Y}_s = \varsigma^{-1}(s)) \subset X$ that support k-cycles in X.

What is homologically substantial?. A family of subsets $Y_s \subset X$ is regarded as homologically substantial if it satisfies some *(co)homology condition* that insures that the subsets $S_{\ni x} \subset S$, $x \in X$, that consist of $s \in S$ such that $Y_s \ni x$, are non-empty for all (some?) $x \in X$.

In the setting of smooth manifolds and smooth families, the simplest such condition is *non-vanishing* of the "*algebraic number of points*" in $S_{\ni x}$ for generic $x \in X$ that make sense if $dim(Y_x) + dimS = dimX$ for these x.

More generally, if $dim(Y_s) + dim(S) \geq dimX$, then the corresponding condition in the *bordisms homology theory* asserts *non-vanishing of the cobordism class of submanifold* $S_{\ni x} \subset X$ (for cobordisms regarded as homologies of a point in the bordism homology theory).

$\mathbb{Z}_2$-*waists.*   One arrives at a particular definition, namely, of what we call $\mathbb{Z}_2$-*waist* if "homologically" refers to homologies with $\mathbb{Z}_2$-*coefficients* (that is, the best understood case), and, accordingly, the above *inf* is taken over all diagrams $D_X = \{X \xleftarrow{\chi} \Sigma \xrightarrow{\varsigma} S\}$ where S and Σ are *pseudomanifolds of dimensions n and $n-k$ with boundaries*[22] (probably, using only *smooth manifolds* S and Σ in our diagrams would lead to the same $\mathbb{Z}_2$-waist) and where ς and χ are continuous maps, such that $\chi : \Sigma \to X$ *respects the boundaries,*[23] i.e., $\partial\Sigma \to \partial X$ and where χ *has nonzero $\mathbb{Z}_2$ degree* that exemplifies the idea of "homological substantiality."[24]

One may render this definition more algebraic by

- admitting an *arbitrary* (decent) topological space for the role Σ (that is, continuously mapped by ς to an m-dimensional *pseudomanifold* S); and
- replacing the "*nonzero degree condition*" by requiring that the fundamental $\mathbb{Z}_2$-homology class $[X]$ of X should lie in the image of homology homomorphism $\chi_* : H_*(\Sigma; \mathbb{Z}_2) \to H_*(X; \mathbb{Z}_2)$, or equivalently, that the cohomology homomorphism $\chi^* : H^*(X; \mathbb{Z}_2) \to H^*(\Sigma; \mathbb{Z}_2)$ does not vanish on the fundamental *cohomology* class of X.

(This naturally leads to a definition of the *waists of an n-dimensional homology and cohomology class h of dimension $N \ge n$*, where one may generalize/refine further by requiring non-vanishing of some cohomology operation applied on $\chi^*(h)$, as in [33].)

Examples of homologically substantial families.   (i) The simplest, yet significant, instances of such families of (virtually k-dimensional) subsets in n-manifolds X are the pullbacks $Y_s = \varsigma^{-1}(s) \subset X$, $s \in \mathbb{R}^{n-k}$, for continuous maps $\varsigma : X \to \mathbb{R}^{n-k}$. (The actual dimension of some among these Y_s may be strictly greater than k.)

(ii) Let S be a subset in the projectivized space[25] P^∞ of continuous maps $X \to \mathbb{R}^m$ and $Y_s = s^{-1}(0) \subset X$, $s \in S$, be the zero sets of these maps.[26] Let $P^\infty_{reg} \subset P^\infty$ consist of the maps $X \to \mathbb{R}^m$ the images of which *linearly span all of $\mathbb{R}^m$*, where, observe, the inclusion $P^\infty_{reg} \subset P^\infty$ is a *homotopy equivalence* in the present (infinite dimensional) case.

Observe that there is a natural map, say G, from P^∞_{reg} to the Grassmanian Gr^∞_m of m-planes in the linear space of functions $X \to \mathbb{R}$, where the linear subspace $G(p)$

---

[22] An "*n-pseudomanifold with a boundary*" is understood here as a simplicial polyhedral space, where all m-simplices for $m < n$ are contained in the boundaries of n-simplices and where the boundary $\partial\Sigma \subset \Sigma$ is comprised of the $(n-1)$-simplices that have *odd* numbers of n-simplices adjacent to them.

[23] Σ *has non-empty boundary only if X does and $\varsigma : \Sigma \to S$, unlike $\chi : \Sigma \to X$, does not have to send $\partial\Sigma \to \partial S$.*

[24] If one makes the definition of waists with the volumes of maps $\chi_{|Y_s} : Y_s \to X$ instead of the volumes of their images $\chi(Y_s) \subset X$ that would lead, a priori, to *larger* waists. However, in the $\mathbb{Z}_2$-case, the waists defined with the volumes of images, probably, equal the ones, defined via the volumes of maps, even for non-manifold targets X. This is obvious under mild regularity/genericity assumptions on the maps ς and χ, but needs verification in our setting of general continuous maps.

[25] *Projectivization $P(L)$ of a linear space L (e.g., of maps $X \to \mathbb{R}^m$) is obtained by removing* zero and dividing $L \setminus 0$ by the action of $\mathbb{R}^\times$.

[26] This P^∞ *is an infinite dimensional projective space, unless X is a finite set.*

in the space of functions $X \to \mathbb{R}$ for a (projectivized) map $p : X \to \mathbb{R}^m$ consists of the compositions $l \circ p : X \to \mathbb{R}$ for all linear functions $l : \mathbb{R}^m \to \mathbb{R}$.

If the cohomology homomorphism $\mathbb{Z}_2 = H^m(P^\infty; \mathbb{Z}_2) \to H^m(S; \mathbb{Z}_2)$ does not vanish, then the S-family $Y_s \subset X$ is $\mathbb{Z}_2$-homologically substantial, provided $S \subset P^\infty_{reg}$, i.e., if the images of $s : X \to \mathbb{R}^m$, $s \in S$, linearly span all of $\mathbb{R}^m$ for all $s \in S$.

Indeed, if S is an m-dimensional pseudomanifold, for $m = n - k$, then the number of points in the subset $S_{\ni x} \subset$ for which $Y_s \ni x$ (that is, defined under standard genericity conditions) does not vanish   mod 2. In fact, it is easy to identify this number with the value of the Stiefel-Whitney class of the complementary bundle to the canonical line bundle over the projective space $P^\infty \approx P^\infty_{reg}$, where this "complementary bundle" is induced from the canonical $(\infty - m)$-bundle over the Grassmanian Gr^∞_m by the above map $G : P^\infty_{reg} \to Gr^\infty_m$.

(iii) In the smooth situation, the above $Y_s = s^{-1}(0) \subset X$ are, generically, submanifolds of codimension m in X with *trivial normal bundles*.

General submanifolds and families of these are obtained by mappings ϕ from X to the Thom space T_m of a universal $\mathbb{R}^m$ bundle V by taking the pullbacks of the zero $\mathbf{0} \subset V \subset T_m$.

And if $\{\phi_s\} : X \to T_m$ is a family of maps parametrized by the m-sphere $S \ni s$ that equals the closure of a fiber of V in $T_m \supset V$, then the family $Y_s = \phi_s^{-1}(\mathbf{0}) \subset X$ is homologically substantial, if the map $S \to T_m$ defined by $s \mapsto \phi_s(x_0) \in T_m$, for some point $x_0 \in X$, has nonzero intersection index with the zero $\mathbf{0} \subset V \subset T_m$.

(iv) The above, when applied to maps between the suspensions of our spaces, $\phi^{\wedge k} : X \wedge S^k \to T_m \wedge S^k$, delivers families of (virtual) submanifolds of dimensions $n - m$, $n = dim(X)$, mapped to X via the projection $X \wedge S^k \to X$, where these families are nomologically substantial under the condition similar to that in (iii).

(v) There are more general (non-Thom) spaces, T, where $((n - m)$-volumes of) pullbacks of m-codimensional subspaces $T_0 \subset T$ are of some interest.

For instance, since the space of m-dimensional Π-cocycles of (the singular chain complex of) X (see [24]) is homotopy equivalent to space of continuous maps $X \to K(\Pi, m)$, it may be worthwhile to look from this perspective at (e.g., cellular) spaces T that represent/approximate *Eilenberg-MacLane's* $K(\Pi, m)$ with the codimension m skeletons of such T taken of $T_0 \subset T$.

Positivity of waists. *The k-dimensional $\mathbb{Z}_2$-waists of all Riemannian n-manifolds X are strictly positive for all $k = 0, 1, 2, \ldots, n$:*

$$[\geq_{non-sharp}]_{\mathbb{Z}_2} \qquad\qquad \mathbb{Z}_2\text{-}waist_k(X) \geq w_{\mathbb{Z}_2} = w_{\mathbb{Z}_2}(X) > 0.$$

About the proof. The waists defined with all of the above "k-volumes" are *monotone under inclusions,*

$$waist_k(U_1) \leq waist_k(U_2) \text{ for all open subsets } U_1 \subset U_2 \subset X,$$

and they also properly behave under λ-Lipschitz maps $f : X_1 \to X_2$ of nonzero degrees,

$$waist_k(X_2 = f(X_1)) \leq \lambda^k waist_k(X_1) \text{ for maps } f : X_1 \to X_2 \text{ with } deg_{\mathbb{Z}_2}(f) \neq 0.$$

Thus, $[\geq_{non-sharp}]_{\mathbb{Z}_2}$ for all X follows from that for the unit (Euclidean) n-ball B^n.

Then the case $X = B^N$ reduces to that of the unit n-sphere S^n,[27] while the lower bound $[\geq_{non-sharp}]_{\mathbb{Z}_2}$ for the k-waist of S^n defined with the *Hausdorff measure* of $Y \subset X = S^n$ is proven in [24] by a reduction to a *combinatorial filling inequality*.

On the other hand, the *sharp values* of $\mathbb{Z}_2$-waists of spheres are known for the *Minkowski volumes* and, more generally, for all δ-$Mink_k$, $\delta > 0$. Namely,

$[waist]_{sharp},$ $\qquad\qquad \delta\text{-}Mink_k\text{-}waist_{\mathbb{Z}_2}(S^n) = \delta\text{-}Mink_k(S^k),$

where $S^k \subset S^n$ is an equatorial k-sphere and

$$\delta\text{-}Mink_k(S^k \subset S^n) =_{def} \frac{vol_n(U_\delta(S^k))}{\delta^{n-k}vol_{n-k}(B^{n-k})}$$

for the δ-neighborhood $U_\delta(S^k) \subset S^n$ of this sphere in S^n and the unit ball $B^{n-k} \subset \mathbb{R}^{n-k}$.

Every $\mathbb{Z}_2$-homologically substantial S-family of "k-cycles" $Y_s \subset S^n$ has a member, say Y_{s_o}, such that

$$\delta\text{-}Mink_k(Y_{s_o}) \geq \delta\text{-}Mink_k(S^k) \text{ for all } \delta > 0 \text{ simultaneously.}$$

In particular, *given an arbitrary continuous map* $\varsigma : S^n \to \mathbb{R}^{n-k}$, *there exist a value* $s_o \in \mathbb{R}^{n-k}$, *such that the δ-neighborhoods of the s_o-fiber* $Y_{s_o} = \varsigma^{-1}(s_o) \subset S^n$ *are bounded from below by the volumes of such neighborhoods of the equatorial k-subspheres* $S^k \subset S^n$,

$$vol_n(U_\delta(Y_{s_o})) \geq vol_n(U_\delta(S^k)) \text{ for all } \delta > 0.$$

Consequently the Minkowski volumes of this Y_{s_o} is greater than the volume of the equator,

$[\circ]_k.$ $\qquad\qquad Mink_k(Y_{s_o}) \geq Mink_k(S^k)$ for this very $s_o \in \mathbb{R}^{n-k}$.

This is shown in [25] by a *parametric homological localization argument*. (See section 7 below; also see section 19 for further remarks, examples and conjectures.)

7   LOW BOUNDS ON VOLUME SPECTRA VIA HOMOLOGICAL LOCALIZATION

Start with a simple instance of homological localization for *the $(n-1)$-volumes of zeros* of families *of real functions* on Riemannian manifolds.

---

[27] In fact, the sphere S^n is bi-Lipschitz equivalent to the double of the ball B^n.

Also the ball $B^n = B^n(1)$ admits *a radial k-volume contracting map onto the sphere $S^n(R)$* of radius R, such that $vol_k(S^k(R)) = vol_k(B^k(1)$; this allows *a sharp evaluation of certain* "*regular k-waists*" of B^n, see [25], [24], [33].

[I] *Spectrum of* $E_{vol_{n-1}}$. Let L be an $(N+1)$-dimensional linear space of functions $s\colon X \to \mathbb{R}$ on a compact N-dimensional Riemannian manifold X and let $U_1, U_2,$ $\ldots, U_N \subset X$ be disjoint balls of radii ρ_N, such that

$$\rho_N^n \sim \odot_n \frac{vol_n(X)}{N} \text{ for large } N \to \infty \, ,$$

where $\odot_n > 0$ is a universal constant (as in the "packing section" 2). By the *Borsuk-Ulam theorem*, there exists a nonzero function $l \in L$ such that the zero set $Y_l = s^{-1}(0) \subset X$ cuts all U_i into equal halves and the Hausdorff measures (volumes) of the intersections of Y_l with U_i and *the isoperimetric inequality* implies the lower bound

$$vol_{n-1}(Y_l \cap U_i) \geq \beta_{n-1}\rho_N^{n-1} - o(\rho_N^{n-1}), \ \ N \to \infty,$$

for

$$\beta_{n-1} = vol_{n-1}(B^{n-1}(1)) = \frac{\pi^{n-1/2}}{\Gamma(\frac{n-1}{2}+1)}.$$

Therefore, *the supremum of the Hausdorff measures of the zeros* $Y_l \subset X$ *of non-identically zero functions* $l\colon X \to \mathbb{R}$ *from an arbitrary* $(N+1)$-*dimensional linear space* L *of functions on* X *is bounded from below for large* $N \geq N_0 = N_0(X)$ *by*

$$\sup_{l \in L \setminus 0} vol_{n-1}(Y_l) \geq \delta_n N^{\frac{1}{n}} vol_n^{\frac{n-1}{n}}(X)$$

for a universal constant $\delta_n > 0$.

[I*] *Homological generalization.* This inequality remains valid for all *non-linear* spaces L (of functions on compact Riemannian manifolds X) that are invariant under scaling $l \mapsto \lambda l$, $\lambda \in \mathbb{R}^\times$, provided the projectivizations $(L \setminus 0)/\mathbb{R}^\times \subset P^\infty$ of L in the projective space P^∞ of all continuous functions on X *support the (only) nonzero cohomology class from* $H^N(P^\infty; \mathbb{Z}_2) = \mathbb{Z}_2$.

[II] $E_{vol_{n-m}}$-*spectra of Riemannian manifolds* X. Let L be an $(mN+1)$-dimensional space of continuous maps $l\colon X \to \mathbb{R}^m$ that is invariant under scaling $l \to \lambda l$, $\lambda \in \mathbb{R}^\times$, and let the projectivized space $S = (L \setminus 0)/\mathbb{R}^\times \subset P^\infty$ in the projective space P^∞ of continuous maps $X \to \mathbb{R}^m$ modulo scaling support non-zero cohomology class from $H^{mN}(P^\infty; \mathbb{Z}_2) = \mathbb{Z}_2$, e.g., L is a linear space of maps $X \to \mathbb{R}^m$ of dimension $mN+1$.

  Then the zero sets $Y_s = Y_l = l^{-1}(0) \subset X$, *for* $s = s(l) \in S \subset P^\infty$ *being nonzero maps* $l\colon X \to \mathbb{R}^m$ *mod* $\mathbb{R}^\times$-*scaling,*

$$[*]_{n-m} \qquad\qquad \sup_{s \in S} vol_{n-m}(Y_s) \geq \delta_n N^{\frac{m}{n}} vol_n^{\frac{n-m}{n}}(X)$$

for large $N \geq N_0 = N_0(X)$ *and a universal constant* $\delta_n > 0$, *and where* vol_{n-m} *stands for the Minkowski* $(n-m)$-*volume,*

$$vol_{n-m}(Y_s \cup U) \leq (n-m)\text{-}waist(U) - \varepsilon.$$

Then, by the definition of waist, the cohomology restriction homomorphism H^m $(P^\infty; \mathbb{Z}_2) \to H^m(S \setminus (U); \mathbb{Z}_2)$ vanishes for all $\varepsilon > 0$.

Apply this to N open subsets $U_i \subset X$, $i = 1, \ldots, N$, and observe that the *non-zero* cohomology class that comes to our $S \subset P^\infty$ from $H^{Nm}(P^\infty; \mathbb{Z}_2)(= \mathbb{Z}_2)$ equals the $\smile$-product of *necessarily nonzero* m-dimensional classes coming from $H^m(P^\infty; \mathbb{Z}_2)$.

This, interpreted as the "simultaneous $\mathbb{Z}_2$-homological substantiality" of the families $Y_s(i) = Y_s \cup U_i \subset U_i$, for all $i = 1, \ldots, N$, shows that *there exists an $s \in S$, such that*

$$vol_{n-m}(Y_s \cup U_i) \geq (n - m)\text{-}waist(U) - \varepsilon,$$

for all $i = 1, \ldots, N$, in agreement with the definition of waists in the previous section.

Finally, take an efficient packing of X by N balls U_i as in the above [**I**] and [**I***] and derive the above $[*]_{n-m}$ from the lower bound on waists (see $[\geq_{non-sharp}]_{\mathbb{Z}_2}$ in the previous section) of U_i. QED.

The lower bound $[\geq_{non-sharp}]_{\mathbb{Z}_2}$ on waists was formulated under technical, probably redundant, assumption on L saying that the restrictions of spaces L to $U_i \subset X$ lie in the subspaces $P^\infty_{reg}(U_i)$ (such that the images of the "regular" $l : U_i \to \mathbb{R}^m$ span $\mathbb{R}^m$) of the corresponding projective spaces $P^\infty = P^\infty(U_i)$ of maps $U_i \to \mathbb{R}^m$. This assumption can be easily removed for $vol_{n-m} = Mink_{n-m}$ by a simple approximation argument applied to δ-$Mink_{n-m}$-waists and, probably, it also seems not hard to remove for $vol_{n-m} = Haumes_{n-m}$ as well.[28]

Further Results, Remarks and Questions

(a) The (projective) space of the above Y_s can be seen, at least if Y_s are "regular," as a (tiny for $n - k > 1$) part of the space $\mathcal{C}_k(X; \mathbb{Z}_2)$ of all k-dimensional $\mathbb{Z}_2$-cycles in X, say for the n-ball $X = B^n$, where the full space $\mathcal{C}_k(B^n; \mathbb{Z}_2)$ of the relative k-cycles mod 2 is Eilenberg-MacLane's $K(\mathbb{Z}_2, n - k)$ by the Almgren-Dold-Thom Theorem, see [2] and section 2.2 in [24].

If $n - k - 1$, then $\mathcal{C}_k(B^n; \mathbb{Z}_2) = P^\infty$ and $H^*(\mathcal{C}_k(B^n; \mathbb{Z}_2); \mathbb{Z}_2)$ is the polynomial algebra in a single variable of degree one, but if $n - m \geq 2$, then the cohomology algebra of $\mathcal{C}_k(X; \mathbb{Z}_2)$ is freely generated by infinitely many monomials in *Steenrod squares* of the generator of $H^{n-k}(\mathcal{C}_k(B^n; \mathbb{Z}_2)) = \mathbb{Z}_2$.

Thus, if $n - k > 2$, the cohomology spectrum of E_{vol_k} is indexed not by integers (that, if $n - k = 1$, correspond to graded ideals of the polynomial algebra in a single variable), but by the graded ideals in a more complicated algebra $H^{n-k}(\mathcal{C}_k(B^n; \mathbb{Z}_2)) = \mathbb{Z}_2$ with the Steenrod algebra acting on it.

(b) *Guth Theorem*. The asymptotic of this "*Morse-Steenrod spectra*" of the spaces $\mathcal{C}_k(B^n; \mathbb{Z}_2)$ of k-cycles in the n-balls were evaluated, up to a, probably redundant, lower order term, by Larry Guth; see [33] where a deceptively simple looking corollary of his results is the following:

Polynomial Bound on the Spectral Homotopy Dimension Cycles for the Volume Energy

The homotopy dimensions (heights) of the sublevels $\Psi_e = E^{-1}(-\infty.e] \subset \Psi$ of the vol_k-energy $E = E_{vol_k} : \Psi \to \mathbb{R}_+$ on the space $\Psi = \mathcal{C}_k(X; \mathbb{Z}_2)$ of k-dimensional rectifiable $\mathbb{Z}_2$-cycles in a compact Riemannian manifold X satisfies

---

[28]This was stated as a problem in section 4.2 of [21] but I do not recall if the major source of complication was the issue of regularity.

$$homdim(\Psi_e) \le ce^\delta$$

where the constant $c = c(X)$ depends on the geometry of X while $\delta = \delta(n)$ depends only on the dimension n of X.

This, reformulated as a lower spectral bound on E_{vol_k}, reads:

*Let X be a compact Riemannian manifold and let $Y_s \subset X$, $s \in S \subset \mathcal{C}_k(X; \mathbb{Z}_2)$, be an S-family of k-cycles (one may think of these as k-pseudo-submanifolds in X) that is **not** contractible to any N-dimensional subset in $\Psi = \mathcal{C}_k(X; \mathbb{Z}_2)$. Then*

$$\sup_{s \in S} vol_k(Y_s) \ge \varepsilon \cdot N^\alpha,$$

where $\varepsilon = \varepsilon(X) > 0$ depends on the geometry of X, while $\alpha = \alpha(n)(=\delta^{-1})$, $n = dim(X)$, is a universal positive constant.

In fact, Guth's results yield a nearly sharp bound

$$\sup_{s \in S} vol_k(Y_s) \ge \varepsilon(X, \alpha) \cdot N^\alpha,$$

for all $\alpha < \frac{1}{k+1}$, where, conjecturally, this must be also true for $\alpha = \frac{1}{k+1}$.

(c) *Is there a direct simple proof of the above inequality with some, let it be non-sharp, α that would bypass fine analysis (due to Guth) of the Morse-Steenrod cohomology spectrum of E_{vol_k} on the space of cycles?*

Does a polynomial lower bound hold for (k-volumes of) $\mathbb{Z}_p$- and for $\mathbb{Z}$-cycles?

Apparently, compactness of spaces of (quasi)minimal subvarieties in X implies discreteness of homological volumes spectra via the Almgren-Morse theory, but this does not seem to deliver even logarithmic lower spectral bounds due to the absence(?) of corresponding bounds on Almgren-Morse indices of minimal subvarieties in terms of their volumes.

(d) The lower bounds for the k-volumes of zeros of families of maps $X \to \mathbb{R}^m$ (see [**II**] above) can probably be generalized in the spirit of Guth's results, at least in the $\mathbb{Z}_2$-setting, to spaces of maps ψ from X to the total spaces of $\mathbb{R}^m$-bundles V and to the Thom spaces of such bundles where $E_{vol_k}(\psi) =_{def} vol_k(\psi^{-1}(\mathbf{0}))$ for the zero sections $\mathbf{0} \subset V$ of these bundles. The Steenrod squares should be replaced by the bordism cohomology operations that in the $\mathbb{Z}_2$-case amount to taking Stiefel-Whitney classes. (A similar effect to taking Thom spaces can be achieved by replacing $\mathbb{R}^\times$-scaling by the action of the full linear group $GL_m(\mathbb{R})$ on $\mathbb{R}^m$ and working with the equivariant cohomology of the space of maps $X \to \mathbb{R}^m$ with the corresponding action of $GL_m(\mathbb{R})$ on it.)

Here, as well as for spaces of maps ψ from X to more general spaces T where $E_{vol_k}(\psi) = vol_k(\psi^{-1}(T_0))$ for a given $T_0 \subset V$, one needs somehow to factor away the homotopy classes of maps ψ with $E_{vol_k}(\psi) = 0$ (compare [43]).

Also it may be interesting to augment the k-volume by other (integral) invariants of $Y = \psi^{-1}(T_0)$, where the natural candidates in the case of k-dimensional (mildly singular) $Y = \psi^{-1}(\mathbf{0})$ would be curvature integrals expressing the k-volumes of the tangential lifts of these $Y \subset X$ to the Grassmann spaces $Gr_k(X)$, $k = dim Y$, of target k-planes in X (compare section 3 in[29]).

(e) One can define spaces of subsets $Y \subset X$ that support k-cycles (or rather *cocycles*), where these Y do not have to be regular in any way, e.g., rectifiable as in Guth's theorem, or even geometrically k-dimensional. But Guth's parametric homological localization along with the bounds on waists from the previous section yield the same lower bounds on the volume spectra on these spaces as in the rectifiable case.

8 VARIABLE SPACES, HOMOTOPY SPECTRA IN FAMILIES AND PARAMETRIC HOMOLOGICAL LOCALIZATION

Topological spaces Ψ with (energy) functions E on them often come in families. In fact, the proofs of the sharp lower bound on the Minkowski waists of spheres (see section 6) and of Guth's lower bounds on the full vol_k-spectra (see the previous section) depend on localizing *not* to (smallish) *fixed* disjoint subsets $U_i \subset X$ but to variable or "*parametric*" ones that may change/move along with the subsets Y_s, $s \in S$, in them.

In general, families $\{\Psi_q\}$ are constituted by "fibers" of continuous maps F from a space $\Psi = \Psi_Q$ to Q where the "fibers" $\Psi_q = F^{-1}(q) \subset \Psi$, $q \in Q$, serve as the members of these families and where the energies E_q on Ψ_q are obtained by restricting functions E from Ψ to $\Psi_q \subset \Psi$.[29]

Homotopy spectra in this situation are defined with continuous families of spaces S_q that are "fibers" of continuous maps $S \to Q$ and where the relevant maps $\phi : S \to \Psi$ send $S_q \to \Psi_q$ for all $q \in Q$ with these maps denoted $\phi_q = \phi_{|S_q}$.

Then the energy of the fibered homotopy class $[\phi]_Q$ of such a fiber preserving map ϕ is defined as earlier as

$$E[\phi]_Q = \inf_{\phi \in [\phi]_Q} \sup_{s \in S} E \circ \phi(s) \leq \sup_{q \in Q} E_q[\phi_q],$$

where the latter inequality is, in fact, an equality in many cases.

Example 1. k-cycles in moving subsets. Let U_q be a Q-family of open subsets in a Riemannian manifold X. An instance of this is the family of the ρ-balls $U_x = U_x(\rho) \in X$ for a given $\rho \geq 0$ where X itself plays the role of Q.

Define $\Psi = \Psi_Q$ as the space $C_k\{U_q; \Pi\}_{q \in Q}$ of k-dimensional Π-cycles[30] $c = c_q$ in U_q for all $q \in Q$, that is, $\Psi = \Psi_Q$ equals the space of pairs $(q \in Q, c_q \in C_k(U_q; \Pi))$, where, as earlier, Π is an Abelian (coefficient) group with a norm-like function; then we take $E(c) = E_q(c_q) = vol_k(c)$ for the energy.

Example 2. Cycles in spaces mapped to an X. Here, instead of subsets in X, we take locally diffeomorphic maps y from a fixed Riemannian manifold U into X and take the Cartesian product $C_k(U; \Pi) \times Q$ for $\Psi = \Psi_Q$.

Example 1+2. Maps with variable domains. One may deal families of spaces U_q (e.g., "fibers" $U_q = \psi^{-1}(q)$ of a map between smooth manifolds $\psi : Z \to Q$) along with maps $y_q : U_q \to X$.

---

[29] Also one may have functions with the range also depending on q, say $a_q : \Psi_q \to R_q$ and one may generalize further by defining families as some (topological) sheaves over Grothendieck sites.

[30] "k-Cycle" in $U \subset X$ means a *relative* k-cycle in $(U, \partial U)$, that is, a k-chain with boundary contained in the boundary of U. Alternatively, if U is an open non-compact subset, "k-cycle" means an *infinite* k-cycle, i.e., with (a priori) *non-compact support*.

On reduction 1 ⇒ 2. There are cases where the spaces $\Psi_Q = C_k \{U_q; \Pi\}_{q \in Q}$ of cycles in moving subsets $U_q \subset X$ topologically split:

$$\Psi_Q = C_k(B; \Pi) \times Q, \text{ for a fixed manifold } U.$$

A simple, yet representative, example is where $Q = X$ for the m-torus, $X = \mathbb{T}^m = \mathbb{R}^m / \mathbb{Z}^m$, where $B = U_0$ is an open subset in $\mathbb{T}^m$ and where $Y = X = \mathbb{T}^m$ equals the space of translates $U_0 \mapsto U_0 + x$, $x \in \mathbb{T}^m$.

For instance, if U_0 is a ball of radius $\varepsilon \leq 1/2$, then it can be identified with the Euclidean ε-ball $B = B(\varepsilon) \subset \mathbb{R}^m$.

Similar splitting is also possible for *parallelizable* manifolds X with injectivity radii $> \varepsilon$ where moving ε-balls $U_x \subset X$ are obtained via the exponential maps $exp_q : T_q = \mathbb{R}^m \to X$ from a fixed ball $B = B(\varepsilon) \subset \mathbb{R}^m$.

In general, if X is *non-parallelizable*, one may take the space of the tangent orthonormal frames in X for Q, where the product space $C_k(B; \Pi) \times Q$, where $B = B(\varepsilon) \subset \mathbb{R}^m$, makes a principle $O(m)$- fibration, $m = dim(X)$, over the space $C_k \{U_x(\varepsilon); \Pi\}_{x \in X}$ of cycles of moving ε-balls $U_x(\varepsilon) \subset X$.[31]

Waists of "variable metric spaces" needed for homological localization of the spectra of the volume energies on *fixed spaces* can be defined as follows.

Let $\mathcal{X} = \{X_q\}_{q \in Q}$ be a family of metric spaces seen as the fibers, i.e., the pullbacks of points, of a continuous map $\varpi : \mathcal{X} \to Q$ and consider S-families of subsets in X_q that are $Y_s \subset X_{q(s)}$ defined with some maps $S \to Q$ for $s \mapsto q = q(s)$.

The k-waist of such a family is defined as

$$waist_k \{X_q\} = \inf_{s \in S} \sup vol_k(Y_s)$$

where vol_k is one of the "volumes" from section 6, e.g., Hausdorff's k-measure and where "inf" is taken over all *"homologically substantial"* families Y_s.

In order to define the latter and to keep the geometric picture in mind, we

- fix a section $Q_0 \subset \mathcal{X}$, that is a continuous family[32] of points $x(q) \in X_q$;
- assume that X is an n-manifold and that the family Y_s is given by fibers of a map $\Sigma = \cup_{s \in S} Y_s \to S$ for Σ being an n-pseudomanifold of dimensions n.

Then the family $Y_s \subset X_{q(s)}$ comes via a map $\Sigma \to \mathcal{X}$ and "homologically substantial" is understood as *non-vanishing* of the intersection index of $Q_0 \subset \mathcal{X}$ with Σ mapped to $\mathcal{X}$.

For instance, if $\varpi : \mathcal{X} \to Q$ is a fibration with *contractible* fibers X_s, there is a homotopically unique section $Q_0 \subset \mathcal{X}$ that makes our index non-ambiguously defined.

Notice that $waist_k \{X_q\}$ may be strictly smaller than $\inf_{q \in Q} waist_k(X_q)$; yet, the argument(s) used for individual X show that the waists of *compact* families of (connected) Riemannian manifolds are *strictly positive*.

Example. "Ameba" penetrating a membrane. Let a domain $U_t \subset \mathbb{R}^3$, $0 \leq t \leq 1$, be composed of a pair of disjoint balls of radii t and $1 - t$ joint by a δ-thin tube. The

---

[31]Vanishing of Stiefel-Whitney classes seems to suffice for (homological) splitting of this vibration in the case $\Pi = Z_2$ as in section 6.3 of my article *"Isoperimetry of Waists. . ."* in GAFA.

[32]In fact, one only needs a distinguished "horizontal (co)homology class" in $\mathcal{X}$.

2-waist of U_t is at least $\pi/4$, that is, the waist of the ball of radius $t/2$ for all $t \in [0, 1]$. But the waist of the "variable domain" U_t equals the area of the section of the tube that is $\pi\delta^2$.

9 RESTRICTION AND STABILIZATION OF PACKING SPACES

There are no significant relations between *individual* packings of manifolds and their submanifolds, but such relations do exist for *spaces of packings*.

For instance, let $X_0 \subset X$ be a closed n_0-dimensional submanifold in an n-dimensional, Riemannian manifold X and let

$$\frown_I : H_*(X^I) \to H_{*-*'}(X_0^I), \quad *' = N(n - n_0), \quad N = card(I),$$

be the homomorphism corresponding to the (generic) intersections of cycles in the Cartesian power X^I with the submanifold $X_0^I \subset X^I$, where the homology groups are understood with $\mathbb{Z}_2 = \mathbb{Z}/2\mathbb{Z}$-coefficients, since we do not assume that X_0 is orientable. Then *the homological packing energies E_* of X and X_0* (obviously) *satisfy*

$$E_*(\frown_I (h)) \leq E_*(h)$$

for all homology classes $h \in H_*(X^I) = H_*(X)^{\otimes I}$, where packings of X_0 are understood with respect to the metric, i.e., distance function, induced from $X \supset X_0$.

Corollary. Let $P(X; I, r) \subset X^I$ be the space of I-packings of X by balls of radii r and let $S \subset P(X; I, r)$ be a K-cycle, $K = N(n - n_0)$ that has a nonzero intersection index with $X_0^I \subset X^I$.

Then X_0 admits a packing by N-balls of radius r.

Next, let us invert the intersection homomorphism $\frown_I$ in a presence of a *projection* also called *retraction* $p : X \to X_0$ of X to X_0, i.e., where p fixes X_0.

If p is a fibration or, more generally it is a generic smooth p, then the pullbacks $Y_i = p_{-1}(x_i)X$ are k-cycles, $k - n - n_0$, that transversally meet X_0 at the points $x_i \in X_0$. It follows that the Cartesian product $N(n - n_0)$-cycle

$$S = \underset{i \in I}{\times} Y_i \subset X^I$$

has nonzero intersection index with $X_0^I \subset X_I$.

And—now geometry enters—if p is a (non-strictly) *distance decreasing* map, then, obviously, this S is positioned in the space $P(X; I, r) \subset X^I$ of I-packing of X by r-balls $U_r(x_i)$ and multiplication of cycles in $C \subset P(X_0; I, r)$ with S, that is, $C \mapsto C \times S$, *embeds*

$$H_*(P(X_0; I, r)) \underset{\times S}{\to} H_{*+Nk}(P(X; I, r)),$$

such that the composed map

$$H_*(P(X_0; I, r)) \underset{\times S}{\to} H_{*+Nk}(P(X; I, r)) \underset{\frown_I}{\to} H_*(P(X_0; I, r))$$

equals the identity. Thus, *the homology packing spectrum of X fully determines such a spectrum of X_0*.[33]

Example. Let $\underline{X}$ be a compact manifold of negative curvature and $X_0 \subset \underline{X}$ be a closed geodesic. Then the above applies to the covering $X = X(X_0)$ of $\underline{X}$ with the cyclic fundamental group generated by the homotopy class of X_0.

Therefore, *the lengths of all the closed geodesics in $\underline{X}$ are determined by the homotopy packing spectrum of $\underline{X}$*.

Questions. How much of the geometry of minimal varieties V in X, that are critical points of the volume energies, can be seen in terms of the above families of packings of X by balls (or by non-round subsets as in [25]) "moving transversally to" V?

Minimal subvarieties V can be approximated by sets of centers of small δ-balls densely packing these V; this suggests looking at spaces of packings of $c\delta$-neighborhoods $U_{c\delta}(V) \subset X$ by $(1 - \varepsilon)\delta$-balls for some $c > 1$.

Symplectic remark. The above relation between, say, individual packings of an $X_0 \subset X$ by N balls and $N(n - n_0)$-dimensional families of packings of X by balls "moving transversally to X_0" is reminiscent of *hyperbolic stabilization of Morse functions* as it is used in the study of *generating functions* in the symplectic geometry (see [49], [16], and references therein).

Is there something more profound here than just a simple-minded similarity?

10   HOMOTOPY HEIGHT, CELL NUMBERS AND HOMOLOGY

The homotopy spectral values $r \in \mathbb{R}$ of $E(\psi)$ are "named" after (indexed by) the homotopy classes $[\phi]$ of maps $\phi : S \to \Psi$, where $r = r_{[\phi]}$ is, by definition, the minimal r such that $[\phi]$ comes from a map $S \to \Psi_r \subset \Psi$ for $\Psi_r = E^{-1}(-\infty, r]$. In fact, such a "name" depends only on the partially ordered set, call it $\mathcal{H}_{\geqslant}(\Psi)$, that is, *the maximal partially ordered reduction* of $\mathcal{H}_\circ(\Psi)$ defined as follows.

Write $[\phi_1] \prec [\phi_2]$ if there is a morphism $\psi_{12} : [\phi_1] \to [\phi_2]$ in $\mathcal{H}_\circ(\Psi)$ and turn this into a partial order by identifying objects, say $[\phi]$ and $[\phi']$, whenever $[\phi] \prec [\phi']$ as well as $[\phi'] \prec [\phi]$.

Perfect example. If X is (homotopy equivalent to) the real projective space P^∞ then the partially ordered set $\mathcal{H}_{\geqslant}(\Psi)$ is isomorphic to the set of non-negative integers $\mathbb{Z}_+ = \{0, 1, 2, 3, \ldots\}$. This is why spectral (eigen) values are indexed by integers in the classical case.

In general the set $\mathcal{H}_{\geqslant}(\Psi)$ may have undesirable(?) "twists." For instance, if Ψ is homotopy equivalent to the circle, then $\mathcal{H}_{\geqslant}(\Psi)$ is isomorphic to set $\mathbb{Z}_+$ with the

---

[33]If p is a *homotopy retraction*, e.g., $p : X \to X_0$ is a vector bundle, then the above homomorphisms come from the Thom *isomorphisms* between corresponding spaces. Yet, there is more to packings of X than what comes from X_0; it is already seen in the case where X is a ball and $X_0 = \{0\}$.

However, ball packings of X_0 in this case properly reflect properties of packing of X by r-neighborhoods of k-cycles $Y_i \subset X$, the intersection indices of which with X_0 equal one.

division order, where $m \succ n$ signifies that m divides n. (Thus, 1 is the maximal element here and 0 is the minimal one.)

Similarly, one can determine $\mathcal{H}_{\geqslant}(\Psi)$ for general Eilenberg-MacLane spaces $\Psi = K(\Pi, n)$. This seems transparent for Abelian groups Π. But if a space Ψ, not necessarily a $K(\Pi, 1)$, has a non-Abelian fundamental group $\Pi = \pi_1(\Psi)$, such as the above space $\Psi_N(X)$ of subsets $\psi \subset X$ of cardinality N, then keeping track of the conjugacy classes of subgroups $\Pi' \subset \Pi$ and maps $\phi : S \to \Psi$ that send $\pi_1(S)$ to these Π' becomes more difficult.

If one wishes (simple-mindedly?) to remain with integer valued spectral values, one has to pass to some numerical invariant that takes values in a quotient of $\mathcal{H}_{\geqslant}$ isomorphic to $\mathbb{Z}_+$, e.g., as follows.

Homotopy height. Define *the homotopy (dimension) height* of a homotopy class $[\phi]$ of continuous map $\phi : S \to \Psi$ as the minimal integer n such that the $[\phi]$ factors as $S \to K \to \Psi$, where K is a cell complex of dimension (at most) n.

"Stratification" of homotopy cohomotopy spectra by height. This "height" or a similar height-like function defines a partition of the homotopy spectrum into the subsets, call them $Hei_n \subset \mathbb{R}$, $n = 0, 1, 2, \ldots$, of the values of the energy $E[\phi] \in \mathbb{R}$ on the homotopy classes $[\phi]$ with homotopy heights n, where either the *supremum* or the *infimum* of the numbers $r \in Hei_n$ may serve as the *"n-th HH-eigenvalue of ψ."*

One also may "stratify" cohomotopy spectra by replacing *"contractibility condition"* of maps $\psi_{|\Psi_r} : \Psi_r \to T$ by $\psi_{|\Psi_r} \leq n$.

In the classical case of $\Psi = P^\infty$ any such "stratification" of homotopy "eigenvalues" leads the usual indexing of the spectrum, where, besides the homotopy height, among other height-like invariants we indicate the following.

Example 1. Total cell number. Define $N_{cell}[\phi]$ as the minimal N such that $[\phi]$ factors as $S \to D \to \Psi$, where D is a cell complex with (at most) N cells in it.

What are, roughly, the total cell numbers of the classifying maps from packing spaces of an X by N balls to the classifying space $B\mathbb{S}_N$?

What are these numbers for the maps between classifying spaces of "classical" finite groups G corresponding to standard injective homomorphisms $G_1 \to G_2$?

Example 2. Homology rank. Define $rank_{H_*}[\phi]$ as the maximum over all fields $\mathbb{F}$ of the $\mathbb{F}$-ranks of the induced homology homomorphisms $[\phi]_* : H_*(S; \mathbb{F}) \to H_*(\Psi; \mathbb{F})$.

On essentiality of homology. There are other prominent spaces, X, besides the infinite dimensional projective spaces $X = P^\infty$, and energy functions on them, such as, *spaces Ψ of loops $\psi : S^1 \to X$ in simply connected Riemannian manifolds X with length(ψ) taken for $E(\psi)$,*[34] where the cell numbers and the homology ranks spectra for $E(\psi) = length(\psi)$ are "essentially determined" by the homotopy height. (This is why the homotopy height was singled out under the name of "essential dimension" in my paper *"Dimension, Non-linear Spectra and Width."*)

However, the homology carries significantly more information than the homotopy height for the k-volume function on the *spaces of k-cycles of codimensions ≥ 2*

---

[34]This instance of essentiality of the homotopy heights is explained in my article *Homotopical Effects of Dilatation*, while the full range of this property among "natural" spaces Ψ of maps ψ between Riemannian manifolds and energies $E(\psi)$ remains unknown.

as it was revealed by Larry Guth in his paper "*Minimax Problems Related to Cup Powers and Steenrod Squares.*"

On height and the cell numbers of Cartesian products. If the homotopy heights and/or cell numbers of maps $\phi_i : S_i \to \Psi_i$, $i = 1, \ldots, k$, can be expressed in terms of the corresponding homology homomorphisms over some filed $\mathbb{F}$ independent of i, then, according to *Künneth formula,* the homotopy height of the Cartesian product of maps,

$$\phi_1 \times \cdots \times \phi_k : S_1 \times \cdots \times S_k \to \Psi_1 \times \cdots \times \Psi_k,$$

is additive,

$$height[\phi_1 \times \cdots \times \phi_k] = height[\phi_1] + \cdots + height[\phi_k],$$

and the cell number is multiplicative,

$$N_{cell}[\phi_1 \times \cdots \times \phi_k] = N_{cell}[\phi_1] \times \cdots \times N_{cell}[\phi_k].$$

What are other cases where these relations remain valid?

Specifically, we want to know what happens in this regard to the following classes of maps:

(a) *maps between spheres* $\phi_i : S^{m_i+n_i} \to S^{m_i}$,
(b) *maps between locally symmetric spaces,* e.g., compact manifolds of *constant negative curvatures,*
(c) *high Cartesian powers* $\phi^{\times N} : S^{\times N} \to A^{\times N}$ of a single map $\phi : S \to \Psi$.

When do, for instance, the limits

$$\lim_{N \to \infty} \frac{height[\phi^{\times N}]}{N} \quad \text{and} \quad \lim_{N \to \infty} \frac{\log N_{cell}[\phi^{\times N}]}{N}$$

not vanish? (These limits exist, since the height and the logarithm of the cell number are sub-additive under Cartesian product of maps.)

Probably, the general question for *"rational homotopy classes* $[\ldots]_\mathbb{Q}$*"* (instead of "full homotopy classes" $[\ldots] = [\ldots]_\mathbb{Z}$) of maps into *simply connected* spaces Ψ_i is easily solvable with *Sullivan's minimal models.*[35] Also, the question may be more manageable for *homotopy classes* mod p.

Multidimensional spectra revisited. Let $h = h^T$ be a cohomotopy class of Ψ, that is, a homotopy class of maps $\Psi \to T$, and let v be a function on homotopy classes of maps $U \to T$ for open subsets $U \subset \Psi$, where the above height-like functions, such as the *homology rank,* are relevant examples of such a v.

Then the values of v on h restricted to open subsets $U \subset \Psi$ define a numerical (set) function, $U \mapsto v(h_{|U})$ and every continuous map $\mathcal{E} : \Psi \to Z$ pushes down this function to open subsets in X.

---

[35]This may follow from the results in the paper *Integral and rational mapping classes* by Manin and Weinberger [44].

For instance, if $v = 0, 1$ depending on whether a map $U \to T$ is contractible or not and if $Z = \mathbb{R}^l$, then this function on the "negative octants" $\{x_1 < e_1, \ldots, x_k < e_k, \ldots, x_l < e_l\}$ in $\mathbb{R}^l$ carries the same message as Σ_h from section 4.

11 GRADED RANKS, POINCARÉ POLYNOMIALS, IDEAL VALUED MEASURES AND SPECTRAL ⌣INEQUALITY

The images as well as kernels of (co)homology homomorphisms that are induced by continuous maps are *graded* Abelian groups and their ranks are properly represented not by individual numbers but by *Poincaré polynomials*.

Thus, sublevel $\Psi_r = E^{-1}(-\infty, r] \subset \Psi$ of energy functions $E(\psi)$ are characterized by the *polynomials* Poincaré$_r(t; \mathbb{F})$ of the inclusion homomorphisms $\phi_i(r)$: $H_i(\Psi_r; \mathbb{F}) \to H_i(\Psi; \mathbb{F})$, that are

$$\text{Poincaré}_r = \text{Poincaré}_r(t; \mathbb{F}) = \sum_{i=0,1,2,\ldots} t^i \text{rank}_{\mathbb{F}} \phi_i(r).$$

Accordingly, the homology spectra, that are the sets of those $r \in \mathbb{R}$ where the ranks of $\phi_*(r)$ change, are indexed by such polynomials with positive integer coefficient. (The semiring structure on the set of such polynomials coarsely agrees with basic topological/geometric constructions, such as taking $E(\psi) = E(\psi_1) + E(\psi_2)$ on $\Psi = \Psi_1 \times \Psi_2$.)

The set function $D \mapsto \text{Poincaré}_D$ that assigns these Poincaré polynomials to subsets $D \subset \Psi$, (e.g., $D = \Psi_r$) has some measure-like properties that become more pronounced for the set function

$$\Psi \supset D \mapsto \mu(D) = \mu^*(D; \Pi) = \mathbf{0}^{\backslash *}(D; \Pi) \subset H^* = H^*(\Psi; \Pi),$$

where Π is an Abelian (homology coefficient) group, e.g., a field $\mathbb{F}$, and $\mathbf{0}^{\backslash *}(D; \Pi)$ is the *kernel* of the cohomology restriction homomorphism for the complement $\Psi \setminus D \subset \Psi$,

$$H^*(\Psi; \Pi) \to H^*(\Psi \setminus D; \Pi).$$

Since the cohomology classes $h \in \mathbf{0}^{\backslash *}(D; \Pi) \subset H^* = H^*(\Psi; \Pi)$ are representable by cochains with the support in D,[36] *the set function*

$$\mu^* : \{subsets \subset \Psi\} \to \{subgroups \subset H^*\}$$

is additive for the sum-of-subsets in H^* and super-multiplicative[37] for the $\smile$-product of ideals in the case where Π is a commutative ring:

[∪ +]
$$\mu^*(D_1 \cup D_2) = \mu^*(D_i) + \mu^*(D_2)$$

---

[36]This property suggests an extension of μ to multi-sheated *domains D over Ψ* where D go to Ψ by non-injective, e.g., locally homeomorphic finite to one, maps $D \to A$.

[37]This, similarly to *Shannon's subadditivity inequality,* implies the existence of "thermodynamic limits" of *Morse Entropies,* see [8].

for *disjoint* open subsets D_1 and D_2 in Ψ, and

$[\cap \smile]$ $\qquad\qquad\qquad\qquad \mu^*(D_1 \cap D_2) \supset \mu^*(D_1) \smile \mu^*(D_2)$

for all open $D_1, D_2 \subset \Psi$.[38]

The relation $[\cap \smile]$ applied to $D_{r,i} = E_i^{-1}(r, \infty) \subset \Psi$ can be equivalently expressed in terms of cohomology spectra as follows.

Spectral $[\min \smile]$-*inequality*.[39] Let $E_1, \ldots, E_i, \ldots, E_N : \Psi \to \mathbb{R}$ be continuous functions/energies and let $E_{min} : \Psi \to \mathbb{R}$ be the minimum of these,

$$E_{min}(\psi) = \min_{i=1,\ldots,N} E_i(\psi), \ \psi \in \Psi.$$

Let $h_i \in H^{k_i}(\Psi; \Pi)$ be cohomology classes, where Π is a commutative ring, and let

$$h \smile \in H^{\sum_i k_i}(\Psi; \Pi)$$

be the $\smile$-product of these classes,

$$h \smile = h_1 \smile \ldots \smile h_i \smile \ldots \smile h_N.$$

Then

$[\min \smile]$ $\qquad\qquad\qquad E_{min}^*(h \smile) \geq \min_{1=1,\ldots,N} E_i^*(h_i).$

Consequently, the value of the *"total energy"*

$$E_\Sigma = \sum_{i=1,\ldots,N} E_i : \Psi \to \mathbb{R}$$

on this cohomology class $h \smile \in H^*(\Psi; \Pi)$ is bounded from below by

$$E_\Sigma^*(h \smile) \geq \sum_{i=1,\ldots,N} E_i^*(h_i).$$

(This has been already used in the homological localization of the volume energy in section 7.)

On multidimensional homotopy spectra. These spectra, as defined in sections 4, 10, and 11 represent the values of the push-forward of the "measure" μ^* under maps $\mathcal{E} : \Psi \to \mathbb{R}^l$ on special subsets $\Delta \subset \mathbb{R}^l$; namely, on complements to $\times_{k=1,\ldots,l}(-\infty, e_k] \subset \mathbb{R}^l$ and the spectral information is encoded by $\mu^*(\mathcal{E}^{-1}(\Delta)) \subset H^*(\Psi)$.

One may generalize this by enlarging the domain of μ^*, say, by evaluating $\mu^*(\mathcal{E}^{-1}(\Delta))$ for some class of simple subsets Δ in $\mathbb{R}^l$, e.g., convex sets and/or their complements.

---

[38] See section 4 of my article *Singularities, Expanders and Topology of Maps, Part 2,* for further properties and applications of these "measures."

[39] This inequality implies the existence of *Hermann Weyl limits* of energies of cup-powers in infinite dimensional projective (and similar) spaces, see [21].

On ∧-product. The (obvious) proof of [∩⌣] (and of [min ⌣]) relies on locality of the ⌣-product that, in homotopy theoretic terms, amounts to factorization of ⌣ via ∧ that is the *smash product* of (marked) Eilenberg-MacLane spaces that represent cohomology, where, recall, the *smash product* of spaces with marked points, say $T_1 = (T_1, t_1)$ and $T_2 = (T_2, t_2)$, is

$$T_1 \wedge T_2 = T_1 \times T_2 / T_1 \vee T_2$$

where the factorization "$/T_1 \vee T_2$" means "*with the subset* $(T_1 \times t_2) \cup (t_1 \times T_2) \subset T_1 \times T_2$ *shrunk to a point*" (that serves to mark $T_1 \wedge T_2$).

In fact, general cohomotopy "measures" (see 9) and spectra defined with maps $\Psi \to T$ satisfy natural (obviously defined) counterparts/generalizations of [∩⌣] and [min ⌣], call them [∩∧] and [min ∧].

On grading cell numbers. Denote by $N_{i\_cell}[\phi]$ the minimal number N_i such that homotopy class $[\phi]$ of maps $S \to \Psi$ factors as $S \to K \to \Psi$ where K is a cell complex with (at most) N_i cells *of dimension i* and observe that the total cell number is bounded by the sum of these,

$$N_{cell}[\phi] \le \sum_{i=0,1,2,\dots} N_{i\_cell}[\phi].$$

Under what conditions on ϕ does the sum $\sum_i N_{i\_cell}[\phi]$ (approximately) equal $N_{cell}[\phi]$?

What are relations between the cell numbers of the covering maps ϕ between (arithmetic) locally symmetric spaces Ψ besides $N_{cell} \le \sum_i N_{i\_cell}$?[40]

12 SYMMETRIES, EQUIVARIANT SPECTRA AND SYMMETRIZATION

If the energy function E on Ψ is invariant under a continuous action of a group G on Ψ—this happens frequently—then the relevant category is that of *G-spaces S*, i.e., of topological spaces S acted upon by G, where one works with *G-equivariant* continuous maps $\phi : S \to \Psi$, *equivariant* homotopies, equivariant (co)homologies, decompositions, etc.

Relevant examples of this are provided by symmetric energies $E = E(x_1, \dots, x_N)$ on Cartesian powers of spaces, $\Psi = X^{\{1,\dots,N\}}$, such as our (ad hoc) packing energy for a metric space X,

$$E\{x_1, \dots, x_i, \dots, x_N\} = \sup_{i \ne j=1,\dots,N} dist^{-1}(x_i, x_j)$$

that is invariant under the symmetric group Sym_N. It is often profitable, as we shall see later on, to exploit the symmetry under certain *subgroups* $G \subset Sym_N$.

---

[40]The identity maps $\phi = id : \Psi \to \Psi$ of locally symmetric spaces Ψ seem quite nontrivial in this regard. On the other hand, general locally isometric maps $\phi : \Psi_1 \to \Psi_2$ between symmetric spaces as well as continuous maps $S \to \Psi$ of positive degrees, where S and Ψ are equidimensional manifolds with only Ψ being locally symmetric, are also interesting.

Besides the group Sym_N, energies E on $X^{\{1,\dots,N\}}$ are often invariant under some groups H acting on X, such as the isometry group $Is(X)$ in the case of packings.

If such a group H is compact, than its role is less significant than that of Sym_N, especially for large $N \to \infty$; yet, if H properly acts on a *non-compact* space X, such as $X = \mathbb{R}^m$ that is acted upon by its isometry group, then H and its action become essential.

MIN-symmetrized energy. An arbitrary function E on a G-space Ψ can be rendered G-invariant by taking a symmetric function of the numbers $e_g = E(g(\psi)) \in \mathbb{R}$, $g \in G$. Since we are mostly concerned with the order structure in $\mathbb{R}$, our preferred symmetrization is

$$E(\psi) \mapsto \inf_{g \in G} E(g(\psi)).$$

Minimization with partitions. This inf-symmetrization does not fully depend on the action of G but rather on the partition of Ψ into orbits of G. In fact, given an arbitrary partition α of Ψ into subsets that we call α-*slices*, one defines the function

$$E_{inf_\alpha} = inf_\alpha E : \Psi \to \mathbb{R}$$

where $E_{inf_\alpha}(\psi)$ equals the infimum of E on the α-slice that contains ψ for all $\psi \in \Psi$. Similarly, one defines $E_{sup_\alpha} = sup_\alpha E$ with E_{min_α} and E_{max_α} understood accordingly.

Example. Energies on Cartesian powers. The energy E on Ψ induces N energies on the space $\Psi^{\{1,\dots,N\}}$ of N-tuples $\{\psi_1, \dots, \psi_i, \dots, \psi_N\}$, that are

$$E_i : \{\psi_1, \dots, \psi_i, \dots, \psi_N\} \mapsto E(\psi_i).$$

It is natural, both from a geometric as well as from a physical perspective, to symmetrize by taking the *total energy* $E_{total} = \sum_i E_i$. But in what follows we shall resort to $E_{min} = \min_i E_i = \min_i E(\psi_i)$ and use it for bounding the total energy from below by

$$E_{total} \geq N \cdot E_{min}.$$

For instance, we shall do it for families of N-tuples of balls U_i in a Riemannian manifold V, thus bounding the k-volumes of k-cycles c in the unions $\cup_i U_i$, where, observe,

$$vol_k(c) = \sum_i vol_k(c \cap U_i)$$

if the balls U_i *do not intersect*. This, albeit obvious, leads, as we shall see later on, to non-vacuous relations between *homotopy/homology spectrum of the vol_k-energy on the space $C_k(X; \Pi)$ and equivariant homotopy/homology of the spaces of packings of X by ε-balls*.

13 EQUIVARIANT HOMOTOPIES OF INFINITE DIMENSIONAL SPACES

If we want to understand homotopy spectra of spaces of "natural energies" on spaces of infinitely many particles in non-compact manifolds, e.g., in Euclidean spaces, we need to extend the concept of the homotopy and homology spectra to infinite dimensional spaces Ψ, where infinite dimensionality is compensated by an additional structure, e.g., by an action of an infinite group Υ on Ψ.

The simplest instance of this is where Υ is a countable group that we prefer to call Γ and $\Psi = B^\Gamma$ is the space of maps $\Gamma \to B$ with the (obvious) *shift action* of Γ on this Ψ. This motivates the following definition (compare [8]). Let H^* be a graded algebra (over some field) acted upon by a countable amenable group Γ. Exhaust Γ by finite *Følner subsets* $\Delta_i \subset \Gamma$, $i = 1, 2, \ldots$, and, given a finite dimensional graded subalgebra $K = K^* \subset H^*$, let $P_{i,K}(t)$ denote the Poincaré polynomial of the graded subalgebra in H^* generated by the γ-transforms $\gamma^{-1}(K) \subset H^*$ for all $\gamma \in \Delta_i$.

Define *polynomial entropy* of the action of Γ on H^* as follows:

$$Poly.ent(H^* : \Gamma) = \sup_K \lim_{i \to \infty} \frac{1}{card(\Delta_i)} \log P_{i,K}(t).$$

Something of this kind could be applied to subalgebras $H^* \subset H^*(\Psi; \mathbb{F})$, such as images and/or kernels of the restriction cohomology homomorphisms for (the energies sublevel) subsets $U \subset \Psi$, IF the following issues are settled.

1. In our example of moving balls or particles in $\mathbb{R}^m$ the relevant groups Υ, such as the group of the orientation preserving Euclidean isometries, are connected and act *trivially* on the cohomologies of our spaces Ψ.

For instance, let $\Gamma \subset \Upsilon$ be a discrete subgroup and Ψ equal the *dynamic Υ-suspension of B^Γ*, that is, $B^\Gamma \times \Upsilon$ divided by the diagonal action of Γ:

$$\Psi = \left(B^\Gamma \times \Upsilon \right) / \Gamma.$$

The (ordinary) cohomology of this space Ψ are bounded by those of B tensored by the cohomology of Υ/Γ that would give *zero* polynomial entropy for finitely generated cohomology algebras $H^*(B)$.

In order to have something more interesting, e.g., the *mean Poincaré polynomial* equal to that of B^Γ, which is the ordinary Poincaré$(H^*(B)) = \sum_i rank(H^*(B))$, one needs a definition of some *mean (logarithm) of the Poincaré polynomial* that might be *far from zero* even if the ordinary cohomology of Ψ vanishes.

There are several candidates for such *mean Poincaré polynomials*, e.g., the one is suggested in section 1.15 of [22].

Another possibility that is applicable to the above $\Psi = \left(B^\Gamma \times \Upsilon \right) / \Gamma$ with *residually finite* groups Γ is using finite i-sheeted covering $\tilde{\Psi}_i$ corresponding to subgroups $\Gamma_i \subset \Gamma$ of order i and taking the limit of

$$\lim_{i \to \infty} \frac{1}{i} \log \text{Poincaré}(H^*(B)).$$

(Algebraically, in terms of actions of groups Γ on abstract graded algebras H^*, this corresponds to taking the normalized limit of logarithms of Γ_i-*invariant* sub-algebras in H^*; this brings to one's mind a possibility of a generalization of the above polynomial entropies to *sofic groups* (compare [5]).)

2. The above numerical definitions of the polynomial entropy and of the mean Poincaré polynomials beg to be rendered in categorical terms similar to the ordinary entropy (see [23]).

3. The spaces $\Psi_\infty(X)$ of (discrete) infinite countable subsets $\psi \subset X$ that are meant to represent infinite ensembles of particles in non-compact manifolds X, such as $X = \mathbb{R}^m$, are more complicated than $\Psi = B^\Gamma$, $\Psi = (B^\Gamma \times \Upsilon)/\Gamma$ and other "product like" spaces studied earlier.

These $\Psi_\infty(X)$ may be seen as limits of finite spaces $\Psi_N(X_N)$ for $N \to \infty$ of N-tuples of points in compact manifolds X_N where one has to choose suitable approximating sequences X_N.

For instance, if $X = \mathbb{R}^m$ acted upon by some isometry group Υ of $\mathbb{R}^m$ one may use either the balls $B^m(R_N) \subset \mathbb{R}^m$ of radii $R_N = const \cdot R^{N/\beta}$ in $\mathbb{R}^m$, $\beta > 0$ for X_N or the tori $\mathbb{R}^m/\Gamma_N$ with the lattices $\Gamma_N = const \cdot M \cdot \mathbb{Z}^m$ with $M = M_N \approx N^{\frac{1}{\beta}}$ for some $\beta > 0$.[41]

Defining such limits and working out functional definitions of relevant structures in the limit spaces, collectively called $\Psi_\infty(X)$, are the problems we need to solve where, in particular, we need to

- incorporate actions of the group Υ coherently with (some subgroup) of the infinite permutations group acting on subsets $\psi \subset X$ of particles in X that represent points in $\Psi_\infty(X)$, and
- define (stochastic?) homotopies and (co)homologies in the spaces $\Psi_\infty(X)$, where these may be associated to limits of families of n-tuples $\psi_{P_N} \subset X_N$ parametrized by some P_N where $dim(P_N)$ may tend to infinity for $N \to \infty$.[42]

4. Most natural energies E on infinite particle spaces $\Psi_\infty(X)$ are everywhere infinite[43] and defining "sublevels" of such E needs attention.

14 SYMMETRIES, FAMILIES AND OPERATIONS ACTING ON COHOMOTOPY MEASURES

Cohomotopy "measures." Let T be a space with a distinguished *marking point* $t_0 \in T$, let $H^\circ(\Psi; T)$ denote the set of homotopy classes of maps $\Psi \to T$ and define the "T-measure" of an open subset $U \subset \Psi$,

$$\mu^T(U) \subset H^\circ(\Psi; T),$$

as the set of homotopy classes of maps $\Psi \to T$ that send the complement $\Psi \setminus U$ to t_0.

---

[41]The natural value is $\beta = m$ that makes the volumes of X_N proportional to N, but smaller values, that correspond to ensembles of points in $\mathbb{R}^m$ of *zero densities*, also make sense as we shall see later on.

[42]We shall meet families of dimensions $dim(P_N) \sim N^{\frac{1}{\gamma}}$ where $\gamma + \beta = m$ for the above β.

[43]In the optical astronomy, this is called *Olbers' dark night sky paradox*.

For instance, if T is the Cartesian product of Eilenberg-MacLane spaces $K(\Pi; n)$, $n = 0, 1, 2, \ldots$, then $H^\circ(\Psi; T) = H^*(\Psi; \Pi)$ and μ^T identifies with the (graded cohomological) ideal valued "measure" $U \mapsto \mu^*(U; \Pi) \subset H^*(\Psi; \Pi)$ from section 14.

Next, given a category $\mathcal{T}$ of marked spaces T and homotopy classes of maps between them, denote by $\mu^{\mathcal{T}}(U)$ the totality of the sets $\mu^T(U)$, $T \in \mathcal{T}$, where the category $\mathcal{T}$ acts on $\mu^{\mathcal{T}}(U)$ via composition $\Psi \overset{m}{\to} T_1 \overset{\tau}{\to} T_2$ for all $m \in \mu^T(U)$ and $\tau \in \mathcal{T}$.

For instance, if $\mathcal{T}$ is a category of Eilenberg-MacLane spaces $K(\Pi; n)$, this amounts to the natural action of the (unary) cohomology operations (such as Steenrod squares Sq^i in the case $\Pi = \mathbb{Z}_2$) on ideal valued measures.

The above definition can be adjusted for spaces Ψ endowed with additional structures.

For example, if Ψ represents a *family of spaces* by being endowed with a partition β into closed subsets—call them β-*slices* or *fibers*—then one restricts to the space of maps $\Psi \to T$ *constant on these slices* (if T is also partitioned, it would be logical to deal with maps sending slices to slices), defines $H_\beta^\circ(\Psi; T)$ as the set of the homotopy classes of these slice-preserving maps and accordingly defines $\mu_\beta^{\mathcal{T}}(U) \subset H_\beta^\circ(\Psi; T)$.

Another kind of a relevant structure is an action of a group G on Ψ. Then one may (or may not) work with categories $\mathcal{T}$ of G-spaces T (i.e., acted upon by G) and perform homotopy, including (co)homology, constructions equivariantly. Thus, one defines equivariant T-measures $\mu_G^T(U)$ for G-invariant subsets $U \subset \Psi$.

(A group action on a space defines a partition of this space into orbits, but this is a weaker structure than that of the action itself.)

Guth Vanishing Lemma. The supermultiplicativity property of the cohomology measures with arbitrary coefficients Π (see 1.5) for spaces Ψ acted upon by finite groups G implies that

$$\mu^* \left(\bigcap_{g \in G} g(U; \Pi) \right) \supset \underset{g \in G}{\smile} \mu^*(g(U; \Pi))$$

for all open subset $U \subset \Psi$.

This, in the case $\Pi = \mathbb{Z}_2$, was generalized by Larry Guth for families of spaces parametrized by spheres S^j as follows.

Given a space Ψ endowed with a partition α, we say that a subset in Ψ is α-*saturated* if it equals the union of some α-slices in Ψ and defines two operations on subsets $U \subset \Psi$,

$$U \mapsto \cap_\alpha(U) \subset U \text{ and } U \mapsto \cup_\alpha(U) \supset U,$$

where $\cap_\alpha(U)$ is the *maximal α-saturated subset that is contained in U* and $\cup_\alpha(U)$ is the *minimal α-saturated subset that contains U*.

Let, as in the case considered by Guth, $\Psi = \Psi_0 \times S^j$ where $S^j \subset \mathbb{R}^{j+1}$ is the j-dimensional sphere, let α be the partition into the orbits of $\mathbb{Z}_2$-action on Ψ by $(\psi_0, s) \mapsto (\psi_0, -s)$ (thus, "α-saturated" means "$\mathbb{Z}_2$-invariant") and let β be the partition into the fibers of the projection $\Psi \to \Psi_0$ (and "β-saturated" means "equal the pullback of a subset in Ψ_0").

Following Guth, define

$$Sq_j : H^{*\geq j/2}(\Psi; \mathbb{Z}_2) \to H^*(\Psi; \mathbb{Z}_2) \text{ by } Sq^j : H^p \to H^{2p-j}$$

and formulate his "Vanishing Lemma" in μ_β-terms as follows:[44]

$$[\cup\cap] \qquad \mu_\beta^*\left(\cup_\beta(\cap_\alpha(U)); \mathbb{Z}_2\right) \supset Sq_j(\mu_\beta^*(U; \mathbb{Z}_2)) \subset H_\beta^*(\Psi; \mathbb{Z}_2),$$

where, according to our notation, $H_\beta^*(\Psi; \mathbb{Z}_2) \subset H^*(\Psi; \mathbb{Z}_2)$ equals the image of $H^*(\Psi_0; \mathbb{Z}_2)$ under the cohomology homomorphism induced by the projection $\Psi \to \Psi_0$.

If $E: \Psi \to \mathbb{R}$ is an energy function, this lemma yields the lower bound on the $maxmin$-energy[45]

$$E_{max_\beta min_\alpha} = max_\beta min_\alpha E$$

evaluated at the cohomology class $St_j(h)$, $h \in H_\beta^*(\Psi; \mathbb{Z}_2)$:

$$[maxmin] \qquad E_{max_\beta min_\alpha}^*(St_j(h)) \geq E^*(h).$$

Question. What are generalizations of $[\cup\cap]$ and $[maxmin]$ to other cohomology and cohomotopy measures on spaces with partitions $\alpha, \beta, \gamma, \ldots$?

15   PAIRING INEQUALITY FOR COHOMOTOPY SPECTRA

Let Ψ_1, Ψ_2 and Θ be topological spaces and let

$$\Psi_1 \times \Psi_2 \overset{\circledast}{\to} \Theta$$

be a continuous map where we write

$$\theta = \psi_1 \circledast \psi_2 \text{ for } b = \circledast(\psi_1, \psi_2).$$

For instance, composition $\psi_1 \circ \psi_2: X \to Z$ of morphisms $X \overset{\psi_1}{\to} Y \overset{\psi_2}{\to} Z$ in a topological category defines such a map between sets of morphisms,

$$mor(X \to Y) \times mor(Y \to Z) \overset{\circledast}{\to} mor(X \to Z).$$

A more relevant example for us is the following.

Cycles $\times$ Packings

Here, Ψ_1 is a space of locally diffeomorphic maps $U \to X$ between manifolds U and X,

Ψ_2 is the space of cycles in X with some coefficients Π,

Θ is the space of cycles in U with the same coefficients,

$\circledast$ stands for *"pullback"*:

---

[44]Guth formulates his lemma in terms of the complementary set $V = \Psi \setminus U$: *if a cohomology class $h \in H_\beta^*(\Psi; \mathbb{Z}_2)$ vanished on V, then $St_j(h)$ vanishes on $\cap_\beta(\cup_\alpha(V))$.*

[45]Recall that $min_\alpha E(\psi)$, $\psi \in \Psi$, denotes the minimum of E on the α-slice containing ψ and max_β stands for similar maximization with β (see 1.12).

$$\theta = \psi_1 \circledast \psi_2 =_{def} \psi_1^{-1}(\psi_2) \in \Theta.$$

This U may equal the disjoint unions of N manifolds U_i that, in the spherical packing problems, would go to balls $B_x(r) \subset X$; since we want these balls *not to intersect*, we take the space of *injective* maps $U \to X$ for Ψ_1.

If the manifold X is *parallelizable* and the balls $B_{x_i}(r) \subset X$ all have some radius r smaller than the injectivity radius of X, then corresponding $U_i = B_{x_i}(r)$ can be identified (via the exponential maps) with the r-ball $B^n(r)$ in the Euclidean space $\mathbb{R}^n$, $n = dim(X)$. Therefore, the space of k-cycles in $U = \bigsqcup_i U_i = B^n(r)$ equals in this case the Cartesian N-th power of this space for the r-ball:

$$\Theta = \mathcal{C}_k(U; \Pi) = (\mathcal{C}_k(B^n(r); \Pi))^N.$$

Explanatory remarks.

(a) Our "*cycles*" are defined as *subsets* in relevant manifolds X and/or U with Π-*valued functions* on these subsets.
(b) In the case of *open* manifolds, we speak of cycles with *infinite supports*, that, in the case of compact manifolds with boundaries or of open subsets $U \subset X$, are, essentially, *cycles modulo the boundaries* ∂X.
(c) "*Pullbacks of cycles*" that preserve their codimensions are defined, following Poincaré for a wide class of smooth *generic* (not necessarily equividimensional) maps $U \to X$ (see [32]).
(d) It is easier to work with *cocycles* (rather than with cycles) where contravariant functoriality needs no extra assumptions on spaces and maps in question (see [24]).

Let h^T be a (preferably nonzero) cohomotopy class in Θ, that is, a homotopy class of non-contractible maps $\Theta \to T$ for some space T (where "*cohomotopy*" reads "*cohomology*" if T is an Eilenberg-MacLane space) and let

$$h^\circledast = \circledast \circ h^T : [\Psi_1 \times \Psi_2 \to T]$$

be the induced class on $\Psi_1 \times \Psi_2$, that is, the homotopy class of the composition of the maps $\Psi_1 \times \Psi_2 \overset{\circledast}{\to} \Theta \overset{h^T}{\to} T$. (Here and below, we do not always notationally distinguish *maps* and *homotopy classes* of maps.)

Let h_1 and h_2 be homotopy classes of maps $S_1 \to \Psi_1$ and $S_2 \to \Psi_2$ for some spaces S_i, $i = 1, 2$. (In the case where h^T is a *cohomology* class, these h_i may be replaced by *homology*—rather than homotopy—classes represented by these maps.)

Compose the three maps,

$$S_1 \times S_2 \overset{h_1 \times h_2}{\to} \Psi_1 \times \Psi_2 \overset{\circledast}{\to} \Theta \overset{h^T}{\to} T,$$

and denote the homotopy class of the resulting map $S_1 \times S_2 \to T$ by

$$[h_1 \circledast h_2]_{h^T} = h^\circledast \circ (h_1 \times h_2) : [S_1 \times S_2 \to T].$$

Let $\chi = \chi(e_1, e_2)$ be a function in two real variables that is monotone unceasing in each variable. Let $E_i : \Psi_i \to \mathbb{R}$, $i = 1, 2$, and $F : \Theta \to \mathbb{R}$ be (energy) functions on the spaces Ψ_1, Ψ_2 and Θ, such that the $\circledast$-pullback of F to $\Psi \times \Theta$ denoted

$$F^\circledast = F \circ \circledast : \Psi_1 \times \Psi_2 \to \mathbb{R}$$

satisfies

$$F^\circledast(\psi_1, \psi_2) \leq \chi(E(\psi_1), E(\psi_2)).$$

In other words, the $\circledast$-image of the product of the sublevels

$$(\Psi_1)_{e_1} = E_1^{-1}(-\infty, e_1) \subset \Psi_1 \text{ and } (\Psi_2)_{e_2} = E_2^{-1}(-\infty, e_2) \subset \Psi_2$$

is contained in the f-sublevel $B_f = F^{-1}(-\infty, f) \subset \Theta$ for $f = \chi(e_1, e_2)$,

$$\circledast \left((\Psi_1)_{e_1} \times (\Psi_2)_{e_2}\right) \subset \Theta_{f=\chi(e_1, e_2)}.$$

$\circledast$-Pairing Inequality

Let $[h_1 \circledast h_2]_{hT} \neq 0$, that is, *the composed map*

$$S_1 \times S_2 \to \Psi_1 \times \Psi_2 \to \Theta \to T$$

is non-contractible. Then the values of E_1 and E_2 on the homotopy classes h_1 and h_2 are *bounded from below* in terms of a lower bound on $F^\circ[h^T]$ as follows:

$$[_{\circ\circ}\geq^\circ] \qquad\qquad \chi(E_{1\circ}[h_1], E_{2\circ}[h_2]) \geq F^\circ[h^T].$$

In other words

$$\left(E_{1_\circ}[h_1] \leq e_1\right) \& \left(E_{2\circ}[h_2] \leq e_2\right) \Rightarrow \left(F^\circ[h^T] \leq \chi(e_1, e_2)\right)$$

for all real numbers e_1 and e_2; thus,

$$\textbf{\textit{upper}} \text{ bound } E_1^\circ[h_1] \leq e_1 \quad + \quad \textbf{\textit{lower}} \text{ bound } F^\circ[h^T] \geq \chi(e_1, e_2)$$

$$\textit{yield}$$

$$\textbf{\textit{upper}} \text{ bound } E_2^\circ[h_2] \geq e_2,$$

where, observe, E_1 and E_2 are interchangeable in this relation. In fact, all one needs for verifying $[_{\circ\circ}\geq^\circ]$ is unfolding the definitions.

Also $[_{\circ\circ}\geq^\circ]$ can be visualized without an explicit use of χ by looking at the $h^\circledast$-*spectral line* in the (e_1, e_2)-plane

$$\Sigma_{h\circledast} = \partial\Omega_{h\circledast} \subset \mathbb{R}^2$$

(we met this Σ in section 1.3) where $\Omega_{h\circledast} \subset \mathbb{R}^2$ consists of the pairs $(e_1, e_2) \in \mathbb{R}^2$ such that the restriction of $h^\circledast$ to the Cartesian product of the sublevels $\Psi_{1e_1} = E_1^{-1}(-\infty, e_1) \subset \Psi_1$ and $\Psi_{2e_2} = E_2^{-1}(-\infty, e_2) \subset \Psi_2$ vanishes,

$$h^\circledast_{|\Psi_{1e_1} \times \Psi_{2e_2}} = 0.$$

16 INEQUALITIES BETWEEN PACKING RADII, WAISTS AND VOLUMES OF CYCLES IN A PRESENCE OF PERMUTATION SYMMETRIES

The most essential aspect of the homotopy/homology structure in the space of packings of a space X by $U_i \subset X$, $i \in I$, is associated with the permutation group $Sim_N = aut(I)$ that acts on these spaces.

A proper description of this needs a use of the concept of "homological substantiality for *variable* spaces" as in section 8. This is adapted to the present situation in the definitions below.

Let

- Ψ_1 be space of I-packings $\{U_i\}$ of X by disjoint open subsets $U_i \subset X$, $i \in I$, for some set I of cardinality N, e.g., by N balls of radii r, $N = card(I)$,
- $\mathcal{U}^\cup$ be the space of pairs $\big(\{U_i\}, u\big)_{i \in I}$, $i \in I$, where $\{U_i\} \in \Psi_1$ and $u \in \bigcup_{i \in I} U_i$,
- $\mathcal{U}^\times$ be the space of pairs $\big(\{U_i\}, \{u_i\}\big)_{i \in I}$, where $\{U_i\} \in \Psi_1$ and $u_i \in U_i$,
- $\mathcal{C}_k^\cup$ be the space of k-cycles with Π coefficient in the unions of U_i for all packings $\{U_i\} \in \Psi_1$ of X,

$$\mathcal{C}_k^\cup = \bigcup_{\{U_i\} \in \Psi_1} \mathcal{C}_k \big(\bigcup_{i \in I} U_i; \Pi\big),$$

- $\mathcal{C}_{Nk}^\times$ be the space of $N \cdot k$-cycles with the N-th tensorial power coefficients in the Cartesian products of U_i for all $\{U_i\} \in \Psi_1$,

$$\mathcal{C}_{Nk}^\times = \bigcup_{\{U_i\} \in \Psi_1} \mathcal{C}_{Nk}^\cup \left(\underset{i \in I}{\times} U_i; \Pi^{\otimes N} \right).$$

The four spaces $\mathcal{U}^\cup, \mathcal{U}^\times, \mathcal{C}_k^\cup$ and $\mathcal{C}_{NK}^\times$ tautologically "fiber" over Ψ_1 by the maps denoted $\varpi^\cup$, $\varpi^\times$, $\varpi_k^\cup$, and $\varpi_{Nk}^\times$; besides, the Cartesian products of cycles $C_i \subset U_i$ defines an embedding

$$\mathcal{C}_k^\cup \hookrightarrow \mathcal{C}_{Nk}^\times.$$

Observe that

- the fibers $\times_i U_i$ of the map $\varpi^\times : \mathcal{U}^\times \to \Psi_1$ for $\big(\{U_i\}, \{u_i\}\big) \overset{\varpi^\times}{\mapsto} \{U_i\}$ are what we call "variable spaces" in section 8 where $\mathcal{U}^\times$ is playing the role of $\mathcal{X}$ and Ψ_1 that of Q from section 8,

- the above four spaces are naturally/tautologically acted upon, along with Ψ_1, by the symmetric group Sym_N of permutations/automorphisms of the index set I and the above four maps from these spaces to Ψ_1 are Sym_N-equivariant,

- the G-factored map $\varpi^\times$, denoted

$$\varpi_{/G}^\times : \mathcal{U}^\times / G \to \Psi_1 / G,$$

has the same fibers as $\varpi^\times$, namely, the products $\times_i U_i$. However, if $card(G) > 1$ and $dim(X) = n > 1$, then $\varpi_{/G}^\times$ is a *nontrivial* fibration, even if all U_i

equal translates of a ball $U = B^n(r)$ in $X = \mathbb{R}^n$, where the $\mathcal{U}^\times = \Psi_1 \times U^I$ and where $\varpi^\times : \Psi_1 \times U^I \to \Psi_1$ equals the coordinate projection.

The "⊛-pairing" described in the previous section via the intersection of cycles in X with $U_i \subset X$ for $\{U_i\}_{i \in I} \in \Psi_1$ followed by taking the Cartesian product of these intersections defines a pairing

$$\Psi_1 \times \Psi_2 \overset{\circledast^\times}{\to} \Theta^\times = \mathcal{C}_{Nk}^\times = \bigcup_{\{U_i\} \in \Psi_1} \mathcal{C}_{Nk}\left(\underset{i \in I}{\times} U_i; \Pi^{\otimes N} \right),$$

where, recall, Ψ_1 is the space of packings of X by $U_i \subset X$, $i \in I$, and $\Psi_2 = \mathcal{C}_k(X; \Pi)$ is the space of cycles in X.

And since this map $\circledast^\times$ is equivariant for the action of Sym_N on $\Theta^\times$ and on the first factor in $\Psi_1 \times \Psi_2$, it descends to

$$\circledast^\times_{/G} : \Psi_1/G \times \Psi_2 \to \Theta^\times/G$$

for all subgroups $G \subset Sym_N$.

Detection of Nontrivial Families of Cycles and of Packings

A family $S_2 \subset \Psi_2$ of cycles in X is called *homologically G-detectable* by (a family of) a packing of X, if there exists a family S_1 of cycles in $\Psi_1/G \times \Psi_2$ such that the corresponding family of product $N \cdot k$-cycles in "variable spaces" $\times_i U_i$, that is, a map from $S_1 \times S_2$ to $\Theta^\times/G$ by $\circledast^\times_{/G}$, is *homologically substantial*.

Recall (see section 8) that this substantiality is non-ambiguously defined if $\circledast^\times_{/G}$ is a fibration with *contractible fibers*, which is the case in our examples where $U_i \subset X$ are topological n-balls and observe that *unavoidably* variable nature of $\times_i U_i$ is due to nontriviality of the fibration $\varpi^\times_{/G} : \mathcal{U}^\times/G \to \Psi_1/G$ for permutation groups $G \neq \{id\}$ acting on packings that is most essential for what we do.

In his "*Minimax-Steenrod*" paper Guth shows that all homology classes h_2 of the space $\Psi_2 = \mathcal{C}_k(B^n; \mathbb{Z}_2)$ of relative k-cycles in the n-ball are G-detectable by families $S_1 = S_1(h_2)$ *of packings* of B^n by sufficiently small δ-balls $U_i = B^n_{x_i}(\delta) \subset B^n = B^n(1)$, for some 2-subgroup $G = G(h_2) \subset Sym_N$.

This is established with the help of the "Vanishing Lemma" stated in section 14; where, if I understand it correctly, the detective power of such a family $S_1 \subset \Psi_1/G$ (of "moving packings" of X by N balls) is due to *non-vanishing* of some cohomology class in Ψ_1/G that comes from the classifying space $\mathcal{B}_{cla}(G)$ via the classifying map $\Psi_1/G \to \mathcal{B}_{cla}(G)$. (Probably, a proper incorporation of the cohomology coming from X would imply similar "detectability" of *all* manifolds X, where it is quite obvious for *parallelizable X*.)

Apparently Guth's "Vanishing Lemma" shows that every $\mathbb{Z}_2$-homology class h_* in $K(\mathbb{Z}_2, m)$ equals the image of a homology class from the classifying space $\mathcal{B}_{cla}(G)$ of some finite 2-groups $G = G(h_*)$ under a map $\mathcal{B}_{cla}(G) \to K(\mathbb{Z}_2, m)$. Equivalently, this means that given *a nontrivial* $\mathbb{Z}_2$-cohomology operation *op* from degree m to n, (that is, necessarily, a polynomial combination of Steenrod squares) there exists a class $h \in H^m(\mathcal{B}_{cla}(G); \mathbb{Z}_2)$ for some finite 2-group G, such that $op(h) \neq 0$. Possibly, this is also true for other primes p (which, I guess, must be known to people working on cohomology of p-groups).

On the other hand it seems unlikely that the (co)homologies of spaces of I-packings for all finite sets I and/or of (the classifying spaces of) all p-groups are fully detectable by the (co)homologies of the space $\mathcal{C}_k(B^n; \mathbb{Z}_p)$ of cycles in the n-ball, where, recall, the space $\mathcal{C}_k(B^n; \mathbb{Z}_p)$ is homotopy equivalent to Eilenberg-MacLane's $K(\mathbb{Z}_p, n-k)$.

Pairing Inequalities Between k-volumes and Packing Radii

Let $S_1 \subset \Psi_1$ be a G-invariant family of I-packings of a Riemannian manifold X by open subsets $U_i = U_{i,s_1}$, $i \in I$, $s_1 \in S$, where G is a subgroup of the group $Sym_N = aut(I)$, and let S_2 be a family of k-cycles (or more general "virtually k-dimensional entities") $Y = Y_{s_2}$ in X.

Let the coupled family that is a map from $S_1/G \times S_2$ to $\Theta^\times/G$ by $\circledast^\times_{/G}$ be *homologically substantial*. Then, by the definition of "the waist of a variable space" (see section 8) the supremum of the volumes of Y_{s_2} is bounded from below by

$$\sup_{s_2 \in S_2} vol_k(Y_{s_2}) \geq \sum_{i \in I} waist_k(U_{i,s_1}).$$

In particular, since small balls of radii r in X have k-$waists \sim r^k$ the above implies

$$\frac{\inf_{s_1 \in S_1} \sum_{i \in I} inrad_k(U_{i,s_1})^k}{\sup_{s_2 \in S_2} vol_k(Y_{s_2})} \leq const(X),$$

where $inrad(U)$, $U \subset X$ denote the radius of the largest ball contained in U.

This inequality, in the case where $U_i \subset B^n = B^n(1) \subset \mathbb{R}^n$ are Euclidean r-balls, is used by Guth in [33] (as it was mentioned earlier) for obtaining a (nearly sharp) *lower bound* on the k-volume spectrum of the n-ball that is on $\sup_{s_2 \in S_2} vol_k(Y_{s_2})$ for families S_2 of k-cycles Y in $X = B^n$ such that a given cohomology class $h \in H^*(\mathcal{C}_k(X; \mathbb{Z}_2); \mathbb{Z}_2)$ does not vanish on this $S_2 \subset \mathcal{C}_k(X; \mathbb{Z}_2)$. This is achieved by constructing a G-invariant family $S_1 = S_1(h)$ of I-packings of X by r-balls for some 2-group $G \subset Sym_N$, $N = card(I)$, such that the coupled family

$$S_1/G \times S_2 \overset{\circledast^\times_{/G}}{\to} \Theta^\times/G$$

is homologically substantial and such that the radii r of the balls $U_i = B_{x_i}(r) \subset X$ are sufficiently large.

Conversely, and this is what we emphasize in this paper, one can *bound from above* the radii r of the balls in a G-invariant family S_1 of I-packings of X by r-balls in terms of a cohomology class h' in the space Ψ_1/G (that is, of packings/G) such that h' does not vanish on S_1, where one can use for this purpose suitable families $S_2 = S_2(h')$ of k-cycles in X with possibly small volumes, e.g., those used by Guth in [33] for his *upper bound* on the vol_k-spectra.

Thus, in the spirit of Guth's paper, one gets such bounds for $N = card(I) \to \infty$ and some $h' \in H^*(\Psi_1/G; \mathbb{Z}_2)$ for certain 2-groups $G \subset Sym_N$, where $ord(G) \to \infty$ and $deg(h') \to \infty$ along with $N \to \infty$. And even though the packing spaces Ψ_1 look quite innocuous being the complements to the $2r$-neighborhoods to the unions of the diagonals in X^I, there is no apparent alternative method to obtain such bounds.[46]

Multidimensional rendition of the pairing inequalities. Let Ψ be a space of pairs $\psi = (\{U_i\}_{i \in I}, Y)$ where $U_i \subset X$ are disjoint open subsets and $Y \subset X$ is a k-cycle, say with $\mathbb{Z}$ or $\mathbb{Z}_2$ coefficients.

Let $inv(U)$ be some geometric invariant of open subsets $U \subset X$ that is monotone under inclusions between subsets, e.g., the n-volume, inradius, some kind of waist of U, etc.

Define $\mathcal{E} = (E_0, E_1, \dots, E_N) : \Psi \to \mathbb{R}^{N+1}$ by

$$E_0(\{U_i\}_{i \in I}, Y) = vol_k(Y) \text{ and } E_i(\{U_i\}_{i \in I}, Y) = inv(U_i)^{-1},$$

assume that Ψ is invariant under the action of some subgroup $G \subset Sym_N$ that permutes U_i, and observe that all of the above can be seen in the kernels of the cohomology homomorphisms from $H^*(\Psi/G)$ to $H^*(\Psi_{e_0,e_1,\dots,e_N}/G)$ for the subsets $\Psi_{e_0,e_1,\dots,e_N} \subset \Psi$ defined by the inequalities $E_i(\psi) < e_i$, $\psi = (\{U_i\}, Y)$, $i = 0, \dots, N$, as in the definition of the multidimensional (co)homology spectra in sections 4, 10, and 11.

For example, besides the pairing inequalities for packings by balls, this also allows an encoding of similar inequalities for *convex partitions* from [25].

17   SUP$_\vartheta$-SPECTRA, SYMPLECTIC WAISTS AND SPACES OF SYMPLECTIC PACKINGS

Let Θ be a set of metrics ϑ on a topological space X and define *sup$_\vartheta$-invariants* of (X, Θ) as *the suprema* of the corresponding invariants (X, ϑ) over all $\vartheta \in \Theta$. (In many cases, this definition makes sense for more general classes Θ of metrics spaces that do not have to be homeomorphic to a fixed X.)

Problem. Find general criteria for finiteness of *sup$_\vartheta$*-invariants.

Two classical examples: Systoles and Laplacians.   (1) Let Θ be the space of Riemannian metrics on the 2-torus X with

$$sup_{\theta \in \Theta} area_\theta(X) \leq 1.$$

Then the *sup$_\vartheta$-systole$_1$ of (X, ϑ) is* $< \infty$.
     In fact,

$$sup_\vartheta\text{-}systole_1(X, \Theta) = \sqrt{\frac{2}{\sqrt{3}}}$$

by *Loewner's torus inequality* of 1949.

This means that all toric surfaces (X, ϑ) of *unit areas* admit closed *non-contractible* curves of lengths $\leq \sqrt{\frac{2}{\sqrt{3}}}$, where, observe, the equality $systole_1 = \sqrt{\frac{2}{\sqrt{3}}}$

---

[46]A natural candidate for such method would be the Morse theory for the distance function to the union of the diagonals in X^I, see [6], but this does not seem to yield such bounds.

holds for $\mathbb{R}^2$ divided by the *hexagonal lattice*. (See Wikipedia article on systolic geometry and references therein for further information.)

(2) Let Θ be the space of Riemannian metrics ϑ on the 2-sphere X with $area_\vartheta(X) \geq 4\pi$ (that is, the area of the unit sphere). Then *the first $\sup_\vartheta$-eigenvalue of the Laplace operator on (X, Θ) is $< \infty$*.

In fact $sup_\vartheta\text{-}\lambda_1(X, \Theta) = 2$, that is, the first eigenvalue of the Laplace operator on the unit sphere, by the *Hersch inequality* of 1970.

Symplectic area spectra and waists.   Let X be a smooth manifold of even dimension $n = 2m$ and let $\omega = \omega(x)$ be a differential 2-form on X.

A *Riemannian metric* ϑ on X, is called *adapted to* or *compatible with* ω, if

$$\bullet_{\geq\omega} \qquad\qquad area_\vartheta(Y) \geq \left| \int_Y \omega \right|$$

for all smooth oriented surfaces $Y \subset X$; and the n-volume $d_\vartheta x$ element satisfies

$$\bullet_{\leq\omega^m} \qquad\qquad d_\vartheta(x) \leq |\omega^m| \text{ at all ponts } x \in X,$$

that is, $vol_\vartheta(U) \leq \int_U \omega^m$ for all open subsets $U \subset X$.[47]

Question. Which part of the (suitably factorized/coarsened) homotopy/homology area spectra of (X, ϑ) remains finite after taking suprema over $\vartheta \in \Theta(\omega)$?

Partial answer provided by the symplectic geometry.   The form ω is called *symplectic* if it is *closed*, i.e., $d\omega = 0$, and $\omega^m = \omega^m(x)$ does not vanish on X. In this case $X = (X, \omega)$ is called *a symplectic manifold*.

The symplectic k-waist of X may be defined as the *supremum of the k-waists*[48] of the Riemannian manifolds (X, ϑ) for all metrics ϑ *compatible* with ω.

It is easy to see that the space $\Theta = \Theta(\omega)$ of metric ϑ compatible with ω is *contractible,* and that ω-compatible metrics are *extendable* from open subsets $X_0 \subset X$ to all of X with the usual precautions at the boundaries of X_0. It follows that the symplectic waists are monotone under equividimensional symplectic embeddings:

$$sympl\text{-}waist_k(X_0) \leq sympl\text{-}waist_k(X)$$

for all open subsets $X_0 \subset X$.

If $k = 2$, then upper bounds on symplectic waists are obtained by proving homological stability of certain families of ϑ-*pseudoholomorphic curves* in X under deformation of compatible metrics ϑ, where "pseudoholomorphic curves" are oriented surfaces $Y \subset X = (X, \omega, \vartheta)$, such that $area_\vartheta(Y) = \int_Y \omega$.

This, along with the symplectomorphism of the ball $B = B^{2m}(1) \subset \mathbb{C}^m$ onto the complement $\mathbb{C}P^m \setminus \mathbb{C}P^{m-1}$, where the symplectic form in $\mathbb{C}P^m$ is normalized in

---

[47]The *inequalities* $area_\vartheta(Y) \geq | \int_Y \omega |$ and $d_\vartheta(x) \leq |\omega^m|$ imply the *equality* $d_\vartheta x = |\omega^m|$ and if $\omega^m(x) \neq 0$, then there is a $\mathbb{R}$-linear isomorphism of the tangent space T_x to $\mathbb{C}^m$, such that $(\sqrt{-1}\omega, \vartheta)_x$ go to the imaginary and the real parts of the diagonal Hermitian form on $\mathbb{C}^m$. But $\bullet_{\leq\omega^m}$ is better adapted for generalizations than $\bullet_{=\omega^m}$.

[48]This may be any kind of a waist defined with families S of "virtually k-dimensional entities" of a particular kind and with a given type of homological substantiality required from S, where the most relevant for the symplectic geometry are $\mathbb{Z}$-waists $waist_k(X; \mathbb{Z})$ defined with families of $\mathbb{Z}$-cycles that are assumed as regular as one wishes.

order to have the area of the projective line $\mathbb{C}P^1 \subset \mathbb{C}P^m$ equal $area(B^2(1)) = \pi$, implies that

$$symp l\text{-}waist_2(\mathbb{C}P^m; \mathbb{Z}) = symp l\text{-}waist_2(B; \mathbb{Z}) = waist_2(B; \mathbb{Z}) = \pi,$$

where $waist_2(\ldots; \mathbb{Z})$ stands for the $\mathbb{Z}$-waist that is defined with homologically substantial families of $\mathbb{Z}$-cycles.

On the other hand, it is probable that $symp l\text{-}waist_k(X; \mathbb{Z}) = \infty$ for all X, unless $k = 0, 2, n = dim(X)$ and this is also what one regretfully expects to happen to *the symplectic k-systoles* $syst_k(X, \omega) =_{def} \sup_\vartheta syst_k(X, \vartheta)$, for $k \neq 0, 2, n$.

For instance, the complex projective space $\mathbb{C}P^m$ may carry Riemannian metrics ϑ_s for all $s > 0$ compatible with the standard symplectic form on $\mathbb{C}P^m$ such that the k-systoles, i.e., the minimal k-volumes of all non-homologous zero k-cycles in $(\mathbb{C}P^m, \vartheta_s)$ of all dimensions k except for $0, 2, 2m$ are $> s$.

Yet, some geometric (topological?) invariants of the functions $\vartheta \mapsto syst_k(X, \vartheta)$ and $\vartheta \mapsto waist_k(X, \vartheta)$ on the space of metrics ϑ compatible with ω may shed some light on the symplectic geometry of (X, ω), where possible invariants of such a function $F(\vartheta)$ may be the asymptotic rate of some kind of "minimal complexity" of the Riemannian manifolds (X, ϑ) (e.g., some integral curvature or something like the minimal number of contractible metric balls needed to cover (X, ϑ) for which $F(\vartheta) \geq s, s \to \infty$.

Let us generalize the above in the spirit of "multidimensional spectra" by introducing the space $\Psi = \Psi(X, I)$ of triples $\psi = (\{U_i\}_{i \in I}, Y, \vartheta)$, where U_i are disjoint open subsets, $Y \subset X$ is an integer 2-cycle and ϑ is a Riemannian metric compatible with ω.

Let $\mathcal{E} : \Psi \to \mathbb{R}^{N+1}$ be the map defined by

$$\mathcal{E}(\{U_i\}_{i \in I}, Y, \vartheta) = (symp l\text{-}waist_2(U_i), area_\vartheta(Y))_{i=1,2,\ldots,N},$$

where, as an alternative to $symp l\text{-}waist_2(U)$, one may use $inrad_\omega(U)$, $U \subset (X, \omega)$, that is, the supremum of the radii r of the balls $B^{2m}(r) \subset \mathbb{C}^m$ that admit symplectic embedding into U. Let $f : \mathbb{R}^{N+1} \to \mathbb{R}$ be a positive function that is symmetric and monotone decreasing in the first N-variables and monotone increasing in the remaining variable (corresponding to $area(Y)$), where the simplest instance of this is $-\sum_{i=1,\ldots,N} z_i + z_{N+1}$.

Let $G \subset Sym_N$ be a permutation group that, observe, naturally acts on Ψ, let S be a topological space with a free action of G and let $[\varphi/G]$ be a homotopy class of maps $\varphi : S/G \to \Psi/G$.

Let $E_f(\psi) = f(\mathcal{E}(\psi))$ and define *the sup ϑ-homotopy spectrum* of E_f (compare section 4)

$$E_f[\varphi/G]_{\sup_\vartheta} = \sup_\vartheta \inf_{\varphi \in [\varphi]} \sup_{s \in S} E_f(\varphi(s)).$$

Playing $\inf_\phi$ against $\sup_\vartheta$. Our major concern here is the possibility of E_f $[\varphi/G]_{\sup_\vartheta} = \infty$ that can be outweighed by enlarging the homotopy class $[\phi]$ to a homology class or to a set of such classes. With this in mind, given a cohomology class h in $H^*(\Psi/G, \Pi)$ with some coefficients Π, one defines

$$E_f[h]_{\sup_\vartheta} = \sup_\vartheta \inf_{\varphi^*(h) \neq 0} \sup_{s \in S} E_f(\varphi(s))$$

where the infimum is taken over all G-spaces S and all maps $\varphi: S/G \to \Psi/G$ such that $\varphi^*(h) \neq 0$.

Another possible measure against $E_f[\varphi/G]_{\sup_\vartheta} = \infty$ is taking $\mathbb{Z}_2$-cycles Y instead of $\mathbb{Z}$-cycles, but this is less likely to tip the balance in our favor.

On the other hand, one may enlarge/refine the outcome of minimization over φ, yet still keeping the final result finite by restricting the topology of Y, e.g., by allowing only Y represented by surfaces of genera bounded by a given number. Also one may incorporate the integral $\int_Y \omega$ into E_f.

But usefulness of all these variations is limited by the means we have at our disposal for proving finiteness of the $\sup_\vartheta$-spectra that are limited to the homological (sometimes homotopical) stability of families of pseudoholomorphic curves.

Packing inequalities. If $X = (X, \omega)$ admits a nontrivial stable family of pseudoholomorphic curves $Y \subset X$, then there are nontrivial constraints on the topology of the space of packings of X by U_i with $inrad_\omega(U_i) \geq r$.

Namely, there are connected G-invariant subsets $S_\circ$ in the space of I-tuples of disjoint topological balls in X for all finite sets $I \ni i$ of sufficiently large cardinalities N and some groups $G \subset Sym_N = aut(I)$, such that *every family S of I-tuples of disjoint subsets $U_{i,s} \subset X$, $s \in S$, that is G-equivariantly homotopic, or just homologous, to $S_\circ$ satisfies*

$$\inf_{s \in S} \sum_{i \in I} sympl\text{-}waist_2(U_{i,s}) \leq \int_Y \omega.$$

Consequently,

$$\inf_{s \in S} \sum_{i \in I} \leq \pi \cdot inrad_\omega(U_{i,s}) \leq \int_Y \omega.$$

Effective and rather precise inequalities of this kind estimates are possible for particular manifolds, say for the projective $\mathbb{C}P^m$ and domains in it where pseudoholomorphic curves are abundant.

But if X is a closed symplectic manifold $X = (X, \omega)$ *with no pseudoholomorphic curves in it*, e.g., the $2m$-torus, one does not know whether there are nontrivial constraints on the homotopy types of symplectic packing spaces. Also it is unclear if such an X must have $syst_2(X, \omega) = \infty$ and/or $sympl\text{-}waist_2(X) = \infty$. (See [4, 10, 11, 13, 14, 17, 18, 35, 38, 45, 46, 49, 50] for what is known concerning *individual* symplectic packings of X and *spaces* of embeddings of *a single* ball into X.)

We conclude this section by observing that spaces of certain symplectic packings can be described entirely in terms of the set $\Theta = \Theta(\omega)$ of ω-compatible metrics ϑ on X with the following definition applicable to general classes Θ of metric spaces.

Θ-packings by balls. A packing of (X, Θ) by I-tuples of r-balls $B^n(r) \subset \mathbb{R}^n$ is a pair $(\vartheta, \{f_i\})$ where $\vartheta \in \Theta$ and an $\{f_i\}$, $i \in I$, is an I-tuple of *expanding maps* $f_i: B^n(r) \to (X, \vartheta)$ with disjoint images.

Problem. Find further (non-symplectic) classes Θ with "interesting" properties of the corresponding spaces of Θ-Packings.

For example, a Riemannian metric ϑ may be regarded as compatible with a *pseudo-Riemannian* (i.e., non-degenerate indefinite) h on a compact manifold X if the h-lengths of all curves are bounded in absolute values by their ϑ-length and if the ϑ-volume of X equals the h-volume.

Are there instances of (X, h) where some Θ-packings and/or sup_ϑ-invariants carry nontrivial information about h?

Are there such examples of other G-*structures* on certain X for groups $G \subset GL(n)$, $n = dim(X)$, besides the symplectic and the orthogonal ones, where metrics ϑ serve for reductions of groups G to their maximal compact subgroups?

18   PACKING MANIFOLDS BY K-CYCLES AND K-VOLUME SPECTRA OF SPACES OF PACKINGS

Define *an I-packing of the space* $\mathcal{C}_*(X; \Pi) = \bigoplus_{k=0,1,\dots} \mathcal{C}_k(X; \Pi)$ *of cycles with Π-coefficients* in a Riemannian manifold X as *an I-tuple* $\{V_i\}$ *of cycles* in X with a given *lower bound* on *some distances*, denoted "*dist*," between these cycles, say

$$\text{``}dist\text{''}(V_i, V_j) \geq d,$$

or, more generally,

$$\text{``}dist\text{''}(V_i, V_j) \geq d_{ij}, \ i, i \in I.$$

Leading example: $dist_X$-packings of the space $\mathcal{C}_*(X; \Pi)$. A significant instance of "distance between cycles in X" is

$$dist_X(V, V') =_{def} \inf_{v \in V, v' \in V'} dist_X(v, v'),$$

where we use here the same notation for cycles V and their supports in X, both denoted $V_i \subset X$.

Question 1. What are the homotopy/homology properties of the spaces

$$\Psi_{d_{ij}} = \mathcal{C}_*(X; \Pi)^I_{<d_{ij}} \subset \mathcal{C}_*(X; \Pi)^I$$

of k-tuples of cycles $V_i \subset X$ that satisfy the inequalities $dist_X(V_i, V_j) \geq d_{ij}$?

(Here, as at the other similar packing occasions, one should think in Sym_N-equivariant terms that make sense since the *totality* of the spaces $\mathcal{C}_*(X; \Pi)^I_{<d_{ij}}$ for all $d_{ij} > 0$ is Sym_N-invariant.)

Volume spectra of $X^I_{>d_{ij}}$. One can approach this question from an opposite angle by looking at the k_N-volume spectra of (the spaces of cycles in the) Riemannian manifolds $X^I_{>d_{ij}}$ for various $k_N \leq N \cdot dim(X)$, where $X^I_{>d_{ij}} \subset X^I$ are defined by the inequalities $dist_X(x_i, x_j) > d_{ij}$, since mutual distances between cycles V_i in X can be seen in terms of locations of their Cartesian products in X^I, as follows:

$$dist_X(V_i, V_j) \geq d_{ij} \Leftrightarrow V^\times = \underset{i \in I}{\times} V_i \subset X^I_{>d_{ij}}.$$

Recall at this point that the spaces $X^I_{>d_{ij}=d}$ represent packings of X by balls of radii $d/2$ and observe that $V^\times \in \mathcal{C}_*(X^I; \Pi^{N\otimes})$ are rather special, namely split, cycles in X^I. Now the above Question 1 comes with the following companion.

Question 2. What are the volume spectra of the spaces $X^I_{>d_{ij}}$ and how do they depend on d_{ij}?

Recall the following relation between volume of cycles W and W' in the Euclidean space and distances between them.

Gehring's linking volume inequality. Let $W \subset \mathbb{R}^n$ be a k-dimensional sub(pseudo)manifold of dimension k.

Suppose, W *is non-homologous to zero in its open d-neighborhood* $U_d(W) \subset \mathbb{R}^n$, or, equivalently, there exists an $(n-k-1)$-dimensional subpseudomanifold $W' \subset \mathbb{R}^n$ that has *a nonzero linking number with W and such that* $dist_{\mathbb{R}^n}(W,W') \geq d$. ("Linking" is understood mod 2 if W is non-orientable.)

Then, according to the *Federer-Fleming inequality* (see next section),

$$vol_k(W) \geq \varepsilon_n d^k, \ \ \varepsilon_n > 0,$$

where, moreover,

$$\varepsilon_n = \varepsilon_k = vol_k(S^k)$$

($S^k = S^k(1) \subset \mathbb{R}^{k+1}$ denotes the unit sphere) by the Bombieri-Simon solution of Gehring's linking problem (see next section).

Proof. Map the Cartesian product $W \times W'$ to the sphere $S^{n-1}(d) \subset \mathbb{R}^n$ of radius d by

$$f : (w, w') \mapsto \frac{d(w - w')}{dist_{\mathbb{R}^m}(w, w')}$$

and observe that

- the family of the f-images of the "slices" $W \times w' \subset W \times W'$, $w' \in W'$, in $S^{n-1}(d)$ is *homologically substantial* (in the sense of section 6) since the degree of the map $f : W \times W' \to S^{n-1}(d)$ equals the linking number between W and W';
- the map f is *distance decreasing, hence, k-volume decreasing* on the "slices" $W \times w'$ for all $w' \in W'$, since $dist(w, w') \geq d$ for all $w \in W$ and $w' \in W'$.

Therefore, by the definition of *waist* (see section 6)

$$vol_k(W) \geq waist_k(S^{n-1}(d))$$

where $waist_k(S^{n-1}(d)) = vol_k(S^k(d)) = d^k vol_k(S^k(1))$ by the sharp spherical waist inequality (see sections 6 and 19). QED.

Linking waist inequality. The above argument also shows that whenever a k-cycle $W \subset \mathbb{R}^n$ is *nontrivially homologically linked* with some $W' \subset \mathbb{R}^N$, and $dist$ $(W, W') \geq d$, then

$$waist_l(W) \geq waist_l(S^n(d)) = d^l vol_l(S^l) \text{ for all } l \leq k.$$

This provides nontrivial constraints on the spaces of packings of $\mathbb{R}^n$ by l-cycles, since $(k-l)$-cycles in the space of l-cycles $Y \subset \mathbb{R}^n$ make l-cycles $W \subset \mathbb{R}^n$.

Another generalization of Gehring inequality concerns several cycles, say Y_i linked to some k-cycle W. In this case the intersections of Y_i with any chain implemented by a subvariety $V = V^{k+1} \subset \mathbb{R}^n$ that fills-in W, i.e., has W as its boundary, $\partial V = W$, makes packings of this V by 0-cycles.

This, applied to the *minimal* V filling-in (spanning) W, suggests a (sharp?) lower bound on distances between such Y_i in terms of what happens to the ordinary packings of the ball $B^{k+1}(r)$ which has $vol_k(\partial B^k) = vol_k(W)$.

Question 3. Let Y_i, $i \in I$, be m-cycles in $\mathbb{R}^{2m+1}$, such that

$$vol_m(Y_i) \leq c \text{ and } dist(Y_i, Y_j) \geq d.$$

What are, roughly, possibilities for the linking matrices $L_{ij} = \#_{link}(Y_i, Y_j)$ of such Y_i depending on c and d?

What are the homotopies/homologies of spaces of such I-tuples of cycles depending on L_{ij}?

19   APPENDIX: VOLUMES, FILLINGS, LINKINGS, SYSTOLES AND WAISTS

Let us formulate certain mutually interrelated fillings, linkings and waists inequalities extending those presented above and earlier in section 6.

1. *Federer-Fleming filling-by-mapping (isoperimetric) inequality.* ([20]) Let $Y \subset \mathbb{R}^n$ be a closed subset with finite k-dimensional Hausdorff measure for an integer $k \leq n$. Then there exists a continuous map $f : Y \to \mathbb{R}^n$ with the following properties.

• $_{k-1}$ The image $f(Y) \subset \mathbb{R}^n$ is *at most* $(k-1)$-*dimensional.* Moreover, $f(Y)$ is contained in *a piecewise linear subset* $\Sigma^{k-1} = \Sigma^{k-1}(X) \subset \mathbb{R}^n$ *of dimension* $k-1$.

• $_{disp}$ *The displacement of* Y *by* f *is bounded in terms of the Hausdorff measure of* Y by

$$\sup_{y \in Y} dist_{\mathbb{R}^n}(f(y), y) \leq const_n Haumes_k(Y)^{\frac{1}{k}}.$$

• $_{vol}$ The $(k+1)$-dimensional measure of *the cylinder* $C_f \subset \mathbb{R}^n$ *of the map* f that is the union of the straight segments $[y, f(y)] \subset \mathbb{R}^n, y \in Y$, satisfies

$$Haumes_{k+1}(C_f) \leq const'_n Haumes_k(Y)^{\frac{k+1}{k}}.$$

(A possible choice of Σ^{k-1} is the $(k-1)$-skeleton of a standard decomposition of the Euclidean n-space $\mathbb{R}^n$ into R-cubes for $R = C_n Haumes_k(Y)^{\frac{1}{k}}$ and a sufficiently large constant C_n, where the map $f : Y \to \Sigma^{k-1}$ is obtained by consecutive radial projections from Y intersected with the m-cubes $\square^m$ to the boundaries $\partial \square^m$ from certain points in $\square^m$ starting from $m = n$ and up-to $m = k$.)

It remains unknown if this holds with $const_k$ instead of $const_n$ but the following inequality with $const_k$ is available.

2. *Contraction inequality.* Let the $Y \subset \mathbb{R}^n$ be a k-dimensional polyhedral subset. Then there exists a continuous map $f : Y \to \mathbb{R}^n$ with the following properties.

$\bullet_{k-1}$ The image $f(Y) \subset \mathbb{R}^n$ is contained in *a piecewise linear subset of dimension* $k - 1$.

$\bullet_{dist}$ The image of f lies within a *controlled distance* from Y. Namely,

$$\sup_{y \in Y} dist_{\mathbb{R}^n}(f(y), Y) \leq const_k vol_k(Y)^{\frac{1}{k}}.$$

$\bullet_{hmt}$ There exists *a homotopy between the identity map and f*, say $F : Y \times [0, 1] \to \mathbb{R}^n$ with $F_0 = id$ and $F_1 = f$, such that the image of F satisfies

$$vol_{k+1}(F(Y \times [0, 1])) \leq const'_k vol_k(Y)^{\frac{k+1}{k}}.$$

This is proven in appendix 2 in [27] for more general spaces X in the place of $\mathbb{R}^n$, including all Banach spaces X.

3. *Almgren's sharp filling (isoperimetric) inequality.* Almgren proved in 1986 [3] the following sharpening of the (non-mapping aspect of) Federer-Fleming inequality. *The volume minimizing $(k + 1)$-chains Z in Euclidean spaces satisfies:*

$$\frac{vol_{k+1}(Z)}{vol_k(\partial Z)^{\frac{k+1}{k}}} \leq \frac{vol_{k+1}(B^{k+1}(1))}{vol_k(S^k(1))^{\frac{k+1}{k}}} = (k+1)^{-\frac{k+1}{k}} vol_{k+1}(B^{k+1}(1))^{-\frac{1}{k}},$$

where $B^{k+1}(1) \subset \mathbb{R}^{k+1}$ is the unit Euclidean ball and $S^k(1) = B^{k+1}(1)$ is the unit sphere.

In fact, *Almgren's local-to-global variational principle* ([3], [24]) reduces filling bounds in Riemannian manifolds X to lower bounds of the suprema of mean curvatures of subvarieties $Y \in X$ in terms of k-volumes of Y, where such a sharp(!) bound for $Y \subset \mathbb{R}^n$ is obtained by Almgren by reducing it to that for *the Gaussian curvature* of the boundary of the convex hull of Y.

4. *Divergence inequality.* Recall that the k-*divergence of a vector field* $\delta = \delta_x$ on a Riemannian manifold X is the function on the tangent k-planes in X that equals *the δ-derivative of the k-volumes of these k-planes.* Thus, *the δ-derivative of the k-volume of each k-dimensional submanifold $V \subset X$ moving by the δ-flow equals* $\int_V div_k(\delta)(\tau_v)dv$, for τ_v denoting the tangent k-plane to $V \subset X$ at v.

For instance, the k-divergence of *the standard Euclidean radial field focused at* zero $\delta_x = \overrightarrow{x} = grad(\frac{1}{2}||x||^2)$ on $\mathbb{R}^n$, where $\overrightarrow{x}$ denotes this very x seen as the tangent vector parallelly transported from $0 \in \mathbb{R}^n$ to x, equals the norm $||x||$. (The δ-flow here equals the homothety $x \mapsto e^t x$, $t \in \mathbb{R}$.)

Let Z be a compact *minimal/stationary* $(k + 1)$-dimensional subvariety with boundary $Y = \partial Z$ and let $\nu = \nu_y$ be the unit vector field tangent to Z, normal to Y and facing outside Z.

Then

$\star=$

$$\int_Z div_k(\delta)(\tau_z)dz = \int_Y \langle \delta_y, \nu_y \rangle.$$

If Z and Y are non-singular, this equality, that follows from the definition of "stationary" and the Gauss-Stokes formula, goes back to the 19th century, while the singular case is, probably, due to Federer-Fleming (Reifenberger? Almgren?).

5. *Isoperimetric corollary.* If $div_k(\delta) > 0$, then every subset $Z_0 \subset Z$ satisfies:

$$\star_{\partial \geq} \qquad vol_k(Y) \geq vol_{k+1}(Z_0) \cdot \frac{\inf_{z \in Z_0} div_{k+1}\delta}{\sup_{y \in Y} ||\delta_y||}.$$

This inequality may be applied to a *radial field* δ focused in $X \supset Z$ at some (say non-singular) point $z_0 \in Z$ (such a field is tangent to the geodesics issuing from z_0) in conjunction with *the coarea inequality* for the intersections Y_r of Z with spheres in X around z_0 of radii r,

$$\int_0^d vol_k(Y_r)dr \geq vol_{k+1}(Z_d),$$

$Z_r \subset Z$ denoting the intersection of Z with the d-ball in X around z_0.

For instance, if $X = \mathbb{R}^n$ and δ is the standard radial field focused at some point $z_0 \in Z$, then one arrives this way at the classical (known since 1960s, 50s, 40s?) monotonicity inequality.

6. *Monotonicity (isoperimetric) inequality.*

$$\frac{vol_{k+1}(Z_r)}{vol_k(Y_r)^{\frac{k+1}{k}}} \leq \frac{vol_{k+1}(B^{k+1}(1))}{vol_k(S^k(1))^{\frac{k+1}{k}}}.$$

Remark. If Z is volume minimizing, rather than being only "stationary," this follows from the above Almgren sharp filling inequality.

Corollary. If the boundary ∂Z lies d-far from z_0, then the volume of Z_d is bounded from below by that of the Euclidean ball $B^{k+1}(d) = B_0(\mathbb{R}^{k+1}, d) \subset \mathbb{R}^{k+1}$ of radius d:

$$vol_{k+1}(Z_d) \geq vol_{k+1}(B^{k+1}(d)),$$

provided the boundary ∂Z lies d-far from z_0.

7. *Application to linking.* The boundary $Y = \partial Z$ (as well as $Y_d = \partial Z_d$) satisfies

$$\bullet_{\partial \geq} \qquad vol_k(Y) \geq vol_k(S^k(d)),$$

where $S^k(d) = \partial B^{k+1}(d)$ is the Euclidean sphere of radius d and where we keep assuming that Y lies *outside the d-ball* in $\mathbb{R}^n$ around z_0.

Indeed, this follows by applying $\star_{\geq}$ to the radial field that is focused at z_0, that equals the standard one (i.e., $\overrightarrow{x - z_0}$) *inside the ball* $B^n_{z_0}(d) \subset \mathbb{R}^n$ and that has *norm (length) equal d everywhere outside* this ball.

Consequently, every (mildly regular) k-cycle $Y \subset \mathbb{R}^n$ *bounds a $(k+1)$-chain Z in its d-neighborhood for d equal the radius of k-sphere with volume equal $vol_k(Y)$.*

Namely, the solution Z of *the Plateau problem* with boundary Y does the job. (This is how Bombieri and Simon solve the Gehring linking problem, see [12].)

8. *Systolic inequality.* A simple adjustment of the above "monotonicity argument" yields the following short-cut in the proof of *the systolic inequality.*[49]

Let $K = K(\Pi, 1)$, $\Pi = \pi_1(K)$, be *an aspherical space* and let $\mathcal{X} = \mathcal{X}(h, R)$, where $h \in H_n(K)$, $R > 0$, be the class of all n-dimensional pseudomanifolds X with piecewise Riemannian metrics on them along with maps $f : X \to K$, such that

$\bullet_h$ the fundamental homology class $[X]_n \in H_n(X)$ of X (defined mod 2 if X is non-orientable) goes to a given nonzero homology class $h \in H_n(K)$;

$\bullet_R$ the restrictions of the map f to the R-balls in X, that are maps $B_x(R) \to K$, are contractible for all $x \in X$.

Then

$$vol(X) \geq \alpha_n R^n \text{ for } \alpha_n = \frac{(2\sigma_{n-1})^n}{n^n \sigma_n^{n-1}},$$

where $\sigma_n = \frac{2\pi^{\frac{n+1}{2}}}{\Gamma(\frac{n+1}{2})}$ is the volume of the unit sphere S^n; thus, $\alpha_n \sim \frac{(2\sqrt{e})^n}{n^n}$, that is, α_n equals $\frac{(2\sqrt{e})^n}{n^n}$ plus a subexponential term. (The expected value of the constant in the systolic inequality is $\sim \frac{c^n}{n^{\frac{n}{2}}}$.)

Proof. Assume that X is *volume minimizing* in $\mathcal{X}$, i.e., under conditions $\bullet_h$ and $\bullet_R$.[50] Then volume of each r-ball $B_x(r) \subset X$ with $r < R$ is bounded by the volume of *the spherical r-cone* $B_{or}(S)$ over the r-sphere $S = S_x(r) = \partial B_x(r) \in X$.[51]

In fact, if you cut $B_x(r)$ from X and attach $B_{or}(S)$ to $X \setminus B_x(r)$ by the boundary $\partial B_{or}(S) = S = \partial B_x(r)$, the resulting space X' will admit map $f' : X' \to K$ with the conditions $\bullet_h$ and $\bullet_R$ satisfied.

Therefore,

$$vol_{n-1}(S_x(r)) = \frac{d}{dr} vol(B_x(r)) \geq \beta_n vol(B_x(r))^{\frac{n-1}{n}} \text{ for } \beta_n = \frac{\sigma_{n-1}}{(\frac{1}{2}\sigma_n)^{\frac{n-1}{n}}},$$

which implies, by integration over $r \in [0, R]$, the bound $vol(B_x(R)) \geq \alpha_n R^n$ for $\alpha_n = (\beta_n/n)^n$.
QED.

Remark. Earlier, Guth [36] suggested a short proof of a somewhat improved systolic inequality, that says, in particular, that the R-ball $B_x(R) \subset X$ at some point $x \in X$ has $vol_n(B_x(R)) \geq (4n)^{-n} R^n$, provided the fundamental class $[X]_n \in H_n(X)$ equals the product of one dimensional classes.

His argument, based on minimal hypersurfaces and induction on dimension, generalizes to minimal hypersurfaces with boundaries as in [28] and yields a similar systolic inequality for spaces X with "sufficiently large" fundamental groups $\Pi = \pi_1(X)$.

---

[49] This was proven in [27] by a reduction to the contraction (filling) inequality generalized to Banach spaces, see [28], [39].

[50] This assumption is justifiable according to [51], or, approximately, that is sufficient for the present purpose, by the formal (and trivial) argument from section 6 in [27].

[51] The spherical r-cone over a piecewise Riemannian manifold S can be seen by isometric embedding from S to the equatorial sphere $S^{N-1}(r) \subset S^N(r) \subset \mathbb{R}^{N+1}$ and taking the geodesic cone over $S \subset S^{N-1}$ from a pole in $S^N \supset S^{N-1}$ for $B_{or}(S)$.

But the proof of the bound $vol_n(B_x(R)) \geq \varepsilon_n R^n$ for general groups Π, also due to Guth (see [37] where a more general inequality is proven), is rather complicated[52] and gives smaller constant ε_n.[53]

9. *Negative curvature and infinite dimensions.* The divergence inequality and its corollaries applies to many non-Euclidean spaces X, such as $CAT(\kappa)$-*spaces* with $\kappa \leq 0$, that are *complete simply connected, possibly infinite dimensional spaces, e.g., Riemannian/Hilbertian manifolds, with non-positive sectional curvatures $\leq \kappa$ in the sense of Alexandrov.*

Albeit vector fields are not, strictly speaking, defined in singular spaces, *radial semigroups of transformations $X \to X$ with controllably positive k-divergence* are available in CAT-spaces. This, along with a solution of Plateau's problem, (let it be only an approximate one) shows that *every (mildly regular) k-cycle Y in a $CAT(\kappa)$-space X is homologous to zero in its d-neighborhood $U_d(Y) \subset X$ for d equal the radius of the corresponding k-sphere in the hyperbolic space with curvature κ, that is, $S^k(\kappa, d) \subset H^{k+1}(\kappa)$, such that $vol_k(S^k(d)) = vol_k(Y)$.*

Remark. If X is *finite dimensional as well as non-singular*, this also follows from the spherical waist inequality that, in fact, does not need k-volume contracting (or expanding) radial fields but rather (controllably) k-volume contracting maps from X to the unit tangent spheres $S_x^{n-1}(1) \subset T_x(X)$, $n = dim(X)$, at all $x \in X$; see appendix 2 in [27]. But proper setting for such an inequality for infinite dimensional spheres and in the presence of singularities remains problematic.

10. *Almgren's inequality for curvature ≥ 0.* Let X be a complete n-dimensional Riemannian manifold X *with non-negative sectional curvatures* and with *strictly positive volume density at infinity*

$$dens_\infty(X) =_{def} \limsup_{R \to \infty} \frac{vol_n(B_{x_0}(X, R))}{vol_n(B_0(\mathbb{R}^n, R))} > 0.$$

Observe, that since $curv(X) \geq 0$, this X admits a *unique tangent cone $T_\infty(X)$ at infinity* and the volume density of X at infinity equals the volume of the unit ball in this cone centered at the apex $o \in T_\infty(X)$, where, recall, the tangent cone $T_\infty(X)$, of a metric space X at infinity is *the pointed Hausdorff limit* of the metric spaces obtained by scaling X by $\varepsilon \to 0$:

$$T_\infty(X) = \lim_{\varepsilon \to 0}(\varepsilon X =_{def} (X, \varepsilon \cdot dist_X)).$$

Let $Y \subset X$ be a compact k-dimensional subvariety such that the distance minimizing segments $[x, y] \subset X$ almost all points $x \in X$ and Y have their Y-endpoints y contained in the $C^{1, Lipschitz}$-regular locus of Y where moreover they are normal to Y. (This is automatic for closed smooth submanifolds $Y \subset X$ but we need it for more general Y.)

---

[52]See *Uryson width and volume* by Panos Papasoglu, arXiv: 1909.03738 [math.DG] and *Filling metric spaces* by Liokumovich, Lishak, Nabutovsky, and Rotman in [41] for recent progress in this direction.

[53]Most *known* examples of fundamental groups of closed aspherical manifolds, e.g., those with non-positive curvatures, are "sufficiently large." But, conjecturally, there are many "non-large" examples.

If the norms of the mean curvatures of Y at almost all of these points $y \in Y$ are bounded by a constant M, then

$$dens_\infty(X) \leq \frac{vol_k(Y)}{vol_k(S^k(k/M))}$$

for the k-sphere of radius k/M in $\mathbb{R}^{k+1}$ that has its mean curvature equal M.

Proof. The volumes of the R-tubes $U_R(Y) \subset X$ around Y are bounded by those of $S^k(k/M) \subset \mathbb{R}^n$ by *the Hermann Weyl tube formula* extended as a (volume comparison) inequality to Riemannian manifolds X with $curv \geq \kappa$ by Bujalo and Heintze-Karcher.

Then it follows by Almgren's variational local-to-global principle, mentioned in the above 3, that *the volume minimizing $(k+1)$-chains Z in X satisfy:*

$[curv \geq 0]_{almg}$ $$\frac{vol_{k+1}(Z)}{vol_k(\partial Z)^{\frac{k+1}{k}}} \leq (k+1)^{-\frac{k+1}{k}}(dens_\infty(X))^{-\frac{1}{k}}.$$

Sharpness of $[curv \geq 0]_{almg}$.   This inequality is *sharp*, besides $X = \mathbb{R}^n$, for certain conical (singular if $\delta < 1$) spaces X, in particular, for $X = (\mathbb{R}^{k+1}/\Gamma) \times \mathbb{R}^{n-k-1}$ for finite isometry group Γ acting on $\mathbb{R}^{k+1}$.

Linking corollary.   The inequality $[curv \geq 0]_{almg}$, combined with the above coarea inequality, yields the following generalization of the Bombieri-Simon (Gehring) linking volume inequality.

The $(k+1)$-volume minimizing chains Z in a complete Riemannian manifold X of non-negative curvature that fill-in a k-cycle Y in X are contained in the d-neighborhood of $Y \subset X$ for d equal the radius of the ball $B_o(T_\infty(X), d) \subset T_\infty(X)$, such that $vol_k(\partial B_o(T_\infty(X), d)) = vol_k(Y)$.

11. *Convex functions and monotonicity inequality for $curv \geq 0$.* Let X be a metric space, let $x_0 \in X$ be a preferred point in X and $\mu_\bullet$, also written as $d\mu_{x_\bullet}$, be a probability measure on X.

Let $h_{x_0}(x, x_\bullet)$, $x, x_\bullet \in X$, be defined as

$$h_{x_0}(x_\bullet, x) = \max(0, -dist(x, x_\bullet) + dist(x_0, x_\bullet))$$

and

$$H_{x_0, \mu_\bullet}(x) = \int_X h^2_{x_0}(x, x_\bullet) d\mu_{x_\bullet}.$$

If X is a complete manifold with $curv(X) \geq 0$ and $\mu_{\bullet,i}$ is a sequence of measures with supports tending to infinity (thus weakly convergent to 0), then the limit $H(x) = H_{x_0}(x)$ of the functions $H_{x_0, \mu_{\bullet,i}}$ for $i \to \infty$, assuming this limit exists, is a *convex* function on X by the old Gromoll-Meyer lemma on convexity of *the Busemann functions.*

Moreover, if $\mu_{\bullet,i} = \rho_i(x_\bullet)dx$ for *radial functions* ρ_i, i.e., $\rho_i(x_\bullet) = \phi_i(dist(x_0, x_\bullet))$, for some real functions ϕ_i, and *if* $dens_\infty(X) > 0$, then the corresponding limit functions $H(x)$ are *strictly convex.*

In fact, this strictness is controlled by the density $\delta = dens_\infty(X) > 0$ as follows.

The second derivatives of these H along all geodesics in X are bounded from below by $\varepsilon = \varepsilon_n(\delta) > 0$.[54]

Now, the strict convexity of the (smoothed if necessary) function $H(x)$ can be seen as a lower bound on the k-divergence of the gradient of H and, as in the above 6, we arrive at *the monotonicity inequality* of intersections of stationary $(k+1)$-dimensional subvarieties $Z \subset X$ with the balls $B_{x_0}(r) \subset X$, $x_0 \in Z$,

$$\frac{vol_{k+1}(Z_r)}{vol_k(\partial Z_r)^{\frac{k+1}{k}}} \leq const = const_n(\delta) \text{ for } n = dim(X) \text{ and } \delta = dens_\infty(X).$$

Remark. A specific evaluation of this *const* depends on a lower bound on the scalar products $\langle s_0, s \rangle$ averaged over subsets $U \subset S^{n-1}$ with spherical measures $\geq \delta \cdot vol_{n-1}(S^{n-1})$ in the tangent spheres $S^{n-1} = S^{n-1}_x \subset \mathbb{R}^n_x = T_x(X)$. But even the best bound

$$\inf_{s_0 \in S^{n-1}} \frac{1}{vol(U)} \int_U \langle s_0, s \rangle ds \geq c(\delta)$$

with sharp $c(\delta)$ does not seem to deliver the sharp constant

$$const_{almgren} = (k+1)^{-\frac{k+1}{k}} (dens_\infty(X))^{-\frac{1}{k}}.$$

11. *On singularities and infinite dimensions with $curv \geq \kappa$.* Since singular Alexandrov spaces X with $curv \geq \kappa$ admit strictly contracting negative "gradient fields," we can *conically fill* cycles in balls: if $Y \subset B_{x_0}(R)$, there is a filling Z such that $vol_{k+1}(Z) \leq cost \cdot R \cdot vol_k(Y)$. According to [27], [51] *the cone inequality* in a space X implies a non-sharp filling inequality in this X.

Probably, these singular spaces enjoy the filling and waist inequalities with similarly *sharp constants* as their non-singular counterparts. (This is easy for *equidimensional Hausdorff limits* of non-singular spaces, since waists and filling constants are Hausdorff continuous in the presence of lower volume bounds.)

In fact, one expects a full-fledged contravariantly Hausdorff continuous (i.e., for collapsing $X_i \to X = X_\infty$) theory of volume minimizing as well as of *quasi-stationary* (similar to quasi-geodesics of Milka-Perelman-Petrunin) subvarieties in X.

Another avenue of possible generalizations is that of infinite dimensional (singular if needed) spaces X with positive curvatures. Here one is encouraged by *the stability of $dens_\infty$:*

$$dens_\infty(X) = dens_\infty(X \times \mathbb{R}^N)$$

that suggests a class of infinite dimensional spaces X with "small positive" curvatures where the differentials of various exponential maps are isometric up to small

---

[54]This, which is obvious once it has been stated, was pointed out to me in slightly different terms by Grisha Perelman along with the following similar observation that constitutes the geometric core of *the Grove-Petersen finiteness theorem.*

If $curv(X) \geq -1$ and if the volume of the unit ball $B_{x_0}(1) \subset X$ around an $x_0 \in X$ has volume $\geq \delta$, then x_0 admits a convex neighborhood $U_\varepsilon \subset B_{x_0}(1)$ that contains the ball $B_{x_0}(\varepsilon)$ for $\varepsilon = \varepsilon_n(\delta) > 0$.

(trace class or smaller) errors. In this case, one may try to define $dens_\infty$ that would allow one to formulate and prove an infinite dimensional counterpart of $[curv \geq 0]$.

12. *Almgren's Morse theory for regular waists.* Recall (see section 6) that k-waists of Riemannian manifolds X are defined via classes $\mathcal{D}$ of diagrams $D_X = \{X \overset{\chi}{\leftarrow} \Sigma \overset{\varsigma}{\rightarrow} S\}$ where S and Σ are pseudomanifolds with $dim(\Sigma) - dim(S) = k$ that represent *homologically substantial* S-families of k-cycles Y_s in X, that are the χ-images of the pullbacks $\varsigma^{-1}(s), s \in S$ and where homological substantiality may be understood as non-vanishing of the image of the fundamental homology class $h \in H_n(Z)$ under the homomorphism $\chi_* : H_*(\Sigma) \rightarrow H_*(X)$. Namely $waist_k(X)$ is defined as

$$waist_k(X) = \inf_{D_X \in \mathcal{D}} \sup_s vol_k(Y_s) \text{ for } Y_s = \chi(\varsigma^{-1}(s)).$$

If a class $\mathcal{D}$ consists of diagrams $D_X = \{X \overset{\chi}{\leftarrow} \Sigma \overset{\varsigma}{\rightarrow} S\}$ with *sufficiently regular* maps ς and χ, e.g., *piecewise real analytic* ones, then the resulting waists, call them *regular*, admit *rather rough* lower bounds in terms of *filling* and of *local contractibility* properties of X, where the latter refers to the range of pairs of numbers (r, R) such that every r-ball in X is contractible in the concentric R-ball.

On the other hand, *the sharp* lower bound on a regular k-waist of the unit n-sphere S^n that (trivially) implies the equality

$$reg\text{-}waist_k(S^n) = vol_k(S^k)$$

can be derived from the Almgren-Morse theory in *spaces of rectifiable cycles with flat topologies*, [48], [34]. This theory implies that

$$reg\text{-}waist_k(X) \geq \inf_{M^k \in \mathcal{MIN}_k} vol_k(M^k)$$

for "inf" taken over all *minimal/stationary* k-subvarieties $M^k \subset X$ and the inequality

$$reg\text{-}waist_k(S^n) \geq vol_k(S^k)$$

follows from the lower volume bound for minimal/stationary subvarieties in the spheres S^n:

$[volmin_k(S^n)]$ $\qquad\qquad\qquad$ $vol_k(M_k) \geq vol_k(S^k).$

Almgren's theory, that preceded the homological localization method from [25], albeit limited to the regular case, has an advantage over the lower bounds on the $\mathbb{Z}_2$-waists indicated in section 6 of being applicable to the integer and to the $\mathbb{Z}_p$-cycles, that allows minimization over the maps $\chi : \Sigma \rightarrow X$ of nonzero *integer degree* in the case of oriented X, S and Σ.[55]

Besides, Almgren's theory plus Weyl-Buyalo-Heintze-Karcher volume tube bound yield the following *sharp waist inequality for closed Riemannian manifolds*

---

[55] I am not certain of all implications of Almgren's theory, since I have not studied the technicalities of this theory in detail.

X with $curv(X) \geq 1$, (see section 3.5 in [24] and [47]):

$$[waist]_{curv\geq1} \qquad\qquad waist_k(X) \geq vol_k(S^k)\frac{vol_n(X)}{vol_n(S^n)}, \ n = dim(X).$$

If X is a manifold with a *convex boundary*, this inequality may be applied to the (smoothed) double of X, but the resulting low bound on $waist(X)$ is non-sharp, unlike those obtained by the convex partitioning argument from [25].

Nevertheless, Almgren's Morse theory seems more suitable for proving the counterpart of $[waist]_{curv\geq1}$ for general singular Alexandrov spaces with curvatures ≥ 1 (where even a properly formulated volume tube bound is still unavailable).

13. *Regularization conjecture.* Probably, general homologically substantial families of virtually k-dimensional subsets $Y_s \subset X$ should admit an ε-approximation, for all $\varepsilon > 0$, by *regular* homologically substantial families $Y_{s,\varepsilon} \subset X$ with

$$Haumes_k(Y_{s,\varepsilon}) \leq Haumes_k(Y_s) + \varepsilon.$$

This would imply that all kinds of k-waists of S^n are equal to $vol_k(S^k)$ with no regularity assumption, that is, for all homologically substantial diagrams with continuous χ and ς and with k-volumes understood as Hausdorff measures. (This remains unknown except for $k+1$ and $k = n-1$.)

14. *Waists at infinity.* Let X be a complete manifold with $curv(X) \geq 0$. Do the waists of the complements to the balls $B_{x_0}(R) \subset X$ satisfy

$$waist_k(X \setminus B_{x_0}(R)) \geq R^k dens_\infty(X) vol_k(S^k)?$$

Are there some nontrivial (concavity, monotonicity) inequalities between these waists for different values of R?

Does the above lower bound hold for waists of the subsets $X \setminus B_x(R)$ seen as variable ones (in the sense of section 8) parametrized by $X \ni x$?

If "yes" this would imply that every m-dimensional subvariety $W \subset X$ that is *not homologous to zero in its R-neighborhood*, satisfies (compare section 18)

$$waist_k(W) \geq R^k dens_\infty(X) vol_k(S^k).$$

Another class of spaces X where evaluation of waists (and volume spectra in general) at infinity may be instructive is that of *symmetric spaces with non-positive curvatures*. (One may define "waists at infinity" via integrals of positive radial functions $\phi(dist(x, x_0))$ over k-cycles in X. For instance, the function $\phi(d) = \exp \lambda d$, $\lambda < 0$, may serve better than $+\infty$ on the ball B_{x_0} and 1 outside this ball which depicts $X \setminus B(R)$.)

15. *Waists of products and fibrations.* Let X be a Riemannian product, $X = \underline{X} \times X_\varepsilon$, or more generally, let $X = X(\varepsilon)$ be fibered over $\underline{X}$.

If X_ε is sufficiently small compared to $\underline{X}$, e.g.,

$$X_\varepsilon = \varepsilon \cdot X_0 =_{def} (X_0, dist_\varepsilon = \varepsilon \cdot dist_0) \text{ for a small } \varepsilon > 0,$$

then, conjecturally,

$$waist_k(X) = waist_k(\underline{X})$$

and something similar is expected for fibrations $X(\varepsilon) \to \underline{X}$ with *small* fibers $X_{\underline{x}} \subset X$ that also for this purpose must vary *slowly* as functions of $\underline{x}$.

Here, conjecturally, the Hausdorff k-waist of $X(\varepsilon)$ equals the maximum of the k-volumes of the fibers of the normal projection $X(\varepsilon) \supset \underline{X}$ plus a lower order term.[56]

This, *the regular case*, follows from the Almgren-Morse theory whenever "small" minimal subvarieties $M^k \subset X(\varepsilon)$ are "vertical," namely, if those with $vol_k(M^k) \le waist_k(X(\varepsilon))$, go to points under $X(\varepsilon) \to \underline{X}$, where a sufficient condition to such "verticality" is a presence of *contracting* vector fields, in sufficiently large balls in $\underline{X}$ and where such fields often come as gradients of convex functions. But it is unclear how to sharply bound from below the Hausdorff waists in the non-regular case; compare 7.4 in [25] and 1.5(B) in [31].

16. *Waists of thin convex sets.* A particular case of interest is where $\underline{X}$ is a compact convex m-dimensional subset in an n-dimensional space, call it Z, of constant curvature, and $X(\varepsilon) \supset \underline{X}$ are n-dimensional convex subsets in Z that are ε-close to this $\underline{X}$.

These $X(\varepsilon)$ *normally* project to $\underline{X}$ with convex ε small k-dimensional fibers $X_{\underline{x}}$ over the interior points $\underline{x} \in int(\underline{X}) = \underline{X} \setminus \partial \underline{X}$ for $k = n - m$[57] and, conjecturally,

$$\frac{waist_k(X(\varepsilon))}{\max_{\underline{x} \in int(\underline{X})} vol_k(X_{\underline{x}})} \to 1 \text{ for } \varepsilon \to 0.$$

This is known for δ-$Mink_k$-waists, and this implies, in conjunction with homological localization, the sharp lower bounds on the Minkowski waists spheres, see [25]. But the case of Hausdorff waists with *no regularity* assumption remains open.

17. *Waists of solids.* Does the regular k-waist of the rectangular solid

$$[0, l_1] \times [0, l_2] \times \cdots \times [0, l_n], \ 0 < l_1 \le l_2 \le \cdots \le l_i \le \cdots \le l_n$$

equal $l_1 \cdot l_2 \cdot \ldots \cdot l_k$?

This is not hard to show for *fast growing* sequences $l_1 << l_2 << \cdots << l_i << \cdots << l_n$ but the case of roughly equal l_i, especially that of the cube $[0, l]^n$, remains problematic.[58]

18. *Waist of the infinite dimensional Hilbertian sphere.* "Homologically substantial" families of k-cycles in S^∞ may be defined with either Fredholm maps $S^\infty \mathbb{R}^\infty$ of index k, (i.e., with virtually k-dimensional fibers) or, with Fredholm maps $Y \times S^\infty \to S^\infty$ of Fredholm degree one, or by bringing the two diagrammatically together as in section 6.

But it is unclear if these waists are not equal to zero.

---

[56]This would imply the sharp lower bound on the Hausdorff waist of spheres by the homological localization argument in [25].

[57]The fibers over the boundary points $\underline{x} \in \partial \underline{X}$ have $dim(X_{\underline{x}}) = n$.

[58]This was solved by Bo'az Klartag [40].

19. *Linking inequalities with δ-$Mink_k$.* Such inequalities provide lower bounds on the volumes of δ-neighborhoods $U_\delta(W) \subset X$ of k-dimensional subvarieties $W \subset X$ that are not homologous to zero in their R-neighborhoods for some $R > \delta$. For instance, if $X = \mathbb{R}^n$, then

$$vol_n(U_\delta(W)) \geq vol_n(U_\delta(S^k(R)))$$

for the R-sphere $S^k(R) \subset \mathbb{R}^{k+1} \subset \mathbb{R}^n$.

Proof. The argument from section 18 (also see section 8 in [27]) with the radial projections from W to the R-spheres with centers $x \in \mathbb{R}^n \setminus W$ applies here, since:

- the spherical waist inequality holds true with δ-$Mink_k$ defined with δ-neighborhoods of subsets $Y \subset S^{n-1} \subset \mathbb{R}^n$ taken in $\mathbb{R}^n \supset S^{n-1}$ (see [25]);
- these projections $W \to S_x^{n-1}(R)$ are not only distance decreasing, but they are obtained from the identity map by a distance decreasing homotopy. Therefore it diminishes the volume of δ-neighborhoods by *Csikós's theorem.* (See [16] and [9] and references therein.)

20. *Geometric linking inequalities.* Let two closed subsets $W, W' \subset \mathbb{R}^n$ of dimensions k and $n - k - 1$ *can not be unlinked*, i.e., moved apart without mutual intersection on the way by a certain class $\mathcal{M}$ of geometric motions.

Can one bound from below the Hausdorff measures of these subsets in terms of $dist(W, W')$?

For instance—this, according to Eremenko [19], was observed by I. Syutric in 1976 or in 1977—if $\mathcal{M}$ consists of homotheties $h_t : x \mapsto x_t = x_0 + (1 - t)(x - x_0)$, $t \in [0, 1]$, for some point $x_0 \in W$, then $Hausmes_1(W) \geq \pi R$ for connected sets W and $Hausmes_1(W) \geq 2\pi R$ for closed curves W.

Indeed, the image of this homothety, that is, the unit cone over W from x_0, intersects W', say at $x' \in W' \cap h_{t'}(W)$ for some $t' \in [0, 1]$; thus the radial projection from W to the R-sphere $S_{x'}^{n-1}(R) \subset \mathbb{R}^n$, that is, distance decreasing, contains two diametrically opposite points, namely the images of x_0 and $x_1 = h_{t'}^{-1}(x') \in W$. QED.

A bound on $vol(W \times W')$. If W and W' can not be unlinked by *parallel translations*, then, obviously, the map $W' \times W \to S_0^{n-1}(R)$ for $(w, w') \mapsto R\frac{w-w'}{dist(w,w')}$ is onto. Hence,

$$Hausmes_k(W) \cdot Hausmes_{n-k-1}(W) \geq const_n R^{n-1}.$$

Thus, either W or W' must have large Hausdorff measure, but one of them may be arbitrarily small. This is seen by taking (large) W that ε-approximates the k-skeleton of a (large) sphere $S^{n-1}(R+r) \subset \mathbb{R}^n$ and where W' much finer, say with $\varepsilon' = 0.1\varepsilon$, approximates the $(n - k - 1)$-skeleton of a (tiny) r'-ball $B_0^n(r') \subset B_0^n$ $(R+r) \subset \mathbb{R}^n$ for $r' \leq r/2$.

If $r' \geq 2\varepsilon$ then W "cages" W' inside $S^{n-1}(R+r)$, provided $r' \geq 10\varepsilon$: W' can not be moved outside of $S^{n-1}(R+r)$ by an isometric motion without meeting W on the way.

Non-accessible articles. There are a dozen or so other papers on the Gehring linking problem but, since they are not openly accessible, one can not tell what is written in there.

Acknowledgments

I am thankful to the referees of this paper for several useful critical remarks and constructive suggestions.

REFERENCES

[1]   A. Adem, R. J. Milgram, Cohomology of finite groups, Springer 2013.

[2]   F. J. Almgren Jr., Homotopy groups of the integral cycle groups, Topology 1 (1962), 257–299.

[3]   F. J. Almgren Jr., Optimal isoperimetric inequalities, Indiana Univ. Math. J. 35 (1986), 451–547.

[4]   S. Anjos, F. Lalonde, M. Pinsonnault, The homotopy type of the space of symplectic balls in rational ruled 4-manifolds, Geometry & Topology 13 (2009), 1177–1227.

[5]   G. Arzhantseva, L. Paunescu, Linear sofic groups and algebras, arXiv:1212.6780.

[6]   Yu. Baryshnikov, P. Bubenik, M. Kahle, Min-type Morse theory for configuration spaces of hard spheres, Int. Math. Res. Not. IMRN (2014), no. 9 2577–2592.

[7]   P. Benevieri, M. Furi, On the uniqueness of degree in infinite dimension, http://sugarcane.icmc.usp.br/PDFs/icmc-giugno2013-short.pdf.

[8]   M. Bertelson, M. Gromov, Dynamical Morse entropy, in Modern Dynamical Systems and Applications, Cambridge Univ. Press, Cambridge (2004), 27–44,

[9]   K. Bezdek, R. Connelly, Pushing disks apart—the Kneser-Poulsen conjecture in the plane. http://arxiv.org/pdf/math/0108098.pdf.

[10]   P. Biran, From symplectic packing to algebraic geometry and back, European Congress of Mathematics, vol. 2, Bareelona, 2000.

[11]   P. Biran, Connectedness of spaces of symplectic embeddings. arXiv.org > dg-ga > arXiv:dg-ga/9603008v1.

[12]   E. Bombieri, An introduction to minimal currents and parametric variational problems, preprint, Institute for Advanced Study, Princeton.

[13]   L. Buhovsky, A maximal relative symplectic packing construction, J. Symplectic Geom. 8 (2010), no. 1, 67–72.

[14]   K. Cieliebak, H. Hofer, J. Latschev, F. Schlenk, Quantitative symplectic geometry. arXiv:math/0506191v1.

[15]   H. Cohn, A conceptual breakthrough in sphere packing, arXiv:1611.01685 (2016).

[16] B. Csikós, On the volume of flowers in space forms, Geom. Dedicata (2001), 59–79.

[17] Y. Eliashberg, M. Gromov, Lagrangian intersection theory: finite-dimensional approach, in Geometry of Differential Equations, Amer. Math. Soc., Providence, RI (1998), 27–118.

[18] M. Entov, M. Verbitsky, Full symplectic packing for tori and hyperkähler manifolds. arXiv.org > math > arXiv:1412.7183.

[19] A. Eremenko, F. Gehring's problem on linked curves, https://mathoverflow. net/questions/8247/one-step-problems-in-geometry/112076#112076.

[20] H. Federer, W. H. Fleming, Normal and integral currents, Ann. of Math. 72:3 (1960), 458–520.

[21] M. Gromov, Dimension, non-linear spectra and width, in Geometric Aspects of Functional Analysis, Springer, Berlin (1988), 132–184.

[22] M. Gromov, Topological invariants of dynamical systems and spaces of holomorphic maps:1, Math. Phys. Anal. Geom. 2 (1999), 323–415.

[23] M. Gromov, In a search for a structure, part 1: on entropy. www.ihes.fr /~gromov/PDF/structre-serch-entropy-july5-2012.pdf.

[24] M. Gromov, Singularities, expanders and topology of maps. Part 2. Geom. Funct. Anal., 19:743–841, 2009.

[25] M. Gromov, Isoperimetry of waists and concentration of maps, Geom. Funct. Anal. 13:1 (2003), 178–215.

[26] M. Gromov, Metric structures for Riemannian and non-Riemannian spaces, Birkhäuser, 1999.

[27] M. Gromov, Filling Riemannin manifolds, http://projecteuclid.org/euclid.jdg /1214509283.

[28] M. Gromov, Systoles and intersystolic inequalities, www.ihes.fr/~gromov /PDF/1[95].ps.

[29] M. Gromov, Crystals, proteins, stability and isoperimetry. www.ihes.fr /~gromov/PDF/proteins-crystals-isoper.pdf.

[30] M. Gromov, H. B. Lawson, Jr., Positive scalar curvature and the Dirac operator on complete Riemannian manifolds, www.ihes.fr/~gromov/PDF/5[89].pdf.

[31] M. Gromov, Plateau-Stein manifolds, www.ihes.fr/~gromov/PDF/Plateau-Stein-Febr-2013.pdf.

[32] M. Gromov, Manifolds: Where do we come from? www.ihes.fr/~gromov/PDF /manifolds-Poincare.pdf.

[33] L. Guth, Minimax problems related to cup powers and Steenrod squares, arxiv.org/pdf/math/0702066.

[34] L. Guth, The waist inequality in Gromov's work, math.mit.edu/~lguth /Exposition/waist.pdf.

[35] L. Guth, Symplectic embeddings of polydisks. arXiv:0709.1957v2.

[36] L. Guth, Systolic inequalities and minimal hypersurfaces, arXiv:0903.5299v2.

[37] L. Guth, Volumes of balls in large Riemannian manifolds, arXiv:math /0610212v1.

[38] R. Hind, S. Lisi, Symplectic embeddings of polydisks, arXiv:1304.3065v1.

[39] M. Katz, Systolic geometry and topology, https://books.google.fr/books? isbn=0821841777.

[40] B. Klartag, Convex geometry and waist inequalities, https://arxiv.org/pdf /1608.04121.pdf.

[41] Y. Liokumovich, B. Lishak, A. Nabutovsky, R. Rotman, Filling metric spaces, arXiv: 1905.06522 [math.DG].

[42] Y. Liokumovich, F. C. Marques, A. Neves, Annals of Math. 187(2018), 1–29.

[43] F. Manin, Volume distortion in homotopy groups, arXiv:1410.3368v3 [math. GT] 12 Nov 2014.

[44] F. Manin, S. Weinberger, Integral and rational mapping classes, arXiv: 1802. 05784 [math.AT].

[45] D. McDuff, Symplectic embeddings and continued fractions: a survey, Notes for the Takagi lectures, Sapporo, Japan, June 2009, available in pdf. http:// www.math.stonybrook.edu/~dusa/capjapaug29.pdf.

[46] D. McDuff, L. Polterovich, Symplectic packings and algebraic geometry. http://link.springer.com/article/10.1007%2FBF01231766.

[47] A. Memarian, A note on the geometry of positively-curved Riemannian manifolds, http://arxiv.org/pdf/1312.0792v1.pdf.

[48] J. T. Pitts, Existence and regularity of minimal surfaces on Riemannian manifolds, Mathematical Notes 27, Princeton University Press, Princeton, NJ (1981).

[49] F. Schlenk, Embedding problems in symplectic geometry, de Gruyter Expositions in Mathematics, vol. 40, Berlin, 2005.

[50] C. Viterbo, Symplectic topology as the geometry of generating functions, Math. Annalen 292(1992), 685–710.

[51] S. Wenger, A short proof of Gromov's filling inequality, arXiv:math/07038 89v1.

Right-Angled Hexagon Tilings of the Hyperbolic Plane

RICHARD KENYON

1 INTRODUCTION

In the spectrum of random planar structures studied in mathematics, the most basic examples are lattice-based systems like random tilings of the Euclidean plane with polyominoes; in these the randomness is encoded in the combinatorial arrangement of tiles. A similar basic setting is in classical lattice statistical mechanics, where the randomness is encoded in the states of "particles" sitting at lattice sites (typically with a finite number of states), and interacting with nearest neighbors.

At the other end of the spectrum of random planar processes, one can consider random metrics on $\mathbb{R}^2$, as in the case of Liouville quantum gravity, where there are no local interactions or combinatorics.

Between these two extremes of either pure combinatorics and fixed geometry, or no combinatorics and random geometry, is the case of non-lattice statistical mechanics, in which particles are free to change position but still interact with each other. Very few examples of this type have been successfully studied.

We discuss here a case of this intermediate type, where the randomness is encoded in the spatial positions of interacting components.

We work in the hyperbolic plane $\mathbb{H}$, and our components are right-angled hexagons (RAHs) of varying shape. The components "interact" in such a way as to form global, edge-to-edge tilings of the hyperbolic plane. One can take a different point of view on the same system and consider the individual components to be bi-infinite geodesics, which interact with each other so as to intersect orthogonally (and so that the complementary components are hexagons).

Other models of random geometric structures in $\mathbb{H}^2$ have been studied, see, e.g., [2]. However in these cases the underlying measures are simpler in a Markovian sense: once a geodesic is determined the left and right half-space structures are independent. Although our model eventually boils down to a similar argument, the analogous condition is a priori much less evident.

There is a periodic tiling with regular (all sides of equal length) RAHs, see Figure 1. It is not at all clear that this structure is flexible in a way that preserves angles.

Let $0 \in \mathbb{H}$ denote the center of the Poincaré disk. Let Ω be the space of edge-to-edge tilings of the hyperbolic plane $\mathbb{H}$ with RAHs. It has a natural topology: two

Research supported by the NSF grant DMS-1208191 and the Simons Foundation award 327929.

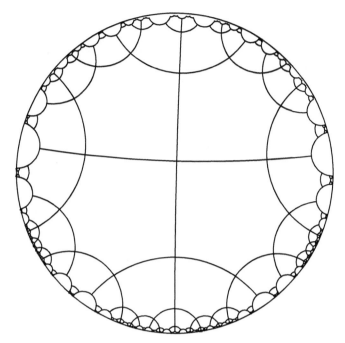

Figure 1: The tiling with regular RAHs.

tilings T_1, T_2 are close if there is a homeomorphism $\phi : \mathbb{H} \to \mathbb{H}$ which is ε-close to the identity ($|\phi(z) - z| < \varepsilon$) on a large neighborhood of 0, and takes one tiling to the other. The space Ω also comes with a natural action of the isometry group of $\mathbb{H}$; it is thus a Riemann surface lamination, whose leaves are generically hyperbolic planes.

One way to explore Ω is to study invariant measures on it. Given a measure μ on RAHs one can try to construct a measure on Ω for which the marginal (the induced measure on tiles) is given by μ. We prove this for a family of measures μ:

Theorem 1. *Given a probability measure ν on $\mathbb{R}_+$ of full support, let μ be the measure on RAHs whose edge lengths ℓ_1, ℓ_3, ℓ_5 are independent and distributed according to ν. Then there is an Isom($\mathbb{H}$)-invariant probability measure on Ω of full support in which the tile containing the origin (and therefore every tile) has measure μ.*

An analogous result holds for tilings with right-angled pentagons (section 4) and presumably for k-gons with $k > 6$ as well. For shapes with fewer sides, such as quadrilaterals with angles $2\pi/5$, say, we have been unable to formulate a similar result, probably because the dimension of the space of shapes is too small.

One way to state the invariance under Isom($\mathbb{H}$) of a measure on Ω is as follows. Take a μ-random tiling. Pick an isometry ϕ of $\mathbb{H}$ independent of the tiling. Applying ϕ to the tiling results in a new μ-random tiling. In fact, since all tiles have the same area, the random tiling "looks the same around any tile," in the following sense. Choose from the tiling any tile in a combinatorial manner, that is, which does not depend on the geometry of the current tiling, except for the information about which tile contains the origin. Translate this tile so that the origin $0 \in \mathbb{H}$ is at a uniform random location in it, and then choose a random rotation fixing the origin. Then the result is another exact sample of the measure.

There are leaves of Ω which are closed Riemann surfaces, corresponding to RAH tilings which are periodic under the action of a cocompact subgroup of isometries. These leaves come in families whose dimension is (as one can show) half the dimension of the Teichmüller space of the associated Riemann surface. Each such family supports many isometry-invariant measures, supported on group-invariant RAH-tilings. We will not address here the question of what natural measures are supported on these subspaces; we are concerned rather with measures of *full support* in Ω.

Another motivation for studying the space Ω comes from integrable systems. There is a surprising connection between the geometry of a RAH and the Yang-Baxter equation for the Ising model. The Yang-Baxter equation, or star-triangle equation, for the Ising model is a local rearrangement of a graph preserving the so-called partition function of the model. The Yang-Baxter equation is a hallmark of an underlying integrable system, and for the Ising model one which has been studied in [4]. One can translate the Ising integrable structure into an integrable structure on certain spaces of surfaces tiled by RAHs. The measures we discuss below are relevant to this system: the "trigonometric" configurations here are in fact fixed points of the system, and the measure μ_0 discussed below is an invariant measure. This will be the subject of a forthcoming work.

Acknowledgments

We thank Oded Schramm and Andrei Okounkov for helpful discussions, and the referee for several suggestions for improvements.

2   MEASURES ON RAHS

2.1   Hexagons

The side lengths $(\ell_i)_{i=1,\dots,6}$ of a right-angled hexagon satisfy the following relation. Let $(a, B, c, A, b, C) = (e^{\ell_1}, \dots, e^{\ell_6})$. If a, b, c are known (but arbitrary in $(0, \infty)$) then the remaining three are determined by

$$\frac{A+1}{A-1} = \sqrt{\frac{(1+abc)(a+bc)}{(b+ac)(c+ab)}}$$

$$\frac{B+1}{B-1} = \sqrt{\frac{(1+abc)(b+ac)}{(a+bc)(c+ab)}} \qquad (1)$$

$$\frac{C+1}{C-1} = \sqrt{\frac{(1+abc)(c+ab)}{(a+bc)(b+ac)}}.$$

It is straightforward to see that the map $\Psi\colon (a, b, c) \mapsto (A, B, C)$ is an involution. In particular the family of RAHs is homeomorphic to $\mathbb{R}^3_+$, parametrized by three non-adjacent edge lengths.

One particularly nice family of hexagons are the *trigonometric* ones, where

$$1 - a - b - c - ab - ac - ab + abc = 0,$$

or equivalently
$$a = \tan\alpha, \quad b = \tan\beta, \quad c = \tan\gamma,$$

and $\alpha + \beta + \gamma = \frac{\pi}{4}$. In this case opposite sides have equal lengths ($A = a$, $B = b$, $C = c$, as one can check from (1)). One can define a measure on trigonometric RAHs by choosing, for example, α, β, γ uniformly with respect to Lebesgue measure on the simplex $\{\alpha + \beta + \gamma = \frac{\pi}{4}\}$.

A more interesting measure μ_0, of full support in $\mathbb{R}^3_+$, is given by choosing ℓ_1, ℓ_3, ℓ_5 independently, each distributed with respect to the measure ν_0 on $(0, \infty)$ with density

$$F(\ell)\, d\ell = \frac{C\, d\ell}{\sinh^{1/3}(\ell)} \tag{2}$$

where

$$C = \frac{\sqrt{3}\,\Gamma(\frac{2}{3})\Gamma(\frac{5}{6})}{2\pi^{3/2}}.$$

In terms of $u = e^\ell \in (1, \infty)$ the density is

$$f(u)\, du = \frac{2^{1/3}C\, du}{u^{2/3}(u^2 - 1)^{1/3}}. \tag{3}$$

The truly remarkable property of this measure is

Theorem 2. *If ℓ_1, ℓ_3, ℓ_5 are chosen i.i.d. with distribution ν_0 then ℓ_2, ℓ_4, ℓ_6 will be i.i.d. with distribution ν_0 as well.*

Proof. This is a computation: one needs to show that

$$f(a)f(b)f(c)\, da\, db\, dc = f(A)f(B)f(C)\, dA\, dB\, dC,$$

that is, that the Jacobian of the mapping from (a, b, c) to (A, B, C) defined by (1) is

$$\left(\frac{\partial_{A,B,C}}{\partial_{a,b,c}}\right) = \frac{f(a)f(b)f(c)}{f(A)f(B)f(C)}. \qquad \square$$

One can "discover" the distribution μ_0 as follows. If we assume a, b, c are i.i.d. with some density $g(x)dx$ then A, B, C will be distributed according to the density

$$\Psi_*(g(a)g(b)g(c)da\, db\, dc) = g(a)g(b)g(c)\left(\frac{\partial_{a,b,c}}{\partial_{A,B,C}}\right) dA\, dB\, dC.$$

Massaging the right-hand side of this expression into a form separating out the A, B and C dependence, one sees that there is a unique choice of g (up to scale), given by (3), for which it can be written as $g(A)g(B)g(C)dA\, dB\, dC$. It follows that μ_0 is the only probability measure with this "independence-preserving" property.

Question

Does this probability measure μ_0 have a geometric significance?

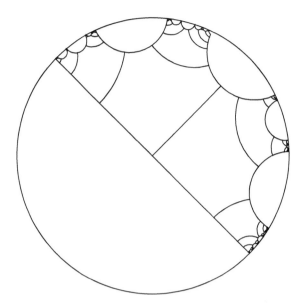

Figure 2: A random hextree in $\mathbb{H}$. For the purposes of illustration this one was constructed using a measure on side lengths which is more tame than ν_0.

One can of course choose any other measure ν on $\mathbb{R}_+$ and define a corresponding measure $\mu = \mu(\nu)$ on RAHs by choosing lengths ℓ_1, ℓ_3, ℓ_5 independent and ν-distributed. The remaining lengths ℓ_2, ℓ_4, ℓ_6 are then determined (and will have the same marginal distribution as each other but will not be independent and typically not distributed according to ν). The resulting measure $\mu = \mu(\nu)$ on RAHs is invariant under rotation by two indices.

3   TILING

3.1   Hextrees

We will build an Isom($\mathbb{H}$)-invariant probability measure λ on Ω with the property that the marginal distribution of a tile (the probability measure restricted to any single tile) is $\mu(\nu)$ for some ν as above. The building blocks of this measure are collections of RAHs glued together in a ternary-tree structure as in Figure 2. We call such sets *hextrees*.

That is, for each edge of the infinite regular degree-3 tree take an independent random real variable distributed according to ν. Then, for each vertex of the tree, use the three adjacent random variables as lengths of edges $1, 3, 5$ of a RAH. These RAHs are glued together so that RAHs at adjacent vertices are glued along the corresponding edge. We refer to these glued edges as the cross-edges of the hextree, since they cross between boundary components.

3.2   Gluing

Now the construction of the tiling in Ω is as follows. Start with a hextree as above; for each of its boundary geodesics construct another hextree, conditioned on having a boundary with the same subdivision points as the first, but otherwise independent

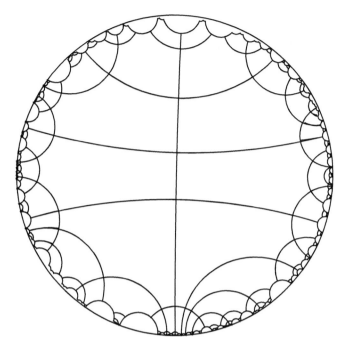

Figure 3: A RAH tiling. For this figure we used the measure ν on edge lengths with density (for $u = e^{\ell}$) $f(u) = C(u-1)e^{a_0(1-u)}du$, with mean 4.

of the first (we discuss how this is accomplished below); then glue these hextrees along their common boundary so that the individual RAHs meet edge-to-edge. We continue gluing hextrees to existing boundaries in this way, always choosing independent new hextrees conditioned to agree along the common boundary.

To show that the result is a tiling in Ω, it remains to show all of $\mathbb{H}$ is covered. This is proved in section 3.3 below.

How does one construct a hextree with the same boundary data (along one boundary geodesic) as a given hextree? Given a hextree R let Y be the boundary data (the sequence of edge lengths) along one boundary of R, and X be the remaining lengths defining the rest of the hextree.

There is a general theorem which applies in this situation [3]: given a joint random variable $(X, Y) \in \Omega_1 \times \Omega_2$ (that is, a random variable in a product space), there is a conditional probability measure $(X, Y)|Y$ obtained by conditioning on Y; a random sample $(X', Y)|Y$ of this measure can be chosen independently of (X, Y) conditional on Y. So to glue a hextree to another we just resample this conditional measure. In practice this resampling is complicated; see section 3.4 below.

The Möbius invariance of the resulting measure is a consequence of the reversibility of the gluing procedure. Each new hextree is distributed according to the same measure μ, and the joint probability measure on the glued object is symmetric: one cannot tell where the construction began and in what order the hextrees were glued. Furthermore, within each hextree each RAH has the same distribution, with edges $1, 3, 5$ which are independent, so there is no information about which RAH was the "first." In particular, as mentioned earlier, if we choose any RAH in the tiling and translate it so that the origin $0 \in \mathbb{H}$ is at a uniform random location in that RAH, the result is another exact sample of λ.

Question

Is the measure on tilings constructed from the above measure μ_0 characterized by some natural property like entropy maximization?

Question

Is there an $\mathrm{Isom}(\mathbb{H})$-invariant measure on Ω supported on trigonometric RAHs?

3.3   Covering property

To show that the union of hextrees covers all of $\mathbb{H}$, we need to show that hextrees are typically "thick": if some cross-edge of a hextree is short then along the geodesic containing this short cross-edge the other cross-edges of nearby hextrees are typically no shorter.

This follows from the reversibility of the construction of the tiling. Let x_0 be a cross-edge. Choose one of the two cross-edges adjacent to it on the same geodesic. This cross-edge x_1 is just as likely to be greater than x_0 as smaller than x_0, by reversibility. Similarly for the next cross-edge x_2, and so on. So the chance of getting a sequence of smaller and smaller cross-edges starting from x_0 goes to zero with the length of this sequence.

Thus along the "cross-edge geodesics" there are no accumulation points of vertices of RAHs, and so all of $\mathbb{H}$ is covered.

3.4   Resampling

We are given a hextree R with boundary data along one of its boundaries given by a sequence of edge lengths $Y = \{\ldots, y_{-1}, y_0, y_1, \ldots\}$, and wish to construct another, independent of R conditional on having the same boundary data. It suffices to construct only the layer of RAHs adjacent to the boundary Y, since the remaining RAHs can be chosen by choosing their cross-edge lengths independently of everything as in the original definition of μ. To construct the RAHs adjacent to the boundary Y, it suffices to construct the sequence x_i of edge lengths of the edges incident to Y (ending on Y and perpendicular to it, so that x_i ends between y_i and y_{i+1}) since the three consecutive edges x_{i-1}, y_i, x_i will determine the RAH at position i.

We construct the x_i using a successive approximation procedure, starting initially from the corresponding lengths x_i' of R, that is, from the edge lengths of the edges of R incident to Y from the other side.

We use the fact that the sequence $\{(x_i, y_i)\}$ is Markovian: given one pair (x_i, y_i) the values (x_j, y_j) for $j > i$ are independent of the values (x_j, y_j) for $j < i$. This follows from the construction of μ from the sequence of independent cross-edge lengths: the values (x_j, y_j) for $j > i$ depend only on the choices of cross-edge lengths on one side of x_i; the values (x_j, y_j) for $j < i$ depend only on the cross-edge lengths on the other side of x_i.

From the sequence $\{x_i'\}_{i \in \mathbb{Z}}$, erase all x_{2i}' for all i. Then for each i, resample x_{2i}' according to the conditional measure defined by $x_{2i-1}', y_{2i-1}, y_{2i}, x_{2i+1}'$. By this we mean, by the Markov property, the marginal distribution of x_{2i}' (given everything else) only depends on these four values. Resampling each x_{2i}' is then a straightforward finite-dimensional computation.

Now repeat for the odd indices: erase each x'_{2i+1} and resample it according to its four neighboring values. Now iterate, resampling successively the even values then the odd values over and over. The resampling map preserves the measure μ conditioned on Y, and the x_i's will quickly become decorrelated from their initial values, that is, after a small number of iterations the state is independent of the initial state. To be a little more precise, if we are interested in functions of $\{x_i\}$ which depend only on indices in the interval $[-m, m]$, then after a polynomial number (in m) of iterations the system is close to uncorrelated. These facts have been established in [1] in a similar setting.

Although it should be possible to show that the dependence of x_i on y_j decays exponentially with $|i - j|$, we have not attempted to prove this here.

4   PENTAGONS

Let $\ell_1, \ldots, \ell_5$ be the side lengths of a right-angled pentagon (RAP). Any two of these determine the rest: we have for example

$$\cosh \ell_4 = \sinh \ell_1 \sinh \ell_2,$$

and its cyclic rotations. Explicitly, given ℓ_1, ℓ_2, which must satisfy $\sinh \ell_1 \sinh \ell_2 > 1$, we have

$$\sinh \ell_3 = \frac{\cosh \ell_1}{\sqrt{\sinh^2 \ell_1 \sinh^2 \ell_2 - 1}}$$

$$\sinh \ell_4 = \sqrt{\sinh^2 \ell_1 \sinh^2 \ell_2 - 1}$$

$$\sinh \ell_5 = \frac{\cosh \ell_2}{\sqrt{\sinh^2 \ell_1 \sinh^2 \ell_2 - 1}}.$$

There is a natural measure τ on right-angled pentagons (RAPs) which is invariant under rotations, defined as follows (one can discover this measure in a similar manner as one discovers the measure ν_0 as discussed after the proof of Theorem 1). If we parametrize by ℓ_1, ℓ_2, the density is

$$f(\ell_1, \ell_2) \, d\ell_1 \, d\ell_2 = \frac{C \, d\ell_1 \, d\ell_2}{\sqrt{\sinh^2 \ell_1 \sinh^2 \ell_2 - 1}} = \frac{C \, d\ell_1 \, d\ell_2}{\sinh \ell_4}$$

where $C = K(1/\sqrt{2})^2$, K being the complete elliptic integral of the first kind. For this measure each edge has expected length $\pi/2$.

In terms of exponentials of the lengths we have (with $a = e^{\ell_1}$, and $b = e^{\ell_2}$)

$$f(a, b) \, da \, db = \frac{\frac{1}{4} C \, da \, db}{\sqrt{(ab + a + b - 1)(ab - a + b + 1)(ab + a - b + 1)(ab - a - b - 1)}}.$$

As in the case of right-angled hexagons, one can construct an $\mathrm{Isom}(\mathbb{H})$-invariant measure on tilings of $\mathbb{H}$ with RAPs whose marginal distribution is τ, as follows.

Gluing four RAPs edge-to-edge around a vertex produces a right-angled octagon after merging four pairs of edges. Its four edge lengths $\ell_1, \ell_3, \ell_5, \ell_7$ opposite to the central vertex (which were originally edges of pentagons) satisfy one relation:

$$\cosh \ell_1 \cosh \ell_5 = \cosh \ell_3 \cosh \ell_7. \tag{4}$$

Given four lengths satisfying this relation, there is a one-parameter family of choices for the central four pentagon edges.

We can glue such octagons together along the ℓ_i for i odd, choosing lengths at each glued edge which are independent subject to (4)—as explained below—to make a thickened 4-valent tree, an "octatree." Then we can proceed as above, gluing octatrees to make a tiling of $\mathbb{H}$.

To choose the edge lengths of the cross-edges of the 4-valent tree, subject to (4) at each vertex, proceed as follows. Take a bi-infinite straight path (not turning left or right at a vertex) in the tree through a fixed vertex v; assign values to its edges which are i.i.d. and distributed according to the cosh of the edge lengths. Assign the edges along the other straight path through v also i.i.d. except that the values on the two edges adjacent to v must have product equal to the product of values on the other two edges adjacent to v. Now proceed outwardly from v; along each new bi-infinite geodesic encountered, assign values on it i.i.d. subject to the condition (4) at each vertex. Since this is a tree the reversibility of the assignments guarantees that the resulting object is Isom($\mathbb{H}$)-invariant.

REFERENCES

[1] M. Balazs, F. Rassoul-Agha, T. Seppalainen, The random average process and random walk in a space-time random environment in one dimension. Comm. Math. Phys. 266 (2006) 499–545.

[2] N. Curien, W. Werner, The Markovian hyperbolic triangulation, J.E.M.S. 15, Issue 4 (2013) 1309–1341.

[3] O. Kallenberg, Foundations of modern probability, 2nd ed. (2002) Springer.

[4] R. Kenyon, R. Pemantle, Double-dimers, the Ising model and the hexahedron recurrence, J. Combin. Theory, Series A **137** (2016), 26–63.

On Balanced Planar Graphs, Following W. Thurston

SARAH KOCH AND TAN LEI

1 INTRODUCTION AND RESULTS

Let S^2 denote the topological 2-sphere, equipped with an orientation. Let $f : S^2 \to S^2$ be an orientation-preserving branched covering map of degree $d \geq 2$, let R_f be the set of ramification values (or critical values) of f, and let C_f be the set of critical points of f. We say that f is *generic* if $|R_f| = 2d - 2$; note that if f is a generic branched cover, then $|C_f| = 2d - 2$ as well. Furthermore, every critical point of a generic map f must be simple.

Let $\Sigma \subseteq S^2$ be an oriented Jordan curve running through the $n := 2d - 2$ critical values of f, where we consider Σ to be an oriented graph with vertex set $V_\Sigma = R_f$; each vertex has valence 2. Then the set $f^{-1}(\Sigma) \subseteq S^2$ is an oriented graph in the domain, with vertex set $V_{f^{-1}(\Sigma)} = f^{-1}(R_f)$. Among the $nd - n$ elements of $V_{f^{-1}(\Sigma)}$ are:

- the vertices with valence 4, which comprise the set of critical points C_f, and
- the vertices with valence 2, which comprise the set $f^{-1}(R_f) - C_f$, often referred to as *cocritical points*.

Forgetting the valence 2 vertices in $f^{-1}(\Sigma)$ gives an oriented graph Γ with vertex set $V_\Gamma = C_f$. We call Γ the *underlying 4-valent graph* of $f^{-1}(\Sigma)$. We will refer to a connected component of $S^2 - \Gamma$ as a *face* of Γ.

The question we address in this article is: which oriented 4-valent graphs on S^2 can be realized as the underlying graph of $f^{-1}(\Sigma)$ for some pair f and Σ? This was asked by W. Thurston as part of a group email discussion, which he initiated in Fall 2010, and which was motivated by his visionary question:

Question (W. Thurston, 2010). What is the shape of a rational map?

Throughout the course of this discussion, he proved the following result.

Theorem 1.1. *(W. Thurston, December 2010) An oriented planar graph Γ with $2d - 2$ vertices of valence 4 is equal to $f^{-1}(\Sigma)$ for some degree d branched cover $f : S^2 \to S^2$ and some Σ if and only if*

1. *every face of Γ is a Jordan domain (in particular Γ is connected),*
2. *(global balance) for any alternating white-blue coloring of the faces of Γ, there are d white faces, and there are d blue faces, and*

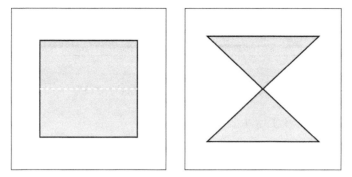

Figure 1: If the dashed curve in the blue region is pinched to a point, the number of blue faces increases by 1, while the number of white faces remains constant.

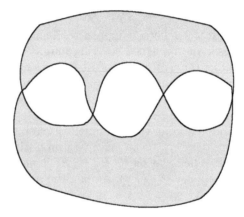

Figure 2: A globally imbalanced graph. There are 4 white faces against 2 blue ones. This figure was made by W. Thurston.

3. *(local balance) for any oriented simple closed curve in* Γ *that is bordered by blue faces on the left and white on the right (except at the corners), there are strictly more blue faces than white faces on the left side of* Γ.

Any oriented 4-valent graph satisfying the three conditions above is called *balanced*. One way to construct arbitrary 4-valent graphs which fail the global balance condition in Theorem 1.1 is to pinch arcs in regions of one color only. The number of faces of that color increases without changing the number of the other faces, see Figure 1. A globally balanced but locally imbalanced graph is shown in Figure 3. Notice that the balance conditions are insensitive to the orientation of the surface, and they are insensitive to the choice of blue and white faces in an alternating coloring of the faces.

Before giving the proof of Theorem 1.1, we will play with some examples and study the graphs arising from maps $f : S^2 \to S^2$ of degree 4 (which is exactly how Bill invited us to participate in the original email discussion where this work began).

Outline

In Section 2, we begin by discussing Hurwitz numbers for generic branched covers $S^2 \to S^2$ of degree d, and we then present a complete list of graphs arising in the

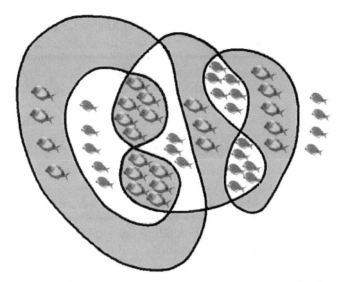

Figure 3: Give the surface an orientation, and orient the graph above. There are equal numbers of blue and white faces, so the graph Γ is globally balanced; however, the right side of the graph violates the local balance condition, regardless of the orientation chosen. Note that in each face of this diagram, the number of fish plus the number of corners is equal to $8 = 2(5) - 2$. This figure is a slightly modified version of one made by W. Thurston.

case $d = 4$. We then establish the proof of Theorem 1.1 in Section 3, closely following Thurston's original proof. We have added some details (mainly for our own understanding). In Section 4, we present some of W. Thurston's work on decomposing balanced planar graphs into "standard pieces." In the appendix, we prove Hurwitz's classical result, that (up to isomorphism) the number of generic degree d branched covers $S^2 \to S^2$ is

$$\frac{(2d - 2)! d^{d-3}}{d!},$$

presenting a geometric proof due to Duchi-Poulalhon-Schaeffer in [DPS].

Acknowledgments

First and foremost, we would like to thank Bill Thurston for starting this email discussion in 2010, engaging a large number of mathematicians in this project, and sharing his work with all of us. It is with great pleasure that we present this part of his work. Thanks to Laurent Bartholdi, Dylan Thurston, and Jerome Tomasini for helping with this manuscript in various ways. The research of Sarah Koch is supported in part by NSF grant DMS 1300315 and a Sloan Research Fellowship, and the research of Tan Lei is supported in part by ANR LAMBDA.

2 EXAMPLES AND A COMPLETE LIST OF QUARTIC DIAGRAMS

2.1 Hurwitz numbers

Let $f : S^2 \to S^2$ be a generic orientation-preserving branched cover of degree $d \geq 2$. Label the critical values $R_f = \{v_1, \dots, v_{2d-2}\}$, and let Σ be an oriented Jordan curve running through the critical values following the order of the labels. Color the left complementary disc of Σ blue, and color the right complementary disk white. Consider Σ as a graph with $V_\Sigma = R_f$, where each $v \in V_\Sigma$ has valence 2.

Then $f^{-1}(\Sigma)$ is an oriented graph with vertices $V_{f^{-1}(\Sigma)} = f^{-1}(R_f)$; there are precisely $2d - 2$ vertices in $V_{f^{-1}(\Sigma)}$ with valence 4 (corresponding to the critical points of f), and the other vertices in $V_{f^{-1}(\Sigma)}$ have valence 2. We label the vertices of $f^{-1}(\Sigma)$ as follows: for each $c \in V_{f^{-1}(\Sigma)}$, label the vertex c with the number j if $f(c) = v_j$. In this way, there are $d - 1$ vertices in $V_{f^{-1}(\Sigma)}$ labeled 1, $d - 1$ vertices in $V_{f^{-1}(\Sigma)}$ labeled 2, and so on. Note that because $f : S^2 \to S^2$ is generic, for each $1 \leq j \leq 2d - 2$, there is a unique vertex labeled j with valence 4. Color each face of $f^{-1}(\Sigma)$ either white or blue according to the color of its image, under f, in $S^2 - \Sigma$. The result is a checkerboard pattern in the domain of $f : S^2 \to S^2$.

Two such labeled graphs are said to be *isomorphic* if there is an orientation-preserving homeomorphism $\phi : S^2 \to S^2$ sending one oriented graph to the other, preserving the labels of the vertices. According to Hurwitz, the number of isomorphic labeled graphs (of degree d) is

$$\frac{(2d - 2)!}{d!} d^{d-3}$$

(see the appendix for further details). When $d = 4$, this number is 120.

2.2   The case $d = 4$: cataloguing all 120 quartic diagrams

We illustrate the complete list of all 120 labeled graphs arising from $f : S^2 \to S^2$, where $d = 4$. They are organized into groups, catalogued by their underlying 4-valent graphs. The data are based on a list, computed by L. Bartholdi with Gap, of 6-tuples of transpositions in S_4, with trivial product, modulo diagonal conjugation by S_4. The number of such tuples is equal to the number of isomorphism classes of generic degree 4 maps $f : S^2 \to S^2$.

2.2.1   Constructing Γ from $f : S^2 \to S^2$

In the case $d = 4$, we recall the steps used to construct Γ, the oriented underlying 4-valent graph of the branched cover $f : S^2 \to S^2$:

1. Take an oriented Jordan curve $\Sigma \subseteq S^2$ in the range running through the 6 critical values $R_f = \{v_1, \dots, v_6\}$, and consider $f^{-1}(\Sigma) \subseteq S^2$ as an oriented graph in the domain.
2. There are 6 vertices in $f^{-1}(\Sigma)$ with valence 4 (corresponding to the critical points of f); there are 12 vertices in $f^{-1}(\Sigma)$ with valence 2 (the cocritical points).
3. Each face of $f^{-1}(\Sigma)$ has 6 vertices on its boundary.
4. The underlying oriented 4-valent graph Γ is the oriented graph $f^{-1}(\Sigma)$ obtained by forgetting the vertices which have valence 2.

2.2.2   Constructing $f : S^2 \to S^2$ from a diagram

For each of the diagrams below, we build a degree 4 branched cover $f : S^2 \to S^2$ in the following way:

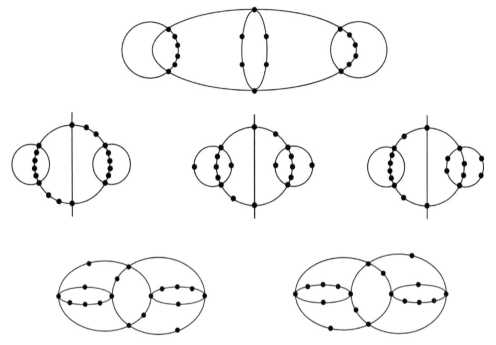

Figure 4: These six graphs represent 36 branched covers; each graph above gives rise to three distinct branched covers $f : S^2 \to S^2$, up to isomorphism.

1. Choose an alternating coloring of the faces (a checkerboard-coloring).
2. Choose one 4-valent vertex (that is, a critical point) to map to the critical value v_1; label this critical point with a "1."
3. Label the remaining cocritical points and critical points so that they appear in counterclockwise order on the boundary of each blue face, see Figure 10 for this labeling scheme.
4. Define the covering f first on the vertices following the labeling, and then on the edges by homeomorphic extension, and finally on the faces by homeomorphic extension (with a little help from the Schöenflies theorem).

Remark 2.1. The recipe above for building a map $f : S^2 \to S^2$ from the labeled diagram does not always work; for example, there is a labeled diagram that satisfies the conditions above, but there is no branched cover $f : S^2 \to S^2$ which gives rise to the labeled graph in Figure 10, or in fact to the underlying unlabeled graph in Figure 6.

Once a diagram is realizable as an underlying oriented 4-valent graph for some f and Σ, there are

- 6 choices of the critical point to label with "1," and
- 2 choices of the set of white/blue faces.

So in general, each diagram below gives rise to 12 distinct labeled diagrams. However, some of these diagrams have extra symmetry, and in these cases, there are fewer than 12 labeled diagrams (up to isomorphism).

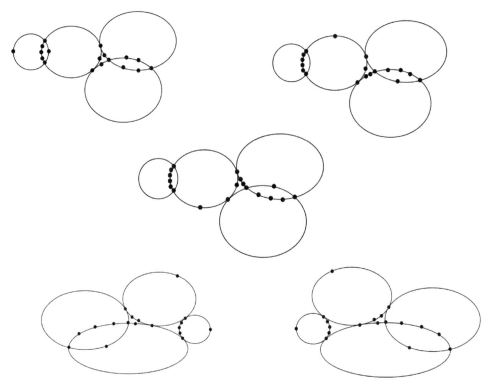

Figure 5: These five graphs represent 60 branched covers; each graph above gives rise to 12 distinct branched covers $f : S^2 \to S^2$, up to isomorphism.

3 PROOF

We will prove Theorem 1.1, following the main ideas in W. Thurston's original proof. Recall the statement:

An oriented planar graph Γ with $2d - 2$ vertices of valence 4 is equal to $f^{-1}(\Sigma)$ for some degree d branched cover $f : S^2 \to S^2$ and some Σ, if and only if

1. *every face of Γ is a Jordan domain (in particular Γ is connected),*
2. *(global balance) for any alternating white-blue coloring of the faces of Γ, there are d white faces, and there are d blue faces, and*
3. *(local balance) for any oriented simple closed curve in Γ that is bordered by blue faces on the left and white on the right (except at the corners), there are strictly more blue faces than white faces on the left side of Γ.*

Proof. The key idea is to first translate the realization problem into finding a pattern of vertices (either 2-valent or 4-valent) so that each face of Γ has precisely $n := 2d - 2$ vertices on the boundary, and then to reduce the problem into a matching problem in graph theory. Indeed, let Σ be an oriented simple closed curve which goes through the critical values of a degree d branched cover $f : S^2 \to S^2$. Consider the oriented graph $f^{-1}(\Sigma)$; it has n vertices of valence 4, and $n(d-1) - n$ vertices of valence 2. Split each 2-valent vertex into two "dots," putting one into each of the two adjacent neighboring regions. Then split each 4-valent vertex (or corner) into 4 "dots," putting one into each of the four neighboring regions. Every face of $f^{-1}(\Sigma)$

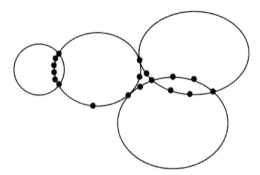

Figure 6: This diagram cannot be realized as the underlying diagram for any $f:$ $S^2 \to S^2$. In fact, any labeling of this diagram will result in a "duplicate critical value." See Figure 10.

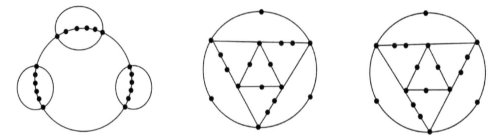

Figure 7: These three graphs represent 6 branched covers; each graph above gives rise to two distinct branched covers $f: S^2 \to S^2$, up to isomorphism.

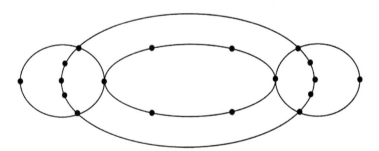

Figure 8: This graph represents 6 branched covers up to isomorphism.

then contains precisely n dots in its interior. Grouping the dots back together at the vertices then becomes a matching problem.

Now take an arbitrary oriented graph Γ with n 4-valent vertices and no other vertices. Color the faces in a blue and white checkerboard pattern. Put n men in each blue face (represented by blue fish plus corners in Figure 3), and n women in each white face (represented by red fish plus corners in Figure 3). Each is trying to find a partner from one of the neighboring faces. In addition, each corner requires one person per neighboring face (that makes two women and two men). Is there a perfect matching? The usual marriage criterion in the marriage theorem from

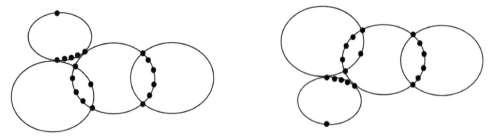

Figure 9: These two graphs represent 12 branched covers; each graph above gives rise to six distinct branched covers $f : S^2 \to S^2$, up to isomorphism.

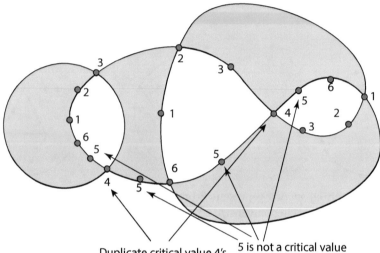

Duplicate critical value 4's   5 is not a critical value

Figure 10: Note how the numbering goes clockwise around each white face and counterclockwise around each blue face. There's a duplicate critical value: the two vertices labeled 4 map to the same point in the codomain, and none of the vertices labeled 5 is a critical point, so there are only five distinct critical values in the codomain, corresponding to the labels: $1, 2, 3, 4, 6$. This figure was made by W. Thurston.

graph theory states that for every set of N women, there are at least N men who are potential mates; this can be reduced to the global balance and local balance conditions in the theorem. Here are the details:

Necessity, condition 1: complementary Jordan domains.
Let $f : S^2 \to S^2$ be a generic orientation-preserving branched cover of degree d, and let Σ be an oriented simple closed curve passing though the n critical values of f. Note that f restricts to a degree d covering map $S^2 - f^{-1}(\Sigma) \to S^2 - \Sigma$; it follows that the connected components of $S^2 - f^{-1}(\Sigma)$ are Jordan domains.

Necessity, condition 2: global balance.
Change coordinates in the range so that Σ becomes the unit circle S^1, and orient it so the unit disk is on the left, and color the unit disk blue. The white face is the

complement. We may also assume that the point 1 is not a critical value. Now the oriented graph we are considering in the domain is $f^{-1}(S^1)$. Each of the blue faces in $S^2 - f^{-1}(S^1)$ contains one zero of f, and each of the white faces in $S^2 - f^{-1}(S^1)$ contains one pole of f. The global balance condition is equivalent to the condition that the number of zeros of f equals the number of poles of f, which is equal to the degree of f.

Necessity, condition 3: local balance.
Choose an alternating coloring of the complementary components of $f^{-1}(\Sigma)$. Let Λ be a Jordan domain bounded by an oriented curve γ in $f^{-1}(\Sigma)$ bordered by blue faces on the left. Suppose that in Λ there are W white faces, B blue faces, E edges in the interior of Λ, and X edges on the boundary. Recall that the number of women in a white face plus the number of corners equal n. It follows that the total number of women in all the white faces is $nW - E$. Similarly, the total number of men in the blue faces is $nB - E - X$, with $X > 0$. For these $nW - E$ women, all of their potential mates are among the $nB - E - X$ men. Necessary and sufficient conditions from the marriage lemma imply $nW - E \leq nB - E - X$, from whence it follows that $B > W$. This establishes the local balance condition of the theorem.

Sufficiency, step I: adding 2-valent vertices to Γ.
Suppose that Γ is an oriented 4-valent graph which satisfies the conditions of the theorem. Every face F of Γ is a Jordan domain, so the number of edges on the boundary of F is equal to the number of corners of F. For any face with k corners, we have $k \leq n$. After matching the corners, there are $n - k$ people left in the face to match. Since the numbers of faces of the two colors are equal, the total number of men who are not matched with corners is equal to the total number of women who are not matched with corners.

In what follows, we will consider the men and women who have not been matched with corners. Let S be any subset of these women. We want to show that the set of potential mates of S is at least as large as S (satisfying the condition in the marriage theorem).

If we augment the set S with all women in every face containing an element of S, this will not increase the number of their potential mates, so we may assume S is actually the union of all women in some collection U of white faces. Let R be the closure of U with all adjacent neighboring blue faces of U. Then the men in R are exactly the potential mates for the women in S. The orientation of the boundary edges of R leaves R on the left, and leaves blue faces on the left as well.

If the interior of R is not connected, then the women in one connected component can only match with men in that same connected component. We will therefore establish that there are enough mates for the elements of S in each connected component of the interior of R, from whence it will follow that there are enough mates in R for all of the elements of S. To this end, we assume that the interior of R is connected, so that every complementary component of R in S^2 is a Jordan domain.

Denote the number of men/women inside/outside R by the following tableau:

	Number of Men	Number of Women
Inside R	m	$w := |S|$
Outside R	m^*	w^*

Assume at first that R is simply-connected, that is, R is bounded by a simple closed curve leaving blue faces on the left.

Denote by:

- X_1 the number of boundary corners of R which are on the boundary of only one face in R (this face is necessarily blue);
- X_3 the number of boundary corners of R that are on the boundary of three faces in R (necessarily two blues and one white);
- v the number of vertices in the interior of R;
- W the number of white faces in R; and
- B the number of blue faces in R.

Note that every boundary corner of R belongs to one of the two types either X_1 or X_3, and

$$w = |S| = nW - (2v + X_3), \quad m = nB - (X_1 + 2X_3 + 2v).$$

By the local balance condition, we have $W \leq B - 1$. Therefore, using $X_1 + X_3 \leq n$, we get

$$w = nW - (2v + X_3) \leq n(B - 1) - (X_3 + 2v)$$
$$= nB - (n + X_3 + 2v)$$
$$\leq nB - (X_1 + 2X_3 + 2v) = m.$$

This is the desired inequality.

We now deal with the case where R is not simply-connected. The number of women in the complement of R is w^*, and the number of men in the complement of R is m^*. If we can show that $m^* \leq w^*$, the fact that $m + m^* = w + w^*$ will imply that $w \leq m$, which is the desired inequality.

Let Y be a complementary component of R. It is bounded by a Jordan curve keeping Y and white faces on the left. The same curve in the opposite direction, enclosing the complement of Y, has blue faces on the left. So by the local balance condition there are more blue faces than white faces outside Y. Using the assumption that there are equal numbers of blue and white faces (the global balance condition), there are more white faces than blue faces in Y. We then apply the same argument above in the case that R is simply-connected (replacing R by Y and reversing the colors) to conclude that the set of men in Y have enough mates (necessarily in Y); this establishes that $m^* \leq w^*$.

We have now proved that for any set S of women, the number of potential mates is at least as large as S. By the marriage theorem, there is a perfect matching. For each matched pair, merge them into a vertex on one common boundary edge of the two faces. Make sure that the vertices are pairwise disjoint. We have now enriched Γ into a graph with $n(d - 1)$ vertices, where n are 4-valent and the rest are 2-valent. Furthermore, every face has exactly n vertices on its boundary.

Sufficiency, step II: consistent labeling of the vertices.

We now want to label the enriched vertices of Γ by the set $\{1, \ldots, n\}$, such that each label appears exactly $d - 1$ times, and the natural order of the labels appears counterclockwise around each blue face (and therefore clockwise around each white face), as shown in Figure 10.

Consider $V_\Gamma \times \mathbb{Z}/n\mathbb{Z}$, where V_Γ denotes the set of vertices of Γ. For every edge e, from vertex v to vertex v' define $h_e : \{v\} \times \mathbb{Z}/n\mathbb{Z} \to \{v'\} \times \mathbb{Z}/n\mathbb{Z}$ by

$$h_e : (v, x) \mapsto (v', x + 1 \bmod n).$$

The collection of h_e forms a cocycle with coefficients in $\mathbb{Z}/n\mathbb{Z}$. Since $H^1(S^2) = 0$, this is the coboundary of a function to $\mathbb{Z}/n\mathbb{Z}$.

More generally: consider a cocycle with values in $S^1 = \mathbb{R}/2\pi\mathbb{Z}$, and parametrize $\mathbb{RP}^1$ by a circle-valued coordinate, then a particular cocycle yields exact positions for the critical values, up to rotations of the circle (constant of integration).

Sufficiency, step III: avoiding duplicate critical values.

It may happen, as we see in the quartic diagrams (see Figure 10) that two critical points have the same label. In this case, we may perturb the graph to obtain matchings and labels so that the n critical points have pairwise distinct labeling.   q.e.d.

3.1   Further comments

It would be nice to have a better proof of sufficiency that can be turned into an algorithm to construct arbitrary rational functions whose critical values are on the unit circle. There should be some kind of cell structure whose lower-dimensional cells correspond to subdivisions of the sphere with vertices of higher even order as critical points coalesce, obtained by collapsing forests contained in the 1-skeleton of the subdivisions in the generic case. Collisions between critical points is related to the strata in Hurwitz's study.

Graph flow

One way to test whether a graph is balanced or not is to directly test a possible matching. There are good algorithms for finding matchings when they exist, or finding an obstruction when they don't exist: this is very related to graph flow, which has fast algorithms.

Here is a graph flow version of the balance condition: let Γ be a connected 4-valent graph with an alternating coloring of its faces. Give each face F a weight $w(F)$ equal to the number of vertices of Γ minus the number of corners of F. Define an abstract graph flow Σ whose vertices are $\{v(F), F$ a face of $\Gamma\}$ and whose directed edges are of the form $\{(v(F), v(F')\}$ for every ordered pair (F, F') of blue-white faces sharing one or more boundary edges in Γ. Add two vertices D (for departure) and A (for arrival) to Σ. Direct an edge from D to every vertex $v(F)$ of blue face F. This edge has capacity $w(F)$. Direct an edge from every vertex $v(F')$ of white face F' to A, with capacity $w(F')$. Now the map is balanced if the maximal flow reaches the maximal capacity $\sum_{F \text{ blue}} w(F)$.

4   DECOMPOSITIONS OF BALANCED PLANAR GRAPHS

Our objective in this section is to understand the structure of balanced graphs from the point of view of decomposing them into standard pieces.

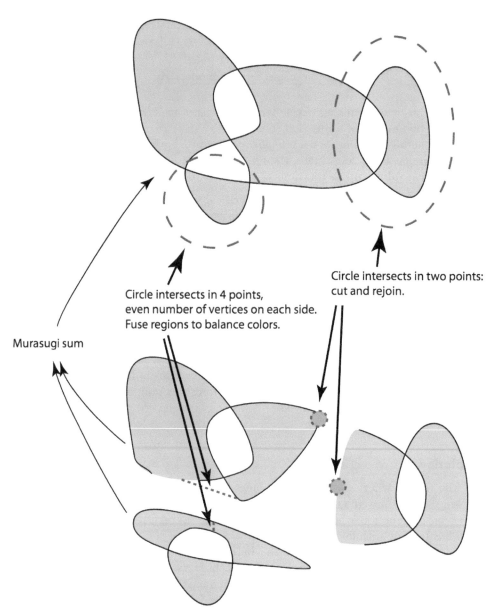

Circle intersects in two points: cut and rejoin.

Circle intersects in 4 points, even number of vertices on each side. Fuse regions to balance colors.

Murasugi sum

Figure 11: The dashed Jordan curve on the right cuts Γ in exactly two points and they are on distinct edges; it is nontrivial. This figure was made by W. Thurston.

4.1   A *22 decomposition

Start with a planar 4-valent graph Γ. Find a nontrivial Jordan curve γ in S^2 that intersects Γ in only two points (here nontrivial means that the two intersection points are on different edges). This happens exactly when two faces of opposite colors have more than two common edges. Cut S^2 along γ and collapse each of the wound curves into a single point (these points will become regular points of the surgered graph). Continue this process on the two new diagrams until there are no longer any nontrivial curves with two intersection points.

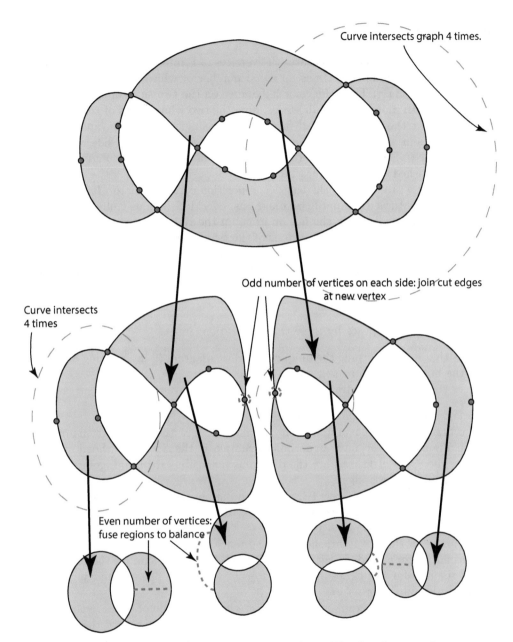

Figure 12: The two types of tangle decompositions. The 2-valent vertices are not taken into consideration; they should be added later. This figure was made by W. Thurston.

4.2   Two types of tangle decompositions

Now look for a nontrivial Jordan curve that intersects Γ in four points; here nontrivial means that the curve does not go around a single vertex, and that the four points are on four distinct edges.

If the curve γ separates the 4-valent vertices of Γ into a pair of odd numbers of vertices, cut S^2 along γ and collapse each of the wound curves into a single point.

The single point becomes a new 4-valent vertex of the surgered graph. See the top right dashed curve in Figure 12.

Assume that the curve γ separates the vertices of Γ into a pair of even numbers of vertices, and assume that Γ has an equal number of white and blue faces. Cut the sphere S^2 along γ. Then the color imbalance on the two sides of the cut is ± 1. For the side with more white (half)-faces, glue the two white edges of the cut copy of γ to connect the two white half faces into one single white face (and fold the two blue edges into two segments), and symmetrically for the other side with more blue (half)-faces. See for example the middle two dashed curves in Figure 12 and the top left dashed curve in Figure 11.

The converse procedure of the last decomposition corresponds to *Murasugi sum* of knots. To do Murasugi sum of two diagrams, consider them as embedded in two spheres. Place one sphere on the left and one on the right (in 3-space). Remove a rectangle from oppositely colored regions of the two diagrams, where the rectangles have two edges on different sides of a region and two edges interior to the region. Then glue them so as to match colors. The two spheres are merged into a single one.

4.3   Indecomposable pieces

Every diagram decomposes by these three operations into

- pieces that are isomorphic the unique quadratic diagram (2 intersecting circles—see the bottom of Figure 12), and
- pieces that are not quadratic diagrams and do not have any cutting curves.

These latter pieces are called *hyperbolic pieces*.

The first hyperbolic example is the octahedron (see the two diagrams on the right in Figure 7). The next hyperbolic example is the 3×4 turkshead, with 8 vertices. The $3 \times n$ turkshead are the diagrams for a standard braid, going around in a circle with $2n$ crossings.

The 3×1 turkshead is $\{|z| = 1 \pm \varepsilon\}$ union an arc in the intersection of the upper half plane with the annulus, connecting $1 + \varepsilon$ to $-1 + \varepsilon$ and the mirror symmetry by complex conjugation, with two vertices $\pm(1 + \varepsilon)$. The $3 \times n$ turkshead is the inverse image of the 3×1 turkshead by the map $z \mapsto z^n$. This is a lacing pattern for a drum, where a cord zigzags back and forth between a skin on the top and a skin on the bottom and closes in a loop after $2n$ segments. The symmetry group is the dihedral group of order $2n$.

Any 4-valent planar graph is a *link projection*: a projection of a link in the 3-ball to the boundary surface in generic positions. There is a nice theory that Martin Bridgeman [Br] worked out for the structure of all possible hyperbolic diagrams, although it doesn't take into account the balancing conditions: they're all generated from $3 \times n$ turkshead examples by choosing an arc cutting across a region between nonadjacent edges but being the first and the third edges of three consecutive edges, and collapsing the arc to a point, thus creating a new vertex. If you do this once in a region of each color, the global balance will be retained.

Problem

(a) Analyze when the balancing conditions are maintained by these various operations and their inverses. (This topic has recently been developed by J. Tomasini, [To].)

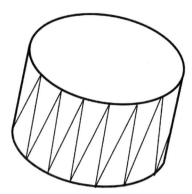

Figure 13: Turkshead.

(b) Analyze the set of valid matchings, and the set of valid labelings under these operations: that is, dots are uniquely partitioned (or not) to decomposed pieces.

APPENDIX A. HURWITZ NUMBERS AS SEEN BY DUCHI-POULALHON-SCHAEFFER

Where does the number $\dfrac{(2d-2)!}{d!}d^{d-3}$ come from? This is a classical result due to Hurwitz, [H]. There are a number of proofs with different approaches, some applicable to larger classes of Hurwitz numbers. See for example [CT, ELSV, St, Va]. Here we sketch a combinatorial approach due to Duchi-Poulalhon-Schaeffer, [DPS].

Take a tree (non-embedded) with d white vertices and $d-1$ edges. Label the edges by $1, \cdots, d-1$, in blue. The number of such edge-labeled trees is known[1] to be d^{d-3}.

For each such tree, put a blue vertex in the middle of every edge and label it by the original edge label. Now label the new set of edges by $1, \ldots, 2d-2$, in red. There are $(2d-2)!$ choices.

So all together there are $(2d-2)! d^{d-3}$ such bipartite edge-labeled and blue-vertex-labeled trees.

Embed (in a unique way) the tree in the plane such that around each white vertex the increasing blue labels of the incident edges appear clockwise. There are $(2d-2)! d^{d-3}$ such bipartite edge-labeled and blue-vertex-labeled embedded trees.

Take a generic branched cover $f : S^2 \to S^2$ of degree d with all critical values on the unit circle, labeled in cyclic order by $1, \ldots, 2d-2$, in red. Color the point 0 in blue and the point ∞ in white. Add in $2d-2$ disjoint rays (in black) from 0 to ∞ so that each complementary region contains a unique critical value. Call this graph S. Now consider the pullback $f^{-1}(S)$ and collapse each face containing a cocritical point to a single edge, obtaining a graph G. It has $2d-2$ faces each with a red

---

[1]Choose a root at a white vertex and label it by d. This root choice together with the edge labeling gives a unique labeling of the vertices by $1, 2, \ldots, d-1, d$ (just label the far end of an edge by the label of the edge). So each root choice gives a Cayley tree and the number of such trees is known to be d^{d-2}.

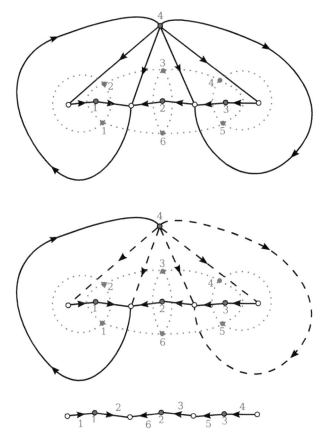

Figure 14: From G to the spanning tree, and then to the chopped spanning tree.

label of a critical point. This is the dual graph to the underlying 4-valent graph $f^{-1}(S^1)$.

The isomorphism class of f is uniquely determined by this face-labeled graph G.

Now label the blue vertices of G by $1, \cdots, d$ (there are $d!$ such choices). Duchi-Poulalhon-Schaeffer show that there is a bijection from these face-labeled and blue-vertex-labeled graphs and the above bipartite edge-labeled and blue-vertex-labeled embedded trees. So the number of isomorphism classes is $\dfrac{(2d-2)!}{d!}d^{d-3}$.

In fact Duchi, Poulalhon, and Schaeffer take the blue vertex labeled d as a root and find a unique spanning tree, as follows:

- Embed G in the plane so that the blue vertex with label d is accessible from the unbounded component. Orient each edge so that the incident face with the greater label is on the left. Every blue vertex has a unique incoming edge and every white vertex has a unique outgoing edge. See Figure 14.
- Now repeatedly reverse orientations of clockwise cycles. This operation does not change the incoming and outgoing degrees of each vertex. A theorem of Felsner in [F] implies that after finitely many such cycle reversals, we obtain an orientation without clockwise cycles, and this orientation is unique.

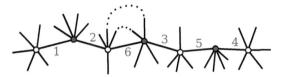

Figure 15: From a hairy tree and matched hairs (only two steps with dotted arcs are shown) in order to obtain $f^{-1}(\Sigma)$.

- Start from the blue vertex labeled d as root and do a rightmost deep-first search in the sense opposite to the edge orientation, while erasing all other incoming edges to each visited blue vertex (including those incident to the root vertex), to get a canonical spanning tree whose edges are oriented toward the root, as done by Bernardi ([Be]). See Figure 14. Chop off the root vertex and its unique incident edge. Label in red each edge of the remaining tree by the label of the white-to-blue right side face. For visual convenience put the label on the same side as the face. Erase the orientation of the tree edges. See Figure 14.

The inverse algorithm from an embedded edge-labeled tree to $f^{-1}(\Sigma)$ is also easy to describe; see Figure 15. Attach some half edges (hairs) to each white and blue vertex to reach the local valence value $2d - 2$ and to match the local cyclic order with the labels of the tree edges (counterclockwise for blue and clockwise for white). Now take a blue-incident half edge whose left edge is a full edge, and walk in clockwise order, while matching each visited blue-incident half edge to the closest available white-incident half edge in the counterclockwise direction. Put an extra blue vertex together with $2d - 2$ half edges in the unbounded face and match them with the remaining $2d - 2$ white-incident half edges.

REFERENCES

[Be]   O. Bernardi, Bijective counting of tree-rooted maps and shuffles of parenthesis systems. Elec. J. of Comb., 14(1):R9, 2007.

[Br]   M. Bridgeman, The structure and enumeration of link projections. Trans. Amer. Math. Soc. 348 (1996), no. 6, 2235–2248.

[CT]   M. Crescimanno and W. Taylor, Large N phases of chiral QCD2, Nuclear Phys. B 437 no. 1 (1995), 3–24.

[DPS] E. Duchi, D. Poulalhon and G. Schaeffer, Uniform random sampling of simple branched coverings of the sphere by itself, Proceedings of the Twenty-Fifth Annual ACM-SIAM Symposium on Discrete Algorithms, 2014, 294–304.

[ELSV] T. Ekedahl, S. Lando, M. Shapiro and A. Vainshtein, Hurwitz numbers and intersections on moduli spaces of curves. Invent. Math. 146 (2001), no. 2, 297–327.

[F]   S. Felsner. Lattice structures from planar graphs. Elec. J. of Comb., 11(1): R15, 2004.

[H] A. Hurwitz, Über die Anzahl der Riemann'schen Flächen mit gegebenen Verzweigungspunkten, Math. Ann. 55 (1902), 53–66.

[St] V. Strehl, Minimal transitive products of transpositions—the reconstruction of a proof of A. Hurwitz. (English summary) Sém. Lothar. Combin. 37 (1996), Art. S37c, 12 pp.

[To] J. Tomasini, On the branched coverings of the sphere: a combinatorial invariant, Preprint 2014.

[Va] R. Vakil, Genus 0 and 1 Hurwitz numbers: recursions, formulas, and graph-theoretic interpretations, Trans. Amer. Math. Soc. 353 (2001), 4025–4038.

Construction of Subsurfaces via Good Pants

Yi Liu and Vladimir Markovic

The good pants technology is a systematic method that has been developed during the past few years to produce surface subgroups in cocompact lattices of $\mathrm{PSL}_2(\mathbf{C})$ or $\mathrm{PSL}_2(\mathbf{R})$ using frame flow. The $\mathrm{PSL}_2(\mathbf{C})$ case leads to a proof of the Surface Subgroup Conjecture, which plays a fundamental role in the resolution of the Virtual Haken Conjecture about 3-manifold topology, and the $\mathrm{PSL}_2(\mathbf{R})$ case leads to a proof of the Ehrenpreis Conjecture on Riemann surfaces. Prior to the work of Jeremy Kahn and Vladimir Markovic, Lewis Bowen first attempted to build surface subgroups by assembling pants subgroups using horocycle flow. This article is intended to guide the readers to these constructions due to Jeremy Kahn and Vladimir Markovic, and survey on some applications of their construction afterwards due to Yi Liu, Hongbin Sun, Ursula Hamenstädt and others.

1   GOOD PANTS AND NICE GLUING

For any cocompact lattice of $\mathrm{PSL}_2(\mathbf{C})$ or $\mathrm{PSL}_2(\mathbf{R})$, any surface subgroup resulting from the Kahn–Markovic construction can be thought of as the subgroup $\pi_1(S)$ corresponding to a π_1-injectively immersed subsurface S of the quotient orbifold M of the underlying symmetric spaces $\mathbb{H}^3$ or $\mathbb{H}^2$ by the lattice. This allows us to discuss the construction from the perspective of hyperbolic geometry. In this section, we set up the general framework of the construction.

Without loss of generality, we may pass to a sublattice of finite index, and assume hereafter that M is a closed orientable hyperbolic manifold of dimension 3 or 2. The construction can be performed for any positive constant ϵ. It gives rise to a closed orientable subsurface S of the closed hyperbolic orbifold M that is $(1+\epsilon)$-bilipschitz equivalent to a finite cover of a model surface $S_0(R)$, as we describe in the following. The positive parameter R can be chosen arbitrarily as long as R is sufficiently large depending on ϵ and the lattice, and $S_0(R)$ has a defining pants decomposition which lifts to be a pants decomposition of the finite cover.

The model surface $S_0(R)$ is an orientable closed surface of genus 2, equipped with a hyperbolic structure described in term of a pants decomposition and its Fenchel–Nielsen parameters. For any positive constant R, a disk with two cone points of order 3 can be endowed with a unique hyperbolic structure to become a hyperbolic 2-orbifold with boundary length R. The unique planar surface cover

---

Supported by NSF grant No. DMS 1308836.

of minimal degree of this 2-orbifold is a hyperbolic pair of pants $\Pi(R)$ with cuff length R which is symmetric under an isometric action of the 3-cyclic group. Take two oppositely oriented copies of the hyperbolic pair of pants, $\Pi(R)$ and $\bar{\Pi}(R)$. We glue $\Pi(R)$ and $\bar{\Pi}(R)$ by identifying their cuffs accordingly, but modify the gluing by a twist of length 1 along each identified curve in the direction induced from boundary of the pants. Note that there is no ambiguity for the twisting direction since the induced direction from $\Pi(R)$ and $\bar{\Pi}(R)$ are opposite to each other. The resulting hyperbolic surface is denoted as $S_0(R)$.

For universally large R, the injectivity radius of $S_0(R)$ is bounded uniformly from 0. It follows that the hyperbolic structure of $S_0(R)$ stays in a compact subset of the moduli space of hyperbolic structures on the underlying topological surface. For any finite cover $\tilde{S}_0(R)$ of $S_0(R)$ to which the pants decomposition lifts, $\tilde{S}_0(R)$ can be constructed by taking copies of the model pants and assembling according to the model gluing. The constant 1 twist in the gluing guarantees that the hyperbolic structure of $\tilde{S}_0(R)$ does not vary significantly under slight perturbation of the cuff length and the twist parameter of the gluing. The intuition leads to the following notion of good pants and nice gluing.

Let M be a closed orientable hyperbolic manifold of dimension 3 or 2. A π_1-injectively immersed pair of pants Π in M can be homotoped so that the cuffs are mutually distinct geodesically immersed curves. Fix a seam decomposition of Π into two hexagons along three arcs, called seams, which join mutually distinct pairs of cuffs. We may further homotope the immersion so that the seams are geodesic of the shortest length. Assuming that none of the seams degenerate to a point, the seams are perpendicular to the cuffs at all the endpoints, and the two hexagons are in fact isometric to each other. The inward unit tangent vectors of the seams at their endpoints are called *feets* of the seams on the cuffs. Therefore, the immersed pair of pants Π has six feets, each cuff carrying a pair of feets as unit normal vectors at an antipodal pair of points. We say that Π is (R, ϵ)-*good* if the length of each cuff of Π is approximately R up to error ϵ, and if the parallel transportation of a foot on each cuff to the antipodal point along the cuff is approximately the other foot up to error $(\epsilon/2)$, measured in the canonical metric on the unit vector bundle of M. Note that the second condition is automatically satisfied if M has dimension 2.

An immersed oriented subsurface S of M is said to be (R, ϵ)-*panted* if it has a decomposition along simple closed curves into pairs of pants, and the immersion restricted to each pair of pants is (R, ϵ)-good. In general, S is not necessarily π_1-injective since we have not controlled the gluing. We say that S is constructed from the (R, ϵ)-good pants by an (R, ϵ)-*nice* gluing, if for each decomposition curve c shared by pairs of pants P and P', the parallel transportation of distance 1 along c of any foot of P on c, in the direction induced from P, is approximately opposite to a foot of P' up to error (ϵ/R).

Theorem 1.1. *Let M be a closed orientable hyperbolic manifold of dimension 3 or 2. The following statement holds for any small positive constant ϵ depending on M and sufficiently large positive constant R depending on ϵ and M.*

Suppose that S is an immersed closed (R, ϵ)-panted subsurface of M constructed by an (R, ϵ)-nice gluing. Then S is π_1-injective and geometrically finite. Furthermore, S can be homotoped so that with respect to the path-induced metric, S is $K(\epsilon)$-bilipschitz equivalent to a finite cover of the model surface $S_0(R)$, where $K(t)$ is a positive function depending on M such that $K(t) \to 1$ as $t \to 0$.

See [Kahn–Markovic 2014, Theorem 2.1] and [Kahn-Markovic 2012, Theorem 2.1].

The construction of Kahn and Markovic produces closed subsurfaces of M by gluing a finite collection of good pants in a nice fashion. Theorem 1.1 describes the geometry of the resulting subsurfaces. To find a finite collection of good pants that admits a nice gluing is the difficult part of their construction, which can be reformulated as a problem of linear programing regarding measures of good pants.

Denote by $\mathbf{\Pi}_{R,\epsilon}$ the collection of the homotopy classes of oriented (R, ϵ)-good pants of M. Any finite collection of (R, ϵ)-good pants can be recorded by a counting measure over the set $\mathbf{\Pi}_{R,\epsilon}$. Denote by $\mathbf{\Gamma}_{R,\epsilon}$ the collection of the homotopy classes of (R, ϵ)-good curves of M, namely, geodesically immersed oriented loops of M of length approximately R and monodromy approximately trivial, up to error ϵ. There is a natural boundary operator between Borel measures over the sets of good pants and good curves:

$$\partial \colon \mathcal{M}(\mathbf{\Pi}_{R,\epsilon}) \to \mathcal{M}(\mathbf{\Gamma}_{R,\epsilon}),$$

which takes the atomic measure supported over any good pants Π to the sum of atomic measures supported over the cuffs of Π. The boundary operator ∂ can be lifted to an operator ranged in Borel measures over unit normal vectors on good curves:

$$\partial^{\sharp} \colon \mathcal{M}(\mathbf{\Pi}_{R,\epsilon}) \to \mathcal{M}(\mathcal{N}(\mathbf{\Gamma}_{R,\epsilon})),$$

which takes the atomic measure supported over Π to the sum of atomic measures supported over its feet. The space $\mathcal{N}(\mathbf{\Gamma}_{R,\epsilon})$ is the disjoint union of unit normal vector bundles $\mathcal{N}(\gamma)$ over good curves γ. To encode the model gluing, denote by

$$\overline{A_1} \colon \mathcal{N}(\mathbf{\Gamma}_{R,\epsilon}) \to \mathcal{N}(\mathbf{\Gamma}_{R,\epsilon})$$

the map that takes any unit normal vector n on a good curve γ to the parallel transportation of $-n$ along the orientation-reversal $\bar{\gamma}$ of distance 1.

Recall that over a metric space X, for any positive constant δ, two Borel measures μ, μ' are said to be δ-equivalent if $\mu(E) \leq \mu'(\mathrm{Nhd}_{\delta}(E))$ holds for all Borel subsets E of X and if $\mu(X) = \mu'(X)$. Being δ-equivalent is a reflexive and symmetric relation which is invariant under rescaling the measures by the same factor. We speak of approximately equivalent measures over $\mathcal{N}(\mathbf{\Gamma}_{R,\epsilon})$ with respect to its canonically induced metric.

Problem 1.2. *Find a probability measure* $\mu \in \mathcal{M}(\mathbf{\Pi}_{R,\epsilon})$ *such that* $\partial^{\sharp}\mu$ *is* (ϵ / R)-*equivalent to* $(\overline{A_1})_{*}\partial^{\sharp}\mu$ *on* $\mathcal{N}(\mathbf{\Gamma}_{R,\epsilon})$.

A solution to Problem 1.2 implies a nontrivial integral measure in $\mathcal{M}(\mathbf{\Pi}_{R,\epsilon})$ with the same property, which gives rise to the finite collection of oriented (R, ϵ)-good pants. It turns out that these (R, ϵ)-good pants can be glued along common cuffs in an (R, ϵ)-nice fashion to create an immersed oriented closed subsurface of M, which meets the requirement of Theorem 1.1. To solve Problem 1.2, one needs to understand the statistics of good pants and the distribution of feet on good curves. The solutions to Problem 1.2 for cocompact lattices of $\mathrm{PSL}_2(\mathbf{C})$ and

$\mathrm{PSL}_2(\mathbf{R})$ lead to affirmative answers to the Surface Subgroup Conjecture and the Ehrenpreis Conjecture respectively.

2   THE EHRENPREIS CONJECTURE

For cocompact lattices of $\mathrm{PSL}_2(\mathbf{R})$, the problem of good pants construction can be resolved through dynamics of hyperbolic surfaces. In this section, we give an introduction to the proof of the Ehrenpreis Conjecture [Kahn–Markovic 2014], as an illustration of how the exponential mixing property of the geodesic flow makes the construction possible.

Let S_1 and S_2 be two closed Riemann surfaces of the same genus. The Ehrenpreis Conjecture asserts that for any positive constant ϵ, there exists a $(1+\epsilon)$-quasiconformal map $f : S_1' \to S_2'$ between some finite covers of S_1 and S_2 accordingly. Leon Ehrenpreis verified the case of genus 1 and proposed the conjecture for genus at least 2, [Ehrenpreis 1970]. In hyperbolic geometry, the conjecture can be equivalently stated as two closed hyperbolic surfaces are virtually $(1+\epsilon)$-bilipschitz equivalent for any positive constant ϵ. There is no direct analog of the conjecture in higher dimensions because two closed hyperbolic manifolds of dimension at least 3 are necessarily commensurable with each other if they are mutually virtually quasi-isometric.

By Theorem 1.1, it suffices to find, for some large R, a finite cover S_i' of S_i that can be constructed by (R, ϵ)-nicely gluing (R, ϵ)-good pants. Therefore, we need to solve Problem 1.2 for cocompact lattices of $\mathrm{PSL}_2(\mathbf{R})$. In this case, a solution can be achieved through dynamics of the geodesic flow.

To illustrate the idea, let us first explain how to construct good curves and good pants in an oriented closed hyperbolic surface M using the mixing property of the geodesic flow.

Recall that the geodesic flow over an oriented closed hyperbolic surface M is a one-parameter family of morphisms of the unit tangent vector bundle

$$g_t : \mathrm{UT}(M) \to \mathrm{UT}(M)$$

which takes any unit tangent vector v at a point p of M to $g_t(v)$, the parallel transportation of v along the unit-speed geodesic ray emanating from p in the direction v for time t. The geodesic flow preserves the Liouville measure m of $\mathrm{UT}(M)$, induced from the Haar measure of $\mathrm{PSL}_2(\mathbf{R})$ by identifying $\mathrm{UT}(M)$ with the left quotient $\pi_1(M) \backslash \mathrm{PSL}_2(\mathbf{R})$. For any functions $\phi, \psi \in C^\infty(\mathrm{UT}(M))$ with integral 1 over $\mathrm{UT}(M)$, the mixing property of the geodesic flow implies:

$$\int_{\mathrm{UT}(M)} (g_t^* \phi)(v)\psi(v)\mathrm{d}m(v) \to \frac{1}{2\pi^2 |\chi(M)|},$$

as t tends to $+\infty$.

For any positive constant δ, we claim that the following Connection Principle holds for sufficiently large L depending only on δ and the injectivity radius of M. For any two unit vectors v_p, v_q at points $p, q \in M$ respectively, there exists a geodesic segment connecting p and q such that the initial and terminal directions are approximately v_p and v_q, respectively, up to error δ. To this end, we may take a C^∞ function $\phi : \mathrm{PSL}_2(\mathbf{R}) \to [0, +\infty)$ with integral 1 over $\mathrm{PSL}_2(\mathbf{R})$, supported in a δ'-neighborhood U of the identity. For sufficiently small δ', the neighborhood $v_p \cdot U$ in $\mathrm{UT}(M)$ is isometric to U so there is an induced function ϕ_p over $\mathrm{UT}(M)$ supported on $v_p \cdot U$ defined by $\phi_p(v_p \cdot u) = \phi(u)$. In a similar fashion we define ϕ_q supported on $v_q \cdot U$. By the mixing property of the geodesic flow, the intersection of $(g_L^* \phi_p)(v_p \cdot U)$ and $v_q \cdot U$ is nonempty for sufficiently large L. In other words, there is a geodesic segment of which the initial and terminal direction vectors are δ'-close to v_p and v_q, respectively. Thus the claim follows since we can choose δ' sufficiently small, for instance, at most $10^{-2}\delta$.

In order to construct an (R, ϵ)-good curve in M for a given ϵ and any sufficiently large R, we apply the Connection Principle by taking p equal to q and v_p equal to v_q. By choosing δ to be $10^{-1}\epsilon$ and a sufficiently large L to be R, we obtain a geodesic segment with coincident endpoints p, which gives rise to a geodesic loop γ by identifying the endpoint and free homotopy. It can be verified by elementary estimation of hyperbolic geometry that γ is an (R, ϵ)-good curve. In order to construct an (R, ϵ)-good pair of pants, we may attach a nearly perpendicular bisecting arc α to γ as follows. Take a pair of antipodal points p, q on γ, and take unit normal vectors n_p, n_q at p, q by rotating the direction vectors of γ counterclockwise. By the Connecting Principle, construct a geodesic segment α connecting p and q with endpoint directions approximately n_p and $-n_q$ of length approximately $\frac{R}{2} + \log(2)$, up to error $10^{-1}\epsilon$. Attaching α to γ gives rise to a θ-shape graph, which is the spine of a unique immersed pair of pants Π up to homotopy. It can be verified, again, that Π is an (R, ϵ)-good pair of pants.

Now let us return to the original Problem 1.2 for cocompact lattices in $\mathrm{PSL}_2(\mathbf{R})$. The strategy is to solve the problem in two steps:

1. Find a probability measure $\mu_0 \in \mathcal{M}(\mathbf{\Pi}_{R,\epsilon})$ so that $\partial^\sharp \mu_0$ is nontrivial and nearly evenly distributed over any (R, ϵ)-good curve γ, which means that $\partial^\sharp \mu_0$ is close to the Lebesgue measure on the unit normal bundle to γ.

2. Resolve the small imbalance of boundary measures between oppositely oriented (R, ϵ)-good curves by a small correction $\mu_\phi \in \mathcal{M}(\mathbf{\Pi}_{R,\epsilon})$.

Then a solution of Problem 1.2 is given by the normalization of $\mu_0 + \mu_\phi$. In the example constructions above, we have not estimated the amount of good curves and good pants in the hyperbolic surface M. The quantitative version can be achieved by knowing the exponential mixing rate of the geodesic flow. Specifically, the following theorems resolve the two steps.

To capture the imbalance of boundary measures, we introduce another boundary operator:

$$\partial_\Delta : \mathcal{M}(\mathbf{\Pi}_{R,\epsilon}) \to \mathcal{M}(\mathbf{\Gamma}_{R,\epsilon}),$$

by defining $(\partial_\Delta \mu)(\gamma) = \max\{0, (\partial \mu)_+(\gamma) - (\partial \mu)_-(\bar{\gamma})\}$. Note that the unit normal vector bundle $\mathcal{N}(\gamma)$ for any (R, ϵ)-good curve γ has two components. Denote by $\mathcal{N}_+(\gamma)$ the unit normal vectors that form a special orthonormal frame with the unit direction vectors of γ. For any measure $\mu \in \mathcal{M}(\mathbf{\Pi}_{R,\epsilon})$, the restriction of the footed boundary $\partial^\sharp \mu$ to any $\mathcal{N}(\gamma)$ is hence supported on $\mathcal{N}_+(\gamma)$ (because $\mathbf{\Pi}_{R,\epsilon}$ consists of oriented good pants which have compatible orientation with the surface).

Theorem 2.1. *For some universal constants q, C and polynomial P, the following statement holds for any small ϵ depending on M and sufficiently large R depending on ϵ and M.*

There exists a probability measure $\mu_0 \in \mathcal{M}(\mathbf{\Pi}_{R,\epsilon})$ with $\partial_\Delta \mu_0$ supported on $\mathbf{\Gamma}_{R,\,\epsilon/2}$ (in fact, one may take μ_0 to be the normalized counting measure on $\mathbf{\Pi}_{R,\epsilon}$). Moreover,

- *for all $\gamma \in \mathbf{\Gamma}_{R,\epsilon}$, the restriction of $\partial^\sharp \mu_0$ to $\mathcal{N}_+(\gamma)$ is $(P(R)e^{-qR})$-equivalent to $(\partial \mu_0)(\gamma)$ times the Lebesgue measure;*
- *for all $\gamma \in \mathbf{\Gamma}_{R,\,\epsilon/2}$, the value $(\partial \mu_0)(\gamma)$ is at least Ce^{-2R}, and $(\partial_\Delta \mu_0)(\gamma)$ is at most $P(R)e^{-(2+q)R}$.*

Theorem 2.2. *For any measure $\nu \in \mathcal{M}(\mathbf{\Gamma}_{R,\epsilon})$ which is contained in the subspace $\partial_\Delta \mathcal{M}(\mathbf{\Pi}_{R,\epsilon})$, there exists a measure $\phi(\nu) \in \mathcal{M}(\mathbf{\Pi}_{R,\epsilon})$ such that $\partial_\Delta \phi(\nu)$ equals ν, and moreover,*

$$\|\phi(\nu)\| \le P(R)e^{-R}\|\nu\|,$$

where $\|\cdot\|$ is understood as the maximum value at individual (R,ϵ)-pants or (R,ϵ)-curves.

See [Kahn–Markovic 2014, Theorems 3.1, 3.4, Lemma 3.3].

Starting with the measure $\mu_0 \in \mathcal{M}(\mathbf{\Pi}_{R,\epsilon})$ of Theorem 2.1, we may construct a correction term $\mu_\phi \in \mathcal{M}(\mathbf{\Pi}_{R,\epsilon})$ using Theorem 2.2, by defining μ_ϕ to be $-\phi(\partial_\Delta \bar\mu_0)$ where $\bar\mu_0$ is the measure induced by the orientation reversion on $\mathbf{\Pi}_{R,\epsilon}$. It can be verified that $\mu_0 + \mu_\phi$ is $(P(R)e^{-qR})$-equivalent to the Lebesgue measure restricted to any unit normal vector bundle $\mathcal{N}(\gamma)$, and that $\partial_\Delta(\mu_0 + \mu_\phi)$ vanishes. By taking R sufficiently large, the normalization of $\mu_0 + \mu_\phi$ provides a solution to Problem 1.2.

To close the present section, we remark that from Theorem 2.2 one can develop a quantitative good correction theory. The qualitative part of the theory is the so-called Good Pants Homology, which has been generalized to the $\mathrm{PSL}_2(\mathbf{C})$ case as we discuss in the next section. The quantitative part of the theory refers to the estimation of $\phi(\nu)$ in Theorem 2.2. The estimation can be achieved through a procedure called randomization that strengthens arguments involved in the study of Good Pants Homology.

3   SURFACE SUBGROUP CONJECTURE

Most techniques of the good pants construction in $\mathrm{PSL}_2(\mathbf{R})$ find their counterparts in $\mathrm{PSL}_2(\mathbf{C})$. In this section, we introduce the proof of the Surface Subgroup Conjecture, which asserts that any cocompact lattice of $\mathrm{PSL}_2(\mathbf{C})$ contains a surface subgroup, [Kahn–Markovic 2012]. Then we discuss the generalization of the qualitative good correction theory for $\mathrm{PSL}_2(\mathbf{C})$, [Liu–Markovic 2014].

The Surface Subgroup Conjecture is related to the Virtual Haken Conjecture in 3-manifold topology. Essentially embedded subsurfaces of closed 3-manifolds are important objects in 3-manifold topology. Such subsurfaces can be constructed via topological methods when the fundamental group of the 3-manifold admits a nontrivial splitting. As a potential approach to the Poincaré Conjecture, Friedhelm Waldhausen conjectured in the 1960s that every closed 3-manifold with an infinite fundamental group has a finite cover which contains an essentially embedded subsurface, or in other words, is virtually Haken [Waldhausen 1968]. By the

Geometrization Conjecture of William P. Thurston, confirmed by Grigori Perelman in 2003, Waldhausen's Virtual Haken Conjecture can be reduced to the case of closed hyperbolic 3-manifolds. Hence the Surface Subgroup Conjecture becomes a natural first step towards Waldhausen's conjecture. In general, a π_1-injectively immersed subsurface in a closed hyperbolic 3-manifold can be either geometrically infinite or quasi-Fuchsian. The only currently known way to construct geometrically infinite subsurfaces invokes the Sutured Manifold Hierarchy as well as the Virtual Special Cubulation, [Agol 2008, Gabai 1983, Wise 2011]. In particular, it requires passage to finite covers and offers no effective control on the geometry of the constructed subsurface. On the other hand, many arithmetic hyperbolic 3-manifolds are known to contain immersed Fuchsian subsurfaces, and hyperbolic surface bundles are known to contain immersed quasi-Fuchsian subsurfaces with arbitrarily thick hull, [Cooper–Long–Reid 1997, Masters 2006]. The Virtual Haken Conjecture has been proved by Ian Agol [Agol 2013] relying on [Kahn–Markovic 2012, Wise 2011].

The subsurfaces constructed through good pants techniques are π_1-injectively immersed and geometrically finite, as one can imply from Theorem 1.1. In fact, one can produce these subsurfaces to be arbitrarily nearly totally geodesic, in the sense that the limit set of a corresponding quasi-Fuchsian subgroup can be required to be contained in any given neighborhood of a round circle in the ideal boundary of $\mathbb{H}^3$. Once again, the task reduces to addressing Problem 1.2 for an oriented closed hyperbolic 3-manifold M. However, this time the solution is even easier since we can pass around the good correction theory. The $\mathrm{PSL}_2(\mathbf{C})$ case was actually resolved earlier than the $\mathrm{PSL}_2(\mathbf{R})$ case.

The first modification to the previous argument for $\mathrm{PSL}_2(\mathbf{R})$ is to employ the frame flow on the special orthonormal frame bundle $\mathrm{SO}(M)$ instead of the geodesic flow on the unit vector bundle $\mathrm{UT}(M)$. The reason is that good curves in closed hyperbolic 3-manifolds must be prescribed to have approximately trivial monodromy besides approximately given length, and similarly, good pants must have small bending in the normal direction. The mixing rate of the frame flow is known to be exponential. A weak analog of Theorem 2.1 is the following:

Theorem 3.1. *For some universal constants q and polynomial P, the following statement holds for any positive constant ϵ and sufficiently large R depending only on ϵ and M.*

There exists a probability measure $\mu_0 \in \mathcal{M}(\mathbf{\Pi}_{R,\epsilon})$ such that for all $\gamma \in \mathbf{\Gamma}_{R,\epsilon}$, the restriction of $\partial^{\sharp}\mu_0$ to $\mathcal{N}(\gamma)$ is $(P(R)e^{-qR})$-equivalent to $(\partial\mu_0)(\gamma)$ times the Lebesgue measure.

See [Kahn–Markovic 2012, Theorem 3.4].

Denote by $\bar{\mu}_0$ the measure induced by the orientation reversion on $\mathbf{\Pi}_{R,\epsilon}$. Then the probability measure $(\mu_0 + \bar{\mu}_0)/2$ yields a solution to Problem 1.2, which completes the proof of the Surface Subgroup Conjecture for cocompact lattices of $\mathrm{PSL}_2(\mathbf{C})$.

Before moving on to the correction theory, we make two technical remarks regarding Theorems 2.1 and 3.1:

1. Current techniques of good pants construction do not apply to non-cocompact lattices in $\mathrm{PSL}_2(\mathbf{R})$ or $\mathrm{PSL}_2(\mathbf{C})$. The argument in the previous section to construct good curves or good pants would fail without the assumption of positive injectivity radius, so the existence of μ_0 is no longer guaranteed.

2. The averaging trick in the $\mathrm{PSL}_2(\mathbf{C})$ case fails in the $\mathrm{PSL}_2(\mathbf{R})$ case essentially because the flip transformation of $\mathcal{N}(\mathbf{\Gamma}_{R,\epsilon})$ that takes any unit normal vector n to $-n$ is orientation reversing. The fact that the flip transformation is central but not contained in the identity component of the isometry group of $\mathcal{N}(\mathbf{\Gamma}_{R,\epsilon})$ obstructs μ_0 from being nearly symmetric under flipping. Similar technical obstructions can be observed for cocompact lattices in $\mathrm{SO}(2m,1)$.

A qualitative correction theory has been developed for cocompact lattices in $\mathrm{PSL}_2(\mathbf{C})$. We state the result in terms of the good panted cobordism group.

Let M be an oriented closed hyperbolic 3-manifold. A possibly disconnected oriented immersed 1-submanifold L of M is called an (R,ϵ)-multicurve, if each component of L is homotopic to an (R,ϵ)-good curve. Two (R,ϵ)-multicurves L, L' are said to be (R,ϵ)-*panted cobordant* if there exists a possibly disconnected (R,ϵ)-panted subsurface bounded by $L \sqcup \bar{L}'$ (an (R,ϵ)-panted subsurface is a subsurface that comes with a decomposition into (R,ϵ) good pants). For any universally small ϵ and sufficiently large R, being (R,ϵ)-panted cobordant is an equivalence relation. Then we define the (R,ϵ)-*panted cobordism group* of M to be the set of (R,ϵ)-panted cobordism classes $[L]_{R,\epsilon}$ of (R,ϵ)-multicurves L, denoted as $\mathbf{\Omega}_{R,\epsilon}(M)$. This is a finitely generated Abelian group with the addition induced by the disjoint union operation and the inverse induced by the orientation reversion.

Theorem 3.2. *Let M be an oriented closed hyperbolic 3-manifold. For any universally small positive ϵ, and any sufficiently large positive R depending only on M and ϵ, there is a canonical isomorphism*

$$\Phi : \mathbf{\Omega}_{R,\epsilon}(M) \longrightarrow H_1(\mathrm{SO}(M); \mathbf{Z}),$$

where $\mathrm{SO}(M)$ denotes the bundle over M of special orthonormal frames with respect to the orientation of M. Moreover, for all $[L]_{R,\epsilon} \in \mathbf{\Omega}_{R,\epsilon}(M)$, the image of $\Phi([L]_{R,\epsilon})$ under the bundle projection is the homology class $[L] \in H_1(M; \mathbf{Z})$.

See [Liu–Markovic 2014, Theorem 5.2].

Since $H_1(\mathrm{SO}(M); \mathbf{Z})$ is a split extension of $H_1(M; \mathbf{Z})$ by $\mathbf{Z}_2$, the only obstruction for a null-homologous good multicurve to bound a good panted subsurface lies in the center $\mathbf{Z}_2$.

For an oriented closed hyperbolic surface M, the Good Pants Homology of M introduced in [Kahn–Markovic 2014] may be equivalently defined as $\mathbf{\Omega}_{R,\epsilon}(M) \otimes \mathbf{Q}$. It has been shown there that $\mathbf{\Omega}_{R,\epsilon}(M) \otimes \mathbf{Q}$ is canonically isomorphic to $H_1(M; \mathbf{Q})$, (see also [Calegari 2009]).

4   FURTHER APPLICATIONS

There have been a number of proceedings of good pants constructions since [Kahn–Markovic 2012, Kahn–Markovic 2014]. Ursula Hamenstädt has considered good pants constructions for cocompact lattices of broader families of Lie groups.

Theorem 4.1 ([Hamenstädt 2014]). *Every cocompact lattice in a rank one simple Lie group other than $\mathrm{SO}(2m,1)$ for positive integers m contains a surface subgroup.*

The complete list of Lie groups in Hamenstädt's result consists of $SO(2m+1, 1)$, $SU(n, 1)$, $Sp(n, 1)$, and F_4^{-20}. It is also conjectured in [Hamenstädt 2014] the existence of surface subgroups for cocompact irreducible lattices in semisimple Lie groups with finite center, without compact factors and without factors locally isomorphic to $SL_2(\mathbf{R})$.

The good correction theory for $PSL_2(\mathbf{C})$ can be applied to construct bounded π_1-injectively immersed subsurface in closed hyperbolic 3-manifolds. The following theorem is the generalization of a result of Danny Calegari in the case of hyperbolic surfaces [Calegari 2009].

Theorem 4.2 ([Liu–Markovic 2014]). *Every π_1-injectively immersed oriented closed 1-submanifold in a closed hyperbolic 3-manifold which is rationally null-homologous admits an equidegree finite cover which bounds an oriented connected compact π_1-injective immersed quasi-Fuchsian subsurface.*

The fact that the good correction theory yields a finite cover of a closed hyperbolic surface can be generalized to a procedure called homological substitution as introduced in [Liu–Markovic 2014]. It replaces any π_1-injectively immersed one-vertexed 2-complex with a good panted complex so that the 1-cells correspond to a collection of good multicurves, and the 2-cells correspond to a collection of good panted subsurfaces. Homological substitution enables us to produce nicely glued good panted subsurfaces with control on the homology class.

Theorem 4.3 ([Liu–Markovic 2014]). *Every rational second homology class of a closed hyperbolic 3-manifold has a positve integral multiple represented by an oriented connected closed π_1-injectively immersed quasi-Fuchsian subsurface.*

Compare the result of Danny Calegari and Alden Walker that in a random group at any positive density, many second homology classes can be rationally represented by quasiconvex (closed) surface subgroups [Calegari–Walker 2013].

A similar idea can be found in Hongbin Sun's work on virtual properties of 3-manifolds. The following results are based on good pants constructions and invoke the separability of quasiconvex subgroups for lattices of $PSL_2(\mathbf{C})$, [Agol 2013].

Theorem 4.4 ([Sun 2014a]). *For any finite Abelian group A, every closed orientable hyperbolic 3-manifold has a finite cover of which the first integral homology contains A as a direct sum component.*

Theorem 4.5 ([Sun 2014b]). *For any closed orientable 3-manifold N, every closed orientable hyperbolic 3-manifold has a finite cover that maps onto N of degree 2.*

In [Sun 2014a], the core construction is a 2-complex which is good panted and nicely glued with certain mild singularity near a 1-submanifold locus. In [Sun 2014b], the core construction involves homological substitution of a carefully chosen one-vertexed 2-complex. In particular, the mapping degree 2 is essentially due to the $\mathbf{Z}_2$ obstruction discovered in Theorem 3.2.

In prospect, there are several directions rather interesting to explore. The first challenge is to generalize good pants constructions to non-cocompact lattices and one-ended word-hyperbolic groups. The next step is to develop qualitative and quantitative versions of the good correction theory. Finally, it remains widely open at this point how to construct higher dimensional submanifolds in locally symmetric spaces through their geometry and dynamics of various flow.

REFERENCES

[Agol 2008] Ian. Agol, *Criteria for virtual fibering*, J. Topol. **1** (2008), 269–284.

[Agol 2013] Ian Agol, *The virtual Haken conjecture*, with an appendix by Ian Agol, Daniel Groves, and Jason Manning, Documenta Mathematica **18** (2013), 1045–1087.

[Calegari 2009] Danny Calegari, *Faces of the scl norm ball*, Geom. Topol. **13** (2009), 1313–1336.

[Calegari–Walker 2013] D. Calegari, A. Walker, *Random groups contain surface subgroups*, Jour. Amer. Math. Soc., to appear, preprint available at arXiv:1304.2188v2, 31 pages, 2013.

[Cooper–Long–Reid 1997] D. Cooper, D. D. Long, A. W. Reid, *Essential closed surfaces in bounded 3-manifolds*, J. Amer. Math. Soc. **10** (1997), 553–563.

[Ehrenpreis 1970] Leon Ehrenpreis, "Cohomology with bounds," *Symposia Mathematica*, Vol. 4, 1970, 389–395.

[Gabai 1983] David Gabai, *Foliations and the topology of 3-manifolds*, J. Diff. Geom. **18** (1983), 445–503.

[Hamenstädt 2014] Ursula Hamenstädt, *Incompressible surfaces in rank one locally symmetric spaces*, preprint available at arXiv:1402.1704v2, 38 pages, 2014.

[Kahn–Markovic 2012] J. Kahn and V. Markovic, *Immersing almost geodesic surfaces in a closed hyperbolic 3-manifold*, Ann. Math. **175** (2012), no. 3, 1127–1190.

[Kahn–Markovic 2014] ———, "The good pants homology and a proof of the Ehrenpreis conjecture," *Proceedings of the ICM 2014*, to appear, preprint available at arXiv:1101.1330v2, 77 pages, 2014.

[Liu–Markovic 2014] Yi Liu and Vladimir Markovic, *Homology of curves and surfaces in closed hyperbolic 3-manifolds*, preprint available at arXiv:1309.7418v2, 66 pages, 2014.

[Masters 2006] Joseph D. Masters, *Thick surfaces in hyperbolic 3-manifolds*, Geometriae Dedicata **119** (2006), 17–33.

[Sun 2014a] Hongbin Sun, *Virtual homological torsion of closed hyperbolic 3-manifolds*, J. Diff. Geom., to appear, preprint available at arXiv:1309.1511v4, 23 pages, 2014.

[Sun 2014b] Hongbin Sun, *Virtual domination of 3-manifolds*, Geom. Topol., to appear, preprint available at arXiv:1401.7049v1, 35 pages, 2014.

[Waldhausen 1968] Friedhelm Waldhausen, *On irreducible 3-manifolds which are sufficiently large*, Ann. Math. (2) **87** (1968), 56–88.

[Wise 2011] Daniel Wise, *The structure of groups with quasiconvex hierarchy*, preprint available at http://www.math.mcgill.ca/wise/papers.html, 186 pages, 2011.

Dehn Surgery on Knots in S^3 Producing Nil Seifert Fibered Spaces

Yi Ni and Xingru Zhang

1 INTRODUCTION

For a knot K in S^3, we denote by $S_K^3(p/q)$ the manifold obtained by Dehn surgery along K with slope p/q. Here the slope p/q is parameterized by the standard meridian/longitude coordinates of K and we always assume $\gcd(p,q) = 1$. In this paper we study the problem of on which knots in S^3 with which slopes Dehn surgeries can produce Seifert fibered spaces admitting the Nil geometry. Recall that every closed connected orientable Seifert fibered space W admits one of 6 canonical geometries: $S^2 \times \mathbb{R}$, $\mathbb{E}^3$, $\mathbb{H}^2 \times \mathbb{R}$, S^3, Nil, $\widetilde{SL_2}(\mathbb{R})$. More concretely if $e(W)$ denotes the Euler number of W and $\chi(\mathcal{B}_W)$ denotes the orbifold Euler characteristic of the base orbifold $\mathcal{B}_W$ of W, then the geometry of W is uniquely determined by the values of $e(W)$ and $\chi(\mathcal{B}_W)$ according to the following table (cf. §4 of [18]):

Table 1: The type of geometry of a Seifert fibered space W.

	$\chi(\mathcal{B}_W) > 0$	$\chi(\mathcal{B}_W) = 0$	$\chi(\mathcal{B}_W) < 0$
$e(W) = 0$	$\mathbb{S}^2 \times \mathbb{R}$	$\mathbb{E}^3$	$\mathbb{H}^2 \times \mathbb{R}$
$e(W) \neq 0$	$\mathbb{S}^3$	Nil	$\widetilde{SL_2}(\mathbb{R})$

Suppose that $S_K^3(p/q)$ is a Seifert fibered space with Euclidean base orbifold. A simple homology consideration shows that the base orbifold of $S_K^3(p/q)$ must be $S^2(2,3,6)$—the 2-sphere with 3 cone points of orders $2, 3, 6$ respectively. The orbifold fundamental group of $S^2(2,3,6)$ is the triangle group $\triangle(2,3,6) = \langle x, y; x^2 = y^3 = (xy)^6 = 1 \rangle$, whose first homology is $\mathbb{Z}/6\mathbb{Z}$. Thus p is divisible by 6. If $p = 0$, then $S_K^3(0)$ must be a torus bundle. By [4], K is a fibered knot with genus one. So K is the trefoil knot or the figure 8 knot. But the 0-surgery on the figure 8 knot is a manifold with the Sol geometry. So K is the trefoil knot. By [18, Table 4.4], there is only one Euclidean 3–manifold with base orbifold $S^2(2,3,6)$, which is the zero surgery on the trefoil knot. So the trefoil knot is the only knot in S^3 and 0 is the only slope which can produce a Seifert fibered space with the Euclidean geometry. Therefore we may assume that $p \neq 0$. Hence $S_K^3(p/q)$ is a Seifert fibered space with the Nil geometry. It is known that on a hyperbolic knot K in S^3, there is at most one surgery which can possibly produce a Seifert fibered space admitting the Nil geometry and if there is one, the surgery slope is integral [2]. In this paper we show

Theorem 1.1. *Suppose K is a knot in S^3 which is not the (right-handed or left-handed) trefoil knot $T(\pm 3, 2)$. Suppose that $S_K^3(p/q)$ is a Seifert fibered space admitting the Nil geometry (where we may assume $p, q > 0$ up to changing K to its mirror image). Then $q = 1$ and p is one of the numbers $60, 144, 156, 288, 300$. Moreover we have the following cases:*

1. $S_K^3(60) \cong -S_{T(3,2)}^3(60/11)$,
2. $S_K^3(144) \cong -S_{T(3,2)}^3(144/23)$ *or* $S_K^3(144) \cong S_{T(3,2)}^3(144/25)$,
3. $S_K^3(156) \cong S_{T(3,2)}^3(156/25)$,
4. $S_K^3(288) \cong S_{T(3,2)}^3(288/49)$,
5. $S_K^3(300) \cong S_{T(3,2)}^3(300/49)$,

where $\cong$ stands for orientation preserving homeomorphism.

Furthermore, under the assumptions of Theorem 1.1, we have the following additional information:

Addendum 1.2. (a) The knot K is either a hyperbolic knot or a cable over $T(3, 2)$ as given in Proposition 4.2.

(b) If Case (1) occurs, then K is a hyperbolic knot and its Alexander polynomial is either

$$\triangle_K(t) = 1 - t - t^{-1} + t^2 + t^{-2} - t^4 - t^{-4} + t^5 + t^{-5} - t^6 - t^{-6} + t^7 + t^{-7} - t^8 - t^{-8}$$
$$+ t^9 + t^{-9} - t^{13} - t^{-13} + t^{14} + t^{-14} - t^{15} - t^{-15} + t^{16} + t^{-16} - t^{22} - t^{-22} + t^{23} + t^{-23},$$

or

$$\triangle_K(t) = 1 - t^2 - t^{-2} + t^4 + t^{-4} - t^7 - t^{-7} + t^9 + t^{-9} - t^{12} - t^{-12} + t^{13}$$
$$+ t^{-13} - t^{16} - t^{-16} + t^{17} + t^{-17} - t^{21} - t^{-21} + t^{22} + t^{-22}.$$

The two Berge knots which yield the lens spaces $L(61, 13)$ and $L(59, 27)$ respectively realize the Nil Seifert surgery with the prescribed two Alexander polynomials respectively. More explicitly these two Berge knots are given in [1], page 6, with $a = 5$ and $b = 4$ in case of Fig. 8, and with $b = 9$ and $a = 2$ in case of Fig. 9, respectively.

(c) If the former subcase of Case (2) occurs, then K is a hyperbolic knot and its Alexander polynomial is

$$\triangle_K(t) = 1 - t - t^{-1} + t^2 + t^{-2} - t^4 - t^{-4} + t^5 + t^{-5} - t^6 - t^{-6} + t^7 + t^{-7} - t^9 - t^{-9}$$
$$+ t^{10} + t^{-10} - t^{11} - t^{-11} + t^{12} + t^{-12} - t^{14} - t^{-14} + t^{15} + t^{-15} - t^{16} - t^{-16} + t^{17} + t^{-17}$$
$$- t^{19} - t^{-19} + t^{20} + t^{-20} - t^{21} - t^{-21} + t^{22} + t^{-22} - t^{24} - t^{-24} + t^{25} + t^{-25} - t^{26} - t^{-26}$$
$$+ t^{27} + t^{-27} - t^{29} - t^{-29} + t^{30} + t^{-30} - t^{34} - t^{-34} + t^{35} + t^{-35} - t^{39} - t^{-39} + t^{40}$$
$$+ t^{-40} - t^{44} - t^{-44} + t^{45} + t^{-45} - t^{49} - t^{-49} + t^{50} + t^{-50} - t^{54} - t^{-54} + t^{55} + t^{-55}.$$

This subcase is realized on the Eudave-Muñoz knot $k(-2, 1, 6, 0)$ of [3, Propositions 5.3 (1) and 5.4 (2)], which is also a Berge knot on which the 143–surgery yields $L(143, 25)$.

(d) If the latter subcase of Case (2) occurs, then $\triangle_K(t) = \triangle_{T(29,5)}(t) \triangle_{T(3,2)}(t^5)$. If Case (4) or (5) occurs, then $\triangle_K(t) = \triangle_{T(41,7)}(t) \triangle_{T(3,2)}(t^7)$ or $\triangle_K(t) = \triangle_{T(43,7)}(t) \triangle_{T(3,2)}(t^7)$ respectively. All these cases can be realized on certain cables over $T(3, 2)$ as given in Proposition 4.2.

(e) If Case (3) occurs, then either $\triangle_K(t) = \triangle_{T(31,5)}(t)\triangle_{T(3,2)}(t^5)$ or

$$\triangle_K(t) = 1 - t^3 - t^{-3} + t^4 + t^{-4} - t^5 - t^{-5} + t^6 + t^{-6} - t^8 - t^{-8} + t^9 + t^{-9} - t^{10}$$
$$- t^{-10} + t^{11} + t^{-11} - t^{13} - t^{-13} + t^{14} + t^{-14} - t^{15} - t^{-15} + t^{16} + t^{-16} - t^{18} - t^{-18}$$
$$+ t^{19} + t^{-19} - t^{20} - t^{-20} + t^{21} + t^{-21} - t^{23} - t^{-23} + t^{24} + t^{-24} - t^{25} - t^{-25} + t^{26}$$
$$+ t^{-26} - t^{28} - t^{-28} + t^{29} + t^{-29} - t^{30} - t^{-30} + t^{31} + t^{-31} - t^{35} - t^{-35} + t^{36} + t^{-36}$$
$$- t^{40} - t^{-40} + t^{41} + t^{-41} - t^{45} - t^{-45} + t^{46} + t^{-46} - t^{50} - t^{-50} + t^{51} + t^{-51} - t^{55}$$
$$- t^{-55} + t^{56} + t^{-56} - t^{60} - t^{-60} + t^{61} + t^{-61}.$$

The former subcase can be realized on the $(31, 5)$-cable over $T(3,2)$, and the latter subcase can be realized on the Eudave-Muñoz knot $k(-3, -1, 7, 0)$, which is also a Berge knot on which the 157–surgery yields $L(157, 25)$.

In other words there are exactly 6 Nil Seifert fibered spaces which can be obtained by Dehn surgeries on non-trefoil knots in S^3 and there are exactly 5 slopes for all such surgeries (while on the trefoil knot $T(3,2)$, infinitely many Nil Seifert fibered spaces can be obtained by Dehn surgeries; in fact by [11], $S_{T(3,2)}(p/q)$ is a Nil Seifert fibered space if and only if $p = 6q \pm 6$, $p \neq 0$). It seems reasonable to raise the following conjecture.

Conjecture 1.3. *If a hyperbolic knot K in S^3 admits a surgery yielding a Nil Seifert fibered space, then K is one of the four hyperbolic Berge knots given in (b) (c) (e) of Addendum 1.2.*

The method of proof of Theorem 1.1 and Addendum 1.2 follows that given in [10] and [8], where similar results are obtained for Dehn surgeries on knots in S^3 yielding spherical space forms which are not lens spaces or prism manifolds. The main ingredient of the method is the use of the correction terms (also known as the d-invariants) for rational homology spheres together with their Spin^c structures, defined in [14]. In fact with the same method we can go a bit further to prove the following theorems.

Theorem 1.4. *For each fixed 2-orbifold $S^2(2,3,r)$ (or $S^2(3,4,r)$), where $r > 1$ is an integer satisfying $\sqrt{6r/Q} \notin \mathbb{Z}$ (resp. $\sqrt{12r/Q} \notin \mathbb{Z}$) for each $Q = 1, 2, \ldots, 8$, there are only finitely many slopes with which Dehn surgeries on hyperbolic knots in S^3 can produce Seifert fibered spaces with $S^2(2,3,r)$ (resp. $S^2(3,4,r)$) as the base orbifold.*

Theorem 1.5. *For each fixed torus knot $T(m,n)$, with $m \geq 2$ even, $n > 1$, $\gcd(m,n) = 1$, and a fixed integer $r > 1$ satisfying $\sqrt{mnr/Q} \notin \mathbb{Z}$ for each $Q = 1, 2 \ldots, 8$, among all Seifert fibered spaces*

$$\left\{ S^3_{T(m,n)}\left(\frac{mnq \pm r}{q}\right) ; q > 0, \gcd(q,r) = 1 \right\}$$

only finitely many of them can be obtained by Dehn surgeries on hyperbolic knots in S^3.

The above results suggest a possible phenomenon about Dehn surgery on hyperbolic knots in S^3 producing Seifert fibered spaces, which we put forward in a form of conjecture.

Conjecture 1.6. *For every fixed 2-orbifold $S^2(k, l, m)$, with all k, l, m larger than 1, there are only finitely many slopes with which Dehn surgeries on hyperbolic knots in S^3 can produce Seifert fibered spaces with $S^2(k, l, m)$ as the base orbifold.*

In the above conjecture we may assume that $\gcd(k, l, m) = 1$.

After recalling some basic properties of the correction terms in Section 2, we give and prove a more general theorem in Section 3. This theorem together with its proof will be applied in the proofs of Theorem 1.1, Addendum 1.2 and Theorems 1.4 and 1.5, which is the content of Section 4.

Acknowledgments

The first author was partially supported by NSF grant numbers DMS-1103976, DMS-1252992, and an Alfred P. Sloan Research Fellowship.

2 CORRECTION TERMS IN HEEGAARD FLOER HOMOLOGY

To any oriented rational homology 3-sphere Y equipped with a Spin^c structure $\mathfrak{s} \in \mathrm{Spin}^c(Y)$, there can be assigned a numerical invariant $d(Y, \mathfrak{s}) \in \mathbb{Q}$, called the *correction term* of $(Y, \mathfrak{s})$, which is derived in [14] from Heegaard Floer homology machinery. The correction terms satisfy the following symmetries:

$$d(Y, \mathfrak{s}) = d(Y, J\mathfrak{s}), \quad d(-Y, \mathfrak{s}) = -d(Y, \mathfrak{s}), \tag{1}$$

where $J \colon \mathrm{Spin}^c(Y) \to \mathrm{Spin}^c(Y)$ is the conjugation.

Suppose that Y is an oriented homology 3-sphere, $K \subset Y$ a knot, let $Y_K(p/q)$ be the oriented manifold obtained by Dehn surgery on Y along K with slope p/q, where the orientation of $Y_K(p/q)$ is induced from that of $Y - K$ which in turn is induced from the given orientation of Y. There is an affine isomorphism $\sigma \colon \mathbb{Z}/p\mathbb{Z} \to \mathrm{Spin}^c(Y_K(p/q))$. See [14, 15] for more details about the isomorphism. We shall identify $\mathrm{Spin}^c(Y_K(p/q))$ with $\mathbb{Z}/p\mathbb{Z}$ via σ but with σ suppressed, writing $(Y_K(p/q), i)$ for $(Y_K(p/q), \sigma(i))$. Note here i is (mod p) defined and sometimes it can appear as an integer larger than or equal to p. The following lemma is contained in [10, 13].

Lemma 2.1. *The conjugation $J : Spin^c(Y_K(p/q)) \to Spin^c(Y_K(p/q))$ is given by*

$$J(i) = p + q - 1 - i, \ for \ 0 \leq i < p + q.$$

For a positive integer n and an integer k we use $[k]_n \in \mathbb{Z}/n\mathbb{Z}$ to denote the congruence class of k modulo n.

Let $L(p, q)$ be the lens space obtained by p/q–surgery on the unknot in S^3. The correction terms for lens spaces can be computed inductively as in [14]:

$$d(S^3, 0) = 0,$$

$$d(L(p, q), i) = -\frac{1}{4} + \frac{(2i + 1 - p - q)^2}{4pq} - d(L(q, [p]_q), [i]_q), \ \text{for } 0 \leq i < p + q. \tag{2}$$

For a knot K in S^3, write its Alexander polynomial in the following standard form:

$$\triangle_K(t) = a_0 + \sum_{i \geq 1} a_i(t^i + t^{-i}).$$

For $i \geq 0$, define

$$b_i = \sum_{j=1}^{\infty} j a_{i+j}.$$

Note that the a_i's can be recovered from the b_i's by the following formula

$$a_i = b_{i-1} - 2b_i + b_{i+1}, \text{ for } i > 0. \tag{3}$$

By [15] [16], if $K \subset S^3$ is a knot on which some Dehn surgery produces an L-space, then the b_i's for K satisfy the following properties:

$$b_i \geq 0, \ b_i \geq b_{i+1} \geq b_i - 1, \ b_i = 0 \text{ for } i \geq g(K) \tag{4}$$

and if $S_K^3(p/q)$ is an L-space, where $p, q > 0$, then for $0 \leq i \leq p-1$,

$$d(S_K^3(p/q), i) = d(L(p,q), i) - 2b_{\min\{\lfloor \frac{i}{q} \rfloor, \lfloor \frac{p+q-i-1}{q} \rfloor\}}. \tag{5}$$

This surgery formula has been generalized in [12] to one that applies to any knot in S^3 as follows. Given any knot K in S^3, from the knot Floer chain complex, there is a uniquely defined sequence of integers V_i^K, $i \in \mathbb{Z}$, satisfying

$$V_i^K \geq 0, V_i^K \geq V_{i+1}^K \geq V_i^K - 1, \quad V_i^K = 0 \text{ for } i \geq g(K) \tag{6}$$

and the following surgery formula holds.

Proposition 2.2. *When $p, q > 0$,*

$$d(S_K^3(p/q), i) = d(L(p,q), i) - 2V_{\min\{\lfloor \frac{i}{q} \rfloor, \lfloor \frac{p+q-i-1}{q} \rfloor\}}^K$$

for $0 \leq i \leq p-1$.

3   FINITELY MANY SLOPES

Theorems 1.1, 1.4 and 1.5 will follow from the following more general theorem and its proof.

Theorem 3.1. *Let L be a given knot in S^3, and r, l, Q be given positive integers satisfying*

$$\sqrt{\frac{rl}{Q}} \notin \mathbb{Z}. \tag{7}$$

Suppose further that l is even. Then there exist only finitely many positive integers q, such that $S_L^3(\frac{lq \pm r}{q})$ is homeomorphic to $S_K^3(\frac{lq \pm r}{Q})$ for a knot K in S^3.

Remark 3.2. The condition that l is even is not essential. We require this condition to simplify our argument. The condition (7) does not seem to be essential either.

We now proceed to prove Theorem 3.1. Let each of ζ and ε denote an element in $\{1, -1\}$, and let $p = lq + \zeta r$. We may assume that p is positive (as long as $q > r/l$). Assume that

$$S_K^3 \left(\frac{p}{Q} \right) \cong \varepsilon S_L^3 \left(\frac{p}{q} \right), \tag{8}$$

where $\epsilon \in \{\pm 1\}$ indicates the orientation and "$\cong$" stands for orientation preserving homeomorphism. Then the two sets

$$\{d(S_K^3(p/Q), i) \mid i \in \mathbb{Z}/p\mathbb{Z}\}, \quad \{d(\varepsilon S_L^3(p/q), i) \mid i \in \mathbb{Z}/p\mathbb{Z}\}$$

are of course equal, but the two parametrizations for Spin^c may not be equal: they could differ by an affine isomorphism of $\mathbb{Z}/p\mathbb{Z}$, that is, there exists an affine isomorphism $\phi \colon \mathbb{Z}/p\mathbb{Z} \to \mathbb{Z}/p\mathbb{Z}$, such that

$$d\left(S_K^3 \left(\frac{p}{Q} \right), i \right) = d\left(\varepsilon S_L^3 \left(\frac{p}{q} \right), \phi(i) \right), \quad \text{for } i \in \mathbb{Z}/p\mathbb{Z}.$$

By Lemma 2.1, the fixed point set of the conjugation isomorphism $J \colon \text{Spin}^c(S_K^3(p/Q)) \to \text{Spin}^c(S_K^3(p/Q))$ is

$$\left\{ \frac{Q-1}{2}, \frac{p+Q-1}{2} \right\} \cap \mathbb{Z}$$

and likewise the fixed point set of $J \colon \text{Spin}^c(\varepsilon S_L^3(p/q)) \to \text{Spin}^c(\varepsilon S_L^3(p/q))$ is

$$\left\{ \frac{q-1}{2}, \frac{p+q-1}{2} \right\} \cap \mathbb{Z}.$$

As J and ϕ commute, we must have

$$\phi\left(\left\{ \frac{Q-1}{2}, \frac{p+Q-1}{2} \right\} \cap \mathbb{Z} \right) = \left\{ \frac{q-1}{2}, \frac{p+q-1}{2} \right\} \cap \mathbb{Z}.$$

It follows that the affine isomorphism $\phi \colon \mathbb{Z}/p\mathbb{Z} \to \mathbb{Z}/p\mathbb{Z}$ is of the form

$$\phi_a(i) = \left[a(i - b) + \frac{(1 - \alpha)p + q - 1}{2} \right]_p \tag{9}$$

where b is an element of $\{\frac{Q-1}{2}, \frac{p+Q-1}{2}\} \cap \mathbb{Z}$, $\alpha = 0$ or 1, and a is an integer satisfying $0 < a < p$, $\gcd(a, p) = 1$. Similarly, replacing a with $p - a$ in (9), we can define ϕ_{p-a}. By (1) and Lemma 2.1, $d(\varepsilon S_L^3(p/q), \phi_a(i)) = d(\varepsilon S_L^3(p/q), \phi_{p-a}(i))$. So we may further assume that

$$0 < a < \frac{p}{2}, \quad \gcd(p, a) = 1. \tag{10}$$

Let

$$\delta_a^\varepsilon(i) = d(L(p, Q), i) - \varepsilon d(S_L^3(p/q), \phi_a(i)). \tag{11}$$

By Proposition 2.2, we have, when $q > r/l$ (so that $p > 0$),

$$\delta_a^\varepsilon(i) = 2V_{\min\{\lfloor \frac{i}{Q} \rfloor, \lfloor \frac{p+Q-1-i}{Q} \rfloor\}}^{\mathrm{K}}. \tag{12}$$

Let $m \in \mathbb{Z}$ satisfy

$$0 \le a + \frac{(1-\alpha)\zeta r + q - 1}{2} - mq < q,$$

then as $0 < a < p/2$, we have $0 \le m \le \frac{l}{2}$ when $q > 2r$.

Let

$$\kappa(i) = \min\left\{\left\lfloor \frac{i}{q} \right\rfloor, \left\lfloor \frac{p+q-1-i}{q} \right\rfloor\right\}.$$

Using Proposition 2.2 and (11), we get

$$\begin{aligned}
\delta_a^\varepsilon(i) &= d(L(p,Q), i) - \varepsilon d(S_L^3(p/q), \phi_a(i)) \\
&= d(L(p,Q), i) - \varepsilon d(L(p,q), \phi_a(i)) + 2\varepsilon V_{\kappa(\phi_a(i))}^{\mathrm{L}}.
\end{aligned} \tag{13}$$

Lemma 3.3. *With the notations and conditions established above, there exists a constant $N = N(r, l, Q, L)$, such that*

$$\left| a - \frac{mp}{l} \right| < N\sqrt{p}$$

for all $q > 2r$.

Proof. It follows from (6) and (12) that

$$\delta_a^\varepsilon(b+1) - \delta_a^\varepsilon(b) = 0 \text{ or } \pm 2. \tag{14}$$

Using (13), (2) and (9), we get

$$\begin{aligned}
&\delta_a^\varepsilon(b+1) - \delta_a^\varepsilon(b) \\
&= \frac{2b + 2 - p - Q}{pQ} - d(L(Q, [p]_Q), [b+1]_Q) + d(L(Q, [p]_Q), [b]_Q) \\
&\quad + 2\varepsilon(V_{\kappa(\phi_a(b+1))}^{\mathrm{L}} - V_{\kappa(\phi_a(b))}^{\mathrm{L}}) \\
&\quad - \varepsilon\left(d\left(L(p,q), a + \frac{(1-\alpha)p + q - 1}{2}\right) - d\left(L(p,q), \frac{(1-\alpha)p + q - 1}{2}\right)\right).
\end{aligned} \tag{15}$$

When $\zeta = 1$, by the recursive formula (2), we have (note that $a + \frac{(1-\alpha)p + q - 1}{2} < p + q$)

$$\begin{aligned}
&d\left(L(p,q), a + \frac{(1-\alpha)p + q - 1}{2}\right) - d\left(L(p,q), \frac{(1-\alpha)p + q - 1}{2}\right) \\
&= \frac{(2a - \alpha p)^2 - (\alpha p)^2}{4pq} - d\left(L(q,r), a - mq + \frac{(1-\alpha)r + q - 1}{2}\right)
\end{aligned}$$

$$+ d\left(L(q,r), \frac{(1-\alpha)r+q-1}{2}\right)$$

$$= \frac{a^2 - a\alpha p}{pq} - \frac{(2a - 2mq - \alpha r)^2 - (\alpha r)^2}{4qr}$$

$$+ d\left(L(r,[q]_r), \left[a - mq + \frac{(1-\alpha)r+q-1}{2}\right]_r\right)$$

$$- d\left(L(r,[q]_r), \left[\frac{(1-\alpha)r+q-1}{2}\right]_r\right)$$

$$= -\frac{l}{pr}\left(a - \frac{mp}{l}\right)^2 + \frac{m^2}{l} - m\alpha$$

$$+ d\left(L(r,[q]_r), \left[a - mq + \frac{(1-\alpha)r+q-1}{2}\right]_r\right)$$

$$- d\left(L(r,[q]_r), \left[\frac{(1-\alpha)r+q-1}{2}\right]_r\right).$$

When $\zeta = -1$,

$$d\left(L(p,q), a + \frac{(1-\alpha)p+q-1}{2}\right) - d\left(L(p,q), \frac{(1-\alpha)p+q-1}{2}\right)$$

$$= \frac{(2a - \alpha p)^2 - (\alpha p)^2}{4pq} - d\left(L(q,q-r), a - mq + \frac{-(1-\alpha)r+q-1}{2}\right)$$

$$+ d\left(L(q,q-r), \frac{-(1-\alpha)r+q-1}{2}\right)$$

$$= \frac{a^2 - a\alpha p}{pq} - \frac{(2a - 2mq + \alpha r - q)^2 - (\alpha r - q)^2}{4q(q-r)}$$

$$+ d\left(L(q-r,r), a - mq + \frac{-(1-\alpha)r+q-1}{2}\right)$$

$$- d\left(L(q-r,r), \frac{-(1-\alpha)r+q-1}{2}\right)$$

$$= \frac{a^2}{pq} - \frac{\alpha a}{q} - \frac{(a - mq + \alpha r - q)(a - mq)}{q(q-r)} + \frac{(a - mq - (1-\alpha)r)(a - mq)}{(q-r)r}$$

$$- d\left(L(r,[q-r]_r), \left[a - mq + \frac{-(1-\alpha)r+q-1}{2}\right]_r\right)$$

$$+ d\left(L(r,[q-r]_r), \left[\frac{-(1-\alpha)r+q-1}{2}\right]_r\right)$$

$$= \frac{l}{pr}\left(a - \frac{mp}{l}\right)^2 + \frac{m^2}{l} - m\alpha$$

$$- d\left(L(r,[q]_r), \left[a - mq + \frac{-(1-\alpha)r+q-1}{2}\right]_r\right)$$

$$+ d\left(L(r,[q]_r), \left[\frac{-(1-\alpha)r+q-1}{2}\right]_r\right).$$

Let

$$
\begin{aligned}
C_0 ={}& \frac{2b+2-p-Q}{pQ} - d(L(Q, [p]_Q), [b+1]_Q) + d(L(Q, [p]_Q), [b]_Q) \\
& + 2\varepsilon(V^{\mathrm{L}}_{\kappa(\phi_a(b+1))} - V^{\mathrm{L}}_{\kappa(\phi_a(b))}) \\
& - \varepsilon\zeta\left(d\left(L(r, [q]_r), \left[a - mq + \frac{\zeta(1-\alpha)r+q-1}{2}\right]_r\right)\right. \\
& \left. - d\left(L(r, [q]_r), \left[\frac{\zeta(1-\alpha)r+q-1}{2}\right]_r\right)\right),
\end{aligned}
$$

then the right-hand side of (15) becomes

$$
\varepsilon\left(\zeta\frac{l}{pr}\left(a - \frac{mp}{l}\right)^2 - \frac{m^2}{l} + m\alpha\right) + C_0.
$$

Using (14), we get

$$
\frac{l}{pr}\left(a - \frac{mp}{l}\right)^2 \le 2 + \left|\frac{m^2}{l} - m\alpha\right| + |C_0|.
$$

Clearly, $|C_0|$ and m are bounded in terms of r, l, Q, L, so the conclusion of the lemma follows. $\diamond$

Lemma 3.4. *Let k be an integer satisfying*

$$
0 \le k < \frac{p - (2l+1)r + l}{2Nl^2\sqrt{p}} - \frac{1}{l}. \tag{16}
$$

Let

$$
i_k = \frac{(1-\alpha)p+q-1}{2} + k(al - mp), \quad j_k = \frac{(1-\alpha)\zeta r+q-1}{2} + k(al - mp).
$$

Then

$$
\delta^\varepsilon_a(b+lk+1) - \delta^\varepsilon_a(b+lk) = Ak + B + C_k,
$$

where

$$
\begin{aligned}
A ={}& \varepsilon\zeta \cdot \frac{2(al-mp)^2}{pr} + \frac{2l}{pQ}, \\
B ={}& \varepsilon\left(\zeta\frac{l}{pr}(a - \frac{mp}{l})^2 - \frac{m^2}{l} + m\alpha\right), \\
C_k ={}& \frac{2b+2-p-Q}{pQ} - d(L(Q, [p]_Q), [b+lk+1]_Q) + d(L(Q, [p]_Q), [b+lk]_Q) \\
& + 2\varepsilon(V^L_{\kappa(\phi_a(b+lk+1))} - V^L_{\kappa(\phi_a(b+lk))}) \\
& - \varepsilon\zeta\big(d(L(r, [q]_r), [a - mq + j_k]_r) - d(L(r, [q]_r), [j_k]_r)\big).
\end{aligned}
$$

Proof. By (16), we have

$$(lk+1)N\sqrt{p} < \frac{p-(2l+1)r+l}{2l} \leq \frac{q-2r+1}{2}. \tag{17}$$

It follows from (10), (17) and Lemma 3.3 that

$$0 \leq i_k < i_k + a < p+q, \qquad 0 \leq j_k, j_k + a - mq < q. \tag{18}$$

For example,

$$\begin{aligned}
j_k + a - mq &= j_k + a - m\frac{p-\zeta r}{l} \\
&= \frac{(1-\alpha)\zeta r + q - 1}{2} + (lk+1)\left(a - \frac{mp}{l}\right) + \frac{m\zeta r}{l} \\
&< \frac{r+q-1}{2} + \frac{q-2r+1}{2} + \frac{r}{2} \\
&= q.
\end{aligned}$$

The other inequalities can be verified similarly.

Using (13), we can compute

$$\begin{aligned}
&\delta_a^\varepsilon(b+lk+1) - \delta_a^\varepsilon(b+lk) \\
&= \frac{2b+2lk+2-p-Q}{pQ} - d(L(Q,[p]_Q),[b+lk+1]_Q) + d(L(Q,[p]_Q),[b+lk]_Q) \\
&\quad + 2\varepsilon(V_{\kappa(\phi_a(b+lk+1))}^{\mathrm{L}} - V_{\kappa(\phi_a(b+lk))}^{\mathrm{L}}) - \varepsilon\big(d(L(p,q),i_k+a) - d(L(p,q),i_k)\big). \tag{19}
\end{aligned}$$

As in the proof of Lemma 3.3, using (18) and the recursion formula (2), when $\zeta = 1$, we can compute

$$\begin{aligned}
&d(L(p,q),i_k+a) - d(L(p,q),i_k) \\
&= \frac{(2i_k+2a+1-p-q)^2 - (2i_k+1-p-q)^2}{4pq} - d(L(q,r),j_k+a-mq) + d(L(q,r),j_k) \\
&= \frac{a(2k(al-mp)+a-\alpha p)}{pq} - \frac{(2j_k+2a-2mq+1-q-r)^2 - (2j_k+1-q-r)^2}{4qr} \\
&\quad + d(L(r,[q]_r),[j_k+a-mq]_r) - d(L(r,[q]_r),[j_k]_r) \\
&= -\frac{2(al-mp)^2}{pr}k - \frac{l}{pr}\left(a - \frac{mp}{l}\right)^2 + \frac{m^2}{l} - m\alpha + d(L(r,[q]_r),[j_k+a-mq]_r) \\
&\quad - d(L(r,[q]_r),[j_k]_r).
\end{aligned}$$

Similarly, when $\zeta = -1$, we get

$$\begin{aligned}
&d(L(p,q),i_k+a) - d(L(p,q),i_k) \\
&= \frac{2(al-mp)^2}{pr}k + \frac{l}{pr}\left(a - \frac{mp}{l}\right)^2 + \frac{m^2}{l} - m\alpha - d(L(r,[q]_r),[j_k+a-mq]_r) \\
&\quad + d(L(r,[q]_r),[j_k]_r).
\end{aligned}$$

So the right-hand side of (19) is $Ak + B + C_k$. $\diamond$

We can now finish the proof of Theorem 3.1. If $S_K^3(p/Q) \cong \varepsilon S_L^3(p/q)$, then (12) holds, so

$$\delta_a^\varepsilon(b + lk + 1) - \delta_a^\varepsilon(b + lk) = 0 \text{ or } \pm 2 \qquad (20)$$

for all k satisfying (16).

Let A, B, C_k be as in Lemma 3.4. By (7), $A \neq 0$. So $Ak + B + C$ is equal to 0 or ± 2 for at most three values of k for any given C. From the expression of C_k, it is evident that there exists a constant integer $M = M(L)$, such that given $p, q, a, \varepsilon, \zeta$, as k varies, C_k can take at most MQr values. Thus $Ak + B + C_k$ can be 0 or ± 2, i.e., (20) holds, for at most $3MQr$ values of k. But if $p \geq 4l^2 N^2 (3lMQr + 2)^2$, then each of k in $\{0, 1, 2, \ldots, 3MQr\}$ satisfies (16) and thus (20) holds for each of these $3MQr + 1$ values of k. This contradiction shows that p is bounded above by $4l^2 N^2 (3lMQr + 2)^2$. Since $p = lq + \zeta r$, we also get a bound for q.

4   SEIFERT SURGERIES

In this section we prove Theorem 1.1, Addendum 1.2 and Theorems 1.4 and 1.5.

Lemma 4.1. *If W is an oriented Seifert fibered space whose base orbifold is $S^2(2, 3, r)$ (or $S^2(3, 4, r)$), $r > 1$, then W is homeomorphic to some surgery on the torus knot $T(3, 2)$ (resp. $T(4, 3)$), i.e.,*

$$W \cong \varepsilon S_{T(3,2)}^3 \left(\frac{6q + \zeta r}{q} \right) \quad \left(resp. \ W \cong \varepsilon S_{T(4,3)}^3 \left(\frac{12q + \zeta r}{q} \right) \right)$$

for some $\varepsilon, \zeta \in \{1, -1\}$ and some positive integer q.

Proof. The proof is a quick generalization of that of [10, Lemma 3.1]. The Seifert space W has three singular fibers of orders $2, 3, r$ (resp. $3, 4, r$). The exterior of the singular fiber of order r in W is homeomorphic (not necessarily orientation preserving) to the exterior of the torus knot $T(3, 2)$ (resp. $T(4, 3)$) in S^3 because there is only one Seifert fibered space (up to homeomorphism) with base orbifold $D^2(2, 3)$ (resp. $D^2(3, 4)$). Now on $T(3, 2)$ (resp. $T(4, 3)$), a surgery gives Seifert fibered space with base orbifold $S^2(2, 3, r)$ (resp. $S^2(3, 4, r)$) if and only if the slope is $\frac{6q + \zeta r}{q}$ (resp. $\frac{12q + \zeta r}{q}$), $\gcd(q, r) = 1$. We may assume $q > 0$ up to change the sign of ζ.   ◇

The following proposition classifies satellite knots in S^3 which admit Nil Seifert surgeries.

Proposition 4.2. *Suppose K is a satellite knot and $S_K^3(p/q)$ is a Nil Seifert fibered space with $p/q > 0$. Then K is a cable over $T(3, 2)$. More precisely, there are four cases for the cable type and the slope:*

Cable Type	p/q
$(29, 5)$	$144/1$
$(31, 5)$	$156/1$
$(41, 7)$	$288/1$
$(43, 7)$	$300/1$

Proof. Let C be a companion knot of K such that C is itself not a satellite knot. Let V be a solid torus neighborhood of C in S^3 such that K is contained in the interior of V but is not contained in a 3-ball in V and is not isotopic to the core circle of V. Let N be a regular neighborhood of K in V, $M_K = S^3 - int(N)$, $M_C = S^3 - int(V)$, and let $V_K(p/q)$ be the p/q-surgery of V along K. Then $S_K^3(p/q) = M_K(p/q) = M_C \cup V_K(p/q)$. Since $S_K^3(p/q)$ does not contain incompressible tori, ∂V must be compressible in $S_K^3(p/q)$ and in fact compressible in $V_K(p/q)$. By [5], it follows that either $V_K(p/q)$ has a connected summand W with $0 < |H_1(W)| < \infty$, or $V_K(p/q)$ is a solid torus. In the former case, by [17] $V_K(p/q)$ contains a lens space as a connected summand, which contradicts the fact that $S_K^3(p/q) = M_K(p/q) = M_C \cup V_K(p/q)$ is a Nil Seifert fibered space. Hence $V_K(p/q)$ is a solid torus. Now by [5], K is a 0 or 1-bridge braid in V with winding number $w > 1$. By [7, Lemma 3.3] the meridian slope of the solid torus $V_K(p/q)$ is $p/w^2 q$ and thus $M_K(p/q) = M_C(p/w^2 q)$. So C is a torus knot by [2] and then C must be the trefoil knot $T(3, 2)$ by [11].

If K is a (s, t)-cable in V (where we may assume $t > 1$ is the winding number of K in V), then by [7, Lemma 7.2], $p = stq + \epsilon_1$, $\epsilon_1 \in \{\pm 1\}$. So $M_K(p/q) = M_C((stq + \epsilon_1)/(t^2 q))$. By [11] we should have $stq + \epsilon_1 = 6t^2 q + \epsilon_2 6$, $\epsilon_2 \in \{\pm 1\}$. So we have $stq - 6t^2 q = \epsilon_1 5$ or $-\epsilon_1 7$, which implies $q = 1$ and $t = 5$ or 7.

If $t = 5$, then $s = 30 + \epsilon_1$ and $p = 5(30 + \epsilon_1) + \epsilon_1$. That is, either K is the $(29, 5)$-cable over $T(3, 2)$, $q = 1$ and $p = 144$ or K is the $(31, 5)$-cable over $T(3, 2)$, $q = 1$ and $p = 156$. Likewise if $t = 7$, K is the $(41, 7)$-cable over $T(3, 2)$, $q = 1$ and $p = 288$ or K is the $(43, 7)$-cable over $T(3, 2)$, $q = 1$ and $p = 300$.

Now suppose that K is a 1-bridge braid in V. By [6, Lemma 3.2], $q = 1$ and $p = \tau w + d$ where w is the winding number of K in V, and τ and d are integers satisfying $0 < \tau < w - 1$ and $0 < d < w$. Hence $M_K(p/q) = M_C(\tau w + d/w^2)$ and by [11] $\tau w + d = 6w^2 \pm 6$. But $6w^2 \pm 6 - \tau w - d \geq 6w^2 - 6 - (w-1)w - w = 5w^2 - 6 > 0$. We get a contradiction, which means K cannot be a 1-bridge braid in V. $\diamond$

Proof of Theorem 1.1 and Addendum 1.2. Let K be any non-trefoil knot in S^3 such that $S_K^3(p/q)$ is a Nil Seifert space. Up to changing K to its mirror image, we may assume that $p, q > 0$. If K is a torus knot, then by [11], no surgery on K can produce a Nil Seifert fibered space. So we may assume that K is not a torus knot. By [2] and Proposition 4.2, $q = 1$. We are now going to give a concrete upper bound for p. As noted in Section 1, the base orbifold of $S_K^3(p)$ is $S^2(2, 3, 6)$. Thus by Lemma 4.1, $S_K^3(p) \cong \varepsilon S_{T(3,2)}^3(p/q)$ with $p = 6q + \zeta 6$, for some $\varepsilon, \zeta \in \{1, -1\}$, and $q > 0$. As $p \neq 0$, $p/q > 1 = g(T(3, 2))$ which implies that $S_K^3(p) \cong \varepsilon S_{T(3,2)}^3(p/q)$ is an L-space by [15, Corollary 1.4]. Therefore we may use surgery formula (5) instead of Proposition 2.2. Now we apply the proof of Theorem 3.1 (and the notations established there) to our current case with $L = T(3, 2)$, $Q = 1$, $l = r = 6$. Then $m \in \{0, 1, 2, 3\}$, $b \in \{0, p/2\}$, $V_i^L = b_i^{T(3,2)}$ (which is 1 if $i = 0$ and 0 if $i > 0$), $V_i^K = b_i^K$ and

$$C_0 = \frac{2b + 2 - p - 1}{p} + 2\varepsilon \left(b_{\kappa(\phi_a(b+1))}^{T(3,2)} - b_{\kappa(\phi_a(b))}^{T(3,2)} \right)$$

$$- \varepsilon\zeta \left(d \left(L(6, [q]_6), \left[a - mq + \frac{\zeta(1 - \alpha)6 + q - 1}{2} \right]_6 \right) \right.$$

$$\left. - d \left(L(6, [q]_6), \left[\frac{\zeta(1 - \alpha)6 + q - 1}{2} \right]_6 \right) \right).$$

Using formula (2) one can compute

$$d(L(6,q),i) = \begin{cases} \left(\frac{5}{4}, \frac{5}{12}, \frac{-1}{12}, \frac{-1}{4}, \frac{-1}{12}, \frac{5}{12}\right), & q=1, i=0,1,\ldots,5, \\ \left(\frac{-5}{12}, \frac{1}{12}, \frac{1}{4}, \frac{1}{12}, \frac{-5}{12}, \frac{-5}{4}\right), & q=5, i=0,1,\ldots,5. \end{cases} \tag{21}$$

Thus $|C_0| \le 1 + 2 + \frac{3}{2} < 5$. Since $|\frac{m^2}{6} - m\alpha| \le 2$ for $m=0,1,2,3$ and $\alpha = 0,1$, we may take $N=3$. Similarly recall the A, B, and C_k in Lemma 3.4, and in our current case, C_k becomes

$$C_k = \frac{2b+2-p-1}{p} + 2\varepsilon\left(b^{T(3,2)}_{\kappa(\phi_a(b+lk+1))} - b^{T(3,2)}_{\kappa(\phi_a(b+lk))}\right)$$
$$-\varepsilon\zeta\left(d(L(6,[q]_6), [a-mq+j_k]_6) - d(L(6,[q]_6), [j_k]_6)\right),$$

which can take at most 18 values as k varies. Thus the bound for p is $4 \cdot 6^2 \cdot 3^2 (3 \cdot 6 \cdot 18 + 2)^2$ when $A \ne 0$.

Now we just need to show that in our current case, A is never zero. Suppose otherwise that $A=0$. Then $\varepsilon\zeta = -1$, $(a - \frac{mp}{6})^2 = 1$, so $a - mq = \zeta m \pm 1$, and by Lemma 3.4

$$\delta^\varepsilon_a(b+lk+1) - \delta^\varepsilon_a(b+lk) = B + C_k$$
$$= -\varepsilon\frac{m^2}{6} + \varepsilon m\alpha + (0 \text{ or } -1) + 2\varepsilon(b^{T(3,2)}_{\kappa(\phi_a(b+lk+1))} - b^{T(3,2)}_{\kappa(\phi_a(b+lk))})$$
$$+d(L(6,[q]_6), [\zeta m \pm 1 + j_k]_6) - d(L(6,[q]_6), [j_k]_6).$$

Thus

$$-\varepsilon\frac{m^2}{6} + d\left(L(6,[q]_6), \zeta m \pm 1 + \left[3\zeta(1-\alpha) + \frac{q-1}{2}\right]_6\right)$$
$$- d\left(L(6,[q]_6), \left[3\zeta(1-\alpha) + \frac{q-1}{2}\right]_6\right) \tag{22}$$

is integer valued. Using (21), we see that for each of $m=0,1,2,3$, $q \equiv 1,5 \pmod 6$, $\alpha \in \{0,1\}$ and $\zeta \in \{1,-1\}$, the expression given in (22) is never integer valued. This contradiction proves the assertion that $A \ne 0$.

Now for the bounded region of integral slopes for p, one can use (5), (2) and computer calculation to locate those possible integral slopes and identify the corresponding Nil Seifert fibered spaces given in Theorem 1.1, so we get Theorem 1.1. One can also recover the possible Alexander polynomials for the candidate knots using formula (3). The rest of Addendum 1.2 follows from [11], Proposition 4.2, [3], and direct verification using SnapPy. $\diamond$

Proof of Theorem 1.4. Let K be a hyperbolic knot in S^3 such that $S^3_K(p/Q)$ is a Seifert fibered space whose base orbifold is $S^2(2,3,r)$ (or $S^2(3,4,r)$). By changing K to its mirror image, we may assume that both p and Q are positive integers. By [9] we have $Q \le 8$. So we just need to show that p is bounded above (independent of hyperbolic K).

By Lemma 4.1,

$$S^3_K(p/Q) \cong \varepsilon S^3_{T(3,2)}\left(\frac{6q+\zeta r}{q}\right) \quad \left(\text{resp. } S^3_K(p/Q) \cong \varepsilon S^3_{T(4,3)}\left(\frac{12q+\zeta r}{q}\right)\right)$$

for some $\varepsilon, \zeta \in \{1, -1\}$ and some positive integer q. Now applying Theorem 3.1 with $l = 6$ and $L = T(3, 2)$ (resp. $l = 12$ and $L = T(4, 3)$), our desired conclusion is true when (7) holds, i.e.,

$$\sqrt{\frac{6r}{Q}} \notin \mathbb{Z} \ \left(\text{resp.} \ \sqrt{\frac{12r}{Q}} \notin \mathbb{Z}\right)$$

for each $Q = 1, \ldots, 8$. $\diamond$

Proof of Theorem 1.5. Let K be a hyperbolic knot in S^3 such that $S_K^3(p/Q) \cong \varepsilon S_{T(m,n)}^3(\frac{mnq + \zeta r}{q})$. Again $Q \leq 8$ and Theorem 3.1 applies with $l = mn$ and $L = T(m, n)$. $\diamond$

REFERENCES

[1] J. Berge, *Some knots with surgeries yielding lens spaces*, preprint, arXiv:1802. 09722.

[2] S. Boyer, *On the local structure of $SL(2, \mathbb{C})$-character varieties at reducible characters*, Topology Appl. 121 (2002), 383–413.

[3] M. Eudave-Muñoz, *On hyperbolic knots with Seifert fibered Dehn surgeries*, Topology and Its Applications 121 (2002), 119–141.

[4] D. Gabai, *Foliations and the topology of 3–manifolds III*, J. Differential Geom. 26 (1987), no. 3, 479–536.

[5] D. Gabai, *Surgery on knots in solid tori*, Topology 28 (1989), 1–6.

[6] D. Gabai, *1–bridge braids in solid tori*, Topology Appl. 37 (1990), no. 3, 221–235.

[7] C. Gordon, *Dehn surgery and satellite knots*, Trans. Amer. Math. Soc. 275 (1983), 687–708.

[8] L. Gu, *Integral finite surgeries on knots in S^3*, preprint (2014), available at arXiv:1401.6708.

[9] M. Lackenby, R. Meyerhoff, *The maximal number of exceptional Dehn surgeries*, Invent. Math. 191 (2013), no. 2, 341–382.

[10] E. Li, Y. Ni, *Half-integral finite surgeries on knots in S^3*, Ann. Fac. Sci. Toulouse Math. (6) 24 (2015), no. 5, 1157–1178.

[11] L. Moser, *Elementary surgery along a torus knot*, Pacific J. Math. 38 (1971), 737–745.

[12] Y. Ni, Z. Wu, *Cosmetic surgeries on knots in S^3*, J. Reine Angew. Math. 706 (2015), 1–17.

[13] B. Owens, S. Strle, *Rational homology spheres and the four-ball genus of knots*, Adv. Math. 200 (2006), no. 1, 196–216.

[14] P. Ozsváth, Z. Szabó, *Absolutely graded Floer homologies and intersection forms for four-manifolds with boundary*, Adv. Math. 173 (2003), no. 2, 179–261.

[15] P. Ozsváth, Z. Szabó, *Knot Floer homology and rational surgeries*, Algebr. Geom. Topol. 11 (2011), 1–68.

[16] J. Rasmussen, *Floer homology and knot complements*, PhD Thesis, Harvard University (2003), available at arXiv:math.GT/0306378.

[17] M. Scharlemann, *Producing reducible 3-manifolds by surgery on a knot*, Topology 29 (1990), 481–500.

[18] P. Scott, *The geometries of 3-manifolds*, Bull. London Math. Soc. 15 (1983), 401–487.

Degree-d-invariant Laminations

William P. Thurston, Hyungryul Baik, Gao Yan,
John H. Hubbard, Kathryn A. Lindsey,
Tan Lei and Dylan P. Thurston

PREFACE

During the last year of his life, William P. Thurston developed a theory of *degree-d-invariant laminations*, a tool that he hoped would lead to what he called "a qualitative picture of [the dynamics of] degree d polynomials." Thurston discussed his research on this topic in his seminar at Cornell University and was in the process of writing an article on this topic, but he passed away before completing the manuscript. Part I of this document consists of Thurston's unfinished manuscript. As it stands, the manuscript is beautifully written and contains a lot of his new ideas. However, he discussed ideas that are not in the unfinished manuscript and some details are missing. Part II consists of supplementary material written by the other authors based on what they learned from him throughout his seminar and email exchanges with him. Tan Lei also passed away during the preparation of Part II. William Thurston's vision was far beyond what we could write here, but, hopefully, this paper will serve as a starting point for future researchers.

Each semester since moving to Cornell University in 2003, William Thurston taught a seminar course titled "Topics in Topology," which was familiarly (and perhaps more accurately) referred to by participants as "Thurston seminar." On the first day of each semester, Thurston asked the audience what mathematical topics they would like to hear about, and he tailored the direction of the seminar according to the interests of the participants. Although the course was nominally a seminar in topology, he discussed other topics as well, including combinatorics, mathematical logic, and complex dynamics. Between 2010 and 2012, a high percentage of the seminar participants were dynamicists, and so (with the exception of one semester) Thurston's seminar during this period primarily focused on topics in complex dynamics. He discussed his topological characterization of rational maps on the Riemann sphere, as well as how to understand complex polynomials via topological entropy and laminations on the circle. During this time, Thurston developed many beautiful ideas, motivated in part by discussions with people in his seminar and in part by email exchanges with others who were at a distance.

His seminar was not an organized lecture series; it was much more than that. He talked about ideas that he was developing at that very moment, as opposed to previously known findings. Thurston invited members of the seminar to be actively involved in the exploration. He often demonstrated computer experiments in class, and he encouraged seminar participants to experiment with his codes. Seminar participants frequently received drafts of Thurston's manuscript as his thinking

evolved. Those fortunate enough to learn from Thurston observed how his understanding of the subject gradually transformed into a beautiful theory.

This paper was the work of many people, written by different people at different
times. We have done our best to make it a unified whole, and apologize for remaining
inconsistencies of notation.

Part I

Degree-d-invariant Laminations

William P. Thurston
February 22, 2012

1   INTRODUCTION

Despite years of strong effort by an impressive group of insightful and hardworking
mathematicians and many advances, our overall understanding and global picture
of the dynamics of degree d rational maps and even degree d complex polynomials
has remained sketchy and unsatisfying.

The purpose of this paper is to develop at least a sketch for a skeletal qualitative picture of degree d polynomials. There are good theorems characterizing
and describing examples individually or in small-dimensional families, but that is
not our focus. We hope instead to contribute toward developing and clarifying the
global picture of the connectedness locus for degree d polynomials, that is, the
higher-dimensional analogues of the Mandelbrot set.

To do this, the main tool will be the theory of degree-d-invariant laminations.

We hope that by developing a better picture for degree d polynomials, we will
develop insights that will carry on to better understand degree d rational maps,
whose global description is even more of a mystery.

2   SOME DEFINITIONS AND BASIC PROPERTIES

A degree d polynomial map $z \mapsto P(z) : \mathbb{C} \to \mathbb{C}$ always "looks like" $z \mapsto z^d$ near ∞.
More precisely, it is known that P is conjugate to $z \mapsto z^d$ in some neighborhood of ∞.

We may as well specialize to monic polynomials such that the center of mass of
the roots is at the origin (so that the coefficient of z^{d-1} is 0), since any polynomial
can be conjugated into that form. In that case, we can choose the conjugating map
to converge to the identity near ∞; this uniquely determines the map. As Douady
and Hubbard noted, we can use the dynamics to extend the conjugacy near ∞
inward toward 0 step by step. If the Julia set is connected, we obtain in this way
a Riemann mapping of the complement of the Julia set to the complement of the
closed unit disk in $\mathbb{C}$ that conjugates the dynamics outside the Julia set (which is
the attracting basin of ∞) to the standard form $z \mapsto z^d$.

It has been known since the time of Fatou and Julia (and easy to show) that the
Julia set is the boundary of the attracting basin of ∞. We want to investigate the
topology of the Julia set, and how this topology varies among polynomial maps of

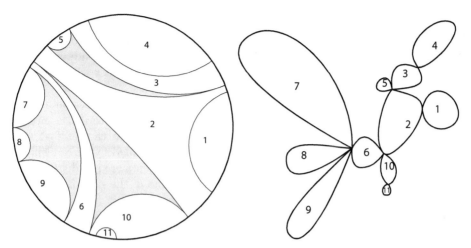

Figure 1: On the left is a finite lamination associated with a treelike equivalence relation with finitely many non-trivial equivalences: the two ends of any leaf are equivalent, and (as a consequence) the vertices of the polygons are all equivalent. The leaves are drawn as geodesics in the Poincaré disk model of the hyperbolic plane. The leaves and shaded areas can be shrunk to points by a pseudo-isotopy of the plane to obtain (topologically) the figure on the right.

degree d. As long as the Julia set is locally connected, the Julia set is the continuous image of the unit circle that is the boundary of the Riemann map around ∞; the key question is to understand the identifications on the circle made by these maps, and the way the identifications vary as the polynomial varies.

Define a *treelike* equivalence relation on the unit circle to be a closed equivalence relation such that for any two distinct equivalence classes, their convex hulls in the unit disk are disjoint. (A relation $R \subset X \times X$ on a topological space X is *closed* if it is a closed subset of $X \times X$.) The condition that the convex hulls of equivalence classes be disjoint comes from the topology of the plane: it translates into the condition that if we take the complement of the open unit disk and make the given identifications on the circle, two simple closed curves that cross the circle quotient using different equivalence classes cannot have intersection number 1, otherwise the quotient space would not embed in the plane.

To develop a geometric understanding of the possible equivalence relations, it helps to translate the concept into the language of laminations. Given a treelike equivalence relation R, there is an associated lamination $\mathrm{Lam}(R)$ of the open disk, where the leaves of $\mathrm{Lam}(R)$ consist of boundaries of convex hulls of equivalence classes intersected with the open disk. The regions bounded by leaves are called *gaps*. Some of the gaps of $\mathrm{Lam}(R)$ are *collapsed gaps*, which touch the circle in a single equivalence class, while other gaps are *intact gaps*. An intact gap necessarily touches the circle in an uncountable set, and its boundary mod R is homeomorphic to a circle.

Conversely, given a lamination λ there is an equivalence relation $\mathrm{Rel}(\lambda)$, usually obtained by taking the transitive closure of the relation that equates endpoints of leaves. However, this relation may not be topologically closed. One can take the closure, which may not be an equivalence relation. It is possible to alternate the operations of transitive closure and topological closure by transfinite induction until it stabilizes to a closed equivalence relation $\mathrm{Rel}(\lambda)$, or simply define $\mathrm{Rel}(\lambda)$ as the intersection of all closed equivalence relations that identify endpoints of leaves of λ.

Often, but not always, the operations Lam and Rel are inverse to each other. If R has equivalence classes that are Cantor sets, then $\mathrm{Rel}(\mathrm{Lam}(R))$ will collapse these only to a circle, not to a point. If λ has chains of leaves with common endpoints, then any leaves in the interior of the convex hull of the chain of vertices will disappear, and new edges may also appear.

A circular order on a set intuitively means an embedding of the set on a circle, up to orientation-preserving homeomorphism. In combinatorial terms, a circular order can be given by a function C from triples of elements to the set $\{-1, 0, 1\}$ (interpreted as giving the sign of the area of the triangle formed by three elements in the plane) such that $C(a, b, c) = 0$ if and only if the elements are not distinct, and C is a 2-cocycle, meaning for any four elements a, b, c and d, the sum of the values for the boundary of a 3-simplex labeled with these elements is 0. (This implies, in particular, that $C(a, b, c) = -C(b, a, c)$, etc.) With this definition, it can be shown that any countable circularly ordered set can be embedded in the circle in such a way that $C(a, b, c)$ is the sign of the area of the triangle formed by the images of a, b and c. One way to prove this assertion is to first observe that a circular order can be "cut" at a point a into a linear order, defined by $b < c \iff C(a, b, c) = 1$. It is reasonably well-known that a countable linear order on a set is induced by an embedding of that set in the interval, from which it follows that a circular ordering is induced by an embedding in S^1.

An *interval* J of a circularly ordered set is a subset with a linear order (=total order) satisfying the condition that

i. for $a, b, c \in J$, $a < b < c \iff C(a, b, c) = 1$, and
ii. for any x and any $a, b \in J$, if $C(a, x, b) = 1$ then $x \in J$.

Usually the linear order is determined by the subset, but there is an exception if the subset is the entire circularly ordered set. In that case the definition is like cutting a circle somewhere to form an interval. Just as for linear orders, any two elements a, b in a circularly ordered set determine a closed interval $[a, b]$, as well as an open interval (a, b), etc.

The *degree* of a map $f : X \to Y$ between two circularly ordered sets is the minimum size of a partition of X into intervals such that on each interval, the circular order is preserved. If there is no such finite partition, the degree is ∞.

A treelike equivalence relation R is *degree-d-invariant*

i. if sRt then $s^d R t^d$, and
ii. if for any equivalence class C, the total degree of the restrictions of $z \mapsto z^d$ to the equivalence classes on the set $C^{1/d}$ is d.

Similarly, a lamination λ is *degree-d-invariant*

i. if there is a leaf with endpoints x and y, then either $x^d = y^d$ or there is a leaf with endpoints x^d and y^d, and
ii. if there is a leaf with endpoints x and y, there is a set of d disjoint leaves with one endpoint in $x^{1/d}$ and the other endpoint in $y^{1/d}$.

These conditions on leaves imply related conditions for gaps, from the behavior of the boundary of the gap. In particular, an equivalence relation R is degree-d-invariant if and only if $\mathrm{Lam}(R)$ is degree-d-invariant. As customary, we will use the word "quadratic" as a synonym for degree 2, "cubic" for degree 3, "quartic" for degree 4, etc.

It is worth pointing out that the map $f_d : z \mapsto z^d$ is centralized by a dihedral group of symmetries of order $2(d-1)$, generated by reflection $z \mapsto \bar{z}$ together with rotation $z \mapsto \zeta z$ where ζ is a primitive $(d-1)$th root of unity. Therefore, the set of all degree-*d*-invariant laminations and the set of all degree-*d*-invariant relations has the same symmetry group.

If R is a degree-*d*-invariant equivalence relation, a *critical class* C is an equivalence class that maps with degree greater than 1. Similarly, a critical gap in $\mathrm{Lam}(R)$ is a gap whose intersection with the circle is mapped with degree greater than 1. The *criticality* of a class or a gap C is its degree minus 1. A critical class may correspond to either a critical leaf, which must have criticality 1, or a critical gap. A critical gap may be a collapsed gap that corresponds to a critical class, or an intact gap.

Proposition 2.1. *For any degree-d-invariant equivalence relation R, the total criticality of equivalence classes of R together with intact critical gaps of $\mathrm{Lam}(R)$ (i.e., the sum of the criticality of the critical classes and the intact critical gaps) equals $d-1$.*

Proof. First, let's establish that if $d > 1$ there is at least one critical leaf or critical gap. We can do this by extending the map $z \mapsto z^d$ to the disk: first extend linearly on each edge of the lamination, then foliate each gap by vertical lines and extend linearly to each of the leaves of this foliation. If there were neither collapsing leaves nor collapsing gaps, this would be a homeomorphism on each leaf and on each gap, so the map would be a covering map, impossible since the disk is simply-connected.

Now consider any critical leaf $\overline{xy}$, the circle with x and y identified is the union of two circles mapping with total degrees d_1 and d_2 where $d_1 + d_2 = d$. Proceed by induction: if the total criticality in these two circles is $d_1 - 1$ and $d_2 - 1$, then the total criticality in the whole circle is $d - 1$.

Now consider any critical gap G. Let X be the union of S^1 with the boundary of G, and X' be its image under $z \mapsto z^d$, extended linearly on each edge. Any leaf in the image of the boundary of G has d preimages, from which it follows readily that the total criticality of the map on the complementary regions of X is $d-1$.   □

Definition 2.2. *The* major *of a degree-d-invariant lamination is the set of critical leaves and critical gaps. The* major *of a degree-d-invariant equivalence relation R is the set of equivalence classes corresponding to critical leaves and critical gaps of $\mathrm{Lam}(R)$.*

3   MAJORS

What are the possibilities for the major of a degree-*d*-invariant lamination?

We'll say that the major of a lamination is *primitive* if each critical gap is a polygon whose vertices are all identified by $z \mapsto z^d$. We will see later that any degree-*d*-invariant lamination is contained in a degree-*d*-invariant lamination whose major is primitive; there are sometimes several possible invariant enlargements, and sometimes uncountably many. The invariant enlargements might not come from invariant equivalence relations, but they are useful nonetheless for understanding the global structure of invariant laminations, invariant equivalence relations, and Julia sets for degree d polynomials.

Even without a predefined lamination, we can define a *primitive degree-d major* to be a collection of disjoint leaves and polygons each of whose vertices are identified

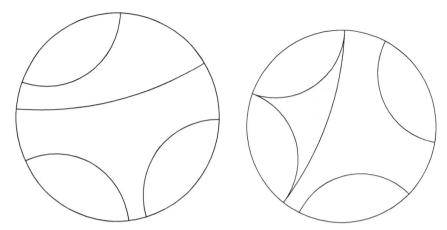

Figure 2: Here are two examples of primitive majors for a quintic-invariant lamination. Each intact gap that touches S^1 touches with total length $2\pi/5$. As a consequence, the two endpoints of any leaf are collapsed to a single point under $z \mapsto z^5$.
Each isolated leaf has criticality 1; the three vertices of the shaded triangular gap in
the example at right map to a single point, so it has criticality 2. If the leaves of the
lamination at left move around so that two of them touch, then a third leaf joining
the non-touching endpoints of the touching leaves is implicit, as in the figure at right.
The example at right can be perturbed so that any one of the three leaves forming
the ideal triangle disappears and the other two become disjoint.

under $z \mapsto z^d$, with total criticality $d - 1$. In Section 4 we will show how to construct
a degree-d-invariant lamination whose major is the given major. First though, we
will analyze the set $\mathrm{PM}(d)$ of all primitive degree-d majors.

An element $m \in \mathrm{PM}(d)$ determines a quotient graph $\gamma(m)$ obtained by identifying each equivalence class to a point. The path-metric of S^1 defines a path-metric on
$\gamma(m)$. In addition, $\gamma(m)$ has the structure of a planar graph, that is, an embedding
in the plane well-defined up to isotopy, obtained by shrinking each leaf and each
ideal polygon of the lamination to a point. These graphs have the property that
$H^1(\gamma(m))$ has rank d, and every cycle has length a multiple of $2\pi/d$. Every edge
must be accessible from the infinite component of the complement, so the metric
and the planar embedding together with the starting point, that is, the image of
$1 \in \mathbb{C}$, is enough to define the major.

The metric $\mathrm{met}(m)$ on the circle induced by the path-metric on $\gamma(m)$ determines
a metric md on $\mathrm{PM}(d)$, defined as the sup difference of metrics,

$$\mathrm{md}(m, m') = \sup_{x,y \in S^1} |\mathrm{met}(m)(x, y) - \mathrm{met}(m')(x, y)| \,.$$

In particular, there is a well-defined topology on $\mathrm{PM}(d)$.

There is a recursive method one can use to construct and analyze primitive
degree-d majors, as follows. Consider any isolated leaf l of a major $m \in \mathrm{PM}(d)$. The
endpoints of l split the circle into two intervals, A of length k/d and B of length
$(d-k)/d$, for some integer k. If we identify the endpoints of A and make an affine
reparametrization to stretch it by a factor of d/k so it fits around the unit circle and
place the identified endpoints at 0, the leaves of m with endpoints in A become a
primitive degree k major. Similarly, we get a primitive degree $(d - k)$ major from B.

If p is any ideal j-gon of an element $m \in \mathrm{PM}(d)$, we similarly can derive a sequence of j primitive majors whose total degree is d.

Theorem 3.1. *The space* $\mathrm{PM}(d)$ *can be embedded in the space of monic degree* d *polynomials as a spine for the set of polynomials with distinct roots, that is, the complement of the discriminant locus. The spine (image of the embedding) consists of monic polynomials whose critical values are all on the unit circle and such that the center of mass of the roots is at the origin.*

A *spine* is a lower-dimensional subset of a manifold such that the manifold deformation-retracts to the spine.

Proof. We'll start by defining a map from the space of polynomials with distinct zeroes to $\mathrm{PM}(d)$. Given a polynomial $p(z)$, we look at the gradient vector field for $|p(z)|$, or better, the gradient of the smooth function $|p(z)|^2$. Any simple critical point for p is a saddle point for this flow. Near infinity, the flow lines align very closely to the flow lines for the gradient of $|z|^{2d}$, so each flow line that doesn't tend toward a critical point goes to infinity with some asymptotic argument (angle). Define a lamination whose leaves join the pairs of outgoing separatrices for the critical points. If there are critical points of higher multiplicity or if the unstable manifolds from some critical points coincide with stable manifolds for others (separatrices collide), then define an equivalence relation that decrees that for any flow line, if the limit of asymptotic angles on the left is not the same as the limit of asymptotic angles on the right, then the two limits are equivalent. For each such equivalence class, adjoin the ideal polygon that is the boundary of its convex hull.

We claim that the lamination so defined is a primitive degree-d major. To see that, notice that the level sets of $|p(z)|$ are mapped under $p(z)$ to concentric circles, so the flow lines for the gradient flow map to the perpendicular rays emanating from the origin. Each equivalence class coming from discontinuities in the asymptotic angles therefore maps to a single point, since there are no critical points for the image foliation except at the origin in the range of p. The total criticality is $d-1$ since the degree of $p'(z)$ is $d-1$.

There are many different polynomials that give any particular major. In the first place, note that for any polynomial, if we translate in the domain by any constant (which amounts to translating all the roots in some direction, keeping the vectors between them constant), the asymptotic angles of the upward flow lines from the critical points do not change, so then lamination does not change. We may as well keep the roots centered at the origin. Algebraically, this is equivalent to saying that the critical points are centered at the origin (by considering the derivative of the defining polynomial).

You can think of the spine this way. In the domain of p, cut $\mathbb{C}$ along each upward separatrix of each critical point. This cuts $\mathbb{C}$ into d pieces, each containing one root of p (that you arrive at by flowing downhill; whenever there is a path from z_1 to z_2 whose downhill flows don't ever meet a critical point, they end up at the same root). For any region, if you glue the two sides of each edge together, starting at the lowest critical point and matching points with equal values for $p(z)$, you obtain a copy of $\mathbb{C}$ which is really just a copy of the image plane; the seams have joined together to become rays.

You can construct other polynomials associated with the same major by varying the length of the cuts. (These cuts are classical branch cuts). Assuming for now that we are in the generic case with no critical gaps, for each leaf choose a positive real number. Take one copy of $\mathbb{C}$ for each region of the complement of the lamination, and for each leaf l $\overline{xy}$ on its boundary, pick a positive real number r_l and make a slit on the ray at angle $x^d = y^d$ from ∞ to the point $r_l x^d$. Now glue the slit copies of $\mathbb{C}$ together so as to be compatible with the parametrization by $\mathbb{C}$. (This is equivalent to forming a branched cover of $\mathbb{C}$, branched over the various critical values, with given combinatorial information or *Hurwitz data* describing the branching.) By uniformization theory, the Riemann surface obtained by gluing the copies of $\mathbb{C}$ together is analytically equivalent to $\mathbb{C}$, and the induced map is a polynomial.

In the nongeneric case (in which there are critical gaps), the set of possibilities branches: it is not parametrized by a product of Euclidean spaces, because of the different possibilities for the combinatorics of saddle connections. Let's look more closely at the possible structure of saddle connections for the gradient flow of $|p(z)|$. A small regular neighborhood of the union of the upward separatrices for the critical points has boundary consisting of $k + 1$ lines (here k is the multiplicity of the critical point); these map to the leaves of the boundary of the corresponding gap of the major. The union of upward separatrices is a tree whose leaves are the vertices of the gap polygon. The height function $\log |p(z)|$ has the property that induces a function with exactly one local minimum on each boundary component of the regular neighborhood projected to the graph.

We can bypass the enumeration of all such structures by observing that there is a direct way to define a retraction to the case when all critical values equal 1. We think of the graph as a metric graph where the height function $log|p|$ has speed 1. Multiply the height function on the compact edges of the graph by $(1 - t)$ while adding the constant times t to the height function on each unbounded component that makes it continuous at the vertices. When $t = 1$, the compact edges all collapse, only the unbounded edges remain, and the polygonal gap contains a single multiple critical point with critical value 1.   □

Corollary 3.2. PM(d) *is a* $K(B_d, 1)$ *where* B_d *is the d-strand braid group. In other words,* $\pi_1(\mathrm{PM}(d)) = B_d$ *and all higher homotopy groups are trivial.*

This follows from the well-known fact that the complement of the discriminant locus is a $K(B_d, 1)$. (See, e.g., [FM12, Section 9.12].) A loop in the space of primitive degree-d majors can be thought of as braiding the d regions in the complement of the union of its leaves and critical gaps.

A primitive quadratic major is just a diameter of the unit circle, so PM(2) is itself a circle. This corresponds to the fact that the 2-strand braid group is $\mathbb{Z}$.

A primitive cubic major is either an equilateral triangle inscribed in the circle, or a pair of chords that each cut off a segment of angle $2\pi/3$. There is a circle's worth of equilateral triangles. To each primitive cubic that consists of a pair of leaves, there is associated a unique diameter that bisects the central region. Thus, the space of two-leaf primitive cubic majors fibers over S^1, with fiber an interval. When the diameter is turned around by an angle of π, the interval maps to itself by reversing the orientation, so this is a Moebius band. The boundary of the Moebius band is attached to the circle of equilateral triangle configurations, wrapping around 3 times, since, given an equilateral triangle, you can remove any of its three edges to get a limit of two-leaf majors.

The space of cubic polynomials can be normalized to make the leading coefficient 1 (monic) and, by changing coordinates by a translation, make the second

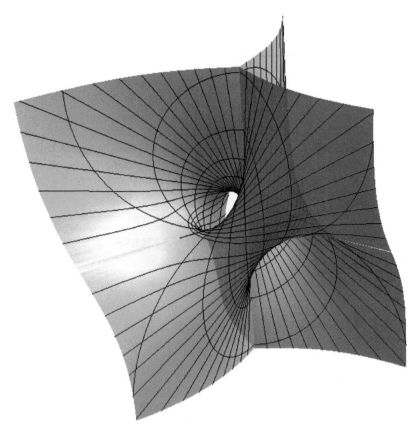

Figure 3: This is the space PM(3) embedded in S^3. It is formed by a 3/2-twisted Moebius band centered on a horizontal circle that grows wider and wider until its boundary comes together in the vertical line, which closes into a circle passing through the point at infinity, and wraps around this circle 3 times. The complement of PM(3) consists of 3 chambers which are connected by tunnels, each tunnel burrowing through one chamber and arriving 2/3 of the way around the post in the middle. A journey through all three tunnels ties a trefoil knot: such a trefoil knot is the locus where the discriminant of a cubic polynomial is 0, and PM(3) is a spine for the complement of the trefoil.

coefficient (the sum of the roots) equal to 0. By multiplying by a positive real constant, one can then normalize so that the other two coordinates, as an element of $\mathbb{C}^2$, are on the unit sphere. The discriminant locus intersects the sphere in a trefoil knot. The spine can be embedded in S^3 as follows: start with a 3/2-twisted Moebius band centered around a great circle in S^3. This great circle can be visualized via stereographic projection of S^3 to $\mathbb{R}^3$ as the unit circle in the xy-plane. The Moebius band can be arranged so that it is generated by geodesics perpendicular to the horizontal great circle, in a way that is invariant by a circle action that rotates the horizontal great circle at a speed of 2 while rotating the z-axis, completed to a great circle by adding the point at infinity, at a speed of 3. Extend the perpendicular geodesics all the way to the z-axis; this attaches the boundary of the Moebius band by wrapping it three times around the vertical great circle, as shown in Figure 3.

The spine gives a graphic description for one presentation of the 3-strand braid group,

$$B_3 = \langle a, b | a^2 = b^3 \rangle ,$$

where a is represented by the core circle of the Moebius band, and b is the circle of equilateral triangles, the two loops being connected via a short arc to a common base point. The presentation is an amalgamated free product of two copies of $\mathbb{Z}$ over subgroups of index 2 and 3. As braids, a is a 180 degree flip of three strands in a line, while b is a 120 degree rotation of 3 strands forming a triangle; the square of a and the cube of b is a 360 degree rotation of all strands, which generates the center of the 3-strand braid group.

4   GENERATING INVARIANT LAMINATIONS FROM MAJORS

Let m be a degree-d primitive major. How can we construct a degree-d-invariant lamination having m for its major?

A leaf of a lamination is defined by an unordered pair of distinct points on the unit circle. The space of possible leaves is topologically an open Moebius band. To see this, consider that any leaf divides the circle into two intervals. The line connecting the midpoints of these two intervals is the unique diameter perpendicular to the leaf. For each diameter, there is an interval's worth of leaves, parametrized by the point at which a leaf intersects the perpendicular diameter. When the interval is rotated 180 degrees, the parameter is reversed, so the space is an open Moebius band. One way to graphically represent the Moebius band is as the region outside the unit disk in $\mathbb{RP}^2$. The pair of tangents to the unit circle at the endpoints of a leaf intersect somewhere in this region, which is homeomorphic to a Moebius band.

Another way to represent the information is by passing to the double cover, the set of ordered pairs of distinct points on a circle, that is, the torus $S^1 \times S^1$ minus the diagonal, which is a $(1,1)$-circle on the torus, going once around each axis. Here $S^1 = \mathbb{R}/\mathbb{Z}$. For $x, y \in S^1$, we use $\overline{xy}$ to represent the geodesic in the unit disc linking $e^{2\pi i x}$ and $e^{2\pi i y}$. Each leaf $\overline{xy}$ is represented twice on the torus, as (x, y) and (y, x).

If a lamination contains a leaf $l = \overline{xy}$, then a certain set $X(l)$ of other leaves are excluded from the lamination because they cross $\overline{xy}$. On the torus, if you draw the horizontal and vertical circles through the two points (x, y) and (y, x), they subdivide the torus into four rectangles having the same vertex set; the remaining two common vertices are (x, x) and (y, y). The leaves represented by points in the interior of two of the rectangles constitute $X(l)$, while the leaves represented by the closures of the other two rectangles are all compatible with the given leaf $l = \overline{xy}$. We will call this good, compatible region $G(l)$, as shown in Figure 4. The compatible rectangles are actually squares, of side lengths $a - b$ mod 1 and $b - a$ mod 1. They form a checkerboard pattern, where the two squares of $G(l)$ are bisected by the diagonal. Another way to express it is that two points p and q define compatible leaves if and only if you can travel on the torus, without crossing the diagonal, from one point to either the other point or the other point reflected

Figure 4: A leaf $\overline{xy}$ of a lamination can be represented by a pair of points $\{(x,y),(y,x)\}$ on a torus. Leaves that are excluded by $\overline{xy}$ because they intersect it are represented in two darker shaded rectangles, and leaves compatible with it are represented in two squares of side length $b-a$ mod 1 and $a-b$ mod 1.

in the diagonal, heading in a direction between north and west or between south and east (using the same conventions as on maps, where up is north, left is west, etc.).

Given a set S of leaves, the excluded region $X(S)$ is the union of the excluded regions $X(l)$ for $l \in S$, and the good region $G(S)$ is the intersection of the good regions $G(l)$ for $l \in S$. If S is a finite lamination, then $G(S)$ is a finite union of rectangles that are disjoint except for corners.

In the particular case of a primitive major lamination $m \in \mathrm{PM}(d)$, each region of the disk minus m touches S^1 in a union of one or more intervals $J_1 \cup \cdots \cup J_k$ of total length $1/d$. This determines a finite union of rectangles $(J_1 \cup \cdots \cup J_k) \times (J_1 \cup \cdots \cup J_k)$ of $G(m)$ whose total area is $1/d^2$ that maps under the degree d^2 covering map $(x,y) \mapsto d \cdot (x,y)$ to the entire torus, as seen in Figures 5 and 6.

Consequently we have (here we use $\mathbb{D}^2$ to denote the unit disc):

Proposition 4.1. *For any primitive degree-d major m, the total area of $G(m)$ is $1/d$. Almost every point $p \in T^2$ has exactly d preimages in $G(m)$ by the degree d^2*

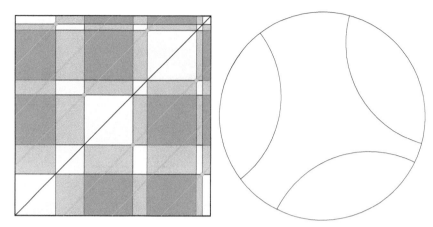

Figure 5: On the left is a plot showing the excluded region $X(m)$, shaded, together with the compatible region $G(m)$ on the torus, where $m \in \mathrm{PM}(4)$ is a primitive degree-4 major. The figure is symmetric by reflection in the diagonal. The quotient of the torus by this symmetry is a Moebius band. Note that $G(m)$ is made up of three $1/4 \times 1/4$ squares (one of them wrapped around) corresponding to the regions that touch the circle in only one edge, together with $3 \cdot 3 = 9$ additional rectangles whose total area is $1/4^2$, corresponding to the region that touches S^1 in 3 intervals. The 6 green dots represent the leaves of the major, one dot for each orientation of the leaf.

covering map $f_d : (x, y) \mapsto d \cdot (x, y)$ with one preimage representing a leaf in each of the d regions of $\mathbb{D}^2 \setminus m$, and all points have at least d preimages in $G(m)$ with at least one preimage representing a leaf in each region of $\mathbb{D}^2 \setminus m$.

For $m \in \mathrm{PM}(d)$ we can now define a sequence of backward-image laminations $b_i(m)$. Let $b_0(m) = m$ and inductively define $b_{i+1}(m)$ to be the union of m with the preimages under f_d of $b_i(m)$ that are in $G(m)$.

Proposition 4.2. *For each i, $b_i(m)$ is a lamination.*[1]

Proof. We need to check that the good preimages under f_d of two leaves of b_i are compatible. If they are in different regions of $\mathbb{D}^2 \setminus m$ they are obviously compatible. For two leaves in a single region of $\mathbb{D}^2 \setminus m$, note that when the boundary of the region is collapsed by collapsing m, it becomes a circle of length $1/d$ that is mapped homeomorphically to S^1. By the inductive hypothesis, the two image leaves are compatible; since the compatibility condition is identical on the small circle and its homeomorphic image, the leaves are compatible. □

---

[1]This proposition from the original text is not correct as stated. Two counter-examples are the cubic lamination $m = \left\{ \left(0, \frac{1}{3}\right), \left(\frac{1}{3}, \frac{2}{3}\right), \left(\frac{2}{3}, 0\right) \right\}$ (for which $b_1(m)$ fails to be a lamination) and the quadratic lamination $m = \left\{ \left(0, \frac{1}{2}\right) \right\}$ (for which $b_2(m)$ fails to be a lamination). Section 14 has a corrected algorithm in the quadratic case.

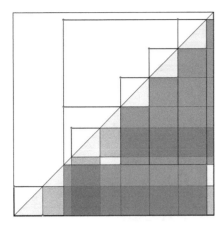

Figure 6: Here is a primitive heptic (degree-7) major with a pentagonal gap, shown in a variation of the torus plot, along with the standard Poincaré disk picture. On the left, half of the torus has been replaced by a drawing that indicates each leaf of the lamination by a path made up of a horizontal segment and a vertical segment. The upper left triangular picture transforms to the Poincaré disk picture by collapsing the horizontal and vertical edges of the triangle to a point, bending the collapsed triangle so that it goes to the unit disk with the collapsed edges going to $1 \in \mathbb{C}$, then straightening each rectilinear path into the hyperbolic geodesic with the same endpoints. Notice how the ideal pentagon corresponds to a rectilinear 10-gon, where two pairs of edges of the 10-gon overlap. The lower right triangle is a fundamental domain for a group Γ that is the covering group of the torus together with lifts of reflection through the diagonal of the torus. The horizontal edge of the triangle is glued by an element $\gamma \in \Gamma$ to the vertical edge. The action of γ on the edge is the same as a 90 degree rotation through the center of the square; γ itself follows this by reflection through the image edge. The quotient space $\mathbb{E}^2/\Gamma$ is topologically a Moebius band. As an orbifold it is the Moebius band with mirrored boundary.

Note that this is an increasing sequence, $b_i(m) \subset b_{i+1}(m)$. By induction, the good region $G(b_i(m))$ has area $1/d^m$.

It follows readily that:

Theorem 4.3. *The closure $b_\infty(m)$ of the union of all $b_i(m)$ is a degree-d-invariant lamination having m as its major.*

The lamination $b_\infty(m)$ may have various issues concerning its quality. In particular, it does not always happen that $b_\infty = \mathrm{Lam}(\mathrm{Rel}(b_\infty(m)))$. We will study the quality of these laminations later, and develop tools for studying more general degree-*d*-invariant laminations by embedding them in $b_\infty(m)$ for some m. The process is illustrated in Figure 7.

Note that as a finite lamination λ varies, a compatible rectangle can become thinner and thinner as two endpoints approach each other, and disappear in the limit. Thus $G(\lambda)$ is not continuous in the Hausdorff topology.

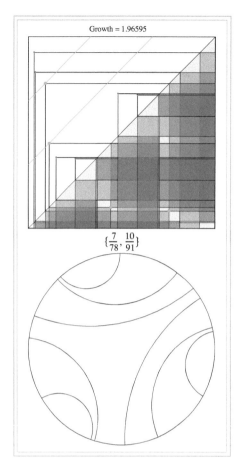
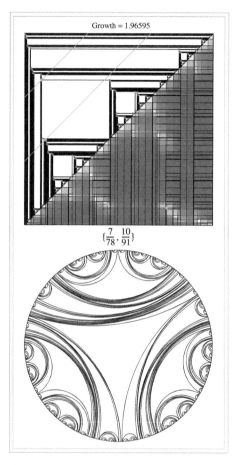

Figure 7: On the left is stage 1 ($b_1(m)$) in building a cubic-invariant lamination for $m = \{(\frac{10}{91}, \frac{121}{273}), (\frac{7}{78}, \frac{59}{78})\} \in \mathrm{PM}(3)$. The two longer leaves of m subdivide the disk into 3 regions, each with two new leaves induced by the map f_3. On the right is a later stage that gives a reasonable approximation of $b_\infty(m)$.

On the other hand,

Proposition 4.4. *The map from* $\mathrm{PM}(d)$ *to the space of compact subsets of* T *endowed with the Hausdorff topology defined by* $m \mapsto X(m)$ *is a homeomorphism onto its image, i.e., the topology of* $\mathrm{PM}(d)$ *coincides with Hausdorff topology on the set* $\{X(m) : m \in \mathrm{PM}(d)\}$.

Proof. Perhaps the main point is that $X(m)$ is a union of fat rectangles, with height and width at least $1/d$, so pieces of $X(m)$ can't shrink and suddenly disappear.

Suppose m and m' are majors that are within ϵ in the metric md, and suppose $p \in X(m)$, so there is some leaf l_m of m intersecting the leaf l_p represented by p. The endpoints of l_m have distance 0 in the quotient graph S^1/m, so there must be a path of length no greater than ϵ on S^1/m' connecting these two points. In particular, assuming $\epsilon < 1/d$, some leaf of m' comes within at most ϵ of intersecting l_p, therefore a leaf of m' intersects a leaf $l_{p'}$ near l_p. Then p' is in $X(m')$. Since this works symmetrically between m and m', it follows that they are close in the Hausdorff metric.

Conversely, given $m \in \mathrm{PM}(d)$ and $\epsilon > 0$, we will show that there exists $\delta > 0$ such that for any $m'' \in \mathrm{PM}(d)$ with the Hausdorff distance between $X(m)$ and $X(m')$ less than δ, the distance $\mathrm{md}(m, m') < \epsilon$. We will do this by induction on d. It is obvious for $d = 2$, since a major is a diameter and $X(m)$ is a union of two squares that intersect at two corners that are the representatives of the single leaf of m.

When $d > 2$, we choose a region of $\mathbb{D}^2 \setminus m$ that touches S^1 in a single interval A of length $1/d$, bounded by a leaf l. Suppose $X(m')$ is Hausdorff-near $X(m)$. Then most of the square $A \times A$ of $G(m)$ is in $G(m')$, and most of the rectangles $A \times (S^1 \setminus A)$ and $(S^1 \setminus A) \times A$ of $X(m)$ is also in $X(m')$. This implies that m' has a nearby l' spanning an interval A' of length $1/d$. Now we can look at the complementary regions, with l or l' collapsed, normalized by the affine transformation that makes the restriction of m or m' a primitive degree-$(d-1)$ major having the collapsed point at 0. Call these new majors m_1 and m'_1. The excluded region $X(m_1)$ is obtained from $X(m)$ intersected with a $(d-1) \times (d-1)$ square on the torus, and similarly for $X(m'_1)$; hence if the Hausdorff distance between $X(m)$ and $X(m')$ is small, so is that between $X(m_1)$ and $X(m'_1)$. By induction, we can conclude that the pseudo-metrics $\mathrm{met}(m)$ and $\mathrm{met}(m')$ on S^1 induced from the quotient graphs are close. $\qquad \square$

5 CLEANING LAMINATIONS: QUALITY AND COMPATIBILITY

We will use the parametrization of the circle by *turns*, that is, numbers interpreted as fractions of the way around the circle. Thus $\tau \in [0, 1]$ corresponds to the point $\exp(2\pi i \tau)$ in the unit circle in the complex plane.

Let's look at the quadratic major $\overline{0, \frac{1}{2}}$. We'll use a variation of the process of backward lifting, depicted in Figure 8, which is really the limit as $\epsilon \to 0$ of the standard process applied to $\{-\epsilon, \frac{1}{2} - \epsilon\}$. The first backward lift adds 2 new leaves, $\overline{0, \frac{1}{4}}$ and $\overline{\frac{1}{2}, \frac{3}{4}}$, and at each successive stage, a new leaf is added joining the midpoint of each interval between endpoints to the last clockwise endpoint of the interval. (The full construction would also join each of these midpoints to the counterclockwise endpoint of the interval.) In the limit, there are only a countable set of leaves, but they are joined in a single tree. The closed equivalence relation they generate collapses the entire circle to a point!

Each finite-stage lamination λ_k can be cleaned up into the standard form $\mathrm{Lam}(\mathrm{Rel}(\lambda_k))$.

6 PROMOTING FORWARD-INVARIANT LAMINATIONS

A degree-d primitive major is a special case of a lamination that is forward invariant. We have seen how to construct a fully invariant degree-d lamination containing it. How generally can forward-invariant laminations be promoted to fully invariant laminations?

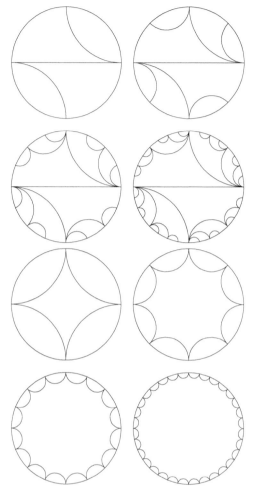

Figure 8: The top four laminations are successive steps of building an invariant quadratic lamination for the major $\{0, \frac{1}{2}\}$. The bottom four laminations are obtained by performing Lam ∘ Rel.

7   HAUSDORFF DIMENSION AND GROWTH RATE

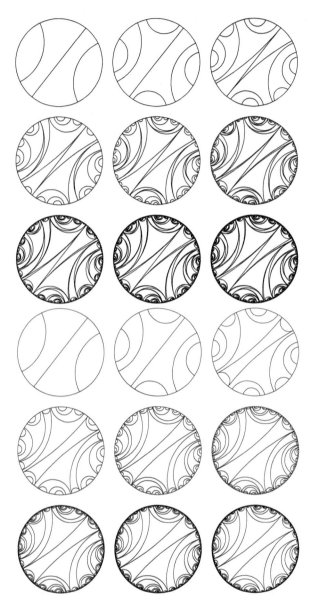

Figure 9: The top nine laminations are successive steps of building an invariant quadratic lamination for the major $\{\frac{1}{7}, \frac{9}{14}\}$. The bottom nine laminations are obtained by performing Lam ∘ Rel (at every third step). It gives the lamination of Douady's rabbit.

Part II

Part II consists of supplementary sections expanding on the material in Part I, which were written by the authors other than W. Thurston. We start by recalling related work on laminations in Section 8. Section 9 gives various dynamical interpretations of PM(d). Section 10 looks at several different parametrizations of PM(d) and describes PM(2). Section 11 gives a complete description of PM(3), and Section 12 discusses PM(4) to some extent.

The remaining sections are devoted to issues around entropy. Section 13 explains and proves correctness of a simple algorithm for computing core entropy extracted from computer code by W. Thurston. Sections 14 and 15 examine different types of laminations related to a primitive major (presumably related to the material intended for Section 6), while Section 16 shows equivalence of several different entropies and Hausdorff dimensions related to a polynomial (presumably related to Section 7).

8   RELATED WORK ON LAMINATIONS

Invariant laminations were introduced as a tool in the study of complex dynamics by William Thurston [Thu09]. Thurston's theory suggests using spaces of invariant laminations as models for parameter spaces of dynamical systems defined by complex polynomials. In degree $d = 2$, Thurston uses QML (quadratic minor lamination) to model the space of 2-invariant laminations. Underpinning this approach in degree 2 is Thurston's No Wandering Triangle Theorem. Thurston conjectured that the boundary of the Mandelbrot set is essentially the quotient of the circle by the equivalence relation induced by QML. The precise relationship between invariant laminations and complex polynomials is even less clear for higher degree.

Not all degree-d-invariant laminations correspond to degree d polynomials. One necessary condition for a lamination to be directly associated to a polynomial is that it be generated by a tree-like equivalence relation on S^1 (see, for example, [BL02, BO04a, Kiw04, Mim10] and Section 2 of Part I). To distinguish between arbitrary invariant laminations and those associated to polynomials, we call the invariant laminations defined by Thurston *geometric invariant laminations*, and the smaller class of laminations defined by tree-like equivalence relations *combinatorial invariant laminations*. (In [BMOV13], such laminations are called q-laminations).

The difference between these two notions can be explained via the following examples. Let a, b, c be three distinct points on S^1 which form an equivalence class under a tree-like equivalence relation. Then in the combinatorial lamination obtained from the given tree-like equivalence, all the leaves $(a, b), (b, c), (a, c)$ are contained. But as a geometric lamination, it may have leaves $(a, b), (b, c)$ without having (a, c) as a leaf. Here is another important case to consider. From the "primitive major" construction, we can end up with leaves $(a, b), (b, d)$, and (c, d), where a, b, c, d are four distinct points on S^1 so that (a, b, c, d) appear in cyclic order. On the other hand, if this were a combinatorial lamination, it would only include the ones around the outside, namely $(a, b), (b, c), (c, d)$, and (d, a). One way to understand both of these examples geometrically is to identify the circle with the ideal boundary of the hyperbolic plane $\mathbb{H}^2$. Then a combinatorial lamination can be understood as the one consisting of the boundary leaves of the convex hulls of finitely many points on the ideal boundary of $\mathbb{H}^2$.

A fundamental global result in the theory of combinatorial invariant laminations is the existence of *locally connected models for connected Julia sets*, obtained by Kiwi [Kiw04]. He associates a combinatorial invariant lamination $\lambda(f)$ to each polynomial f that has no irrationally neutral cycles and whose Julia set is connected. Then the topological Julia set $J_{\sim_f} := S^1/_{\sim_f}$ is a locally connected continuum, where $\sim_f$ is the equivalence relation generated by $x \sim_f y$ if x and y are connected by a leaf of $\lambda(f)$, and $f|_{J_f}$ is semi-conjugate to the induced map $f_{\sim_f}$ on $J_{\sim_f}$ via a monotone map $\phi\colon J_f \to J_{\sim_f}$ (by monotone we mean a map whose point preimages are connected). Kiwi characterizes the set of combinatorial invariant laminations that can be realized by polynomials that have no irrationally neutral cycles and whose Julia sets are connected. In [BCO11], Blokh, Curry and Oversteegen present a different approach, one based upon continuum theory, to the problem of constructing locally connected dynamical models for connected polynomial Julia sets J_f; their approach works regardless of whether or not f has irrational neutral cycles. These locally connected models yield nice combinatorial interpretations of connected quadratic Julia sets that themselves may or may not be locally connected.

The No Wandering Triangle Theorem is a key ingredient in Thurston's construction [Thu09] of a locally connected model $\mathcal{M}_c^2$ of the Mandelbrot set. The theorem asserts the non-existence of wandering non-(pre)critical branch points of induced maps on quadratic topological Julia sets. Branch points of $\mathcal{M}_c$ correspond to topological Julia sets whose critical points are periodic or preperiodic. Thurston posed the problem of extending the No Wandering Triangle Theorem to the higher-degree case. Levin showed [Lev98] that for "unicritical" invariant laminations, wandering polygons do not exist. Kiwi proved [Kiw02] that for a combinatorial invariant lamination of degree d, a wandering polygon has at most d edges. Blokh and Levin obtained more precise estimates on the number of edges of wandering polygons [BL02]. Soon after, Blokh and Oversteegen discovered that some combinatorial invariant laminations of higher-degree ($d \geq 3$) do admit wandering polygons [BO04b, BO09].

Extending Thurston's technique of using invariant laminations to construct a combinatorial model $\mathcal{C}_d$ of the connectedness locus for polynomials of degree $d > 2$ remains an area of inquiry. In [BOPT14] and [Pta13], A. Blokh, L. Oversteegen, R. Ptacek and V. Timorin make progress in this direction. They establish two necessary conditions of laminations from the polynomials in the *Main Cubioid CU*, i.e., the boundary of the principal hyperbolic component of the cubic connectedness locus $\mathcal{M}_3$; CU is the analogue of the main cardioid in the quadratic case. They propose this set of laminations as the *Combinatorial Main Cubioid CU^c*, a model for CU.

9 INTERPRETATIONS OF THE SPACE OF DEGREE d PRIMITIVE MAJORS

9.1 Recalling definitions

We begin by recalling some concepts from Part I. A *critical class* of a degree-d-invariant equivalent relation is subset of $S^1 = \partial\mathbb{D} \subset \mathbb{C}$ that consists of all elements of an equivalence class and that maps under $z \mapsto z^d$ with degree greater than 1; the associated subsets of $\mathbb{D}$ are the *critical leaves* and *critical gaps* of the lamination. The *criticality* of a critical class is defined to be one less than the degree of the restriction of the map to that subset. Per Definition 2.2, the *major* of a degree-d-invariant lamination (resp. equivalence relation) is the set of critical leaves and critical gaps (resp. equivalence classes corresponding to critical leaves and critical gaps). Such a major is said to be *primitive* if every critical gap is a "collapsed polygon" whose vertices are identified under the map $z \mapsto z^d$. (The restriction of $z \mapsto z^d$ to an intact gap of a primitive degree-d-invariant lamination is necessarily injective.) A critical leaf may be thought of as a critical gap defined by a polygon with precisely two vertices, and those vertices are identified by $z \mapsto z^d$. By Proposition 2.1, the sum over the critical classes of their criticalities equals $d - 1$.

9.2   Metrizability of PM(d)

As described in Part I, an element $m \in \mathrm{PM}(d)$ determines a quotient graph $\gamma(m)$ obtained by identifying each equivalence class to a point. The path-metric of S^1 defines a path-metric on $\gamma(m)$. In addition, $\gamma(m)$ has the structure of a planar graph, that is, an embedding in the plane well-defined up to isotopy, obtained by shrinking each leaf and each ideal polygon of the lamination to a point. These graphs have the property that $H^1(\gamma(m))$ has rank d, and every cycle has length a multiple of $1/d$. Every edge must be accessible from the infinite component of the complement, so the metric and the planar embedding together with the starting point, that is, the image of $1 \in \mathbb{C}$, is enough to define the major.

The pseudo-metric $\mathrm{met}(m)$ on the circle induced by the path-metric on $\gamma(m)$ determines a continuous function on $S^1 \times S^1$. The sup-norm on the space of continuous functions on $S^1 \times S^1$ induces a metric md on PM(d):

$$\mathrm{md}(m, m') = \sup_{(x,y) \in S^1 \times S^1} |\mathrm{met}(m)(x, y) - \mathrm{met}(m')(x, y)| \, .$$

For the sake of completeness, we give a proof of the fact that md is indeed a metric.

Lemma 9.1. *The function* md *is a metric on* PM(d).

Proof. Non-negativity and symmetry follow automatically from the definition. The rest of the proof is also pretty straightforward.

Suppose $\mathrm{md}(m, m') = 0$ for some $m, m'' \in \mathrm{PM}(d)$. Since $\sup_{x,y} |\mathrm{met}(m)(x, y) - \mathrm{met}(m')(x, y)| = 0$, this means $\mathrm{met}(m)(x, y) = \mathrm{met}(m')(x, y)$ for all $x, y \in S^1$. This implies indeed $m = m'$. In order to see this, suppose $m = \{W_1, \ldots, W_n\}$ and $m'' = \{W_1', \ldots, W_k'\}$ are different. Then one can pick two distinct points p, q such that $p, q \in W_i$ for some i but there is no W_j' which contains both p and q. Clearly one has $\mathrm{met}(m)(p, q) = 0$ while $\mathrm{met}(m')(p, q) > 0$. This proves that $\mathrm{md}(m, m') = 0$ if and only if $m = m'$.

It remains to prove the triangular inequality. Let m_1, m_2, m_3 be three elements of PM(d). Then

$$\mathrm{md}(m_1, m_3)$$
$$= \sup_{x,y} |\mathrm{met}(m_1)(x, y) - \mathrm{met}(m_3)(x, y)|$$
$$\leq \sup_{x,y} \left(|\mathrm{met}(m_1)(x, y) - \mathrm{met}(m_2)(x, y)| + |\mathrm{met}(m_2)(x, y) - \mathrm{met}(m_3)(x, y)| \right)$$

$$\le \sup_{x,y} |\mathrm{met}(m_1)(x,y) - \mathrm{met}(m_2)(x,y)| + \sup_{x,y} |\mathrm{met}(m_2)(x,y) - \mathrm{met}(m_3)(x,y)|$$

$$= \mathrm{md}(m_1, m_2) + \mathrm{md}(m_2, m_3). \qquad \square$$

9.3 A spine for the complement of the discriminant locus

Let $\mathcal{P}_d$ be the space of monic centered polynomials of degree d, and $\mathcal{P}_d^0 \subset \mathcal{P}_d$ be the space of polynomials with distinct roots.

There is a natural map $f : \mathcal{P}_d^0 \to \mathrm{PM}(d)$ defined as follows: for $p \in P_d^0$ consider the meromorphic 1-form

$$\frac{1}{2d\pi i} \frac{d}{dz} \log p(z) = \frac{1}{2d\pi i} \frac{p'(z)}{p(z)} dz.$$

Denote by $Z(p)$ the set of roots of p. This 1-form gives $\mathbb{C} - Z(p)$ a Euclidean structure. Near infinity, we see a semi-infinite cylinder of circumference 1 (∞ is a simple pole of $d \log p$ with residue d) and near all the zeroes of p we see a semi-infinite cylinder with circumference $1/d$.

We restate Theorem 3.1, and give another proof here.

Theorem 9.2. *The map $f : \mathcal{P}_d^0 \to \mathrm{PM}(d)$ is a homotopy equivalence. More specifically, there exists a section $\sigma : \mathrm{PM}(d) \to \mathcal{P}_d^0$ which is a deformation retract.*

Proof. For $m \in \mathrm{PM}(d)$ consider the half-infinite cylinder $X'_m = (S^1 \times [0, \infty))/\sim$ where

$$(\theta_1, t_1) \sim (\theta_2, t_2) \iff (\theta_1, t_1) = (\theta_2, t_2) \quad \text{or} \quad t_1 = t_2 = 0 \text{ and } \theta_1 \sim_m \theta_2.$$

The graph quotient of $S^1 \times \{0\}$ by $\sim_m$ consists of d closed curves of length $1/d$. Glue to each of these closed curves a copy of $(\mathbb{R}/\frac{1}{d}\mathbb{Z}) \times (-\infty, 0)$, to construct X_m, which is a Riemann surface carrying a holomorphic 1-form ϕ_m given by $d\theta + idt$ on the upper cylinder. The integral of this 1-form around any of the d punctures at $-\infty$ is $1/d$, so we can define a function $p_m : X_m \to \mathbb{C}$

$$p_m(x) = e^{2\pi i d \int_{x_0}^x \phi_m},$$

well-defined up to post-multiplication by a constant depending on x_0. (Since the integral of ϕ_m along a loop around one of the lower cylinders $(\mathbb{R}/\frac{1}{d}\mathbb{Z}) \times (-\infty, 0)$ is $1/d$, p_m is well-defined.) But the endpoint compactification $\overline{X}_m$ of X_m is homeomorphic to the 2-sphere, and the complex structure extends to the endpoints, so $\overline{X}_m$ is analytically isomorphic to $\overline{\mathbb{C}}$. With this structure we see that p_m is a polynomial of degree d with distinct roots at the finite punctures. We can normalize p_m to be centered and monic by requiring that $1 \times [0, \infty]$ is mapped to a curve asymptotic to the positive real axis. The map $m \mapsto p_m$ gives the inclusion $\sigma : \mathrm{PM}(d) \to \mathcal{P}_d^0$.

We need to see that σ is a deformation retract. For each $p \in \mathcal{P}_d^0$, we consider the manifold with 1-form $(\mathbb{C} - Z(p), \phi_p)$, and adjust the heights of the critical values until they are all 0. $\qquad \square$

9.4   Polynomials in the escape locus

Again let $\mathcal{P}_d$ be the space of monic centered polynomials of degree d; this time they will be viewed as dynamical systems. For $p \in \mathcal{P}_d$ let G_p be the *Green's function* for the filled Julia set K_p.

Let $Y_d(r) \subset \mathcal{P}_d$ be the set of polynomials p such that $G_p(c) = r$ for all critical points of p. The set $Y_d(0)$ is the degree d *connectedness locus*; it is still poorly understood for all $d > 2$.

For $r > 0$, there is a natural map $Y_d(r) \to \mathrm{PM}(d)$ that associates to each polynomial $p \in \mathcal{P}_d$ the equivalence relation m_p on S^1 where two angles θ_1 and θ_2 are equivalent if the external rays at angles θ_1 and θ_2 land at the same critical point of p.

Theorem 9.3. *For $r > 0$ and $p \in Y_d(r)$, the equivalence relation m_p is in $\mathrm{PM}(d)$, and the map $p \mapsto m_p$ is a homeomorphism $Y_d(r) \to \mathrm{PM}(d)$.*

The above theorem is a combination of a theorem of L. Goldberg [Gol94] and a theorem of Kiwi [Kiw05]. To see a more recent proof using quasiconformal surgery, the readers are referred to [Zen14].

9.5   Polynomials in the connectedness locus

For $m \in \mathrm{PM}(d)$, we geometrically identify it as the unions of convex hulls within $\overline{\mathbb{D}}$ of non-trivial equivalence classes of m. Let us denote by $J_1(m), \ldots, J_d(m)$ the open subsets of S^1 that are intersections with S^1 of the components of $\mathbb{D} \setminus m$; each $J_i(m)$ is some finite union of open intervals in S^1. An *attribution* will be a way of attributing each non-empty intersection $\overline{J_i(m)} \cap \overline{J_j(m)}$ to $J_i(m)$, or to $J_j(m)$, or to both. Call A such an attribution, and denote by J_i^A the interval J_i together with all the points attributed to it by A. Define the equivalence relation $\sim_{(m,A)}$ to be

$$\theta_1 \sim_{m,A} \theta_2 \iff d^k\theta_1 \text{ and } d^k\theta_2 \text{ belong to the same } J_i^A \text{ for all } k \geq 0.$$

Suppose that some $p \in \mathcal{P}_d$ belongs to the connectedness locus without Siegel disks, and that K_p is locally connected, so that there is a Carathéodory loop $\gamma_p : \mathbb{R}/\mathbb{Z} \to \mathbb{C}$. Then γ_p induces the equivalence relation $\sim_p$ on $\mathbb{R}/\mathbb{Z}$ by $\theta_1 \sim_p \theta_2$ if and only if $\gamma_p(\theta_1) = \gamma_p(\theta_2)$.

Theorem 9.4. *There exists $m \in \mathrm{PM}(d)$ and an attribution A such that the equivalence relation $\sim_p$ is precisely $\sim_{m,A}$.*

This theorem is essentially proved in [Zen15, Theorem 1.2]. Understanding when different $\sim_{m,A}$ correspond to the same polynomial is a difficult problem, even for quadratic polynomials.

Proof. (Sketch) Choose an external ray landing at each critical value in J_p, and an external ray landing at the root of each component of $\overset{\circ}{K}_p$ containing a critical value if the critical value is attracted to an attracting or parabolic cycle. Then for each critical point $c \in J_p$, the angles of the inverse images of the chosen rays landing at c form an equivalence class for (m, A).

For each critical point $c \in \overset{\circ}{K_p}$, the angles of the inverse images of the chosen rays landing on the component of $\overset{\circ}{K_p}$ containing c form the other equivalence classes. These are the ones that need to be attributed carefully.   □

9.6   Shilov boundary of the connectedness locus

Conjecture 9.5. *All stretching rays through $Y_d(r)$ land on the Shilov boundary of the connectedness locus. Furthermore, if $r \to 0$, $Y_d(r)$ accumulates to the Shilov boundary.*

If this is true, it gives a description of the Shilov boundary of $Y_d(0)$, which is probably the best description of the connectedness locus we can hope for.

10   PARAMETRIZING PRIMITIVE MAJORS

As part of his investigations into core entropy, William P. Thurston wrote numerous Mathematica programs. This section presents an algorithm found in W. Thurston's computer code which cleverly parametrizes primitive majors using starting angles.

Denote by I the circle of unit length. We will interpret I as the fundamental domain $[0, 1)$ in $\mathbb{R}/\mathbb{Z}$ with the standard ordering on $[0, 1)$. Simultaneously, we will think of I as the space of angles of points in the boundary of the unit disk.

Throughout this section, we will let $m = \{W_1, \cdots, W_s\} \in \mathrm{PM}(d)$ denote a generic primitive major. By "generic," we mean that the associated lamination $\mathcal{M}$ consists of $d - 1$ leaves and has no critical gaps. Each leaf ℓ_i in $\mathcal{M}$ has two distinct endpoints in I. We will call the lesser of these two points the *starting point* of ℓ_i and denote it s_i, and we will call the greater the *terminal point* of ℓ_i and denote it t_i. We will adopt the labelling convention that the labels of the leaves are ordered so that

$$s_1 < s_2 < \cdots < s_{d-1}.$$

Since, for each i, $d \cdot s_i \pmod 1 = d \cdot t_i \pmod 1$, there exists a unique natural number $k_i \in \{1, ..., d-1\}$ such that $t_i = s_i + \frac{k_i}{d}$; we will show how to find the k_i.

Each leaf $\ell_i \in \mathcal{M}$ determines two open arcs of I: $I_i = (s_i, t_i)$ and $I_i^0 = I \setminus [s_i, t_i]$. The complement in $\overline{\mathbb{D}}$ of the lamination $\mathcal{M}$ consists of d connected sets. We will adopt the notation that C_i is the connected component whose boundary contains ℓ_i and has non-empty intersection with the arc I_i, for $1 \leq i \leq d$, and C_0 is the connected set whose boundary contains an arbitrarily small interval $(1 - \epsilon, 1) \subset I$. For each connected set C_i, denote by $\mu(C_i)$ the Lebesgue measure of the boundary of C_i in $I = \partial \overline{\mathbb{D}}$.

Lemma 10.1. *Let $m \in \mathrm{PM}(d)$ be a generic primitive major. Then $\mu(C_i) = 1/d$ for all i.*

(This lemma is used implicitly several times in Part I.)

Proof. Since for every leaf ℓ_i the lengths of the arcs I_i and I_i^0 are both integer multiples of $1/d$, $\mu(C_i)$ is also an integer multiple of $1/d$. Thus, we can write $\mu(C_i) = m_i/d$

for a unique natural number m_i. Then, since the C_i are pairwise disjoint,

$$1 = \sum_{i=0}^{d-1} \mu(C_i) = \frac{1}{d} \sum_{i=0}^{d-1} m_i.$$

Consequently, $m_i = 1$ for all i.   □

Lemma 10.2. *Let $m \in \mathrm{PM}(d)$ be a generic primitive major. Then $s_{d-1} < \frac{d-1}{d}$ and $t_{d-1} = s_{d-1} + \frac{1}{d}$.*

Proof. First, we will prove that $s_{d-1} < \frac{d-1}{d}$. To see this, suppose $s_{d-1} \in [\frac{d-1}{d}, 1)$. We know t_{d-1} is the fractional part of $(s_{d-1} + \frac{k_{d-1}}{d})$ (mod 1) for some $k_{d-1} \in \mathbb{N}$. Consequently, $t_{d-1} \notin (\frac{d-1}{d}, 1)$, contradicting the fact that $s_{d-1} < t_{d-1}$.

Now, suppose $s_{d-1} + \frac{1}{d} < t_{d-1}$. Since s_{d-1} is the biggest of the s's, there is no leaf in $\mathcal{M}$ whose starting point lies in the arc I_{d-1}. Therefore the boundary of C_{d-1} contains the entire arc I_{d-1}, and so $\mu(C_{d-1}) > 1/d$, a contradiction.   □

Definition 10.3. *Let $m \in \mathrm{PM}(d)$ be a generic primitive major. The derived primitive major m' is an equivalence relation on I that is the image of m under the following process: collapse the interval $[s_{d-1}, t_{d-1}]$ in I to a point, and then affinely reparametrize the quotient circle so that it has unit length, keeping the point 0 fixed.*

Lemma 10.4. *For any generic primitive major $m \in \mathrm{PM}(d)$, the derived primitive major m' is in $\mathrm{PM}(d-1)$.*

Proof. By Lemma 10.2, the arc $[s_{d-1}, t_{d-1}]$ has length $\frac{1}{d}$, so the reparametrization affinely stretches the quotient circle by a factor of $\frac{d}{d-1}$. For $i \leq d-2$, denote the image of s_i and t_i in $\mathcal{M}'$ by s_i' and t_i'. If $t_i - s_i = \frac{k_i}{d}$, then

$$t_i' - s_i' = \frac{k_i'}{d} \cdot \frac{d}{d-1} = \frac{k_i'}{d-1},$$

where $k_i' = k_i - 1$ if $[s_{d-1}, t_{d-1}] \subset [s_i, t_i]$ and $k_i' = k_i$ otherwise. In either case, $(d-1) \cdot (t_i' - s_i') = 0$ (mod 1). Hence $m' = \{(s_i', t_i') \mid i = 1, \cdots, d-2\}$ is in $\mathrm{PM}(d-1)$.   □

Lemma 10.5. *Let $m \in \mathrm{PM}(d)$ be a generic primitive major. Then $s_i < \frac{i}{d}$ for all i.*

Proof. Repeatedly deriving the major m yields a sequence of primitive majors

$$m^{(0)}(:=m), m^{(1)}(:=m'), m^{(2)}, \dots, m^{(d-3)}.$$

The major $m^{(j)}$ consists of $d-1-j$ leaves, $\ell_1^{(j)}, \dots, \ell_{d-1-j}^{(j)}$, that are the images under j derivations of the leaves $\ell_1^{(0)}, \dots, \ell_{d-1-j}^{(0)}$ of the original major m. We will denote the starting point of the leaf $\ell_k^{(j)}$ by $s_k^{(j)}$.

We wish to show, for any fixed i, that $s_i^{(0)} < \frac{i}{d}$. The major $m^{(d-1-i)}$ consists of i leaves, of which $\ell_i^{(d-1-i)}$ has the largest starting point. Hence by Lemma 10.2, we have

$$s_i^{(d-1-i)} < \frac{i}{i+1}.$$

When deriving $m^{(j-1)}$ (which is in $\mathrm{PM}(d-j+1)$) to form $m^{(j)}$, for any j, we collapse an interval of length $\frac{1}{d-j+1}$ and rescale by a factor of $\frac{d-j+1}{d-j}$. Thus,

$$s_i^{(d-1-i)} = s_i^{(0)} \cdot \frac{d}{d-1} \cdot \frac{d-1}{d-2} \cdot \dots \cdot \frac{d-(d-1-i)+1}{d-(d-1-i)}$$

$$= s_i^{(0)} \cdot \frac{d}{i+1}.$$

Hence

$$s_i^{(0)} \cdot \frac{d}{i+1} < \frac{i}{i+1},$$

implying

$$s_i^{(0)} < \frac{i}{d}. \qquad \square$$

Theorem 10.6. *Given any increasing sequence $0 \le s_1 < s_2 < \dots < s_{d-1} < 1$ such that $s_i < \frac{i}{d}$ for all i, there is a unique degree-d-invariant primitive major m whose starting points are $s_1, \dots, s_{d-1}$. There is an algorithm to find m.*

Proof. Let $m \in \mathrm{PM}(d)$ be any primitive major whose leaves have starting points $s_1, \dots, s_{d-1}$; we will find possible values for the terminal points of each leaf of m are uniquely determined.

When we derive m n times, we collapse a union of arcs, namely $\bigcup_{i=1}^{n} cl(I_{d-i})$. The starting point $s_{d-1-n}^{(n)}$ is the biggest starting point of the resulting major, $m^{(n)}$. By Lemma 10.2, $t_{d-1-n}^{(n)} - s_{d-1-n}^{(n)} = \frac{1}{d-n}$. Reversing the rescaling process which at each derivation rescales the quotient circle to have unit length, an interval of length $\frac{1}{d-n}$ in $\mathcal{M}^{(n)}$ corresponds to an interval of length $\frac{1}{d}$ in the original major $\mathcal{M}$ when we measure the arcs which get collapsed as having length 0.

Thus, for each natural number $n < d-1$, we can choose t_{d-1-n} to be the smallest number in $[0, 1)$ such that

$$m \left([s_{d-1-n}, t_{d-1-n}] \setminus \bigcup_{i=1}^{n} I_{d-i} \right) = \frac{1}{d}, \qquad (1)$$

where m is Lebesgue measure. For $n = 0$, we have $t_{d-1} = s_{d-1} + \frac{1}{d}$, and thus $I_{d-1} = (s_{d-1}, s_{d-1} + \frac{1}{d})$, by Lemma 10.2. Inductively, if $t_{d-1}, \dots, t_{d-n}$ are known, Equation (1) gives t_{d-n-1}.

The resulting union of leaves might not be a primitive major, since in non-generic cases it might not come from an equivalence relation. Once we add in the missing leaves as necessary to make ideal polygons, the result is unique. (There might be more than one value for t_{d-1-n} satisfying Equation (1), but they are identified by the rest of the lamination.) $\qquad \square$

Theorem 10.6 describes an algorithm which associates a primitive major to any ordered sequence of points $\{s_i\}_{i=1}^{i=d-1}$ such that $s_i < \frac{i}{d}$ and $s_i < s_{i+1}$ for all i. We now describe an algorithm used in W. Thurston's code for constructing such sequences of points from arbitrary collections of points.

Definition 10.7. *For any sequence* $X = \{x_1, \dots, x_{d-1}\}$ *of* $d-1$ *distinct points* x_i *in* I, *define* A *to be the map*

$$A : X \mapsto Y$$

where $Y = \{y_1, \dots, y_{d-1}\}$ *is the sequence of* $d-1$ *points* y_i *in* I *defined by the following process:*

1. Reorder and relabel (if necessary) the elements of the sequence X *so that*

$$x_1 < x_2 < \cdots < x_{d-1}.$$

2. Set

$$y_i = \begin{cases} x_i & \text{if } x_i < \frac{i}{d} \\ x_i - \frac{1}{d} & \text{if } x_i \geq \frac{i}{d}. \end{cases}$$

Theorem 10.8. *Let* $X = \{x_1, \dots, x_{d-1}\}$ *be any sequence of* $d-1$ *points* x_i *in* I. *Then there exists* $n \in \mathbb{N}$ *such that* $A^n(X) = A^m(X)$ *for all* $m \geq n$. *For such an* n, $A^n(X) = \{a_1, \dots, a_{d-1}\}$ *is a sequence of* $d-1$ *distinct points* a_i *in* I *such that*

$$a_1 < a_2 < \cdots < a_{d-1}$$

and $a_i < \frac{i}{d}$ *for all* i.

Proof. For convenience, denote by $B^j = \{b_1^j, \dots, b_{d-1}^j\}$ the result of applying Step 1 of the definition of the map A to $A^j(X)$. Thus, B^j is just the sequence $A^j(X)$ reordered and relabeled so that $b_1^j < \cdots < b_{d-1}^j$. Notice that when passing from $A^j(X)$ to $A^{j+1}(X)$ (with B^j as an intermediate step), if we do not subtract $1/d$ from at least one of the elements b_i^j, then $B^j = A^m(X)$ for all $m > j$.

Since we only subtract $1/d$ from b_i^j if $b_i^j \geq i/d \geq 1/d$, every b_i^j is non-negative. Since $\sum_i b_i^0 < \infty$, we can only subtract $1/d$ finitely many times from elements of B^0 without some b_i^j becoming negative. Hence, there are only finitely many integers j such that we subtract $1/d$ from at least one element of B^j when passing from $A^j(X)$ to $A^{j+1}(X)$. Hence, there exists $n_0 \in \mathbb{N}$ such that $B^{n_0} = A^m(X)$ for all $m > n_0$. Set $n = n_0 + 1$. Then $B^{n_0} = A^n(X) = A^m(X)$ for all $m \geq n$.

By assumption, $b_1^{n_0} < \cdots < b_{d-1}^{n_0}$. Since we do not subtract $1/d$ from any $b_i^{n_0}$ when passing from $A^{n_0}(X)$ to $A^n(X)$, this means $b_i^{n_0} < i/d$ for all i.   □

11   UNDERSTANDING PM(3)

11.1   Topology of PM(3)

We now explicitly describe PM(3). For brevity, we will say simply *a major* to denote a cubic primitive major throughout this section. One can represent points on the circle by their angle from the positive real axis. This angle is measured as the number

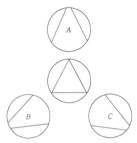

Figure 10: The three generic majors A, B and C are close to the degenerate major m shown in the center.

of turns we need to get to that point, i.e., as a number between 0 and 1. Let $m \in \mathrm{PM}(3)$. In the generic case, m has two leaves, each of which bounds one third of the circle. Assume that we start with a generic choice of a major and rotate it counterclockwise. We get a new major at each angle until we make one full turn. But there are two types of special cases to look at.

One case is when two major leaves share an endpoint. Putting an extra leaf connecting the non-shared endpoints of the major leaves, we get a regular triangle. In fact, which two sides of this regular triangle you choose as the major does not matter. Therefore, we get an extra symmetry in this case; if we rotate such a major, then it does not take a full turn to see the same major again: only a 1/3-turn is enough. Let's call the set of all majors of this type the *degeneracy locus*. One can think of this space as the space of regular triangles inscribed in the unit circle.

Another special case is when two major leaves are parallel (if drawn as straight lines). We call the set of all majors of this type the *parallel locus*. In this case, after making a half-turn, we see the same major again.

Starting from these two types of singular loci allows us to understand the topology of the space $\mathrm{PM}(3)$. Note that both the degeneracy locus and the parallel locus are topological circles. Let's see what a neighborhood of a point on the degeneracy locus looks like. From a major m on the degeneracy locus, there are three different ways to move into the complement: remove one of the sides of the regular triangle and open up the shared endpoint of the remaining two, as illustrated in Figure 10. They are all nearby, but there is no short path connecting two of A, B, C without crossing the degeneracy locus.

On the other hand, if you move along a locus consisting of majors that are distance ϵ away from the degeneracy locus for a small enough positive number ϵ and look at the closest degenerate major at each moment, then you will see each degenerate major exactly three times, as in Figure 11.

We can thus obtain a neighborhood of the degeneracy locus from a tripod cross an interval by gluing the tripod ends with a 2/3-turn. (We choose to rotate by 2/3 to agree with the other part of $\mathrm{PM}(3)$ below.) The boundary is again a topological circle which is embedded in $\mathbb{R}^3$ as a trefoil knot, as on the right side of Figure 12.

Now, we move on to the local picture around the parallel locus. Almost the same argument works, except now the situation is somewhat simpler and the neighborhood is homeomorphic to a Moebius band. One can embed this space into $\mathbb{R}^3$ so that the boundary is again a trefoil knot. (See the image on the left side of Figure 12.) Now the whole space $\mathrm{PM}(3)$ is obtained from these two spaces by gluing along the boundary. Figure 3 illustrates what the space looks like after this gluing.

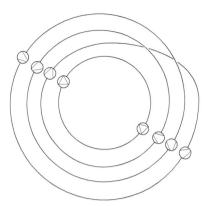

Figure 11: The degeneracy locus (the inner-most circle) and the locus of majors which are distance ϵ away from it.

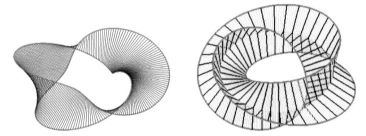

Figure 12: Neighborhoods of the parallel locus and the degeneracy locus.

Visualizing PM(3) by dividing into neighborhoods of two singular loci in this way allows one to see that PM(3) is a $K(B_3, 1)$-space, as follows.

We first construct the universal cover of PM(3). Write PM(3) as $A \cup B$, where A is the closure of the neighborhood of the degeneracy locus and B is the closure of the neighborhood of the parallel locus. We glue them in a way that $A \cap B = \partial A = \partial B$.

Now it is easy to see that $\widetilde{A}$ is just the product of a tripod with $\mathbb{R}$, and $\widetilde{B}$ is simply an infinite strip. $\widetilde{A}$ has three boundary lines each of which is glued to a copy of $\widetilde{B}$, and each end of each copy of $\widetilde{B}$, one needs to glue a copy of $\widetilde{A}$, and so on. To get $\widetilde{\mathrm{PM}(3)}$, we need to do this infinitely many times and finally get the product of an infinite trivalent tree with $\mathbb{R}$, which is obviously contractible. Therefore, all the higher homotopy groups of PM(3) vanish.

On the other hand, the Seifert–van Kampen theorem says that

$$\pi_1(\mathrm{PM}(3)) = \pi_1(A) *_{\mathbb{Z}} \pi_1(B)$$
$$= \langle \alpha, \beta \mid \alpha^3 = \beta^2 \rangle$$

which is one presentation for B_3.

11.2   Parametrization of PM(3) using the angle bisector

Let m be a non-degenerate cubic major. The endpoints of leaves of m divide the circle into four arcs, two of them with length 1/3 and the other two have length

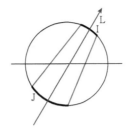

Figure 13: A cubic major (red) and its angle bisector (blue).

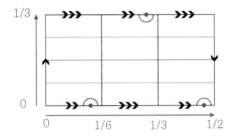

Figure 14: The parameter space for cubic majors using the angle bisector.

between 0 and $1/3$. Call these other two arcs I and J. Draw a line L passing through the midpoint of I and the midpoint of J, as in Figure 13. Let θ be the angle from the positive real axis to L. Relabeling I and J if necessary, let I be the interval that L meets at the angle θ, and let a be the length of the interval I. Our parameters are a and θ. Note that we can choose θ from $[0, 1/2]$, since (a, θ) represents the same major as $(1/3 - a, \theta - 1/2 \mod 1)$. Also note that a runs from 0 to $1/3$. Hence, the set $\{(a, \theta) : 0 \leq a \leq 1/3, 0 \leq \theta \leq 1/2\}$ with appropriate identifications on the boundary gives a parameter space of PM(3) (see Figure 14).

When a is either 0 or $1/3$, either I or J becomes a single point, and this corresponds to the degeneracy locus. The locus where $a = 1/6$ is the parallel locus (the red line in Figure 14).

It is easy to see that $(a, 0)$ and $(1/3 - a, 1/2)$ represent the same major. Also observe that, for any $0 \leq \theta \leq 1/6$, the pairs $(0, \theta)$, $(1/3, \theta + 1/6)$, and $(0, \theta + 1/3)$ are just different choices of two edges of the regular triangle whose vertices are at $(\theta, \theta + 1/6, \theta + 1/3)$. Hence they must be identified. Similarly, for $1/6 \leq \theta \leq 1/3$, $(1/3, \theta - 1/6)$, $(0, \theta)$, and $(1/3, \theta + 1/6)$ represent the same point. For instance, the three large dots in Figure 14 should be identified.

11.3   Embedding of PM(3) into S^3

We can visualize how PM(3) embeds into S^3. Consider the decomposition of S^3 into two solid tori glued along the boundary. Put the degeneracy and parallel locus as central circles of the solid tori. Seeing S^3 as $\mathbb{R}^3$ with a point at infinity, one may assume that the parallel locus coincides with the unit circle on xy-plane and the degeneracy locus is the z-axis with the point at infinity. Then one can view PM(3) with one point removed as a 2-complex in $\mathbb{R}^3$.

We already know that PM(d) embeds as a spine of the complement of the discriminant locus, so it would be instructive to see the discriminant locus in this picture. A monic centered cubic polynomial is written as $z^3 + az + b$ for some complex numbers a, b. Hence the space of all such polynomials can be seen as $\mathbb{C}^2$. The unit sphere is the locus $|a|^2 + |b|^2 = 1$, and the discriminant locus is $4a^3 + 27b^2 = 0$. The intersection of these two loci is a trefoil knot. We will embed PM(3) into S^3 so that it forms a spine of the complement of this trefoil knot, and thus also as a spine of the complement of the discriminant locus in $\mathbb{C}^2$.

Consider the stereographic projection $\phi : S^3 \setminus \{(0, 0, 0, 1)\} \to \mathbb{R}^3$ defined by

$$\phi(x_1, x_2, x_3, x_4) = \left(\frac{x_1}{1 - x_4}, \frac{x_2}{1 - x_4}, \frac{x_3}{1 - x_4} \right).$$

Instead of taking the line segment connecting a point on the unit circle of xy-plane and the z-axis, we first take the preimages of these two points under ϕ and consider the great circle passing through them in S^3. Then we take the image of this great circle under ϕ. While one wraps up the parallel locus twice and the degeneracy locus three times, we construct a surface as the trajectory of the image of the great circle passing through the preimages of the points on the parallel and the degeneracy locus. Now it is guaranteed to be an embedded 2-complex (not a manifold, since the degeneracy locus is singular) by construction.

Let's return to the space of normalized cubic polynomials

$$\left\{ z^3 + az + b \mid a, b \in \mathbb{C}, |a|^2 + |b|^2 = 1 \right\}.$$

We can identify the degeneracy locus inside this space as the subset cut out by $a = 0$, and the parallel locus as the subset cut out by $b = 0$. To connect the degeneracy locus to the parallel locus by spherical geodesics, running three times around the degeneracy locus while running twice around the parallel locus, we look at the subset

$$\mathrm{PM}(3) = \left\{ z^3 + s\theta^2 + t\theta^3 \mid s, t \in \mathbb{R}_{\geq 0}, \theta \in S^1, |s\theta^2|^2 + |t\theta^3|^2 = 1 \right\}$$

$$= \left\{ z^3 + az^2 + bz^3 \mid a, b \in \mathbb{C}, |a|^2 + |b|^2 = 1, \frac{a^3}{b^2} \in [0, \infty] \right\}.$$

This is clearly disjoint from the discriminant locus.

Recall that in the proof of Theorem 9.2, we constructed a section $\sigma : \mathrm{PM}(d) \to \mathcal{P}_d^0$ where $\mathcal{P}_d^0$ is the space of all monic centered polynomials of degree d with distinct roots. The above embedding of PM(3) gives another section into $\mathcal{P}_3^0$, but it is not exactly the same as σ in Theorem 9.2. For instance, σ has the property that for each polynomial f in the image of PM(3) under σ, all the critical values of f have the same modulus, while the above embedding does not have this property.

11.4   Other parametrizations

It is not a priori clear that the angle bisector parametrization of PM(3) can be generalized to the parametrization of PM(d) for higher d. In this section, we briefly survey several different ways of parametrizing PM(3) and see the advantages and disadvantages of each method.

11.4.1 Starting point method

Here we start with the most naive way to parametrize the space of primitive majors, from Section 10. Start from angle 0 and walk around the circle until you meet an end of a major leaf, and say the angle is x_1. Keep walking until you meet an end of different major leaf, say the angle is x_2. The numbers x_1, x_2 are regarded as the starting points of the major leaves of given cubic major. We have two cases: either $x_2 - x_1 < 1/3$ and the leaves are $\{(x_1, x_1 - 1/3 \mod 1), (x_2, x_2 + 1/3)\}$ or $x_2 - x_1 \geq 1/3$ and the leaves are $\{(x_1, x_1 + 1/3), (x_2, x_2 + 1/3)\}$. As you see, it is fairly easy to get a neat formula for leaves. x_1 has the range from 0 to 1/3 and x_2 has the range from 0 to 2/3. But not every point in the rectangle $[0, 1/3] \times [0, 2/3]$ is allowed. First of all, there is a restriction $x_2 \geq x_1$, and sometimes even $x_2 \geq x_1 + 1/3$. So, this parametrization method is not as neat as the rectangular parameter space obtained by the angle bisector method. Another issue is that there are many more combinatorial possibilities in higher degrees. We will see this in more detail while we discuss the next method.

11.4.2 Sum and difference of the turning number

Given a cubic lamination, start from angle 0 and walk around the circle until the first time you meet two consecutive ends $x < y$ belonging to distinct leaves. We call these numbers x, y turning numbers of the given lamination. Let $S = y + x$ and $D = y - x$ (S stands for the "sum" and D stands for the "difference"). Then one can easily get $x = (S - D)/2$ and $y = (S + D)/2$, and the leaves are $\{x, x - 1/3 \mod 1\}, \{y, y + 1/3 \mod 1\}$. Note that D runs from 0 to 1/3 and for a given D, S runs from D to $4/3 - D$. Hence we get a trapezoid shape domain for the parameter space and one can figure out which points on the boundary are identified as we did for the other models.

It is pretty clear what each parameter means and one gets a neat formula for the leaves. On the other hand, it has some drawbacks when one tries to generalize to higher-degree cases. Even for degree 4, the complement of the degeneracy locus in PM(4) is not connected. (This is what we postponed discussing in the last subsection. See Figure 15.) Hence, however one defines the turning numbers, it is hard to determine which configuration one has. One can divide the domain into pieces, each of which represents one combinatorial configuration, and give a different formula for each such piece. This requires understanding the different possible configurations. In particular, one can start with counting the number of connected components of the complement of the degeneracy locus in PM(d). Tomasini counted the number of components in his thesis (see Theorems 4.3.1 and 4.3.2, pp. 118–121, in [Tom14]).

11.4.3 Avoiding the Moebius band

One way to avoid the Moebius band is this: if you take the quotient of the set of majors by the symmetry $z \leftrightarrow -z$ that conjugates a cubic polynomial to a dynamically isomorphic polynomial, then the Moebius band folds in half to an annulus. One boundary component of the annulus is wrapped three times around a circle (laminations with a central lamination), and the other boundary component consists of laminations whose majors are parallel. The parameter transverse to the annulus is the shortest distance between endpoints of the majors, in the interval $[0, 1/6]$. The other parameter is the most clockwise endpoint of this shortest distance interval.

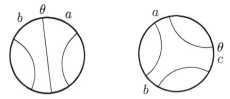

Figure 15: The two topological types of generic primitive majors in PM(4), with the parameters marked.

12   UNDERSTANDING PM(4)

We now give some brief comments on the shape of PM(4). Unlike in PM(2) and PM(3), there is more than one way that a generic primitive major can be arranged topologically: the complement of the degeneracy locus is not connected. The two possibilities are illustrated in Figure 15.

PM(4) is a 3-complex, and each of these topological types of generic majors contributes a top-dimensional stratum of the 3-complex. The *parallel stratum* is the piece corresponding to the stratum on the left, containing the part where the three leaves are parallel. Topologically, the parallel stratum may be parametrized by triples (θ, a, b) with θ in the circle and $a, b \in (0, 1/4)$. Here θ is the angle to one endpoint of the central leaf and $\theta - a$ and $\theta + b$ are the angles to adjacent endpoints of the other two leaves, so that the three leaves have endpoints

$$(\theta - a, \theta - a - 1/4), \quad (\theta, \theta + 1/2), \quad \text{and} \quad (\theta + b, \theta + b + 1/4)$$

(with coordinates interpreted modulo 1). There is an equivalence relation:

$$(\theta, a, b) \equiv (\theta + 1/2, 1/4 - b, 1/4 - a).$$

Therefore, the parallel stratum topologically is a square cross an interval, with the top glued to the bottom by a half-twist. As a manifold with corners, this stratum has 2 codimension-1 faces, each an annulus.

The other stratum, the *triangle stratum*, can be parametrized by quadruples (θ, a, b, c) with $a, b, c > 0$, $a + b + c = 1/4$, and θ in the circle. Here a, b, and c are the lengths of the three intervals on the boundary of the central gap, and θ is the angle to the start of one leaf, so that the three leaves have endpoints

$$(\theta, \theta + 1/4), \quad (\theta + 1/4 + a, \theta + 1/2 + a), \quad \text{and} \quad (\theta - 1/4 - c, \theta - c).$$

(The last leaf is also $(\theta + 1/2 + a + b, \theta + 3/4 + a + b)$.) Again, there is an equivalence relation:
$$(\theta, a, b, c) \equiv (\theta + 1/4 + a, b, c, a) \equiv (\theta + 1/2 + a + b, c, a, b).$$

This stratum is therefore topologically a triangle cross an interval, with the top glued to the bottom by a 1/3 twist. As a manifold with corners, it has only one codimension-1 face, an annulus.

We next turn to the codimension-1 degeneracy locus. Here there is only one topological type (the stratum is connected), consisting of a triangle and another leaf. As illustrated in Figure 16, the major can be uniquely parametrized by an angle θ,

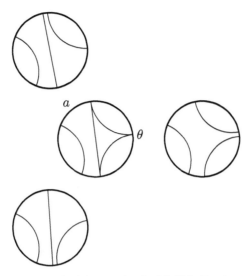

Figure 16: A codimension-1 primitive major in PM(4) (in the center) and three per-turbations to generic majors.

the angle to the vertex of the triangle opposite the leaf, and a number $a \in (0, 1/4)$, the length of one of the intervals on the boundary of the gap between the triangle and the leaf. The codimension-1 degeneracy locus is therefore an annulus.

There are three ways to perturb a codimension-1 degenerate major into generic majors. Note that two of the generic majors are in the parallel stratum and one is in the triangle stratum. Thus, all three annuli that we found on the boundary of the top-dimensional strata are glued together.

13   THURSTON'S ENTROPY ALGORITHM AND ENTROPY ON THE HUBBARD TREE

To any rational angle θ (mod 1), Douady-Hubbard [DH84] associated a unique postcritically finite quadratic polynomial $f_\theta : z \mapsto z^2 + c_\theta$. This polynomial induces a Markov action on its Hubbard tree. The topological entropy [AKM65] of the polynomial f_θ on its Hubbard tree is called the *core entropy* of f_θ.

In order to combinatorially encode and effectively compute the core entropy, W. Thurston developed an algorithm that takes θ as its input, constructs a non-negative matrix A_θ (bypassing f_θ), and outputs its Perron-Frobenius leading eigenvalue $\rho(A_\theta)$. We will prove that $\log(\rho(A_\theta))$ is the core entropy of f_θ.

More precisely, we will define the notions in the following diagram and establish the equality on the right column (Theorem 13.10):

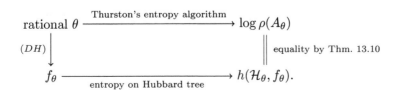

13.1   Thurston's entropy algorithm

Set $\mathbb{S} = \mathbb{R}/\mathbb{Z}$. All angles in this section are considered to be mod 1, i.e., elements of $\mathbb{S}$.

Let $\tau : \mathbb{S} \to \mathbb{S}$ denote the angle doubling map. An angle θ is *periodic* under the action of τ if and only if it is rational with odd denominator, and (strictly) *preperiodic* if and only if it is rational with even denominator.

Fix a rational angle $\theta \in \mathbb{S} \setminus \{0\}$. If θ is periodic, exactly one of $(\theta+1)/2$ and $\theta/2$ is periodic and the other is preperiodic. If θ is preperiodic, both $(\theta+1)/2$ and $\theta/2$ are preperiodic. Set $2^{-1}\theta$ to be the periodic angle, if it exists, among $(\theta+1)/2$ and $\theta/2$, and otherwise set it to be $\theta/2$. Define the set

$$O_\theta := \Big\{ \{2^n\theta, 2^l\theta\} \,\Big|\, l, n \geq -1 \text{ and } 2^n\theta \neq 2^l\theta \Big\},$$

with the convention that the pairs $\{2^n\theta, 2^l\theta\}$ that constitute O_θ are unordered sets. We divide the circle $\mathbb{S}$ at the points $\{\theta/2, (\theta+1)/2\}$, forming two closed half circles, with the boundary points belonging to both halves.

Define Σ_θ to be the abstract linear space over $\mathbb{R}$ generated by the elements of O_θ. Define a linear map $\mathcal{A}_\theta : \Sigma_\theta \to \Sigma_\theta$ as follows. For any basis vector $\{a, b\} \in O_\theta$, if a and b are in a common closed half circle, set $\mathcal{A}_\theta(\{a, b\}) = \{2a, 2b\}$; otherwise set $\mathcal{A}_\theta(\{a, b\}) = \{2a, \theta\} + \{\theta, 2b\}$. Denote by A_θ the matrix of $\mathcal{A}_\theta$ in the basis O_θ; it is a non-negative matrix. Denote its leading eigenvalue, which exists by the Perron-Frobenius theorem, by $\rho(A_\theta)$. It is easy to see that A_θ is not nilpotent, so $\rho(A_\theta) \geq 1$.

Definition 13.1. Thurston's entropy algorithm *is the map*

$$(\mathbb{Q} \cap \mathbb{S} \setminus \{0\}) \ni \theta \mapsto \log \rho(A_\theta).$$

We will relate the output of Thurston's entropy algorithm, $\log \rho(A_\theta)$, to quadratic polynomials in Subsection 13.3.

Example 13.2. Set $\theta = \frac{1}{5}$. The abstract linear space Σ_θ has basis

$$O_{\frac{1}{5}} = \{ \{\tfrac{1}{5}, \tfrac{2}{5}\}, \{\tfrac{1}{5}, \tfrac{3}{5}\}, \{\tfrac{1}{5}, \tfrac{4}{5}\}, \{\tfrac{2}{5}, \tfrac{3}{5}\}, \{\tfrac{2}{5}, \tfrac{4}{5}\}, \{\tfrac{3}{5}, \tfrac{4}{5}\} \}.$$

We divide the circle $\mathbb{S}$ by the pair $\{\frac{1}{10}, \frac{3}{5}\}$. The linear map $\mathcal{A}_{\frac{1}{5}}$ acts on the basis vectors as follows:

$$\left\{\tfrac{1}{5}, \tfrac{2}{5}\right\} \mapsto \left\{\tfrac{2}{5}, \tfrac{4}{5}\right\}, \quad \left\{\tfrac{1}{5}, \tfrac{3}{5}\right\} \mapsto \left\{\tfrac{2}{5}, \tfrac{1}{5}\right\}, \qquad \left\{\tfrac{1}{5}, \tfrac{4}{5}\right\} \mapsto \left\{\tfrac{1}{5}, \tfrac{2}{5}\right\} + \left\{\tfrac{1}{5}, \tfrac{3}{5}\right\},$$

$$\left\{\tfrac{2}{5}, \tfrac{3}{5}\right\} \mapsto \left\{\tfrac{4}{5}, \tfrac{1}{5}\right\}, \quad \left\{\tfrac{2}{5}, \tfrac{4}{5}\right\} \mapsto \left\{\tfrac{4}{5}, \tfrac{1}{5}\right\} + \left\{\tfrac{1}{5}, \tfrac{3}{5}\right\}, \quad \left\{\tfrac{3}{5}, \tfrac{4}{5}\right\} \mapsto \left\{\tfrac{1}{5}, \tfrac{3}{5}\right\}.$$

We then compute $\log \rho(A_\theta) = 0.3331$.

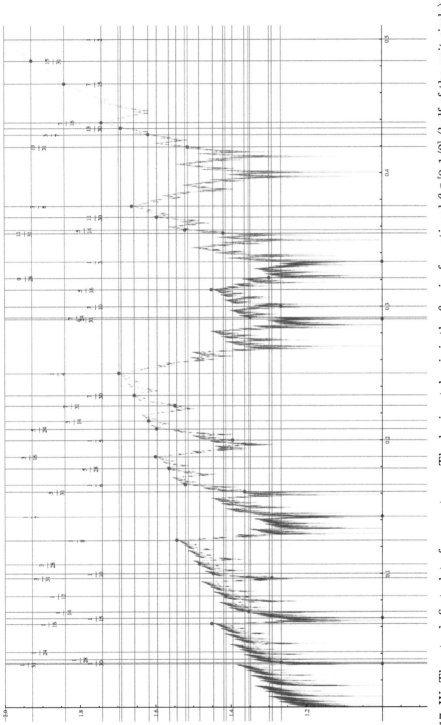

Figure 17: W. Thurston's first plot of core entropy. The horizontal axis is the θ-axis, for rational $\theta \in [0, 1/2]$ (half of the unit circle), and the vertical axis is core entropy, or $\log \rho(A_\theta)$.

13.2   Hubbard trees following Douady and Hubbard

We now recall background material about Hubbard trees, used in later sections to justify Thurston's entropy algorithm as computing core entropy. (See, for example, [DH84, Poi09, Poi10] for additional information about Hubbard trees.)

Let f be a postcritically finite polynomial, i.e., a polynomial all of whose critical points have a finite (and hence periodic or preperiodic) orbit under f. By classical results of Fatou, Julia, Douady, and Hubbard, the *filled Julia set*

$$\mathcal{K}_f = \{z \in \mathbb{C} \mid f^n(z) \not\to \infty\}$$

is compact, connected, locally connected and locally arc-connected. These conditions also hold for the *Julia set* $\mathcal{J}_f := \partial \mathcal{K}_f$. The Fatou set $\mathcal{F}_f := \overline{\mathbb{C}} \setminus \mathcal{J}_f$ consists of one component $U(\infty)$ which is the basin of attraction of ∞, and at most countably many bounded components constituting the interior of $\mathcal{K}_f$. Each of the sets $\mathcal{K}_f, \mathcal{J}_f, \mathcal{F}_f$ and $U(\infty)$ is fully invariant by f; each Fatou component is (pre)periodic (by Sullivan's no-wandering domain theorem, or by hyperbolicity of the map); and each periodic cycle of Fatou components contains at least one critical point of f (counting ∞).

There is a system of Riemann mappings

$$\left\{ \phi_U : \mathbb{D} \to U \;\middle|\; U \text{ Fatou component} \right\}$$

each extending to a continuous map on the closure $\overline{\mathbb{D}}$, so that for all U and some d_U, the following diagram commutes:

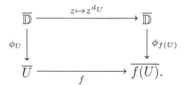

In particular, on every periodic Fatou component U, including $U(\infty)$, the map ϕ_U realizes a conjugacy between a power map and the first return map on U. The image in U under ϕ_U of radial lines in $\mathbb{D}$ are, by definition, *internal rays* on U if U is bounded and *external rays* if $U = U(\infty)$. Since a power map sends a radial line to a radial line, the polynomial f sends an internal/external ray to an internal/external ray.

If U is a bounded Fatou component, then $\phi_U : \overline{\mathbb{D}} \to \overline{U}$ is a homeomorphism, and thus every boundary point of U receives exactly one internal ray from U. This is in general not true for $U(\infty)$, where several external rays may land at a common boundary point.

Definition 13.3 (Supporting rays). *We say that an external ray R supports a bounded Fatou component U if*

1. *the ray lands at a boundary point q of U, and*
2. *there is a sector based at q delimited by R and the internal ray of U landing at q which does not contain other external rays landing at q.*

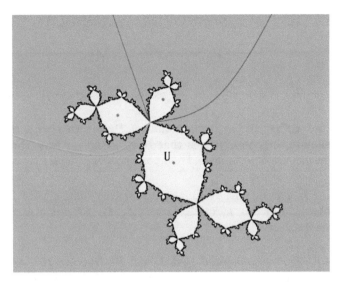

Figure 18: The green (left) and red (right) rays are supporting rays of the Fatou component U, but the blue (middle) one is not.

It follows from Definition 13.3 that for any bounded Fatou component U and point $z \in \partial U$ there are at most two external rays which support U and land at z. Start from the internal ray in U which lands at z and turn in the counterclockwise direction centered at z. The first (resp. last) encountered external ray landing at z is called the *right-supporting ray* (resp. *left-supporting ray*) of U at z. See Figure 18.

The system of internal/external rays does not depend on the possible choices of ϕ_U. If $f : z \mapsto z^2 + c$ and f is postcritically finite, there is actually a unique choice of ϕ_U for each Fatou component U. In particular, $\phi_{U(\infty)}$ conjugates z^2 to f and ϕ_U conjugates z^2 to f^p if U is a bounded periodic Fatou component and p is the minimal integer such that $f^p(0) = 0$ (if no such p exists, $\mathcal{K}_f = \mathcal{J}_f$). In this case, for any $x \in \mathbb{S}$ we use $\mathcal{R}_f(x)$ or simply $\mathcal{R}(x)$ to denote the image under $\phi_{U(\infty)}$ of the ray $\{re^{2\pi i x}, 0 < r < 1\}$ and will call it the *external ray of angle* x. Angles of internal rays can be defined similarly. We also use $\gamma(x) = \phi_{U(\infty)}(e^{2\pi i x})$ to denote the *landing point* of the ray $\mathcal{R}(x)$.

Any pair of points in the closure of a bounded Fatou component can be joined in a unique way by a Jordan arc consisting of (at most two) segments of internal rays. We call such arcs *regulated* (following Douady and Hubbard). Since $\mathcal{K}_f$ is arc-connected, given two points $z_1, z_2 \in \mathcal{K}_f$, there is an arc $\gamma : [0,1] \to \mathcal{K}_f$ such that $\gamma(0) = z_1$ and $\gamma(1) = z_2$. In general, we will not distinguish between the map γ and its image. It is proved in [DH84] that such arcs can be chosen in a unique way so that the intersection with the closure of every Fatou component is regulated. We still call such arcs regulated and denote them by $[z_1, z_2]$. We say that a subset $X \subset \mathcal{K}_f$ is *allowably connected* if for every $z_1, z_2 \in X$ we have $[z_1, z_2] \subset X$.

Definition 13.4. *We define the* regulated hull *of a subset X of $\mathcal{K}_f$ to be the minimal closed allowably connected subset of $\mathcal{K}_f$ containing X.*

Proposition 13.5. *For a collection of $z_1, \ldots, z_n$ finitely many points in $\mathcal{K}_f$, their regulated hull is a finite tree.*

Definition 13.6. *Let f be a postcritically finite polynomial. The* postcritical set *$\mathcal{P}_f$ is defined to be[2]*

$$\mathcal{P}_f = \left\{ f^n(c) \,\middle|\, f'(c) = 0, n \geq 0 \right\}.$$

The Hubbard tree *$\mathcal{H}_f$ is defined to be the regulated hull of the finite set $\mathcal{P}_f$.*

The *vertex set* $V(\mathcal{H}_f)$ of $\mathcal{H}_f$ is the union of $\mathcal{P}_f$ together with the branching points of $\mathcal{H}_f$, namely the points p such that $\mathcal{H}_f \setminus \{p\}$ has at least three connected components. The closure of a connected component of $\mathcal{H}_f \setminus V(\mathcal{H}_f)$ is called an *edge*.

Lemma 13.7. *For a postcritically finite polynomial f, the set $\mathcal{H}_f$ is a tree with finitely many edges. Moreover $f(\mathcal{H}_f) \subset \mathcal{H}_f$ and $f : \mathcal{H}_f \to \mathcal{H}_f$ is a Markov map (as defined in Appendix A).*

Definition 13.8. *For a polynomial f such that its Hubbard tree exists and is a finite tree, the* core entropy *of f is the topological entropy of the restriction of f to its Hubbard tree, $h(\mathcal{H}_f, f)$.*

Using Proposition A.6, we may relate the topological entropy of f on $\mathcal{H}_f$ to the spectral radius of a transition matrix D_f constructed from f by $h(\mathcal{H}_f, f) = \log \rho(D_f)$.

We remark that if a polynomial f has a connected and locally connected filled Julia set such that the postcritical set lies on a finite topological tree, then its Hubbard tree exists. This condition is called "topologically finite" by Tiozzo [Tio15], and is more general than postcritically finite.

13.3   Relating Thurston's entropy algorithm to polynomials

Thurston's entropy algorithm effectively computes the topological entropy $h(\mathcal{H}_f, f)$ for any postcritically finite polynomial without actually computing the Hubbard tree. We will see how to relate the quadratic version of the algorithm given in Section 13.1 to quadratic polynomials.

On one hand, Thurston's entropy algorithm produces a quantity, $\log \rho(A_\theta)$, from any given rational angle θ. On the other hand, Douady-Hubbard defined a finite-to-one map $\mathbb{Q} \ni \theta \mapsto c_\theta$ so that the quadratic polynomial $z \mapsto z^2 + c_\theta$ is postcritically finite. More precisely:

Theorem 13.9 (Douady-Hubbard). *If $\theta \in \mathbb{Q}$ is preperiodic (resp. p-periodic) under the angle doubling map, there is a unique parameter c_θ such that for $f : z \mapsto z^2 + c_\theta$ both external rays $\mathcal{R}(\frac{\theta}{2})$ and $\mathcal{R}(\frac{\theta+1}{2})$ land at 0, and 0 is preperiodic (resp. support the Fatou component containing 0, and 0 is p-periodic). Furthermore, every postcritically finite quadratic polynomial arises in this way.*

Our objective is to establish:

Theorem 13.10. *For θ a rational angle, $\log \rho(A_\theta) = h(\mathcal{H}_f, f)$ for $f : z \mapsto z^2 + c_\theta$.*

---

[2]For the purpose of this section, we include critical points in the postcritical set.

Proof. The idea of the proof is the following, inspired by [Gao13, Jun14]:

1. Construct a topological graph G and a Markov action $L : G \to G$ such that the spectral radius of the transition matrix $D_{(G,L)}$ is equal to the spectral radius of A_θ.
2. Construct a continuous, finite-to-one, and surjective semi-conjugacy Φ from $L : G \to G$ to $f : \mathcal{H}_f \to \mathcal{H}_f$.

Then we may conclude that

$$\log \rho(A_\theta) \overset{\text{same matrix}}{=} \log \rho(D_{(G,L)}) \overset{\text{Prop. A.6}}{=} h(G, L) \overset{\text{Prop. A.3}}{=} h(\mathcal{H}_f, f).$$

Let G be a topological complete graph whose vertex set is the forward orbit after identifying the diagonal angles $\frac{\theta}{2}$ and $\frac{\theta+1}{2}$:

$$V_G = \{2^n \theta, n \geq -1\} / \left(\frac{\theta}{2} \sim \frac{\theta+1}{2} \right).$$

The set of edges of G is

$$E_G = \{ e(x, y) \mid x \neq y \in V_G \}$$

(with $e(x, y) = e(y, x)$). Being a topological graph means that G is a topological space and each edge $e(x, y)$ is homeomorphic to a closed interval with ends x and y. Thus E_G is in bijection with the set O_θ indexing A_θ.

Mimicking the action of the linear map A_θ, we can define a piecewise monotone map $L : G \to G$ as follows. On vertices, $L = \tau$. Let x, y be two distinct vertices in V_G. If x, y belong to the same closed half circle, i.e., the closure of a complete component of $\frac{\theta}{2}, \frac{\theta+1}{2}$, then $\tau(x) \neq \tau(y) \in V_G$. In this case, let L map the edge $e(x, y)$ homeomorphically onto the edge $e(\tau(x), \tau(y))$. If x, y belong to distinct open half circles, subdivide the edge $e(x, y)$ into non-trivial arcs $e(x, z)$ and $e(z, y)$, and let L map the arc $e(x, z)$ (resp. $e(z, y)$) homeomorphically onto $e(\tau(x), \theta)$ (resp. $e(\theta, \tau(y))$).

It is easy to see that the transition matrix of (G, L) is exactly A_θ. By Proposition A.6, the topological entropy $h(G, L)$ is always equal to $\log \rho(A_\theta)$, regardless of the precise choices of L as homeomorphisms on the edges.

However, to relate $h(G, L)$ to $h(\mathcal{H}_f, f)$, we will redefine the homeomorphic action of L on each edge by lifting corresponding actions of f via suitably defined conjugacies.

We will treat the periodic and preperiodic cases separately.

Case 1: θ is periodic

In this case the rays of angles $\frac{\theta}{2}$ and $\frac{\theta+1}{2}$ both land at the boundary of the Fatou component U containing 0 and support that component. If $\frac{\theta}{2}$ is periodic then the rays are right-supporting rays, and if it is $\frac{\theta+1}{2}$, that is, periodic then the rays are left-supporting rays. For simplicity we will only treat the former case and thus right-supporting rays.

The angles in V_G form a periodic cycle. For each $x \in V_G$, the external ray $\mathcal{R}(x)$ of angle x may support one or two Fatou components, but it right-supports a unique periodic Fatou component, denoted by U_x. Define a map $\Phi : V_G \to \mathcal{P}_f$ such that $\Phi(x)$

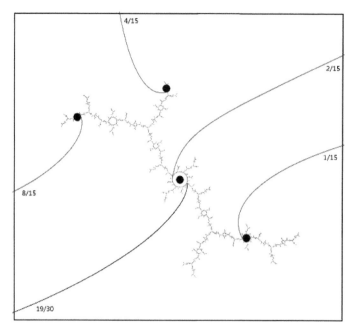

Figure 19: In this periodic example, $\theta = 4/15$ and $\theta/2 = 2/15$. All the rays in this figure are right-supporting rays.

is the center of the Fatou component U_x. Since external rays with distinct angles of V_G right-support distinct Fatou components, we have $U_x \neq U_y$ if $x \neq y \in V_G$. Thus the map $\Phi : V_G \to \mathcal{P}_f$ is bijective.

Extend Φ to a map, also denoted by Φ, from G to $\mathcal{H}_f$ such that Φ maps the edge $e(x,y)$ homeomorphically to the regulated arc $[\Phi(x), \Phi(y)] \subset \mathcal{H}_f$. We now assert and justify several facts about Φ:

1. Φ is finite-to-one.

   It follows directly from the fact that $\Phi : V_G \to \mathcal{P}_f$ is a bijection.

2. Φ is surjective.

   Since $\Phi(V_G) = \mathcal{P}_f$ and G is a complete graph, for any $p, q \in \mathcal{P}_f$, $[p,q] \subset \Phi(G)$. So we only need to evoke the fact that each edge of $\mathcal{H}_f$ is contained in a regulated arc $[p,q]$ with $p, q \in \mathcal{P}_f$.

3. $f \circ \Phi = \Phi \circ L$ after suitable modification of L on each edge.

   If two distinct vertices $x, y \in V_G$ belong to the same closed half circle, the interior of the regulated arc $[\Phi(x), \Phi(y)]$ does not contain the critical point of f. Its f-image is

$$f([\Phi(x), \Phi(y)]) = [f(\Phi(x)), f(\Phi(y))] = [\Phi(2x), \Phi(2y)]$$
$$= \Phi(e(2x, 2y)) = \Phi(L(e(x, y))).$$

So we can redefine L on the edge by lifting f, i.e., by setting

$$L = \Phi^{-1} \circ f \circ \Phi$$

on the edge of $e(x,y)$.

If two vertices $x, y \in V_G$ belong to distinct open half circles, the interior of the regulated arc $[\Phi(x), \Phi(y)]$ contains the critical point 0. Its f-image is

$$f([\Phi(x), \Phi(y)]) = [f(\Phi(x)), c_\theta] \cup [c_\theta, f(\Phi(y))]$$
$$= [\Phi(2x), \Phi(\theta)] \cup [\Phi(\theta), \Phi(2y)]$$
$$= \Phi(e(2x, \theta) \cup e(\theta, 2y)) = \Phi(L(e(x, y))).$$

So we can redefine L on $e(x, y)$ such that $\Phi \circ L = f \circ \Phi$ on each of the two segments of $e(x, y)$ subdivided by $\Phi^{-1}(0)$.

Thus the maps L, Φ and f have been shown to satisfy the properties of Proposition A.3, so the equation

$$h(G, L) = h(\mathcal{H}_f, f)$$

holds, as desired.

Case 2: θ is preperiodic

In this case, the filled Julia set is equal to the Julia set. We can also define a map $\Phi: V_G \to \mathcal{P}_f$ such that $\Phi(x)$ is the landing point of $\mathcal{R}(x)$. It is easy to see that Φ is surjective. However, if we extend Φ piecewise monotonically on the edge of G as we did in the periodic case, the map Φ will lose the property of being finite-to-one, because some rays with distinct angles in V_G may land at the same point, which means that Φ collapses some edges of G to points. So we may no longer apply Proposition A.3 directly. To overcome this difficulty, let us define a subgraph Γ of G as follows:

$$V_\Gamma := V_G \quad \text{and} \quad E_\Gamma := \{e(x, y) \in E_G | \ \Phi(x) \neq \Phi(y)\}.$$

We will check that the graph Γ satisfies the following properties.

1. Γ is connected.

   This is because for any $x \in V_\Gamma$, the edge $e(\theta/2, x)$ belongs to Γ.
2. Γ is L-invariant.

   First, we observe that for any two distinct vertices $x, y \in V_G$ belonging to the same closed half circle, if $\mathcal{R}(x)$ and $\mathcal{R}(y)$ land at distinct points, then $\mathcal{R}(2x)$ and $\mathcal{R}(2y)$ also land at distinct points.

   Let $e(x, y)$ be an edge of Γ. Then the rays $\mathcal{R}(x)$ and $\mathcal{R}(y)$ land at distinct points $\Phi(x)$ and $\Phi(y)$ respectively. If x, y belong to the same closed half circle, by this observation the image edge

$$L(e(x, y)) = e(2x, 2y)$$

belongs to Γ. If x, y belong to distinct open half circles, then $(x, \theta/2)$ and $(y, \theta/2)$ belong to the same closed half circle and the image

$$L(e(x, y)) = e(2x, \theta) \cup e(\theta, 2y)$$

belongs to Γ.

3. $h(G, L) = h(\Gamma, L)$.

We claim that the set $G \setminus \Gamma$ is L-invariant. In fact, an edge $e(x, y)$ belongs to $E_G \setminus E_\Gamma \overset{\text{definition}}{\Longleftrightarrow} \Phi(x) = \Phi(y) \Longleftrightarrow$ the rays $\mathcal{R}(x), \mathcal{R}(y)$ land at a common periodic point in $\mathcal{P}_f$. To see the last implication, we only need to show the "$\Rightarrow$" part. Since x, y are both in the forward orbit of θ, we may assume $y = 2^k x, k \geq 1$. Then

$$\Phi(y) = \Phi(2^k x) = f^k(\Phi(x)).$$

So $\Phi(x)$ is a periodic point, receives both rays $\mathcal{R}(x), \mathcal{R}(y)$, and belongs to $\mathcal{P}_f$. In this case the angles x, y must belong to the same half circle and $\Phi(2x) = \Phi(2y)$. It follows that

$$L(e(x, y)) = e(2x, 2y)$$

also belongs to $E_G \setminus E_\Gamma$.

The argument above also shows that L maps an edge in $E_G \setminus E_\Gamma$ homeomorphically onto an edge in $E_G \setminus E_\Gamma$. According to the definition of the topological entropy, we have

$$h(G \setminus \Gamma, L) = 0.$$

So by Proposition A.1, we obtain

$$h(G, L) = \max \left\{ h(\Gamma, L), h(G \setminus \Gamma, L) \right\} = h(\Gamma, L).$$

By these properties, it is enough to prove that $h(\Gamma, L) = h(\mathcal{H}_f, f)$. In this case, the map $\Phi|_\Gamma : \Gamma \to \mathcal{H}_f$ is finite-to-one. By the same argument as in the periodic case, we also obtain that $\Phi|_\Gamma$ is surjective and

$$f \circ \Phi = \Phi \circ L$$

on Γ. So the equality

$$h(\Gamma, L) = h(\mathcal{H}_f, f)$$

holds.

This finishes the proof of Theorem 13.10.   □

Thurston's entropy algorithm and Theorem 13.10 are generalized to higher-degree maps in [Gao].

14   COMBINATORIAL LAMINATIONS AND POLYNOMIAL LAMINATIONS

The action of a degree d postcritically finite polynomial on its Julia set can be combinatorially encoded by the action of a degree d expanding map on an invariant lamination on the circle $\mathbb{S} = \mathbb{R}/\mathbb{Z}$ or on the torus $\mathbb{T} = \mathbb{S} \times \mathbb{S}$. Here we illustrate this connection.

14.1 Combinatorial laminations

We denote the diagonal Δ of $\mathbb{T}$ by

$$\Delta = \{(x,x), x \in \mathbb{S}\}.$$

For $x \neq y \in \mathbb{S}$, we use $\overline{xy}$ to denote the closure in $\overline{\mathbb{D}}$ of the hyperbolic chord in $\mathbb{D}$ connecting $e^{2\pi ix}$ and $e^{2\pi iy}$, which is called a *leaf*. This leaf is represented (twice) on the torus $\mathbb{T} \setminus \Delta$ as (x,y) and (y,x).

Two leaves $\overline{xy}$ and $\overline{x'y'}$ are said to be *compatible* if they are either equal or do not cross inside $\mathbb{D}$. We say also that the two points (x,y) and (x',y') on the torus are compatible. In this case the four points $(x,y), (y,x), (x',y'), (y',x')$ are pairwise compatible.

A *lamination in* $\mathbb{D}$ is a collection of leaves which are pairwise compatible and whose union is closed in $\mathbb{D}$. A *lamination on the torus* is a subset L closed in $\mathbb{T} \setminus \Delta$ and symmetric with respect to the diagonal Δ so that points in L are pairwise compatible.

14.2 Polynomial laminations

Let f be a monic degree d polynomial with a connected and locally connected Julia set. There is a unique Riemann mapping $\phi : \overline{\mathbb{C}} \setminus \overline{\mathbb{D}} \to \overline{\mathbb{C}} \setminus \mathcal{K}_f$ tangent to the identity at ∞; it extends continuously to the closure by the Carathéodory theorem. External rays are parametrized by external angles. The set of pairs of external rays landing at a common point gives another combinatorial characterization of the polynomial dynamics.

Two distinct external rays $\mathcal{R}(x)$ and $\mathcal{R}(y)$ landing at the same point are said to be a *ray-pair of f*. We also say that $\{x,y\}$ is an *angle-pair of f*. If the rays land at z, we say that $\{x,y\}$ is an angle-pair at z.

Let $\{x,y\}$ be an angle-pair at z. Then the union of the two rays $\mathcal{R}(x)$ and $\mathcal{R}(y)$ together with $\{z\}$ divides the plane into two regions. If at least one of the two regions does not contain other rays landing at z, we say that $\{x,y\}$ is an *adjacent angle-pair*.

A polynomial f induces a lamination $L(f)$ whose leaves (i.e., points in $\mathbb{T}$) consist of geodesics connecting adjacent angle-pairs. $L(f)$ is said to be the *polynomial lamination* of f.

14.3 Good and excluded regions of a lamination

We recall here some notions introduced in Section 4. If a lamination on $\mathbb{T}$ contains a point $l = (x,y)$, then a certain set $X(l)$ of other points are excluded from the lamination because they are not compatible with (x,y). If you draw the horizontal and vertical circles through the two points (x,y) and (y,x), they divide the torus into four rectangles having the same vertex set; the remaining two common vertices are (x,x) and (y,y). The two rectangles bisected by the diagonal are actually squares, of side lengths $a - b \mod 1$ and $b - a \mod 1$. Together, they form the compatible region $G(l)$. The leaves represented by points in the interior of the remaining two rectangles constitute the excluded region $X(l)$. See Figure 4, where the blue region is $X(l)$ and the tan region is $G(l)$.

Given a set S of leaves, the *excluded region* $X(S)$ is the union of the excluded regions $X(l)$ for $l \in S$, and the *good region* $G(S)$ is the intersection of the good regions $G(l)$ for $l \in S$. If S is a finite lamination, then $G(S)$ is a finite union of closed rectangles that are disjoint except for corners.

14.4 Invariant laminations from majors

Define the map $F : \mathbb{T} \to \mathbb{T}$ by

$$(x, y) \mapsto (\tau(x), \tau(y)),$$

where τ is the angle-doubling map defined in Section 1.1. A lamination $L \subset \mathbb{T} \setminus \Delta$ is said to be *F-invariant* if, for every $(x, y) \in L$,

1. either $F(x, y) \in \Delta$ or $F(x, y) \in L$; and
2. there exist two preimages of (x, y) in L, which have different x components and different y components.

The polynomial lamination $L(f)$ for a postcritically finite quadratic polynomial $f : z \mapsto z^2 + c$ is an example of an F-invariant lamination.

Given an angle $\theta \in \mathbb{S}$, there are several more or less natural ways to define an F-invariant lamination, mimicking the polynomial lamination of $f : z \mapsto z^2 + c_\theta$. And as we shall see, they differ by at most a countable set. Here we choose one that mimics the fact that preimages of the critical point of f accumulate on every Julia point. Other definitions will be given later (see Section 16).

Set, inductively,

$$b_0 := \left\{ \left(\frac{\theta}{2}, \frac{\theta+1}{2} \right), \left(\frac{\theta+1}{2}, \frac{\theta}{2} \right) \right\} \quad \text{the major leaves}$$

$$b_{i+1} := \left(F^{-1}(b_i) \cap G(b_i) \right) \cup b_i$$

$$\mathrm{Pre}_\theta := \bigcup_{i \geq 0} b_i \quad \text{the set of pre-major leaves}$$

$$\mathrm{cluster}(\mathrm{Pre}_\theta) := \text{the set of cluster points in } \mathbb{T} \text{ of the pre-major leaves.}$$

$$L_\theta := \text{the set of cluster points in } \mathbb{T} \setminus \Delta \text{ of the pre-major leaves.}$$

(Cluster points are also known as accumulation points.)

14.5 Relating combinatorial laminations to polynomial laminations

Proposition 14.1. *For a rational angle θ, the set L_θ is equal to the polynomial lamination of $f : z \to z^2 + c_\theta$. In particular, L_θ is an F-invariant lamination.*

Proof. This says that every accumulation point (x, y), $x \neq y$, of Pre_θ is an adjacent angle-pair for $f : z \mapsto z^2 + c_\theta$, and conversely every adjacent angle-pair is the accumulation point of a sequence $(x_n, y_n) \in \mathrm{Pre}_\theta$.

If θ is preperiodic, the ray-pair of angles $\frac{\theta}{2}$ and $\frac{\theta+1}{2}$ land at the critical point 0. So each point in Pre_θ is a (not necessarily adjacent) angle-pair at a precritical point. If a sequence of distinct leaves $(x_n, y_n) \in \mathrm{Pre}_\theta$ converges to (x, y), one can find a subsequence converging from one side of the limit leaf (x, y). It follows that (x, y) is an angle-pair, and is an adjacent angle-pair. Conversely, one just needs to consider an adjacent angle-pair at a point z of the Hubbard tree, and then use expansions on the tree to see that z is approximated by preimages of 0 on the tree. Details are presented in Appendix B.

The case that θ is periodic is considerably more complicated. The ends of a leaf in Pre_θ are not an angle-pair, but correspond to a pair of rays supporting a common Fatou component with dyadic internal angles, as in Figure 19. They would form a cutting line if we add the two internal rays. See details in Appendix B.   □

15 COMBINATORIAL HUBBARD TREE FOR A RATIONAL ANGLE θ

We have now seen the combinatorial encodings of Julia sets in $\mathbb{S}$ and $\mathbb{T}$. Let us turn now to the corresponding encodings of Hubbard trees. There are actually two characterizations of a Hubbard tree: one by looking at ray-pairs separating the postcritical set, the other by looking at forward images of ray-pair landing points.

15.1 Combinatorial Hubbard tree

Note that the Julia points of the Hubbard tree $\mathcal{H}_f$ can be considered as the set of Julia points that "separate" the postcritical set $\mathcal{P}_f$. Such Julia points receive necessarily at least two external rays. One can therefore define a combinatorial counterpart of the postcritical set and the Hubbard tree for quadratic maps as follows.

Let $\theta \in \mathbb{S}$ be a rational angle. Define the post-major angle set to be:

$$P_\theta^{\mathbb{S}} = \{\tau^n \theta \mid n \geq -1\} \quad \text{and} \quad P_\theta = \big\{(\alpha, \alpha) \in \mathbb{T} \mid \alpha \in P_\theta^{\mathbb{S}}\big\}.$$

As θ is rational, the set P_θ is a finite F-forward-invariant set in the diagonal of the torus (also $P_\theta^{\mathbb{S}}$ is a finite τ-forward-invariant set).[3]

For $(x, y) \in \mathbb{T}$, we say that (x, y) *separates* P_θ if $x \neq y$ and both components of $\Delta \setminus \{(x, x), (y, y)\}$ contain points of P_θ, or equivalently, both components of $\mathbb{S} \setminus \{x, y\}$ contain points of $P_\theta^{\mathbb{S}}$. Visually, this is equivalent to assert that both open tan squares in $\mathbb{T}$ in Figure 4 generated by (x, y) contain points of P_θ in the interior. We say that (x, y) *intersects* P_θ if either (x, x) or (y, y) is in P_θ.

Recall that

$$L_\theta = \mathrm{cluster}(\mathrm{Pre}_\theta) \setminus \Delta = \text{the cluster set in } \mathbb{T} \setminus \Delta \text{ of the pre-major leaves.}$$

Set

$$X_\theta = \big\{\, (x, y) \in L_\theta \mid (x, y) \text{ intersects or separates } P_\theta \big\}.$$

Definition 15.1 (Combinatorial Hubbard tree). *The set $H_\theta := X_\theta \cup P_\theta$ is called the combinatorial Hubbard tree of θ.*

---

[3]More formally, P_θ should be written $P_\theta^{\mathbb{T}}$. Since we work more often on the torus model, we omit the superscript $\mathbb{T}$.

It is known that the Hubbard tree of a postcritically finite polynomial is an attracting core of the landing points of ray-pairs, in the sense that some forward iterate of any such point will be in the Hubbard tree and stay there under further iterations. Here is the combinatorial counterpart.

Proposition 15.2. *Let $\theta \in \mathbb{S}$ be a rational angle. The set X_θ is an attracting core of L_θ, in the sense that $X_\theta \subset L_\theta$ and for any $(x,y) \in L_\theta$, there is $n \geq 0$ so that $F^n(x,y) \in X_\theta$.*

Proof. By definition, $X_\theta \subset L_\theta$.

For a pair of points $x, y \in \mathbb{S}$, let us define the distance of x, y, denoted by $|x - y|$, to be the arc-length of the shortest of the two arcs in $\mathbb{S}$ determined by x, y. Let $(x,y) \in L_\theta$. Note that if $|x - y| \leq \frac{1}{4}$, then

$$|\tau(x) - \tau(y)| = 2|x - y|.$$

So there exists a minimal $n \geq 0$ such that $|\tau^n(x) - \tau^n(y)| > \frac{1}{4}$. In this case,

$$\frac{1}{2} \geq |\tau^n(x) - \tau^n(y)| > \frac{1}{4}.$$

It follows that the shorter closed arc between $\tau^{n+1}(x)$ and $\tau^{n+1}(y)$ must contain the point θ. Since the leaf $\overline{\tau^{n+1}(x)\tau^{n+1}(y)}$ belongs to the closure of a component of

$$\mathbb{D} \setminus \overline{\tfrac{\theta}{2} \tfrac{\theta+1}{2}},$$

it follows that the leaf $\overline{\tau^{n+1}(x)\tau^{n+1}(y)}$ separates or intersects

$$P_\theta^{\mathbb{D}} := \{e^{2\pi i t} \mid t \in P_\theta^{\mathbb{S}}\}. \qquad \square$$

15.2 Relation between combinatorial trees and pre-major leaves

Fix a rational angle $\theta \in \mathbb{S}$. It is relatively easy to find in the countable set Pre_θ of points those that separate or intersect the post-major angle set. We will see in Proposition 15.4 that, starting from these points, we can recover the combinatorial Hubbard tree. Set

$$S_\theta := \big\{(x,y) \in \mathrm{Pre}_\theta \mid (x,y) \text{ separates or intersects } P_\theta\big\}$$
$$= \big\{\text{pre-major leaves separating or intersecting post-major angles}\big\}$$
$$\mathrm{cluster}(S_\theta) := \text{the set of cluster points of } S_\theta \text{ in } \mathbb{T}.$$

Lemma 15.3. *Any $(x,y) \in \mathrm{cluster}(S_\theta)$ intersects or separates P_θ.*

Proof. This is an easy consequence of the fact that P_θ is finite. $\qquad \square$

Proposition 15.4. *The combinatorial Hubbard tree H_θ and the set $\mathrm{cluster}(S_\theta)$ are both compact and F-forward-invariant. Moreover the two sets differ by a finite set.*

Proof. The compactness of cluster(S_θ) is obvious since $\mathbb{T}$ is compact. We proceed to prove that it is F-forward-invariant.

We show first that the union of the post-major set P_θ with the set S_θ is F-forward-invariant. As Pre$_\theta$ is the set of pre-major leaves, if $(x,y) \in$ Pre$_\theta$, either $F(x,y) \in$ Pre$_\theta$ or $F(x,y) \in P_\theta$. So we are left to prove that if (x,y) separates or intersects P_θ, so does $(\tau(x), \tau(y))$.

Denote, as before, the set

$$P_\theta^{\mathbb{D}} = \{e^{2\pi i t} \mid t \in P_\theta^{\mathbb{S}}\},$$

which is the post-major angle set in the unit disc model. Note that (x,y) separates or intersects P_θ if and only if $\overline{xy}$ separates or intersects $P_\theta^{\mathbb{D}}$ in $\overline{\mathbb{D}}$. If $\overline{xy}$ intersects $P_\theta^{\mathbb{D}}$, the leaf $\overline{\tau(x)\tau(y)}$ must intersect $P_\theta^{\mathbb{D}}$. If $\overline{xy}$ separates (in $\overline{\mathbb{D}}$) but does not intersect $P_\theta^{\mathbb{D}}$, the leaf $\overline{xy}$ belongs to a component W of

$$\overline{\mathbb{D}} \setminus \overline{\tfrac{\theta}{2} \tfrac{\theta+1}{2}}.$$

Then, $\overline{xy}$ must separate a point $e^{2\pi i t}$ of $P_\theta^{\mathbb{D}}$ and the point $e^{2\pi i \frac{\theta}{2}}$. It follows that $\overline{\tau(x)\tau(y)}$ separates the two points $e^{2\pi i \tau(t)}, e^{2\pi i \theta} \in P_\theta^{\mathbb{D}}$. This shows that $S_\theta \cup P_\theta$ is F-forward-invariant.

Second, we will show that cluster(S_θ) is F-forward-invariant. Take $\{(x_n, y_n)\}$ a sequence of distinct points in S_θ such that

$$\lim_{n \to \infty} (x_n, y_n) = (x,y) \in \text{cluster}(S_\theta).$$

According to the F-invariant property of $S_\theta \cup P_\theta$ proved above, the sequence of points $\{F(x_n, y_n)\}$ belong to $S_\theta \cup P_\theta$. As P_θ has only finitely many F-preimages, countably many of $\{F(x_n, y_n)\}$ must belong to S_θ. Since

$$\lim_{n \to \infty} F(x_n, y_n) = F(x,y),$$

we have $F(x,y) \in \text{cluster}(S_\theta)$.

The proof that H_θ is compact and F-forward-invariant is similar and left as an exercise.

By Lemma 15.3 we have cluster(S_θ) $\subset H_\theta$.

Let us show that $H_\theta \setminus \text{cluster}(S_\theta)$ is also finite. Note that if a point $(x,y) \in X_\theta$ separates P_θ, then $(x,y) \in L_\theta$, which is the cluster set in $\mathbb{T} \setminus \Delta$ of Pre$_\theta$. One can thus find a sequence $\{(x_n, y_n)\}$ in Pre$_\theta$ converging to (x,y). So for all large n, the point (x_n, y_n) must separate the finite post-major set P_θ. It follows by definition that $(x_n, y_n) \in S_\theta$, and therefore $(x,y) \in \text{cluster}(S_\theta)$. So a point of $X_\theta \setminus \text{cluster}(S_\theta)$ must intersect but not separate P_θ.

On the other hand, since L_θ is the polynomial lamination of f_θ (by Proposition 14.1), each angle belongs to at most two adjacent angle-pairs. In particular, for any point $(p,p) \in P_\theta$, there are at most two points of L_θ intersecting (p,p). So $X_\theta \setminus \text{cluster}(S_\theta)$, and therefore $H_\theta \setminus \text{cluster}(S_\theta)$, is a finite set. $\square$

16 ENTROPIES

The objective is to prove that several quantities in the combinatorial and dynamical worlds all encode the core entropy of a quadratic polynomial.

16.1 Various entropies and Hausdorff dimensions

Let $\theta \in \mathbb{S}$ be a rational angle. Recall that for the polynomial $f_\theta : z \mapsto z^2 + c_\theta$, we use $\gamma(\eta)$ to denote the landing point of the external ray of angle η, and $\mathcal{H}_\theta$ to denote its Hubbard tree.

We have several natural actions (τ on the circle $\mathbb{S}$, F on the torus $\mathbb{T}$, and f_θ on the Hubbard tree $\mathcal{H}_\theta$). We have also defined the polynomial lamination L_θ and the combinatorial Hubbard tree H_θ, both in $\mathbb{T}$.

The set H_θ is compact and F-forward-invariant, and thus has a topological entropy $h(H_\theta, F)$.

The polynomial f_θ induces two more combinatorial sets[4]

- the angle-pair set

$$B_\theta = \{(\beta, \eta) \in \mathbb{T} \mid \beta \neq \eta \text{ and } \gamma(\beta) = \gamma(\eta)\}$$

- and its projection to the circle

$$B_\theta^{\mathbb{S}} = \{\beta \in \mathbb{S} \mid \exists \; \eta \neq \beta \text{ s.t } \gamma(\beta) = \gamma(\eta)\}.$$

There is also a non-escaping set NE, defined later.
We can then consider entropies or Hausdorff dimensions of these objects.

The Actions	$F : (\eta, \zeta) \mapsto (2\eta, 2\zeta)$	f_θ	$\tau : \eta \mapsto 2\eta$	On a Graph
Entropies On Trees		$h(H_\theta)$ $h(\mathcal{H}_\theta)$		
Dimensions On trees		$HD(H_\theta)$		
Dimensions On Bi-acc Sets	$HD(L_\theta)$ $HD(NE \setminus \Delta)$	$HD(B_\theta)$	$HD(B_\theta^{\mathbb{S}})$	
Algorithm				$\rho(A_\theta)$

Theorem 16.1. *For a rational angle $\theta \in \mathbb{S}$, all the quantities in the above table (as well as several other quantities detailed in the proof) are related by*

$$\log \rho(A_\theta) = h(\cdot) = HD(*) \cdot \log 2.$$

---

[4]These sets are sometimes referred to in the literature as the set of bi-accessible angles.

Proof. We will establish the equalities following the schema:

$$\log \rho(A_\theta)$$

$$\Big\| \text{ by Thm 13.10}$$

$$h(H_\theta, F) = \quad h(\mathcal{H}_\theta, f_\theta) \qquad \text{(entropies)}$$

$$\uparrow \cdot \log 2 \qquad \uparrow \cdot \log 2$$

$$\text{HD}(H_\theta) \qquad \text{HD}(B_\theta^S) \qquad \text{(dimensions)}$$

$$\Big\|$$

$$\text{HD}(L_\theta) = \text{HD}(NE \setminus \Delta) = \text{HD}(B_\theta)$$

$$\text{(by Prop. 16.5 below)}$$

For simplicity set $f = f_\theta$ and $\mathcal{H} = \mathcal{H}_\theta$.

- $h(H_\theta, F) = h(\mathcal{H}, f)$.

  By Proposition 15.4, the set H_θ is compact F-invariant. By Proposition 14.1 the set L_θ is equal to the polynomial lamination of f. By the definition of H_θ, there is a continuous surjective and finite-to-one map $\varphi : H_\theta \to \mathcal{H} \cap \mathcal{J}$ realizing a semi-conjugacy: $f \circ \varphi = \varphi \circ F$. By Proposition A.3, we have that

$$h(H_\theta, F) = h(\mathcal{H} \cap \mathcal{J}, f).$$

If θ is preperiodic, $\mathcal{H} \cap \mathcal{J} = \mathcal{H}$ so

$$h(H_\theta, F) = h(\mathcal{H}, f).$$

If θ is periodic, the points in $\mathcal{H} \setminus \mathcal{J}$ are attracted to the critical orbit $\mathcal{P}_f$, on which f has 0 entropy. So

$$h(\mathcal{H}, f) = h\Big((\mathcal{H} \cap \mathcal{J}) \cup \mathcal{P}_f, f\Big) \quad \text{by Proposition A.2}$$

$$= \max\Big(h(\mathcal{H} \cap \mathcal{J}, f), h(\mathcal{P}_f, f)\Big) \quad \text{by Proposition A.1}$$

$$= h(\mathcal{H} \cap \mathcal{J}, f) = h(H_\theta, F) \quad \text{by Proposition A.3.}$$

Note that this entropy is also equal to

$$h(\text{cluster}(S_\theta), F),$$

as $\text{cluster}(S_\theta)$ and H_θ differ by a finite set (Proposition 15.4).

- $\text{HD}(H_\theta) \cdot \log 2 = h(H_\theta, F)$.

  This follows directly from Proposition 15.4 and Lemma A.4.

- $\text{HD}(B_\theta^S) \cdot \log 2 = h(\mathcal{H}, f)$.

  This fact was first noticed by Tiozzo [Tio15, Tio16]. It is based on the fact that the set B_θ^S has an attracting core which is the set H_θ^S of angles whose rays land at the Hubbard tree, so has the same dimension. One can then relate the dimension of H_θ^S to entropy of τ on H_θ^S by Lemma A.4, and then relates to

the entropy on the Hubbard tree via the ray-landing semi-conjugacy which is finite-to-one (so one can apply Proposition A.3).
- $HD(L_\theta) = HD(X_\theta)$ since X_θ is an attracting core of L_θ (Proposition 15.2).
- $HD(L_\theta) = HD(NE \setminus \Delta) = HD(B_\theta)$ by Proposition 16.5 below. $\square$

16.2   Non-escaping sets on the torus

Let us fix an angle $\theta \in \mathbb{S}$ (not necessarily rational). We define two invariant subsets mimicking filled Julia sets as follows:

Definition (Non-escaping sets). Set inductively

$$\Omega_0 := \text{ the interior of the good region } G\left(\frac{\theta}{2}, \frac{\theta+1}{2}\right)$$
$$\Omega_{i+1} := F^{-1}(\Omega_i) \cap \Omega_0$$
$$NE_1(\theta) := \bigcap_{n \geq 0} \Omega_n$$
$$NE_2(\theta) := \bigcap_{n \geq 0} \overline{\Omega_n}.$$

In other words, $NE_1(\theta)$ consists of the points in $\mathbb{T}$ whose orbits never escape Ω_0.

Lemma 16.2. *We have* $\Delta \subset \overline{NE_1}(\theta) \subset NE_2(\theta)$.

Proof. We omit θ for simplicity.

For each $n \geq 0$, the diagonal Δ is contained in $\overline{\Omega_n}$, and the set $\Delta \setminus \Omega_n$ is finite. It follows that $\Delta \subset NE_2$ and $\Delta \setminus NE_1$ is a countable set and hence $\Delta \subset \overline{NE_1}$. The inclusion $\overline{NE_1} \subset NE_2$ is obvious. $\square$

For $\theta \neq 0$, the diameter $\overline{\frac{\theta}{2} \frac{\theta+1}{2}}$ subdivides the unit circle into two half open halves. Denote by S_0 the half circle containing the angle 0 and by S_1 the other one. Set

$$\text{Pre}_\theta^{\mathbb{S}} = \bigcup_{n \geq 1} \tau^{-n}(\theta).$$

For any angle $\alpha \in \mathbb{R}/\mathbb{Z} \setminus \text{Pre}_\theta^{\mathbb{S}}$, we can assign a sequence $\epsilon_0 \epsilon_1 \ldots \in \{0,1\}^{\mathbb{N}}$ such that $\epsilon_i = \delta$ if and only if $\tau^i(\alpha) \in S_\delta$ ($\delta = 0$ or 1). This sequence is said to be the *itinerary of α relative to θ*, denoted by $\iota_\theta(\alpha)$.

Let θ be a rational angle. There is a parallel description in the dynamical plane of $f : z \mapsto z^2 + c_\theta$.

If θ is preperiodic, the external rays $\mathcal{R}\left(\frac{\theta}{2}\right)$ and $\mathcal{R}\left(\frac{\theta+1}{2}\right)$ both land at the critical point 0. In this case, we set

$$\mathcal{R}\left(\frac{\theta}{2}, \frac{\theta+1}{2}\right) := \mathcal{R}\left(\frac{\theta}{2}\right) \cup \{0\} \cup \mathcal{R}\left(\frac{\theta+1}{2}\right).$$

This set is called the *major cutting line*.

If θ is periodic, say of period p, the rays $\mathcal{R}\left(\frac{\theta}{2}\right)$ and $\mathcal{R}\left(\frac{\theta+1}{2}\right)$ support the Fatou component U that contains 0. The period of U is also p. The first return map

$f^p : \overline{U} \to \overline{U}$ is conjugated to the action of z^2 on $\overline{\mathbb{D}}$, and thus admits a unique fixed point on the boundary, called the root of U. This root point has internal angle 0, and its unique f^p-preimage on ∂U has internal angle $1/2$. Denote by $r_U(0)$ and $r_U(1/2)$ the corresponding internal rays in U. They land at $\gamma\left(\frac{\theta}{2}\right)$, $\gamma\left(\frac{\theta+1}{2}\right)$ respectively (or vice versa, depending on which of $\frac{\theta}{2}$, and $\frac{\theta+1}{2}$ is periodic). In this case, we define the *major cutting line* to be

$$\mathcal{R}\left(\frac{\theta}{2}, \frac{\theta+1}{2}\right) := \mathcal{R}\left(\frac{\theta}{2}\right) \cup \overline{r_U(0) \cup r_U(1/2)} \cup \mathcal{R}\left(\frac{\theta+1}{2}\right).$$

In both cases, $\mathcal{R}\left(\frac{\theta}{2}, \frac{\theta+1}{2}\right)$ is a simple curve which subdivides the complex plane into two open regions. Assume $\theta \neq 0$. Denote by V_0 the region containing the ray $\mathcal{R}(0)$ and by V_1 the other region. For an angle

$$\alpha \in \mathbb{R}/\mathbb{Z} \setminus \mathrm{Pre}_\theta^{\mathbb{S}},$$

it is easy to see that $\iota_\theta(\alpha) = \epsilon_0 \epsilon_1 \ldots$ if and only if

$$f_\theta^i(\gamma_{c_\theta}(\alpha)) \subset V_{\epsilon_i},$$

for all $i \geq 0$.

Set

$$\mathrm{Pre\text{-}cr} = \bigcup_{n \geq 1} f^{-n}(\gamma(\theta)).$$

For $z \in \mathcal{J} \setminus \mathrm{Pre\text{-}cr}$, the orbit of z does not meet the major cutting line. So we can define its itinerary relative to this cutting line:

$$\iota_\theta(z) = \epsilon_0 \epsilon_1 \epsilon_2 \ldots \qquad \text{if} \quad f^i(z) \in V_{\epsilon_i}.$$

It is easy to see that if $z \in \mathcal{J} \setminus \mathrm{Pre\text{-}cr}$ has an external angle α, then $\iota_\theta(\alpha) = \iota_\theta(z)$.

The following known result will be useful for us (see Appendix C for a proof):

Lemma 16.3. *Let θ be a rational angle, and let $\alpha \neq \beta \in \mathbb{R}/\mathbb{Z} \setminus \mathrm{Pre}_\theta^{\mathbb{S}}$. Then the rays $\mathcal{R}(\alpha)$ and $\mathcal{R}(\beta)$ land together if and only if $\iota_\theta(\alpha) = \iota_\theta(\beta)$.*

Corollary 16.4. *For $z_1, z_2 \in \mathcal{J} \setminus \mathrm{Pre\text{-}cr}$, then $z_1 \neq z_2$ if and only if $\iota_\theta(z_1) \neq \iota_\theta(z_2)$.*

Proposition 16.5. *For θ a rational angle, the four sets L_θ, $NE_1(\theta) \setminus \Delta$, $NE_2(\theta) \setminus \Delta$ and B_θ are pairwise different by a set that is at most countable. Consequently, they have the same dimension.*

Proof. Fix a rational angle θ. Let us introduce a new set

$$Z = \{(x, y) \in \mathbb{T} \mid \{x, y\} \text{ is an angle-pair (not necessarily adjacent)}$$

$$\text{at a point in } \mathcal{J} \setminus \mathrm{Pre\text{-}cr}\}.$$

We will compare this set with the four sets L_θ, $NE_1 \setminus \Delta$, $NE_2 \setminus \Delta$ and B_θ. The notation "a.e" below on top of the equal sign means that the symmetric difference between the two sets has measure zero.

1. $NE_1 \setminus \Delta = Z$.

By the definition of NE_1, a point $(x, y) \in NE_1 \setminus \Delta \iff x \neq y$ and the orbit of (x, y) stays in $\Omega_0 \iff x \neq y$ and $\iota_\theta(x) = \iota_\theta(y) \overset{\text{Lemma } 16.3}{\iff}$ the rays $\mathcal{R}(x)$ and $\mathcal{R}(y)$ land at a common point $z \in \mathcal{J} \setminus \text{Pre-cr} \iff (x, y) \in Z$.

2. $NE_1 \setminus \Delta \overset{a.e}{=} NE_2 \setminus \Delta$.

We already know $NE_1 \setminus \Delta \subset NE_2 \setminus \Delta$. Let us fix a point (x, y) in $NE_2 \setminus (NE_1 \cup \Delta)$. By the definition of NE_2 and NE_1, we have that

$$(x, y) \in \left(\bigcap_{n \geq 0} \overline{\Omega}_n \right) \setminus \left(\bigcap_{k \geq 0} \Omega_k \right)$$

so

$$(x, y) \in \overline{\Omega}_n \ \forall n \quad \text{and} \quad (x, y) \notin \Omega_N \text{ for some } N \geq 0.$$

But Ω_k is decreasing with respect to k (it is the set of points that remain in Ω_0 up to k iterates). So for every $n \geq N$,

$$(x, y) \in \overline{\Omega}_n \setminus \Omega_n = \partial \Omega_n.$$

We can then select a point $(x_n, y_n) \in \Omega_n$ such that $\{(x_n, y_n), n \geq 1\}$ converges to (x, y) as n goes to infinity. Set

$$z_n = \gamma(x_n), \quad w_n = \gamma(y_n), \quad z = \gamma(x), \quad w = \gamma(y).$$

We obtain that

$$\lim_{n \to \infty} z_n = z \quad \text{and} \quad \lim_{n \to \infty} w_n = w.$$

It is known that f_θ is uniformly expanding on a neighborhood of $\mathcal{J}$ with respect to the orbifold metric of f_θ (see [DH84, McM94, Mil06]). Since $(x_n, y_n) \in \Omega_n$, the first n-entries of $\iota_\theta(x_n)$ and $\iota_\theta(y_n)$ are identical. By the argument in the proof of Lemma 16.3, the orbifold distance of z_n and w_n is less than $C\lambda^{-n}$ for some constant C and $\lambda > 1$. It follows that $z = w$ and thus (x, y) is an angle-pair.

On the other hand, by an inductive argument, one can see that for a point $(p, q) \in \partial \Omega_n$, either p or q belongs to the set $\bigcup_{i=1}^{n+1} \tau^{-i}(\theta)$. So, in our case, either x or y belongs to $\text{Pre}_\theta^{\mathbb{S}}$. It follows that

$$\gamma(x) = \gamma(y) \in \text{Pre-cr}.$$

It is known that the angle-pairs at points in Pre-cr form a countable set, so we obtain that $NE_1 \setminus \Delta \overset{a.e}{=} NE_2 \setminus \Delta$.

3. $L_\theta \overset{a.e}{=} Z$.

By Proposition 14.1, the set L_θ is the lamination of f_θ. Then a point $(x, y) \in L_\theta \setminus Z$ represents an angle-pair at a common point of Pre-cr, so

$$\text{cluster}(\text{Pre}_\theta) \setminus (\Delta \cup \setminus Z)$$

is a countable set. On the other hand, the set $Z \setminus L_\theta$ consists of angle-pairs that are not adjacent and that land at non-precritical points. Then the landing points each receive at least 4 external rays and the number of rays remain constant under forward iteration. But each point receiving at least two rays will be mapped eventually into the Hubbard tree and stay there. So a landing point of angle-pairs in $Z \setminus L_\theta$ is mapped eventually to a branching point of the Hubbard tree. As the tree is finite, there are only finitely many branching points forming finitely many forward-invariant orbits. So the set of such landing points is countable, and each point receives finitely many ray-pairs. It follows that the set $Z \setminus L_\theta$ is countable.

4. Finally, $B_\theta \supset Z$ and $B_\theta \setminus Z$ is countable as it concerns only angle-pairs whose rays land at the countable set Pre-cr.   □

APPENDIX A. BASIC RESULTS ABOUT ENTROPY

The following can be found in [Dou95, dMvS93]. Let X be a compact topological space, $f : X \to X$ a continuous map.

Proposition A.1. *If $X = X_1 \cup X_2$, with X_1 and X_2 compact, $f(X_1) \subset X_1$ and $f(X_2) \subset X_2$, then*

$$h(X, f) = \sup\big(h(X_1, f), h(X_2, f)\big).$$

Proposition A.2. *Let Z be a closed subset of X such that $f(Z) \subset Z$. Suppose that for any $x \in X$, the distance of $f^n(x)$ to Z tends to 0 uniformly on any compact set in $X \setminus Z$. Then $h(X, f) = h(Z, f)$.*

In the setting of Proposition A.2, we say that Z is an *attracting core* of f.

Proposition A.3. *Assume that π is a surjective semi-conjugacy*

$$
\begin{array}{ccc}
Y & \xrightarrow{\ q\ } & Y \\
\pi \downarrow & & \downarrow \pi \\
X & \xrightarrow{\ f\ } & X.
\end{array}
$$

We have $h(X, f) \leq h(Y, q)$. Furthermore, if $\sup\limits_{x \in X} \#\pi^{-1}(x) < \infty$, then $h(X, f) = h(Y, q)$.

The following can be found in [Fur67].

Lemma A.4. *Let A be a compact τ-invariant subset of $\mathbb{S}$. Then*

$$e^{h(A,\tau)} = 2^{HD(A)}.$$

Similarly, if A is a compact F-invariant subset of $\mathbb{T}$,

$$e^{h(A,F)} = 2^{HD(A)}.$$

Definition A.5. (Finite connected graph). *A finite connected graph is a connected topological space G consisting of the union of the finite vertex set V_G and the edge set E_G with the properties that*

- *each edge $e \in E_G$ is homeomorphic to a closed interval and connects two points of V_G; these two vertices are called the ends of e;*
- *the interior of an edge does not contain the vertices and two edges can intersect only at their ends.*

A *finite tree* is a finite connected graph without cycles.

Let T be a finite tree. A point p in T is called an *endpoint* if $T \setminus \{p\}$ is connected and is called a *branching point* if $T \setminus \{p\}$ has at least 3 connected components. An important property of the finite tree is that for any two points $p, q \in T$, there is a unique arc in T that connects p and q. This arc is denoted by $[p, q]_T$. For a subset $X \subset T$, denote by $[X]_T$ the connected hull of X in T.

Let X be a finite connected graph. We call $f : X \to X$ a *Markov map* if each edge of X admits a finite subdivision into segments and f maps each segment continuously monotonically onto some edge of X.

Enumerating the edges of X by γ_i, $i = 1, \ldots, k$, we obtain a *transition matrix* $D_f = (a_{ij})_{k \times k}$ of (X, f) such that $a_{ij} = \ell$ ($\ell \geq 1$) if $f(\gamma_i)$ covers ℓ times γ_i and $a_{ij} = 0$ otherwise. Note that for different enumerations of the edges, the obtained transition matrices are pairwise similar. Denote by $\rho(D_f)$ the leading eigenvalue of D_f. By the Perron-Frobenius theorem, $\rho(D_f)$ is a non-negative real number and it is also the growth rate of $\|D_f^n\|$ for any matrix norm.

Since the matrix D_f has integer coefficients, we have $\rho(D_f) \geq 1$ unless there exists n such that $D_f^n = 0$ (nilpotent). But the latter case cannot happen because every edge is mapped onto at least one edge. It follows that we have always $\rho(D_f) \geq 1$.

The following is classical:

Proposition A.6. *The topological entropy $h(X, f)$ is equal to $\log \rho(D_f)$.*

APPENDIX B. PROOF OF PROPOSITION 14.1

For a rational angle $\theta \neq 0$, we want to prove that L_θ, the cluster set in $\mathbb{T} \setminus \Delta$ of the pre-major lamination, is equal to the polynomial lamination of $f : z \mapsto z^2 + c_\theta$.

Proof. To get a good geometric intuition of an invariant lamination, we choose to use the unit disk model. Let

$$\wp : \mathbb{T} \to \{\text{closures of hyperbolic geodesics in } \mathbb{D}\}$$

map a point (x, y) to $\overline{xy}$, which is called a *leaf* (or a point in the case $x = y$). Denote the image of Pre_θ and L_θ under $\wp$ by $\mathrm{Pre}_\theta^{\mathbb{D}}$ and $L_\theta^{\mathbb{D}}$, respectively. For each $i \geq 0$, define

$$L_0 = \wp(b_0) \qquad \text{and} \qquad L_i = \wp(b_i \setminus b_{i-1}), \quad i \geq 1.$$

Then we have

$$L_0 = \left\{ \overline{\tfrac{\theta}{2} \tfrac{\theta+1}{2}} \right\}$$

and $\tau : L_{i+1} \to L_i$ is at least 2 to 1. (Here by abuse of notation we set $\tau(\overline{xy}) = \overline{\tau(x)\tau(y)}$.)

The leaf $\overline{\tfrac{\theta}{2} \tfrac{\theta+1}{2}}$ corresponds in the dynamical plane to the major cutting line $\mathcal{R}\left(\tfrac{\theta}{2}, \tfrac{\theta+1}{2} \right)$.

For any $n \geq 1$, let $\overline{x_n y_n}$ be a leaf of L_n. Set

$$(x_{n-i}, y_{n-i}) := F^i(x_n, y_n), \quad i \in [0, n].$$

Then $\overline{x_i y_i}$ is a leaf of L_i. For each i, the set $\{x_i, y_i\}$ belongs to $\overline{S_0}$ or $\overline{S_1}$. (Recall that S_0 denotes the half open circle containing 0 bounded by $\tfrac{\theta}{2}, \tfrac{\theta+1}{2}$.) The ray-pair $\mathcal{R}(x_i) \cup \mathcal{R}(y_i)$ belongs to the corresponding component of $\overline{V_0}, \overline{V_1}$. Inductively, this follows:

- If θ is preperiodic, each $\{x_i, y_i\}$ is an angle-pair at a point z_i with $f_\theta(z_i) = z_{i-1}$. In this case, we define a cutting curve corresponding to $\overline{x_i y_i}$ by

$$\mathcal{R}(x_i, y_i) := \mathcal{R}(x_i) \cup \{z_i\} \cup \mathcal{R}(y_i)$$

- If θ is periodic, each $\{x_i, y_i\}$ is a pair of angles whose rays support a common Fatou component U_i, and $f(U_i) = U_{i-1}$. In this case, we define a cutting curve corresponding to $\overline{x_i y_i}$ by

$$\mathcal{R}(x_i, y_i) := \mathcal{R}(x_i) \cup \overline{r_{U_i}(\alpha_i) \cup r_{U_i}(\beta_i)} \cup \mathcal{R}(y_i)$$

where $r_{U_i}(\alpha_i)$ (resp. $r_{U_i}(\beta_i)$) is the internal ray in U_i landing at $\gamma(x_i)$ (resp. $\gamma(y_i)$).

For $i \geq 0$, set

$$\Gamma_i = \left\{ \mathcal{R}(x, y) \,\middle|\, \overline{xy} \in L_i \right\} \quad \text{and} \quad \Gamma_\infty = \bigcup_{i \geq 0} \Gamma_i.$$

Then there is a natural bijection $l_i : L_i \to \Gamma_i$ which maps a leaf $\overline{xy} \in L_i$ to the curve $\mathcal{R}(x, y)$ and satisfies the commutative diagram:

$$
\begin{array}{ccc}
L_i & \xrightarrow{\tau} & L_{i-1} \\
\downarrow{\scriptstyle l_i} & & \downarrow{\scriptstyle l_{i-1}} \\
\Gamma_i & \xrightarrow{f_\theta} & \Gamma_{i-1}.
\end{array}
$$

We will talk a little more about the curves in Γ_∞ for a periodic angle θ. In this case, there is the unique Fatou component cycle of degree 2 which contains the critical point 0. Then the internal angle for each Fatou component can be uniquely defined. Using the fact that the two internal rays contained in $\mathcal{R}\left(\tfrac{\theta}{2}, \tfrac{\theta+1}{2} \right)$ have angles 0 and 1/2, together with the construction of L_i $(i \geq 1)$, the correspondence between L_i and Γ_i, and the definition of internal angles of a Fatou component, it is not difficult to check the following results:

- Every curve $\mathcal{R}(x, y) \in \Gamma_\infty$ passes through a single Fatou component U, and the rays $\mathcal{R}(x)$, $\mathcal{R}(y)$ support the component U.
- The two internal rays of U contained in $\mathcal{R}(x, y)$ have angles $\frac{j}{2^n}, \frac{j+1}{2^n}$ for some $n \geq 1, 0 \leq j \leq 2^n - 1$.
- For any Fatou component U, any integer $n \geq 1$ and $0 \leq j \leq 2^n - 1$, there exists a unique curve $\mathcal{R}(x, y) \in \Gamma_\infty$ such that it contains the two internal rays $r_U\left(\frac{j}{2^n}\right)$ and $r_U\left(\frac{j+1}{2^n}\right)$. In fact, there exists a minimal k such that f_θ^k sends $r_U(\frac{j}{2^n})$ and $r_U(\frac{j+1}{2^n})$ to the internal rays contained in $\mathcal{R}(\frac{\theta}{2}, \frac{\theta+1}{2})$. Suppose that $f_\theta^k(r_U(\frac{j}{2^n}))$ is the periodic one. Then $\mathcal{R}(x, y)$ is a lift by f_θ^k of $\mathcal{R}(\frac{\theta}{2}, \frac{\theta+1}{2})$ based at $r_U(\frac{j}{2^n})$.

With these preparations, we are ready to prove the proposition. First, we will show that each leaf in $L_\theta^{\mathbb{D}}$ represents an adjacent angle-pair.

Let $\overline{xy}$ be a leaf in $L_\theta^{\mathbb{D}}$. Then there exists a sequence of leaves $\overline{x_i y_i} \in L_i$ such that $\{\overline{x_i y_i}, i \geq 1\}$ converges to $\overline{xy}$ in the Hausdorff topology. Without loss of generality, we may assume that

$$\lim_{n \to \infty} x_n = x, \qquad \lim_{n \to \infty} y_n = y.$$

To see the dynamical explanation of the point (x, y), we need to discuss the set of accumulation points R of the corresponding curves $\mathcal{R}(x_i, y_i)$. At first, we only consider the case that $\overline{x_i y_i}$ converges to $\overline{xy}$ from one side. We distinguish 3 basic cases.

1. No curves in $\{\mathcal{R}(x_i, y_i) \mid i \geq 1\}$ pass through Fatou components. In this case, the external rays $\mathcal{R}(x_i)$ converge to $\mathcal{R}(x)$ and the rays $\mathcal{R}(y_i)$ converge to $\mathcal{R}(y)$. Consequently, the common landing point z_i of $\mathcal{R}(x_i)$ and $\mathcal{R}(y_i)$ converge to a point $z \in \mathcal{J}_f$, which must be the common landing point of $\mathcal{R}(x)$ and $\mathcal{R}(y)$. Then we have

$$R := \lim_{i \to \infty} \mathcal{R}(x_i, y_i) = \mathcal{R}(x) \cup \{z\} \cup \mathcal{R}(y).$$

We claim that $\{x, y\}$ is an adjacent angle-pair. Otherwise, each component of $\mathbb{C} \setminus \mathcal{R}(x, y)$ contains at least two components of $\mathcal{K}_f \setminus z$. Since $\lim_{i \to \infty}(x_i, y_i) = (x, y)$, for sufficiently large i, the points z_i belong to a common component of $\mathcal{K}_f \setminus z$. Set

$$\{z_i\}_{i \geq 1} \subset K \subset U,$$

where K is a component of $\mathcal{K}_f \setminus z$ and U is a component of $\mathbb{C} \setminus \mathcal{R}(x, y)$. Let $\{v, w\}$ be the angle-pair at z bounding K. On one hand, we know

$$[v, w] \subsetneqq [x, y],$$

since, by assumption, the pair $\{x, y\}$ is not adjacent. On the other hand, the angles x_i, y_i $(i \geq 1)$ belong to $[v, w]$ and hence the interval $[x, y]$ is a subset of $[v, w]$. This leads to a contradiction.

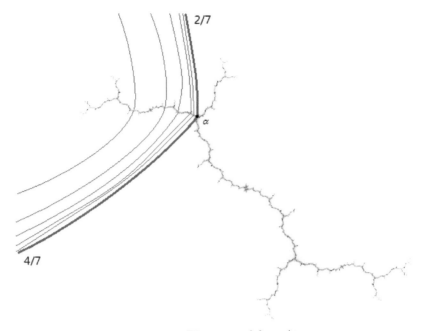

Figure 20: The case of $\theta = 1/4$.

Example B.1. Set $\theta = 1/4$. The red curves in Figure 20 are cutting curves $\{\mathcal{R}(x_i, y_i)\}$ satisfying case (1) such that $x_i \to 2/7, y_i \to 4/7$ as $n \to \infty$. It is seen that $\{\mathcal{R}(2/7), \mathcal{R}(4/7)\}$ is an adjacent ray-pair landing at the α-fixed point of $f_{1/4}$ and $R = \mathcal{R}(2/7) \cup \{\alpha\} \cup \mathcal{R}(4/7)$.

2. For each $i \geq 1$, the curve $\mathcal{R}(x_i, y_i)$ passes through a Fatou component U_i and the diameters of U_i ($i \geq 1$) converge to 0. In this case, the external rays $\mathcal{R}(x_i)$ converge to $\mathcal{R}(x)$, the rays $\mathcal{R}(y_i)$ converge to $\mathcal{R}(y)$ and the Fatou components U_i converge to a point $z \in \mathcal{J}_f$. This point must be the common landing point of $\mathcal{R}(x)$ and $\mathcal{R}(y)$. Then we have

$$R = \lim_{i \to \infty} \mathcal{R}(x_i, y_i) = \mathcal{R}(x) \cup \{z\} \cup \mathcal{R}(y).$$

Using the same argument as in (1) above, we see that the angle-pair $\{x, y\}$ is an adjacent angle-pair.

3. For each $i \geq 1$, the curve $\mathcal{R}(x_i, y_i)$ passes through a Fatou component U_i and the infimum of the diameters of U_i is bounded from 0. In this case, for sufficiently large i, and passing to a subsequence if necessary, all U_i coincide, and hence may be denoted by U. It follows that the two internal rays of U contained in the curves $\mathcal{R}(x_i, y_i)$ tend to a single internal ray as $i \to \infty$. So the two limit external rays $\mathcal{R}(x)$ and $\mathcal{R}(y)$ must land together. In fact,

$$R = \lim_{i \to \infty} \mathcal{R}(x_i, y_i) = \mathcal{R}(x) \cup \mathcal{R}(y) \cup r_U(\alpha) \cup \{z\}$$

where $z = \gamma(x) = \gamma(y)$ and $r_U(\alpha)$ is the internal ray in U landing at z. Since each pair of rays $\mathcal{R}(x_i)$, $\mathcal{R}(y_i)$ support the component U, the angle-pair $\{x, y\}$ must be an adjacent angle-pair.

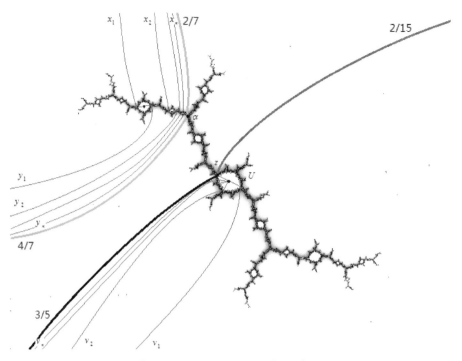

Figure 21: The case of $\theta = 4/15$.

Example B.2. Set $\theta = 4/15$. The blue curves in Figure 21 are cutting curves $\{\mathcal{R}(x_i, y_i)\}$ satisfying case (2) such that $x_i \to 2/7, y_i \to 4/7$ as $n \to \infty$. It is seen that $\{\mathcal{R}(2/7), \mathcal{R}(4/7)\}$ is an adjacent ray-pair landing at the α-fixed point of $f_{4/15}$ and $R = \mathcal{R}(2/7) \cup \{\alpha\} \cup \mathcal{R}(4/7)$. The red curves in Figure 21 are cutting curves $\{\mathcal{R}(2/15, v_i)\}$ satisfying case (3) such that $v_i \to 3/5$ as $n \to \infty$. It is seen that $\{\mathcal{R}(2/15), \mathcal{R}(3/5)\}$ is an adjacent ray-pair landing at z and $R = \mathcal{R}(2/15) \cup \overline{r_U(0)} \cup \mathcal{R}(3/5)$.

Generally, we allow the leaves $\overline{x_i y_i}$ to converge to $\overline{xy}$ from two sides. Then the sequence $\{(x_i, y_i)\}_{i \geq 1}$ may contain subsequences satisfying 1, 2 or 3 of the basic cases described above from either side of $\overline{xy}$. So the set of accumulation points R is formed by the possible combinations of the set of accumulation points in the basic cases from two sides. That is, R consists of an adjacent ray-pair $\{\mathcal{R}(x), \mathcal{R}(y)\}$ together with eventual 0, 1 or 2 internal rays. Anyway, a leaf $\overline{xy} \in L_\theta^{\mathbb{D}}$ corresponds to the unique adjacent angle-pair $\{x, y\}$.

Next, we will show the opposite implication. That is, any adjacent angle-pair $\{x, y\}$ is contained in $L_\theta^{\mathbb{D}}$ as a leaf $\overline{xy}$.

Let $\{x, y\}$ be an adjacent angle-pair at z. Let V be one of the regions bounded by $\mathcal{R}(x)$ and $\mathcal{R}(y)$ without other rays landing at z, set $K = V \cap \mathcal{K}$. Choose any point $w \in K$, denote by $[w, z]_K$ the regulated arc-connecting z and w in K. We distinguish two cases:

1. $[w, z]_K \cap \mathcal{J}$ clusters at z. Then two further cases may happen.

   - $[w, z]_K$ passes through an ordered infinite sequence of Fatou components $\{U_i, i \geq 1\}$ from w to z so that U_i converges to z as $i \to \infty$. We claim that for any $i \geq 2$, we can pick up a curve $\mathcal{R}(x_i, y_i) \in \Gamma_\infty$ such that it separates U_{i-1} and z (in Figure 21, we have $\{x, y\} = \{2/7, 4/7\}$ and the blue curves are what we want). By this claim, the sequence of curves $\mathcal{R}(x_i, y_i)$ $(i \geq 2)$ converge to the ray-pair of $\{x, y\}$ in the Hausdorff topology, so $\overline{xy} \in L_\theta^{\mathbb{D}}$.

Proof of this Claim: Fix U_i. The intersection of $U_i \cap [w, z]_K$ consists of two internal rays $r_{U_i}(\alpha)$ and $r_{U_i}(\beta)$ of U_i. We assume that $r_{U_i}(\alpha)$ lies in between z and $r_{U_i}(\beta)$. If α is not equal to $\frac{j}{2^n}$ for any $n \geq 1, j \geq 0$, we can choose integers n_0 and j_0 so that

$$\alpha \in \left(\frac{j_0}{2^{n_0}}, \frac{j_0 + 1}{2^{n_0}} \right) \quad \text{and} \quad \beta \notin \left[\frac{j_0}{2^{n_0}}, \frac{j_0 + 1}{2^{n_0}} \right].$$

Then the curve $\mathcal{R}(x_i, y_i) \in \Gamma_\infty$ that contains $r_{U_i}\left(\frac{j_0}{2^{n_0}}\right)$ and $r_{U_i}\left(\frac{j_0+1}{2^{n_0}}\right)$ separates U_{i-1} and z. If $\alpha = \frac{j_0}{2^{n_0}}$ for some integers n_0, j_0, we can choose a sufficiently large n and a suitable integer j such that

$$\alpha = j/2^n \quad \text{and} \quad \beta \notin \left[\frac{j-1}{2^n}, \frac{j+1}{2^n} \right].$$

In this case, either the curve (in Γ_∞) containing $r_{U_i}\left(\frac{j-1}{2^n}\right), r_{U_i}\left(\frac{j}{2^n}\right)$ or the curve (in Γ_∞) containing $r_{U_i}\left(\frac{j-1}{2^n}\right), r_{U_i}\left(\frac{j}{2^n}\right)$ separates U_{i-1} and z. Denote this curve by $\mathcal{R}(x_i, y_i)$.

- No subsequence of the Fatou components (if any) passed through by $[w, z]_K$ converges to z. Then, replacing w by a point closer to z if necessary, we may assume $[w, z]_K \subset \mathcal{J}$. In this case, we can pick up a sequence of points

$$\{z_i, \ i \geq 1\} \subset [w, z]_K \cap (\mathcal{J} \setminus \text{Pre-cr})$$

from w to z such that z_i $(i \geq 1)$ converges to z as $n \to \infty$. We claim that for any $i \geq 1$, we can pick up a curve $\mathcal{R}(x_i, y_i) \in \Gamma_\infty$ so that it separates z_i and z. Thus, we also obtain that $\overline{xy} \in L_\theta^{\mathbb{D}}$ (in Figure 20, we have $\{x, y\} = \{2/7, 4/7\}$ and the red curves are what we want).

Proof of this Claim: We only need to prove that each $\mathcal{R}(x_i, y_i)$ separates z_i and z_{i+1}. Fix $i \geq 1$. By Corollary 16.4, $\iota_\theta(z_i) \neq \iota_\theta(z_{i+1})$. Denote by p_i the index of the first distinct entries of $\iota_\theta(z_i)$ and $\iota_\theta(z_{i+1})$. Then the points $f^{p_i}(z_i)$ and $f^{p_i}(z_{i+1})$ are separated by the curve $\mathcal{R}\left(\frac{\theta}{2}, \frac{\theta+1}{2}\right)$. Lifting this curve along the orbit of the pair $\{z_i, z_{i+1}\}$, we obtain a curve $\mathcal{R}(x_i, y_i) \in \Gamma_\infty$ that separates z_i and z_{i+1}.

2. The set $[w, z]_K \cap \mathcal{J}$ does not cluster z. In this case, the point z is on the boundary of a Fatou component $U \subset V$ and the angle-pair $\{x, y\}$ bounds U. Denote by $r_U(\alpha)$ the internal ray of U landing at z. If α is not equal to $\frac{j}{2^n}$ for any $n \geq 1, j \geq 0$, we can choose a sequence of integers $\{j_n, n \geq 1\}$ such that $\alpha \in \left(\frac{j_n}{2^n}, \frac{j_n+1}{2^n} \right)$ for any $n \geq 1$. Let $\mathcal{R}(x_n, y_n)$ be the curve in Γ_∞ that contains $r_U\left(\frac{j_n}{2^n}\right)$ and $r_U\left(\frac{j_n+1}{2^n}\right)$. Then this sequence of curves converge to the ray-pair of $\{x, y\}$ as $n \to \infty$. If $\alpha = j_0/2^{n_0}$ for some integers n_0, j_0, then α can be expressed as $j_n/2^n$ for every $n \geq n_0$. Note that one of x and y, say x, belongs to the set $\text{Pre}_\theta^{\mathbb{S}}$. Then we obtain

two sequences of curves,

$$\{\mathcal{R}(x, y_n)\}_{n \geq n_0} \quad \text{and} \quad \{\mathcal{R}(x, s_n)\}_{n \geq n_0},$$

belonging to Γ_∞ such that each ray $\mathcal{R}(x, y_n)$ contains the internal rays $r_U\left(\frac{j_n - 1}{2^n}\right)$, $r_U\left(\frac{j_n}{2^n}\right)$ and each ray $\mathcal{R}(x, s_n)$ contains the internal rays $r_U\left(\frac{j_n + 1}{2^n}\right), r_U\left(\frac{j_n}{2^n}\right)$. By the construction of the curves in Γ_∞, we can see that one sequence of curves converge to a single ray $\mathcal{R}(x)$ and the other sequence converges to the ray-pair of $\{x, y\}$. So $\overline{xy} \in L_\theta^{\mathbb{D}}$ (in Figure 21, we have $x = 2/15, y = 3/5$ and the red curves are what we want). $\qquad\qquad\qquad\qquad\qquad\qquad\qquad\qquad\qquad\qquad\qquad\qquad\qquad\quad\square$

APPENDIX C. PROOF OF LEMMA 16.3

For a rational angle $\theta \in \mathbb{S} \setminus \{0\}$, and $f : z \mapsto z^2 + c_\theta$, this lemma claims that two non pre-major angles α and β form an angle-pair (i.e., the external rays $\mathcal{R}(\alpha)$ and $\mathcal{R}(\beta)$ land together) if and only if they have the same itinerary relative to the major leaf.

The necessity is obvious because α and β have the same itineraries as that of z, where z is the common landing point of $\mathcal{R}(\alpha)$ and $\mathcal{R}(\beta)$.

For the sufficiency, we only need to prove the following result: if the first n-th entries of $\iota_\theta(\alpha)$ and $\iota_\theta(\beta)$ are identical (see Subsection 16.2 for the definition of $\iota_\theta(\alpha)$), then the distance of z, w is less than $C \cdot \lambda^{-n}$ with some metric, where $\lambda > 1$, C are constants and z, w are the landing points of $\mathcal{R}(\alpha), \mathcal{R}(\beta)$ respectively.

We will prove this by distinguishing between the following two cases: i) the case that c_θ is (strictly) periodic and ii) the case that c_θ is (strictly) preperiodic.

i) Case c_θ is periodic:

In this case, $\mathcal{P}_f$ consists of the orbit of c_θ. For a point $p \in \mathcal{P}_f$, set

$$B_p = \left\{ z \in U_p \,\middle|\, |\phi_{U_p}(z)| < e^{-1} \right\}$$

and set

$$B_\infty = \left\{ z \in U_\infty \,\middle|\, |\phi_{U_\infty}(z)| > e \right\}.$$

Define $\mathcal{L}$ the complement of the union

$$B_\infty \cup \left(\bigcup_{p \in \mathcal{P}_f} B_p \right)$$

in $\mathbb{C}$. Then $\mathcal{L}$ is a compact connected neighborhood of the Julia set $\mathcal{J}$ on which f is uniformly expanding with respect to the hyperbolic metric ρ in a neighborhood U of $\mathcal{L}$, i.e., $\exists \lambda > 1$ such that if $\gamma \subset \mathcal{L}$ is an arc and $f : \gamma \to f(\gamma)$ is a homeomorphism, then

$$\operatorname{diam}_\rho(f(\gamma)) > \lambda \cdot \operatorname{diam}_\rho(\gamma).$$

Define

$$\mathcal{R}(P_\theta) := \bigcup_{n \geq 1} f^n \left(\mathcal{R}\left(\frac{\theta}{2}, \frac{\theta + 1}{2}\right) \right).$$

For any $p, q \in \mathcal{L}$, we can find an arc in $\mathcal{L}$ connecting p, q such that it doesn't cross the rays in $\mathcal{R}(P_\theta)$. So we can define a number

$$M_{p,q} = \inf \left\{ \text{length}_\rho(\gamma) \mid \gamma \subset \mathcal{L} \text{ is an arc connecting } p \text{ and } q \right.$$

$$\left. \text{but not crossing the rays in } \mathcal{R}(P_\theta) \right\}.$$

It is not difficult to check that $M_{p,q}$ is uniformly bounded for $p, q \in \mathcal{L}$. That is, there is a $C > 0$ so that $M_{p,q} < C$ for any $p, q \in \mathcal{L}$.

Now, suppose the first n-th entries of $\iota_\theta(\alpha)$ and $\iota_\theta(\beta)$ are identical. Set z, w to be the landing points of $\mathcal{R}(\alpha), \mathcal{R}(\beta)$, respectively, and z_i, w_i the i-th iteration of z, w by f. Choose an arc $\gamma_n \in \mathcal{L}$ with

$$\text{length}_\rho(\gamma_n) < C$$

such that it connects z_n, w_n and doesn't cross the rays in $\mathcal{R}(P_\theta)$. Lift γ_n to the arc γ_{n-1} with starting point z_{n-1}. Since γ_n doesn't cross the rays in $\mathcal{R}(P_\theta)$, then so doesn't γ_{n-1} and it implies that γ_{n-1} is contained in the same component of

$$\mathcal{L} \setminus \mathcal{R}\left(\frac{\theta}{2}, \frac{\theta + 1}{2}\right).$$

Note that the i-th entries of $\iota_\theta(\alpha)$ and $\iota_\theta(\beta)$ are equal if and only if z_i, w_i belongs to the same component of $\mathcal{L} \setminus \mathcal{R}\left(\frac{\theta}{2}, \frac{\theta+1}{2}\right)$. Then w_{n-1} is the unique preimage of w_n contained in the component of $\mathcal{L} \setminus \mathcal{R}\left(\frac{\theta}{2}, \frac{\theta+1}{2}\right)$ that contains z_{n-1}. It follows that γ_{n-1} connects w_{n-1} and z_{n-1}. As

$$\gamma_{n-1} \subset f^{-1}(\mathcal{L}) \subset \mathcal{L},$$

by the uniform expansion of f in $\mathcal{L}$, we have

$$\text{length}_\rho(\gamma_{n-1}) < \lambda^{-1} \text{length}_\rho(\gamma_n).$$

Repeating this process n times, we obtain an arc γ_0 which is the n-th lift of γ_n with the starting point z. With the same argument as before, the ending point of γ_0 is w and

$$\text{dist}_\rho(z, w) \leq \text{length}_\rho(\gamma_0) < \lambda^{-n} \text{length}_\rho(\gamma_n) < C\lambda^{-n}.$$

ii) Case c_θ is preperiodic:

In this case, $\mathcal{J} = \mathcal{K}$ and f is uniformly expanding with respect to an *admissible metric (orbifold metric)* ρ in a compact neighborhood of $\mathcal{J}$ (see [DH84, McM94, Mil06]).

Since $\mathcal{J}$ is compact, the length of the regulated arc $[z, w]$ (with respect to the metric ρ) is uniformly bounded by a constant C for any $z, w \in \mathcal{J}$. Using the same notions as the periodic case, if the first n-th entries of $\iota_\theta(\alpha)$ and $\iota_\theta(\beta)$ are equal, then for $0 \leq i \leq n - 1$, the polynomial f maps the regulated arc $[z_i, w_i]$ homeomorphically to the regulated arc $[z_{i+1}, w_{i+1}]$. By the uniform expansion of f on $\mathcal{J}$, we have that

$$\text{dist}_\rho(z, w) \leq \text{length}_\rho([z, w]) \leq \lambda^{-n}\text{length}_\rho([z_n, w_n]) < C \cdot \lambda^{-n}.$$

We note that the result of Lemma 16.3 is covered by a more general theorem in [Zen15].

APPENDIX D. ADDITIONAL IMAGES

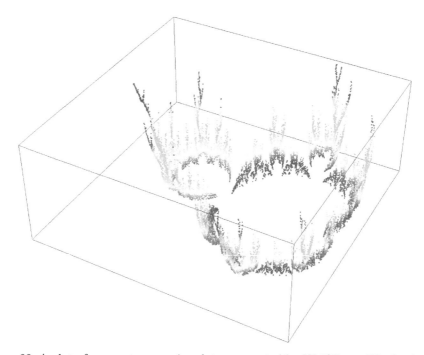

Figure 22: A plot of core entropy, using data computed by Wolf Jung. The horizontal plane is $\mathbb{C}$ and the vertical axis is the interval $[0, \log 2] \subset \mathbb{R}$. A plotted point $(c, h) \in \mathbb{C} \times \mathbb{R}$ represents the data that the core entropy of the polynomial $z \mapsto z^2 + c$ is h. Data is shown for a selection of parameters c in the boundary of the Mandelbrot set with rational external angle. Points are color-coded according to the h-coordinate, core entropy.

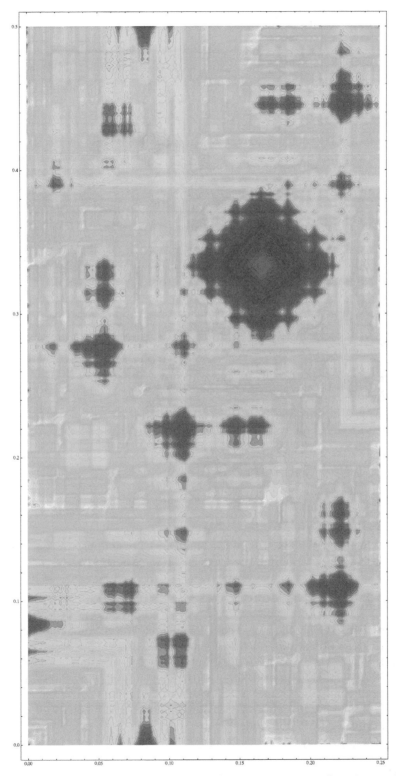

Figure 23: A contour plot by W. Thurston of core entropy as a function on PM(3), using the starting point parametrization of Section 11.4.1.

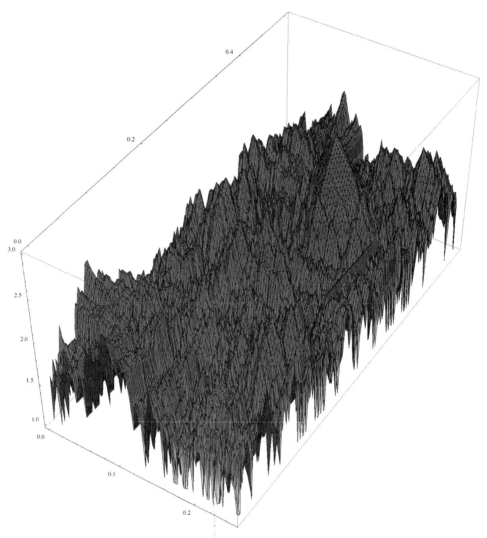

Figure 24: A plot by W. Thurston of the exponential of core entropy as a function on PM(3), using the starting point parametrization of Section 11.4.1.

Acknowledgments

The authors thank the anonymous referee for their careful reading and helpful comments. During the preparation of this manuscript, Kathryn Lindsey received support from the NSF via a Graduate Research Fellowship and, later, a Postdoctoral Research Fellowship. Hyungryul Baik was partially supported by Samsung Science & Technology Foundation grant No. SSTF-BA1702-01. Yan Gao was partially supported by NSFC grant No. 11871354. Dylan Thurston was partially supported by NSF Grant Number DMS-1507244.

REFERENCES

[AKM65]   R. L. Adler, A. G. Konheim, and M. H. McAndrew, *Topological entropy*, Trans. Amer. Math. Soc. **114** (1965), 309–319.

[BCO11] Alexander M. Blokh, Clinton P. Curry, and Lex G. Oversteegen, *Locally connected models for Julia sets*, Adv. Math. **226** (2011), no. 2, 1621–1661. MR 2737795 (2012d:37106).

[BL02] A. Blokh and G. Levin, *An inequality for laminations, Julia sets and "growing trees,"* Ergodic Theory Dynam. Systems **22** (2002), no. 1, 63–97. MR 1889565 (2003i:37045).

[BMOV13] Alexander M. Blokh, Debra Mimbs, Lex G. Oversteegen, and Kirsten I. S. Valkenburg, *Laminations in the language of leaves*, Trans. Amer. Math. Soc. **365** (2013), no. 10, 5367–5391. MR 3074377.

[BO04a] Alexander Blokh and Lex Oversteegen, *Backward stability for polynomial maps with locally connected Julia sets*, Trans. Amer. Math. Soc. **356** (2004), no. 1, 119–133 (electronic). MR 2020026 (2005c: 37081).

[BO04b] ———, *Wandering triangles exist*, C. R. Math. Acad. Sci. Paris **339** (2004), no. 5, 365–370. MR 2092465 (2005g:37081).

[BO09] ———, *Wandering gaps for weakly hyperbolic polynomials*, Complex Dynamics: Families and Friends, A. K. Peters, Wellesley, MA, 2009.

[BOPT14] Alexander Blokh, Lex Oversteegen, Ross Ptacek, and Vladlen Timorin, *The main cubioid*, Nonlinearity **27** (2014), no. 8, 1879–1897. MR 3246159.

[DH84] Andrien Douady and John H. Hubbard, *Étude dynamique des polynômes complexes*, Publications Mathématiques d'Orsay, vol. 84–2, Université de Paris-Sud, Département de Mathématiques, Orsay, 1984, available in English as *Exploring the Mandelbrot set. The Orsay notes* from http://math.cornell.edu/~hubbard/OrsayEnglish.pdf.

[dMvS93] Welington de Melo and Sebastian van Strien, *One-dimensional dynamics*, Ergebnisse der Mathematik und ihrer Grenzgebiete (3) [Results in Mathematics and Related Areas (3)], vol. 25, Springer-Verlag, Berlin, 1993. MR 1239171.

[Dou95] A. Douady, *Topological entropy of unimodal maps: monotonicity for quadratic polynomials*, Real and complex dynamical systems (Hillerød, 1993), NATO Adv. Sci. Inst. Ser. C Math. Phys. Sci., vol. 464, Kluwer Acad. Publ., Dordrecht, 1995, pp. 65–87.

[FM12] Benson Farb and Dan Margalit, *A primer on mapping class groups*, Princeton Mathematical Series, vol. 49, Princeton University Press, Princeton, NJ, 2012.

[Fur67] Harry Furstenberg, *Disjointness in ergodic theory, minimal sets, and a problem in Diophantine approximation*, Math. Systems Theory **1** (1967), 1–49.

[Gao] Gao Yan, *On Thurston's core entropy algorithm*, To appear in Transactions of the AMS.

[Gao13]   ———, *Dynatomic curve and core entropy for iteration of polynomials*, Ph.D. thesis, Université d'Angers, France, April 2013.

[Gol94]   L. R. Goldberg, *On the multiplier of a repelling fixed point*, Invent. Math. **118** (1994), 85–108.

[Jun14]   Wolf Jung, *Core entropy and biaccessibility of quadratic polynomials*, Preprint, 2014, arXiv:1401.4792.

[Kiw02]   Jan Kiwi, *Wandering orbit portraits*, Trans. Amer. Math. Soc. **354** (2002), no. 4, 1473–1485. MR 1873015 (2002h:37070).

[Kiw04]   ———, $\mathbb{R}$*eal laminations and the topological dynamics of complex polynomials*, Adv. Math. **184** (2004), no. 2, 207–267. MR 2054016 (2005b: 37094).

[Kiw05]   J. Kiwi, *Combinatorial continuity in complex polynomial dynamics*, Proc. London Math. Soc. **91** (2005), no. 3, 215–248.

[Lev98]   G. Levin, *On backward stability of holomorphic dynamical systems*, Fund. Math. **158** (1998), no. 2, 97–107. MR 1656942 (99j:58171).

[McM94]   Curtis T. McMullen, *Complex dynamics and renormalization*, Annals of Mathematics Studies, vol. 135, Princeton University Press, Princeton, NJ, 1994. MR 1312365.

[Mil06]   John Milnor, *Dynamics in one complex variable*, third ed., Annals of Mathematics Studies, vol. 160, Princeton University Press, Princeton, NJ, 2006. MR 2193309.

[Mim10]   Debra L. Mimbs, *Laminations: A topological approach*, ProQuest LLC, Ann Arbor, MI, 2010, Thesis (Ph.D.), University of Alabama at Birmingham. MR 2801683.

[Poi09]   Alfredo Poirier, *Critical portraits for postcritically finite polynomials*, Fund. Math. **203** (2009), no. 2, 107–163.

[Poi10]   ———, *Hubbard trees*, Fund. Math. **208** (2010), no. 3, 193–248.

[Pta13]   Ross M. Ptacek, *Laminations and the dynamics of iterated cubic polynomials*, ProQuest LLC, Ann Arbor, MI, 2013, Thesis (Ph.D.), University of Alabama at Birmingham. MR 3187504.

[Thu09]   William P. Thurston, *Polynomial dynamics from combinatorics to topology*, Complex Dynamics: Families and Friends, A. K. Peters, Wellesley, MA, 2009, pp. 1–109.

[Tio15]   Giulio Tiozzo, *Topological entropy of quadratic polynomials and dimension of sections of the Mandelbrot set*, Adv. Math. **273** (2015), 651–715.

[Tio16]   ———, *Continuity of core entropy of quadratic polynomials*, Invent. Math. **203** (2016), no. 3, 891–921.

[Tom14]   J. Tomasini, *Géométrie combinatoire des fractions rationnelles*, Ph.D. thesis, Université d'Angers, 2014.

[Zen14]   Jinsong Zeng, *On the existence of shift locus for given critical portrait*, Preprint, 2014.

[Zen15]   ———, *Criterion for rays landing together*, Preprint online at https:// arxiv.org/abs/1503.05931, March 2015.

An Invitation to Coherent Groups

DANIEL T. WISE

Research supported by NSERC.

1 INTRODUCTION TO COHERENCE

Definition 1.1. *A group G is* coherent *if every finitely generated subgroup of G has a finite presentation.*

There are several known extensive classes of coherent groups. Fundamental groups of 3-manifolds were shown to be coherent independently by Scott [Sco73b] and Shalen (unpublished). Ascending HNN extensions of free groups were shown to be coherent by Feighn-Handel [FH99]. Finally, "perimeter groups" were shown to be coherent by McCammond-Wise [MW05].

The most evident coherent groups are small groups like finite groups or polycyclic groups. In the other extreme, groups with very few relations like free groups and surface groups are coherent. Our focus will be on the coherence of the latter such "sparsely related" groups.

The main goal in this topic is to understand exactly which groups are coherent, and more realistically, to find useful characterizations of coherent groups.

The general approach towards proving coherence of a group G is to represent G as $\pi_1 X$ where X is a topological space with a very tractable structure. This approach underlies every theorem stated here except for the characterization of coherent solvable groups in Theorem 7.1. There has not yet been a successful approach using "coarse geometric group theory."

1.a Some simple algebra

Lemma 1.2. *Coherence is a commensurability invariant.*

Sketch. If G is coherent, then so is any subgroup of G. We now show that G is coherent if $[G:G'] < \infty$ and G' is coherent. Let N be the intersection of the conjugates of G' in G, and observe that $[G:N] < \infty$, since the intersection of two finite index subgroups has finite index. Then N is coherent, and G/N is coherent, and $N \cap H$ is f.g. for any f.g. subgroup $H \subset G$. Hence G is coherent by Lemma 1.3. $\square$

It is unknown whether coherence is a quasi-isometry invariant. See Problem 9.16.

Lemma 1.3. *Let* $1 \to N \to G \to Q \to 1$ *be a short exact sequence. Then* G *is coherent provided that* Q *and* N *are coherent and every f.g. subgroup of* G *has f.g. intersection with* N.

We say that G is N-by-Q when it has such a short exact sequence.

Proof. Let H be a f.g. subgroup of G. Then there is a short exact sequence: $1 \to N \cap H \to H \to \bar{H} \to 1$, where $\bar{H}$ denotes the image of H in Q. Hence H is f.p.-by-f.p., so a finite presentation for H can be built from finite presentations of $N \cap H$ and $\bar{H}$. $\qquad\square$

We use the term *incoherent* to mean not coherent.

The incoherent group $F_2 \times F_2$ shows that Lemma 1.3 does not hold for arbitrary extensions of coherent groups. We discuss $F_2 \times F_2$ in Example 9.22 and Example 10.7. Another incoherent group to consider in this vein is the wreath product $\mathbb{Z} \wr \mathbb{Z}$ discussed in Example 7.3.

I have not noticed any incoherent groups that are f.g. coherent-by-cyclic, and refer to Problem 7.2.

1.b   Related properties

Definition 1.4. G *is* noetherian *if every subgroup of* G *is f.g.*

The most prominent examples of noetherian groups are polycyclic groups.

Problem 1.5. Is every coherent noetherian group virtually polycyclic?

Definition 1.6. *Let* G *act properly and cocompactly on a geodesic metric space* Υ, *and let* $\upsilon \in \Upsilon$ *be a basepoint. The motivating case is where* Υ *is the Cayley graph of* G *with respect to a finite generating set, and* $\upsilon = 1_G$.

A subgroup $H \subset G$ *is* κ-quasiconvex *with respect to this action if for any geodesic* $\gamma \subset \Upsilon$ *whose endpoints lie in* $H\upsilon$, *we have* $\gamma \subset \mathcal{N}_\kappa(H\upsilon)$. *We say* H *is* quasiconvex *if it is* κ-quasiconvex *for some* $\kappa > 0$.

G *is* locally-quasiconvex *if each f.g. subgroup* H *is quasiconvex.*

Being quasiconvex highly depends on the choice of Υ. In particular, $\mathbb{Z}^2$ is not locally-quasiconvex, and for any finite generating set, there are only finitely many commensurability classes of subgroups that are quasiconvex with respect to the associated Cayley graph.

We refer to [Sho91] for an introduction to quasiconvexity, including a proof of the following:

Proposition 1.7. *Let* H *have Cayley graph* Υ *and let* $H \subset G$ *be a quasiconvex subgroup. Then* H *is f.g.*

Definition 1.8. *A f.g. group* G *is* locally quasi-isometrically embedded *if the inclusion* $H \subset G$ *is a local isometry, whenever* H *is f.g.*

It is easy to see that locally-quasiconvex implies locally quasi-isometrically embedded. However, $\mathbb{Z}^2$ has the latter property, which is an intrinsic property that does not depend on the choice of generators. For hyperbolic groups, being locally-quasiconvex and being locally quasi-isometrically embedded are equivalent.

Definition 1.9 (f.g.i.p.). *G has the* finitely generated intersection property *if $A \cap B$ is f.g. whenever A and B are f.g. subgroups.*

A subgroup $H \subset G$ has the finitely generated intersection property, *if $H \cap K$ is f.g. whenever K is f.g.*

The f.g.i.p. is also known as the *Howson property* since Howson proved that free groups have this property. It is tricky to put quantitative estimates on $\text{rank}(A \cap B)$ in terms of $\text{rank}(A)$ and $\text{rank}(B)$. This led to the Hanna Neumann conjecture for free groups (see Section 15.b) and there is more work to be done here for more general classes of groups.

The following is also proven in [Sho91]:

Proposition 1.10. *Let G act on the geodesic metric space Υ. Suppose A and B are quasiconvex. Then $A \cap B$ is quasiconvex.*

In particular, local-quasiconvexity implies the f.g.i.p.

Problem 1.11. Give an example of a f.g. subgroup $H \subset G$ such that G is f.p. and H is quasi-isometrically embedded, but H is not f.p.

Definition 1.12. *A group G is n-free if H is free whenever $H \subset G$ and $\text{rank}(H) \leq n$.*

A simple example is the surface group $\langle a_1, b_1, \ldots, a_g, b_g \mid [a_1, b_1] \cdots [a_g, b_g] \rangle$ which is $(2g - 1)$-free, since any low rank subgroup has infinite index, and is hence free. We review some other classes in Section 19.

Definition 1.13. *A group G is* locally indicable *if for each f.g. subgroup $H \subset G$, either H is trivial, or H has an infinite cyclic quotient.*

The connection with coherence is currently circumstantial since we do not yet know the resolution of Conjecture 12.11. However, all currently known examples of coherent groups G with $\text{cd}(G) = 2$ have the property that G is locally indicable. Indeed, all such groups fall within the rubric of nonpositive immersions (see Definition 1.15) and so Theorem 12.6 applies.

Definition 1.14 (Compact core). *Let X be a space with a universal cover. We say X has the* compact core property *if for each covering space $\widehat{X} \to X$ with $\pi_1 \widehat{X}$ f.g. there exists a compact subspace $Y \subset \widehat{X}$ such that $\pi_1 Y \to \pi_1 \widehat{X}$ is an isomorphism.*

The compact core property holds in many cases where coherence holds. It was proven when X is a 3-manifold in [Sco73a], and it was proven for 2-complexes satisfying the perimeter reduction hypothesis in [MW05] and having negative sectional curvature in [Wis04a].

When X is a complex or a manifold, compactness ensures that $\pi_1 Y$ and hence $\pi_1 \widehat{X}$ is f.p. and so the compact core property is a strong form of coherence. However, the compact core property is not a property of π_1 and depends on the choice of X. See Example 20.4 for a simple example illustrating this.

A useful strengthening of this condition is to require that for each compact subspace $C \subset \widehat{X}$, we can choose Y above so that $C \subset Y$. A natural weakening of the condition asks for a compact core under the additional assumption that $\pi_1 \widehat{X}$ has finite homotopy type. This is interesting and under-explored in the 2-dimensional case.

A central theme in this text is the conjectured relationship between coherence and the following property which is the focus of Section 12.

Definition 1.15 (Nonpositive immersions). *A 2-complex X has nonpositive immersions provided that for each immersion $Y \to X$ with Y compact and connected either $\chi(Y) \leq 0$ or $\pi_1 Y$ is trivial.*

I do not know of a coherent group G with cohomological dimension 2 such that $G \neq \pi_1 X$ where X is a 2-complex with nonpositive immersions. On the other hand, I believe that when X has nonpositive immersions then $\pi_1 X$ is coherent. See Conjecture 12.11.

1.c   Overview

We offer some motivation for the subject of coherence in §2. The circumstantial low-dimensionality of coherence is discussed in §3. We sketch the proof of the coherence of 3-manifold groups in §4, and also explain how 3-manifold coherence is connected to free-by-cyclic coherence. The coherence of certain graphs of groups is discussed in §5. The coherence of ascending HNN extensions of free groups is proven in §6, and the characterization of solvable coherence groups is stated in §7. The perimeter approach to coherence and local quasiconvexity is discussed in §8, where we also describe basic results of small-cancellation theory and one-relator groups with torsion, and show how they are employed towards this subject. A sampling of incoherent groups are described in §9. The finitely generated intersection property is discussed in §10. A general form of nonpositive and negative sectional curvature is described in §11, where a variety of examples are given and the connection to coherence and local quasiconvexity is described. The 2-complexes with nonpositive immersions are described in §12, where towers, an alternate to foldings, are described, and connections are made towards locally indicable groups and asphericity. Moreover, coherence is obtained from negative immersions which is an abstraction of negative sectional curvature. The test case of (p, q, r)-complexes are described in §13. One-relator groups and staggered 2-complexes are described in §14, and then §15 explores two methods for proving the nonpositive immersion property in these cases. Connections with L^2-betti numbers are described in §16. Virtual algebraic fibering is discussed in §17. Some general statements about locally quasiconvex groups are mentioned in §18. The n-free groups are described briefly in §19, and the compact core property is described in §20. We conclude with a problem list in §21.

2   WHY?

Dylan Thurston insisted that I give some explanations and justifications of why this topic might be of interest. I thank him for his encouragement.

Max Dehn advocated the fundamental nature of the word problem, conjugacy problem, and isomorphism problem for finitely presented groups [Deh11]. His vision and influence here was extraordinary. Ultimately these problems have all been shown to be algorithmically undecidable in general, yet solvable within the ubiquitous context of Gromov's hyperbolicity. The urgency of these problems for Dehn was their topological applicability, as they were certainly motivated by his interest in knot complements. The problems are algebraizations of the problems of determining whether a closed circle can be homotoped to a point, whether two closed circles can be homotoped to each other, and whether two knot complements can be recognized as different by virtue of their fundamental groups. In this direction, the topic has triumphed in the sense that all these problems are decidable in their original 3-manifold contexts; their solutions are decisive and have led to many amplifications and generalizations.

From an algorithmic viewpoint, among Dehn's problems, the study of coherence is most closely related to the word problem, since it is connected to the membership problem in a f.g. subgroup. From a topological viewpoint, it is related to our desire to build and grasp the most approachable covering spaces of a complex. From the viewpoint of a combinatorial group theorist, who is fascinated and delighted by the surprises and mysteries packed into a finite presentation, it is especially natural to be curious about the nature of groups whose f.g. subgroups admit such condensed descriptions themselves.

A primary motivation for me stems from the desire to develop a class of groups that are close in nature to the fundamental groups of 3-manifolds, and echo one of their most salient properties.

I first started assembling a collection of problems and research directions about coherent groups around 2003, and was asked by Mladen Bestvina to present some of them at a meeting at the AIM a few years later. Bestvina was eager that they be disseminated because of the "fresh" problems that indicated novel research veins. But in addition I feel that there is also a continuity here with the combinatorial group theory of the 1960s.

At that time, I was mostly focused on my other research program: using cube complexes to elucidate geometric group theory and 3-manifolds. Although it took some time until others came aboard, it was transparent for me to justify that topic to myself. Cubes had a specific technical pulse, naturally interacted with a variety of topics, and the techniques were aimed at specific open problems, most notably those concerning 3-manifolds. There, an agenda had already been set by an acknowledged giant—William Thurston—and so one could justify the interest in cubes by the hope that they would play a role in offering additional explanations on a pre-accepted target.

There is a tremendous desire in mathematics to understand and classify all possible forms of some family of objects, and our human condition makes 3-manifolds appear rather quickly on the list of priorities. Nonetheless, I don't think 3-manifolds are "for" something else in mathematics, and I am unconvinced about how frequently they appear as crucial objects outside their domain—in contrast to surfaces which are comparatively ubiquitous in mathematics. Instead, 3-manifolds are a topological ultimate goal, and I likewise hope coherent groups are an algebraic telos, that might motivate significant and useful techne.

In terms of scope, coherent groups form a larger class of objects. Indeed, this is arguably already the case for the subclass of free-by-cyclic groups. However, whereas

3-manifolds have shown themselves to be a jungle, exhibiting many opportunities for different viewpoints and techniques, I suppose the class of coherent groups will form more of a garden—perhaps a bit less natural, a bit more contained and with fewer strata, but it is a beautiful garden nonetheless.

The interest in coherence comes from a variety of factors:

- The aesthetics of the subject.
- The challenge of simple-looking but perplexing questions that permeate the topic. For example, is $\langle a, b \mid a^{a^b} = a^2 \rangle$ coherent?
- The preference of combinatorial group theorists for groups that are finitely presented.
- The desire to generalize from free groups and surface groups towards a class of groups that are tractable in much the same way.
- A testing ground for the development of tools to engage with infinite group theory that isn't covered by the methods of hyperbolicity.
- Conversely, quasiconvexity has been the main subgroup notion in geometric group theory, and a crucial enabling hypothesis in many studies. It is interesting to consider the extreme scenario where all f.g. subgroups are quasiconvex—a salient source of coherence.
- A recognition that there are shrouded patterns relating coherence to geometric hypotheses on presentations, but their exact nature is not yet explicit.
- The mystery of the variety of possible subgroups lurking within a group defined through a simple recipe.

Whether there is ultimately a unified subject here, or whether this is a miscellany, has yet to be determined, but this document is in favor of a certain unity to the class of coherent groups.

3   DIMENSION

The known examples suggest that coherence is a ≤ 3 dimensional phenomenon and local-quasiconvexity is a ≤ 2 dimensional phenomenon. The degenerate counterexamples to this phenomenon for coherence are the virtually polycyclic groups, and more generally, virtual graphs of other coherent groups with virtually polycyclic edge groups (see Theorem 5.3). Indeed, virtually polycyclic groups themselves virtually arise as iterated HNN extensions of smaller polycyclic groups.

Problem 3.1. Let M be a closed aspherical manifold of dimension ≥ 4. Can $\pi_1 M$ be coherent but not polycyclic? Can $\pi_1 M$ be word-hyperbolic and coherent?

See Section 9.g for a discussion of the noncoherence of π_1 of an arithmetic hyperbolic manifold of dimension ≥ 4.

Problem 3.2. Does there exist a locally-quasiconvex torsion-free word-hyperbolic group G of cohomological dimension ≥ 3?

Does there exist a closed negatively curved manifold of dimension ≥ 4 such that $\pi_1 M$ is locally-quasiconvex? (I don't think so.)

Note that local-quasiconvexity fails for $\pi_1 M$ where M is a hyperbolic 3-manifold because M virtually fibers in both the finite-volume case and the closed cases [Wisc, Ago13].

Pansu's *conformal dimension* for a metric space S is the infimal Hausdorff dimension of all metric spaces quasiconformally homeomorphic to S. This provides an interesting quasi-isometry invariant of the boundary of a hyperbolic group with its visual metric.

Problem 3.3. Let G be a hyperbolic group. Suppose the conformal dimension of ∂G is at most 2. Is G coherent? Suppose the conformal dimension of ∂G is strictly less than 2. Is G locally-quasiconvex?

4   3-MANIFOLDS

The fact that fundamental groups of 3-manifolds are coherent was proven independently by Scott [Sco73b] and Shalen (unpublished), employing an idea of Swarup [Swa73]. Another exposition of the proof is given in [Sta77].

Swarup applied Delzant's T-invariant to reprove an important element of the proof of the coherence of 3-manifolds [Del95, Swa04]. Namely:

Proposition 4.1. *For each f.g. noncyclic indecomposable group G, there is a f.p. noncyclic indecomposable group K and a surjection $K \to G$, such that for any intermediate group H satisfying $K \twoheadrightarrow H \to G$, the group H is also indecomposable.*

The following combines the loop and disk theorem of Papakyriakopoulos [Rol90]:

Theorem 4.2. *Let S be a component of ∂M where M is a 3-manifold. Suppose $S \to M$ is not π_1-injective. Then there exists a proper embedding $(D, \partial D) \to (M, S)$ of a 2-disk, with ∂D an essential closed curve in S.*

Theorem 4.3. *Let M be a 3-manifold, if $\pi_1 M$ is f.g. then $\pi_1 M$ is f.p.*

Note that Theorem 4.3 implies that $\pi_1 M$ is coherent, since every f.g. subgroup of $\pi_1 M$ is itself equal to $\pi_1 \widehat{M}$ where $\widehat{M} \to M$ is a covering space.

Proof. The argument is by induction on $\operatorname{rank}(\pi_1 M)$, where the result obviously holds when $\operatorname{rank} = 0$ or $\operatorname{rank} = 1$. Since $\operatorname{rank}(A * B) = \operatorname{rank}(A) + \operatorname{rank}(B)$ by Grushko's theorem, and the free product of f.p. groups is f.p., it suffices to prove the statement when $\pi_1 M$ is indecomposable.

By Proposition 4.1, there is a surjection $K \to \pi_1 M$ such that $\bar{K}$ is indecomposable for any composition $K \to \bar{K} \to \pi_1 M$ with $K \to \bar{K}$ surjective.

Choose a π_1-surjective submanifold $M_1 \subset M$. For each i, let $M_i' = M - \operatorname{Int}(M_i)$. If $M_i \subset M$ is not π_1-injective, then either $\partial M_i \to M_i$ is not π_1-injective, or $\partial M_i \to M_i'$ is not π_1-injective (on some component). For otherwise, $\pi_1 M$ splits as a graph of groups along $\{\pi_1 \partial M_i\}$, in which case the vertex group $\pi_1 M_i$ injects. By Theorem 4.2, there is a properly embedded disk D_i in either M_i or M_i' with $S_i = \partial D_i$ a simple closed curve representing an element in the kernel. Let R_i be a regular neighborhood of D_i. In the M_i' case, we let $M_{i+1} = M_i \cup R_i$. In the M_i case, using the hypothesis that $\pi_1 M$ is freely indecomposable, we let M_{i+1} be the component of $M_i - \big(\mathrm{Int}(R_i) \cup (\partial M_i \cap \partial R_i)\big)$ that contains a conjugate of the image of K.

Finally, this procedure terminates after finitely many steps because ∂M_i is getting smaller with each compression disc. For instance, letting $\partial M_i = \sqcup_j S_{ij}$ be the disjoint union of surfaces, $\sum_j \mathrm{genus}(S_{ij})^2$ is decreasing. $\qquad \square$

The proof actually produces a compact core in the indecomposable case, and it is shown in [Sco73a] that compact cores exist in general.

Many hyperbolic 3-manifolds have locally-quasiconvex fundamental groups because of Theorem 18.2.

4.a   3-manifold coherence from free-by-cyclic coherence

Given our modern understanding of 3-manifolds—the geometrization theorem and virtual fibering of most of the pieces—and in view of Theorem 5.1, the coherence of 3-manifold groups is essentially a consequence of Theorem 6.1 and Theorem 5.1. This was explained in detail in [Wis19], and we sketch that explanation here.

We emphasize that this section depends on Sections 5 and 6. We are certainly not suggesting that this is an easier proof of the coherence than that of Theorem 4.3, but rather explaining that once we understand the nature of the groups, the coherence has a 2-dimensional character after all.

Let M be a compact 3-manifold. It suffices to understand the irreducible case (since the coherence of a free product is implied by coherence of its free factors—a trivial case of Theorem 5.1). By doubling M along its boundary, we may assume $\partial M = \emptyset$. The JSJ decomposition of M provides a splitting of $\pi_1 M$ as a graph of groups where each edge group is $\mathbb{Z}^2$ and each vertex group is geometric. Thus by Theorem 5.1 it suffices to show that each vertex group is coherent. Each non-hyperbolic piece has coherent π_1 by Lemma 1.3. Each hyperbolic piece is either a finite-volume hyperbolic manifold with one or more cusps, i.e., it is compact atoroidal and boundary consisting of one or more tori, hence its π_1 is virtually free-by-cyclic by [Wisc]. Note that when M is a mixed 3-manifold, it is virtually free-by-cyclic anyhow [PW18]. When M is a graph manifold, its coherence holds by Theorem 5.1, since each Seifert fibered piece has π_1 that is Z-by-S, where Z is a normal cyclic subgroup and S is virtually a surface group, and any Z-by-S group is coherent by Lemma 1.3.

When the JSJ decomposition of M is trivial, either $\pi_1 M$ is polycyclic, and hence coherent by Theorem 1.3, or $\pi_1 M$ is finite, or $\pi_1 M$ is coherent since M is Seifert-fibered as above. In the remaining case, M is a closed hyperbolic manifold. But then if $H \subset \pi_1 M$, either H is of finite index and thus also f.p., or H is of infinite index, in which case H is (either a surface group or) virtually free-by-cyclic, since M is virtually surface-by-cyclic by [Ago13]. Hence H is f.p. by Theorem 6.1.

5   GRAPHS OF GROUPS

The following theorem is essentially due to Karrass-Solitar [KS70]:

Theorem 5.1. *A graph of groups with coherent vertex groups and noetherian edge groups is coherent.*

Definition 5.2. *G has the* finitely generated intersection property *relative to the subgroup H, if $H \cap K$ is f.g. for each f.g. subgroup K of G.*

Theorem 5.1 is a special case of:

Theorem 5.3. *Suppose G splits as a graph of groups such that G has the finitely generated intersection property relative to each edge group, and each vertex group is coherent. Then G is coherent.*

Proof. Let X be a graph of spaces with underlying graph Γ corresponding to the graph of groups decomposition of G. Let $\widehat{X}$ be the based covering space of X corresponding to a f.g. subgroup of G, and let $\widehat{\Gamma}$ be the underlying graph of the graph of space structure of $\widehat{X}$.

As $\pi_1 \widehat{X}$ is f.g., there is a finite subgraph $\Lambda \subset \widehat{\Gamma}$, such that the graph of spaces Y corresponding to Λ has the property that $\pi_1 Y \to \pi_1 \widehat{X}$ is a π_1-isomorphism.

Consider the splitting of $\pi_1 \widehat{X}$ associated to Λ. By hypothesis, each vertex group of Λ is f.p., and each edge group of Λ is f.g. This gives a finite presentation for $\pi_1 \widehat{X}$.   □

We caution that Theorem 5.3 requires that G itself have the f.g.i.p. relative to each edge group; it does not suffice for each vertex group to have the f.g.i.p. relative to its own edge groups. See Remark 10.4.

Theorem 5.1 implies that the coherence property is determined by the rigid subgroups in a JSJ decomposition for G. In particular, for a 3-manifold M (which satisfies the geometrization conjecture), coherence is actually a consequence of coherence of hyperbolic 3-manifolds. In Section 4.a we explain that coherence is essentially a consequence of virtual fibering and Theorem 6.1. It would be interesting if there were an explanation that was entirely hyperbolic.

An immediate consequence of Theorem 5.3 is the following:

Corollary 5.4 (Abelian Hierarchies). *Every group in $\mathcal{AH}$ is coherent, where $\mathcal{AH}$ is the smallest class of groups such that:*

1. *$\mathcal{AH}$ contains the trivial group;*
2. *an amalgamated product $A *_C B \in \mathcal{AH}$ whenever $A, B \in \mathcal{AH}$, and C is abelian; and*
3. *an HNN extension $A*_{C^t=D} \in \mathcal{AH}$ whenever $A \in \mathcal{AH}$, and C, D are abelian.*

A natural notable class of groups that lie in $\mathcal{AH}$ are the fully residually free groups, which are a class of groups that are the quintessential generalizations of free groups and genus 2 surface groups. A group G is *fully residually free* if for each finite subset $S \subset G - \{1_G\}$, there is a free quotient $G \to F$ such that no element of S maps to 1_F. F.g. fully residually free groups, which are called *limit groups* by Sela [Sel01] were shown to be f.p. in [KM99], and indeed, they lie in $\mathcal{AH}$. We refer the reader to [Kap02] for a comprehensive study of their subgroup properties.

6   ASCENDING HNN EXTENSIONS OF FREE GROUPS

In the 1970s, Baumslag raised the question of whether every free-by-cyclic group is coherent. This was answered by Feighn-Handel in [FH99] who proved the following:

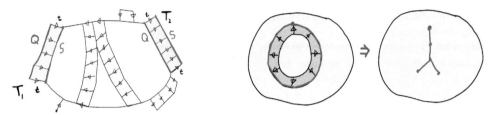

Figure 1: Inward and outward t-strips T_1 and T_2 are depicted on the left. The right illustrates why a minimal diagram cannot contain an annular t-strip.

Theorem 6.1. *Ascending HNN extensions of free groups are coherent.*

Remark 6.2. Let H be a f.g. subgroup of G. Either H is a subgroup of the base group of G whence H is free, or H itself splits as an ascending HNN extension of a free group. It thus suffices to prove that a f.g. ascending HNN of a free group is f.p.

   A version of the Feighn-Handel proof is described in Theorem 6.8. It relies on the change in "relative rank" between a graph and a subgraph, where the graph corresponds to the generators in the vertex group, and the subgraph corresponds to an edge group.

Definition 6.3. *Let $\phi: B \to B$ be a map from a graph to itself, and suppose $\phi(B^0) \subset B^0$. The mapping torus M of ϕ is the 2-complex:*

$$M = \left(B \cup (B \times [0,1]) \right) \,/\, \left\{ (b,0) \sim b, \ (b,1) \sim \phi(b) \ : b \in B \right\}.$$

When B is a bouquet of circles and ϕ induces a monomorphism of $\pi_1 B$, then M is the standard 2-complex of the presentation $\langle b_1, b_2, \ldots \,|\, t^{-1} b_i t = \phi(b_i) \rangle$.

Definition 6.4. *Let M be the mapping torus of Definition 6.3. We will use a graph of spaces structure of M. Its vertex space is B and its open edge space is $B \times (0,1)$.*

   Let $D \to M$ be a disk diagram (see Definition 8.1). Note that D has an induced graph of space structure, whose vertex spaces are components of the preimage of B, and whose open edge spaces are the components of the preimage of $B \times (0,1)$.

   A t-strip in D is an open edge space. In the motivating case, each t-strip corresponds to a sequence of rectangles $t^{-1} b_i t \phi(b_i)^{-1}$ that are glued together along successive t-edges. See the left of Figure 1.

   Let T be a t-strip with $\partial_p T = t^{-1} Q t S$. Then T is inward if $t^{-1} Q t$ is a subpath of $\partial_p D$, and T is outward if $t S t^{-1}$ is a subpath of $\partial_p D$.

   The following variant of the normal form theorem holds by choosing an outermost t-strip:

Lemma 6.5. *Suppose ϕ is π_1-injective. Every nontrivial spurless minimal disk diagram $D \to M$ has an inward or outward pointing t-strip.*

Proof. Suppose D is nontrivial, and has no spur. We use the decomposition of D as a graph Γ of spaces whose open edge spaces are t-strips. Its vertex spaces are components of the preimage of B in D. Each vertex space is a tree, since the inclusion $B \to M$ is π_1-injective. Indeed, if there were a cycle σ in D such that σ

maps to B, then σ maps to a tree in D, for otherwise we could replace the sub-diagram bounded by σ with a tree, as $\sigma \to B$ is nullhomotopic. Consequently each t-strip of D starts and ends on ∂D. In particular, if a tree with boundary path σ has an outgoing t-strip around it, then $\phi(\sigma)$ is also nullhomotopic in B, and so we could replace its interior by a tree, and remove that entire t-strip. And likewise for an incoming t-strip. Thus each t-strip has both t-edges on ∂D. See the right of Figure 1.

Since D is simply-connected, Γ is a tree. If Γ consists of a single vertex, then D would lie in a single vertex space mapping to B, and so by minimality, D would be a finite tree, so either D would be a single 0-cell or D would have a spur. Hence we may assume Γ has an edge, so by finiteness, we may consider a spur of Γ. Let T be the associated t-strip of D, let $t^{-1}QtS = \partial_{\mathsf{p}}T$ be its boundary path, and note that the two t-edges of $\partial_{\mathsf{p}}T$ lie in ∂D. Either Q or S lies in a degree 1 vertex space of Γ, and since D has no spur, Q or S is an arc equal to that entire vertex space and T will be inward or outward pointing, respectively. $\qquad\square$

Definition 6.6. *Let $Z \subset Y$ be a subgraph where Y and Z are connected. Their relative rank is defined by:*

$$\mathsf{rr}(Y, Z) = \mathrm{rank}(Y) - \mathrm{rank}(Z)$$

where $\mathrm{rank}(J) = \mathrm{rank}(\pi_1 J) = 1 + V - E$, where V and E are the numbers of vertices and edges of J.

A map between cell complexes is *combinatorial* if open cells are mapped homeomorphically to open cells. See Definition 11.4.

Let $Y \to B$ be a combinatorial map of graphs. Suppose e_1, e_2 are edges of Y that meet at a vertex of Y and map to the same edge of B. We can *fold* by forming a new graph $Y' = Y/\{e \sim e'\}$, and note that there is an induced map $Y' \to B$. See [Sta83] for more about foldings.

Lemma 6.7. *Let $Z \subset Y$ be a connected subgraph of a connected graph. Let $\phi : Y \to B$ be a combinatorial map. Let (Y', Z') be obtained by folding a pair of edges. Then $\mathsf{rr}(Y, Z) \geq \mathsf{rr}(Y', Z')$.*

Suppose $\phi : Z \to B$ is π_1-injective but $\phi : Y \to B$ is not π_1-injective. Let $(\bar{Y}, \bar{Z})$ be obtained by repeatedly folding until $\bar{Y} \to B$ is an immersion. Then $\mathsf{rr}(Y, Z) > \mathsf{rr}(\bar{Y}, \bar{Z})$.

Proof. The first part holds by considering the various cases for a pair of folded edges e_1, e_2. We refer to Figure 2. Note that rr stays the same when there is no change in homotopy type for Y or Z. However, there is a decrease when there is a fold of a bigon or pair of loops that lie in Y but not in Z, as $\mathrm{rank}(Y')$ increases but $\mathrm{rank}(Z')$ stays the same. There is also a decrease of rr when two vertices of Z are identified without edges of Z being identified since then $\mathrm{rank}(Y') = \mathrm{rank}(Y)$ but $\mathrm{rank}(Z') > \mathrm{rank}(Z)$.

To see the second part, we first fold the edges of Z, to assume that $\phi : Z \to B$ is an immersion, without affecting rr. Subsequently, as we proceed through the folding process, either the homotopy type of Z changes and rr decreases, or at some point the homotopy type of Y changes without affecting the homotopy type of Z, so rr decreases. $\qquad\square$

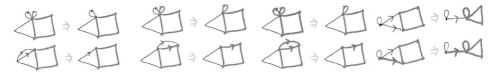

Figure 2: Folding and rr: The subgraph $Z \subset Y$ is indicated in blue.

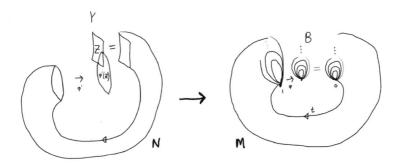

Figure 3: A heuristic depiction of the combinatorial immersion $N \to M$ occurring in the proof of Theorem 6.8.

Theorem 6.8. *A f.g. ascending HNN of a free group is f.p.*

Proof. As in Definition 6.3, consider a mapping torus M with $\pi_1 M$ f.g.

Let $\psi : N \to M$ be a map such that:

1. ψ is a combinatorial immersion;
2. $\pi_1 N \to \pi_1 M$ is surjective;
3. N is compact;
4. $\psi^{-1}(B)$ is a connected graph Y;
5. $\psi^{-1}(B \times (0,1)) \cong (Z \times (0,1))$ where Z is a connected graph; and
6. letting $Z \subset Y$ be the subgraph associated to the 0-end of $Z \times (0,1)$ we have: $\mathrm{rr}(N) = \mathrm{rr}(Y, Z)$ is minimal with the above properties.

N arises as follows: There is an immersion $Y \to B$ of a compact connected graph, a connected subgraph $Z \subset Y$, and a map $\phi' : Z \to Y$ induced from $\phi : B \to B$. Generalizing Definition 6.3, N is the mapping torus of the partial endomorphism ϕ'.

$$N = \big(Y \cup Z \times (0,1)\big) \, / \, \big\{ (z,0) \sim z, \ (z,1) \sim \phi'(z) \ : \ z \in Z \big\}$$

To see that N exists, let $b_o \in B^0$ be the basepoint, let Y be a finite connected subgraph of B containing $\{b_o, \phi(b_o)\}$, let $Z = \{b_o\}$, let $\phi' : Z \to Y$ be induced by ϕ, and let $N = Y \cup (b_o \times (0,1))$ where $(b_o \times (0,1))$ maps to the open edge t of M. As $\pi_1 M$ is f.g., a sufficiently large finite subgraph Y will ensure π_1-surjectivity of $N \to M$.

We now show that if $\mathrm{rr}(N)$ is minimal, then $N \to M$ is π_1-injective. Otherwise, there is an essential path $P \to N$ whose image $P \to M$ bounds a disk diagram D. Choose such a pair (D, P) so that D is minimal among all such choices with essential P. Note that D is nontrivial since P is nontrivial, and that D has no spur since otherwise removing the spur and the associated backtrack in P would provide a smaller diagram.

By Lemma 6.5, D has a t-strip T with $\partial_{\mathsf{p}} T = t^{-1} Q t S$. Either T is outward and $t S t^{-1}$ is a subpath of $P = \partial_{\mathsf{p}} D$, or T is inward and $t^{-1} Q t$ is a subpath of P. We treat these two cases in turn.

Outward case: Let $Y' = Y \cup_{\partial Q} Q$, and $Z' = Z \cup_{\partial Q} Q$. Observe that $\mathsf{rr}(Y', Z') = \mathsf{rr}(Y, Z)$, and although $Z' \to B$ might not be immersed, it is π_1-injective since $Z' \xrightarrow{\phi} Y$ and $Y \to B$ are π_1-injective. Let $N' = N \cup_{t S t^{-1}} T$, so there is a map $N' \to M$, and folding of Y' induces folding of N'. Note that $Y' \to B$ must be π_1-injective or else folding will result in a decrease of rr by Lemma 6.7.

Inward case: $Y' = Y \cup_{\partial S} S$ and $Z' = Z \cup_{\partial Q} Q$, so $Z' \subset Y$. Let $N' = N \cup_{t^{-1} Q t} T$, so there is a map $N' \to M$. As $Z' \subset Y$ is a subgraph and $Y \to B$ is an immersion, the map $Z' \to B$ is π_1-injective. Moreover, as $Y' = Y \cup S$ is the union of an immersed graph with a path, either S will fold entirely into Y as we fold $Y' \to B$ to become an immersion, in which case $\mathsf{rr}(\bar{Y}', \bar{Z}') < \mathsf{rr}(Y', Z') = \mathsf{rr}(Y, Z)$ by Lemma 6.7, or disjoint initial and terminal parts of S will fold into Y. In the latter case, the associated folding of N' to $\bar{N}$ is likewise a homotopy equivalence.

Starting with $N_1 = N$ and $D_1 = D$, and proceeding in this way, we obtain a sequence N_i associated to (Y_i, Z_i). At each stage we use an inward or outward t-strip T_i in a diagram $D_i \to M$ for a path $P_i \to N_i$ that is homotopic to $P \to N \to N_i$. Note that D_i can be formed from D_{i-1} by absorbing into N_i as many cells from D_{i-1} as possible. At some point, rr must decrease, since at a last stage, the innerpath S_i of the inward strip T_i (which is all of D_i) has the property that S_i is a subpath of $\partial_{\mathsf{p}} D_i$, so S_i would get folded completely into Z_i, and rr decreases. $\square$

7   SOLVABLE GROUPS

The following characterization was given in [BS79]. See also [Gro78]:

Theorem 7.1. *Let G be a f.g. solvable group. The following are equivalent:*

1. *G is coherent.*
2. *G is an ascending HNN extension of a polycyclic group.*

Problem 7.2. Is there an example of a f.g. coherent group H and an automorphism $\phi : H \to H$, such that $H \rtimes_\phi \mathbb{Z}$ is not coherent?

Similarly, is there an incoherent ascending HNN extension for a monomorphism $\phi : H \to H$ where H is f.g. and coherent?

The following example shows that Problem 7.2 has a negative solution if the f.g. assumption on H is omitted:

Example 7.3. The wreath product $\mathbb{Z} \wr \mathbb{Z}$ is f.g. but not f.p., since its 2nd homology is not f.g. Nevertheless, $\mathbb{Z} \wr \mathbb{Z} \cong \mathbb{Z}^\infty \rtimes_\phi \mathbb{Z}$ where $\phi : \mathbb{Z}^\infty \to \mathbb{Z}^\infty$ is given by $\phi(x_i) = (x_{i+1})$ where

$$\mathbb{Z}^\infty = \langle\, x_i : i \in \mathbb{Z} \mid [x_i, x_j] : i, j \in \mathbb{Z} \,\rangle.$$

While I expect Problem 7.2 to have a negative solution, I believe the following special case holds:

Conjecture 7.4. *If H is word-hyperbolic and coherent, then $H \rtimes_\phi \mathbb{Z}$ is coherent for any automorphism ϕ of H.*

This is probably also true for any ascending HNN extension of H. Conjecture 7.4 presumably holds because of the limited structure of automorphisms of word-hyperbolic groups. It is largely proven in [BW03], but there remains a gap when H is freely decomposable—a case which is understood already for H free by Theorem 6.1.

8   THE PERIMETER APPROACH

In [MW05], McCammond-Wise developed the notion of perimeter, and used it to prove the halting of an algorithm that produces a compact 2-complex whose fundamental group equals the desired subgroup H. In [MW08] we incorporated results of [MW02] with [MW05] to examine local-quasiconvexity of small-cancellation groups.

In Section 8.a we review definitions about disk diagrams. In Section 8.b we define the notion of "perimeter." In Section 8.c we give a quick statement of some of the results obtained using perimeter. In Section 8.d we discuss the addition of perimeter-reducing missing shells. In Section 8.e we prove the main results using perimeter to deduce coherence and local-quasiconvexity but without specifying the context. In Section 8.f we review the spelling theorem and its variants for one-relator groups with torsion. In Section 8.g we review the definitions and main theorems of small-cancellation theory. In Section 8.h we discuss small-cancellation criteria for determining that maps are π_1-injective and have quasiconvex image. In Section 8.i, we describe the applications towards coherence and local-quasiconvexity of small-cancellation groups.

8.a   Disk diagrams

We now review the basic definitions concerning disk diagrams. We refer to [LS77] or [MW02] for a more detailed account.

Definition 8.1. *A disk diagram D is a planar, simply- connected 2-complex. Given a 2-complex X we will often study a map $D \to X$ and refer to this map as the disk diagram. Let $P \to X$ be a closed path which factors as $P \to D \to X$. The disk diagram $D \to X$ is a disk diagram for $P \to X$ provided that P maps onto ∂D, and furthermore for each 1-cell e of ∂D, the preimage of e in P must consist of one or two 1-cells according to whether or not e lies on the boundary of a 2-cell of D. The path P is a boundary path for D. It is well-known that a disk diagram $D \to X$ exists for each nullhomotopic closed path $P \to X$ (see, for instance, [LS77]).*

A disk diagram $D \to X$ is minimal if there does not exist a disk diagram $D' \to X$ with $\partial_p D' = \partial_p D$ but D' has fewer cells than D. In many contexts it suffices to work with the weaker notion of reduced diagram. See Definition 8.32.

8.b   Perimeter

Definition 8.2 (Perimeter). *Let X be a locally-finite 2-complex. Let $\phi : Y \to X$ be a combinatorial map. Define $\mathbf{P}(Y) = \mathbf{P}(Y \to X)$ as follows: For each 1-cell e in Y, let $\mathbf{P}(e) = |\operatorname{link}(\phi(e)) - \phi(\operatorname{link}(e))|$, which equals the number of sides of 2-cells along*

Figure 4: Sides in a 2-cell and at a 1-cell on the left. On the right, the sides of quadrilaterals at $\phi(e)$ are missing at e. The other sides are present.

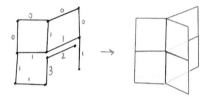

Figure 5: The map above has $\mathbf{P} = 11$. The number of missing sides is indicated along each 1-cell.

$\phi(e)$ *that are* missing *at e in the sense that they do not lift to sides of 2-cells along e. See Figure 4. Define the* perimeter *of Y to be:*

$$\mathbf{P}(Y) = \sum_{e \in\ Edges(Y)} \mathbf{P}(e). \tag{1}$$

There is a more flexible notion called weighted perimeter. *A* weighting *on a 2-complex X is an assignment of nonnegative integer* weights w_i *to the points i in* link(d) *for each 1-cell d of X. Regard this as a weighting on the sides of the 2-cells of X. Given a map $\phi : Y \to X$, for each edge e in Y define:*

$$\mathbf{P}(e) = \sum_{i \in \left(\text{link}(f(e)) - f(\text{link}(e)) \right)} w_i.$$

Define $\mathbf{P}(Y)$ as in Equation (1). This generalizes the unit perimeter *defined above which uses the weighting $w_i = 1$ for each i.*

The attaching map $\partial R \to X^1$ of each 2-cell R in X can be expressed as W^n where W is a path in X^1 that is not periodic. We let $\mathbf{Wt}(R) = \mathbf{P}(W) = \sum_{i \in Sides(R)} w_i$, and the number n is the exponent *of R. We will generally require that a weighting has the property that $\mathbf{Wt}(R) > 0$ for each 2-cell R.*

Generally speaking, one can use rational weights as long as finitely many values occur. However, a crucial consequence of nonnegativity of the weighting is that $\mathbf{P}(Y) \geq 0$ for each map $Y \to X$.

We encourage the reader to ignore this generalization, by assuming all weights are 1.

Lemma 8.3 (Folding and perimeter). *Let $Y \to Y' \to X$ be a composition of combinatorial immersions with $Y \to Y'$ surjective. Then $\mathbf{P}(Y) \geq \mathbf{P}(Y')$.*

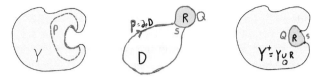

Figure 6: Y has an essential path P bounding a disk diagram $D \to X$ with a shell R. We form $Y^+ = Y \cup_Q R$.

Proof. Each missing side in Y' lifts to a missing side in Y.   □

Lemma 8.4 (Adding a shell). *Let R be a 2-cell with $\partial_p R = QS$, and suppose R is missing along a lift $Q \to Y$. Let $Y^+ = Y \cup_Q R$. Then $\mathbf{P}(Y^+) = \mathbf{P}(Y) + \mathbf{P}(S) - |R|$.*

Proof. This follows from the definition of $\mathbf{P}$ since R is missing along $\partial_p R$ in $Y \cup S$.   □

The desire to make $\mathbf{P}(Y^+) < \mathbf{P}(Y)$ in Lemma 8.4 leads to the following:

Definition 8.5 (Perimeter reducing shell). *A shell $R = r^n$ with innerpath S is perimeter reducing if $\mathbf{P}(S) < n|r|$.*

Definition 8.6 (Perimeter reduction hypothesis). *The weighted 2-complex X satisfies the* perimeter reduction hypothesis *if each spurless reduced disk diagram $D \to X$ is either trivial or has a perimeter reducing shell.*

To give some intuition about how we treat torsion, and why it is to our advantage, consider the following:

Example 8.7 (Why torsion helps). Let X_n denote the standard 2-complex of the following presentation for $n \in \mathbb{N}$:

$$\langle \, a, b, c \mid (aba^{-1}b^{-2})^n, (bcb^{-1}c^{-2})^n, (cac^{-1}a^{-2})^n \, \rangle.$$

Observe that all pieces in X_1 have length 1, which considerably simplifies the discussion. Moreover, this remains the case for arbitrary n, since a length 2 subword is determined uniquely, up to a $\mathbb{Z}_n$-symmetry of its relator. The $C(6)$ small-cancellation condition is satisfied when $n \geq 2$. (In fact, so is the $C(10)$ condition.)

We now discuss how torsion interacts with out perimeter wishes. Similar reasoning holds if we vary the exponents of the relators independently.

In X_1, we have $\mathbf{P}(e) = 5$ for each edge e. If we counted perimeter in the obvious fashion, then in X_n, we would have $\mathbf{P}(e) = 5n$, since there would be n times as many 2-cells attached along e in X_n as in X_1.

Consider a map $Y \to X_n$, and suppose we are extending it to a map $Y \cup_Q R \to X_n$ for some 2-cell R. Moreover, assume that a 2-cell lies in Y if and only if its "associates" lie in Y. Here if $\partial R = r^n$ then we regard R' to be an associate of R if it is a copy of R, but its basepoint differs from R by a path of the form r^p in Y. When we attach a 2-cell to Y, we will actually attach all n associates. (There are n of them if r has order n in $\pi_1 X_n$.)

If $Q \to R$ wraps entirely around ∂R and R is missing along $Q \to Y$, then we certainly have $\mathbf{P}(Y \cup_Q R) < \mathbf{P}(Y)$. So let us consider the case where Q does not circumnavigate ∂R, so $QS = \partial_\mathsf{p} R$ for some nontrivial path S. Then when we form $Y \cup_Q R$, we are first attaching S to Y, and then adding n associates of R together. Hence $\mathbf{P}(Y \cup_Q R) = \mathbf{P}(Y) + \mathbf{P}(S \to X_n) - n|r^n|$.

To guarantee that R is perimeter reducing, we would want $n|r^n| > \mathbf{P}(S \to X_n)$. In a small-cancellation setting, where our 2-cell R has innerpath S that is at most 3 pieces long, there is a huge advantage toward perimeter reduction as n grows, since S is bounded. Hence n makes a quadratic contribution to $n|r^n| = n^2|r|$ but only a linear contribution to $\mathbf{P}(S \to X_n) = n\mathbf{P}(S \to X_1)$. It thus suffices to choose n so that $n^2|r| > n|\mathbf{P}(S)|$, or equivalently, $n > \frac{|\mathbf{P}(S)|}{|r|}$ for each S that is the concatenation of three pieces of r.

For our example, we have $\mathbf{P}(S \to X_1) \leq 5\mathbf{P}(e) = 15$, and $|r| = 5$. Hence X_n has perimeter reducing shells for $n \geq 4$.

Before continuing towards applications of these definitions, we remark that the perimeter reduction hypothesis appears to imply the nonpositive immersion property, but I have not proven this in general:

Conjecture 8.8. *Any compact aspherical 2-complex X satisfying the perimeter reduction hypothesis is either collapsible to a point or has $\chi(X) \leq 0$.*

One has to use an appropriate generalization of Euler characteristic to accommodate the case where X has duplicate 2-cells or 2-cells attached by proper powers. In particular, Conjecture 8.8 would imply the nonpositive immersion hypothesis for 2-complexes satisfying the perimeter reduction hypothesis (provided there aren't duplicate 2-cells in the above sense).

Progress on Conjecture 8.8 is made through a counting argument in [Wis18], and another argument using L^2-betti numbers made in [Wisa]; see Section 16.

8.c   Preview of coherence and quasiconvexity results

The following are proven in Section 8.i. The $C(p)$-$T(q)$ and $C'(\frac{1}{6})$ small-cancellation conditions are defined in Section 8.g.

Theorem 8.9 (Coherence using i-shells). *Let X be a weighted 2-complex which satisfies $C(6)$-$T(3)$ $[C(4)$-$T(4)]$. Suppose $\mathbf{P}(S) \leq n\mathbf{Wt}(R)$ for each 2-cell $R \to X$ and path $S \to \partial R$ which is the concatenation at most three [two] consecutive pieces in the boundary of R. Then $\pi_1 X$ is coherent.*

Theorem 8.10 (Local quasiconvexity). *Let X be a compact weighted 2-complex that satisfies $C'(\frac{1}{6})$-$T(3)$ $[C'(\frac{1}{4})$-$T(4)]$. Suppose $\mathbf{P}(S) < n\mathbf{Wt}(R)$ for each 2-cell $R \to X$ and path $S \to \partial R$ which is the concatenation at most three [two] consecutive pieces in the boundary of R. Then $\pi_1 X$ is locally-quasiconvex.*

Example 8.11 (Surface groups). Let X be the standard 2-complex of the presentation $\langle a_1, \ldots, a_g \mid a_1^2 a_2^2 \cdots a_g^2 \rangle$. Then X is homeomorphic to a nonorientable surface of genus g. Clearly for any $g \geq 2$, X satisfies $C(2g)$-$T(4)$, and the pieces are of length 1. Using the unit perimeter, we have $\mathbf{P}(e) = 2$ for each edge (and hence each piece), and the weight of the 2-cell is $2g$. Thus by Theorem 8.9, $\pi_1 X$ is coherent for $g \geq 2$ and by Theorem 8.10, $\pi_1 X$ is locally-quasiconvex for $g > 2$.

A similar result holds with X the standard 2-complex of the presentation

$$\langle a_1, b_1, \ldots, a_g, b_g \mid [a_1, b_1][a_2, b_2] \cdots [a_g, b_g] \rangle$$

so that X is an orientable surface of genus g. For $g \geq 1$ the 2-complex X satisfies $C(4g)$-$T(4)$, the pieces are of length 1, the weight of each piece is 2, and the weight of the 2-cell is $4g$. Thus by Theorem 8.9, $\pi_1 X$ is coherent for $g \geq 1$ and by Theorem 8.10, $\pi_1 X$ is locally-quasiconvex for $g > 1$.

The fact that these methods apply to surface groups is to be expected since the boundary of a 2-manifold inspired the notion of perimeter. Here is a more novel application of Theorem 8.9.

Theorem 8.12. *Let* $G = \langle a_1, \ldots \mid R_1, \ldots \rangle$ *be a finite presentation that satisfies* $C'(\frac{1}{n})$. *If each* a_i *occurs at most* $n/3$ *times among the* R_j, *then* G *is coherent and locally-quasiconvex.*

A "ladder" form of Theorem 8.27 obtained in [HW01], supports the following application of Theorem 8.23. The analogous statement for $n \geq 3|W|$ was proven in [MW05]. Although we now know how to prove coherence whenever $n \geq 2$ (see Corollary 14.14) quasiconvexity is not yet known.

Theorem 8.13. *Let* $G = \langle a_1, \ldots \mid W^n \rangle$ *be a one-relator group with* $n \geq |W|$. *Then* G *is locally-quasiconvex.*

The following theorem generalizes the one-relator situation by showing that finite presentations with sufficient torsion will always be coherent and locally-quasiconvex.

Theorem 8.14 (Power theorem). *Let* $\langle a_1, \ldots, a_r \mid W_1, \ldots, W_s \rangle$ *be a finite presentation, where each* W_i *is a cyclically reduced word that is not a proper power. If* W_i *is not freely conjugate to* $W_j^{\pm 1}$ *for* $i \neq j$, *then there exists a number* N *such that for all choices of integers* $n_i \geq N$ *the group* $G = \langle a_1, \ldots, a_r \mid W_1^{n_1}, \ldots, W_s^{n_s} \rangle$ *is locally-quasiconvex and word-hyperbolic.*

Theorem 8.14 can easily be applied to obtain the following which distinguishes the results in [MW05] from other results described in this text.

Corollary 8.15. *There exist perfect groups that satisfy the perimeter reduction hypothesis and are thus coherent.*

8.d   Reducing perimeter by attaching shells

Showing that a 2-complex has a weighting satisfying the perimeter reduction hypothesis leads to the consideration of strong forms of Dehn's Lemma for word-hyperbolic groups.

Given a "compression disc" D for $\bar{Y} \to X$ a natural choice is to let $Y' = \bar{Y} \cup_P D$, and hope that perhaps after folding, $\mathbf{P}(Y') < \mathbf{P}(Y)$. Unfortunately, it is complicated to keep track of the exact effect of attaching the disk diagram and then folding (see Problem 8.25). However, the hypothesis that D has minimal area among all possible such choices implies that all the 2-cells along ∂D are missing in $\bar{Y}$. We therefore attempt to add a single 2-cell R of D to $\bar{Y}$.

Definition 8.16 (Perimeter reducing 2-cell). *Let $D \to X$ be a disk diagram with boundary path P and let R be a 2-cell in D, and let n be the exponent of R as 2-cell of X. Suppose ∂R is the concatenation QS where the outerpath Q of R is the subpath Q of P that lies in ∂R, and S is the innerpath of R. Say R is weakly perimeter reducing if the inequality in Equation (2) holds, R is perimeter reducing if the equation holds with a strict inequality.*

$$\mathbf{P}(S) \leq n \cdot \mathbf{Wt}(R). \tag{2}$$

Remark 8.17 (Torsion). The argument is slightly simpler when $n = 1$. In general, we will add n-distinct copies of R, unless $\partial_{\mathsf{p}} R$ is already a closed path in Y. When $\partial_{\mathsf{p}} R = \gamma^n$, and γ has order n in $\pi_1 X$, then n copies of R are attached along the same boundary cycle in $\widetilde{X}$.

Because of the way that R will arise in our arguments, we will always be able to assume that when one is missing in Y, then all n of these copies are missing. This justifies the multiplicative "n" in Equation (2).

8.e   Coherence, cores, and quasiconvexity via perimeter

Definition 8.18 (Shells, outerpaths and innerpaths). *A shell R in a disk diagram D is a 2-cell such that $\partial_{\mathsf{p}} R = QS$ where Q is a subpath of $\partial_{\mathsf{p}} D$, and S is "small" in a sense that will depend on the context (see Remark 8.19). The path Q is the outerpath of R and S is the innerpath of R.*

Remark 8.19.

1. When D is a $C(6)$ diagram, S is the concatenation of at most 3 pieces.
2. When D is a $C(4)$-$T(4)$ diagram, S is the concatenation of at most 2 pieces.
3. When D is a disk diagram arising from a one-relator group with torsion $\langle a, b \ldots \mid W^n \rangle$, we require that $|S| < |W|$. More generally, this is suitable for a staggered 2-complex with torsion. (See Section 14.a.)

Theorem 8.20 (Compact core). *Let X be a weighted 2-complex satisfying the perimeter reduction hypothesis. Then X has the compact core property.*

Specifically, let $\widehat{X} \to X$ be a finite cover, and let $C \subset \widehat{X}$ be a compact subcomplex. Then there is a compact connected subcomplex Y with $C \subset Y$, and such that the inclusion $Y \to \widehat{X}$ is a π_1-isomorphism.

Proof. Since $\pi_1 \widehat{X}$ is f.g. we can pass to a compact connected subcomplex Y containing C such that $\pi_1 Y \to \pi_1 \widehat{X}$ is surjective. Choose Y with this property so that $\mathbf{P}(Y)$ is minimal. Suppose $\pi_1 Y \to X$ is not π_1-injective. Choose an essential closed path $P \to Y$ that bounds a disk diagram $D \to \widehat{X}$. And assume (D, P) are chosen so that $D \to \widehat{X}$ has minimal area among all such choices. Note that we can assume that $P \to Y$ is immersed by removing backtracks and hence spurs in D. By hypothesis, D has a perimeter reducing shell R with outerpath $Q \subset P$. Note that Q must be missing in Y along $Q \to Y$, for otherwise we can produce a smaller area diagram D' with $\partial_p D' \to Y$ homotopic to $P \to Y$, and thus essential. Let $Y^+ = Y \cup R$, and note that $\mathbf{P}(Y^+) < \mathbf{P}(Y)$ which is impossible.   □

Theorem 8.21 (Presentation algorithm and membership problem). *Let X be a compact 2-complex. Suppose every minimal diagram $D \to X$ is either trivial, or has a spur, or has a perimeter reducing shell. Then there is an algorithm to compute finite presentations of f.g. subgroups of $\pi_1 X$.*

Moreover, suppose each such D is either trivial or a single 2-cell, or has at least two spurs and/or shells. Then there is an algorithm to solve the membership problem for an element in a f.g. subgroup.

Proof. Let H be a f.g. subgroup of $\pi_1 X$, so $H = \langle h_1, \ldots, h_r \rangle$ for some elements of $\pi_1 X$. For each i, let σ_i be a based combinatorial path at the basepoint of X. Let $Y_1 = \vee_{i=1}^r h_i$, so there is a based combinatorial map $Y_1 \to X$ with $\pi_1 Y_1$ mapping to H.

For each i, we either

1. obtain Y_{i+1} from Y_i by folding two edges,
2. obtain Y_{i+1} from Y_i by identifying two 2-cells sharing the same boundary path,
3. or $Y_i \to X$ is an immersion and $Y_{i+1} = Y_i \cup_Q R$ is obtained by adding a missing perimeter reducing shell along its outerpath.

Note that both folding/identifying and finding missing shells are algorithmic (they take linear time in the size of Y_i). The folding step reduces the number of edges but does not increase $\mathbf{P}$ by Lemma 8.3, the 2-cell identification step makes the complex smaller without affecting $\mathbf{P}$, and the step that adds shells actually reduces $\mathbf{P}$; the algorithm terminates at an immersion $Y_t \to X$ of a compact 2-complex after finitely many steps.

We claim that the final result $Y_t \to X$ is a π_1-injection. To see this, suppose $P \to Y_t$ is a closed essential path that is nullhomotopic in X, and let $D \to X$ be a disk diagram for P. And assume (D, P) are chosen as above, so that D is minimal among all possible choices. By hypothesis, D has a perimeter reducing shell, and it must be missing along its outerpath, or there would be a small choice. This contradicts that the algorithm terminates.

Now consider the strengthened hypothesis, that there are at least two spurs and/or shells. Then the final complex Y_t has the property that the map $\widetilde{Y_t} \to \widetilde{X}$ is an injection. For suppose two distinct vertices of $\widetilde{Y_t}$ map to the same vertex of $\widetilde{X}$, and let $P \to \widetilde{Y_t}$ be a path between them. Let $D \to X$ be a disk diagram for P, and suppose (D, P) are chosen as above so that D is minimal.

We can assume that D is nontrivial. Observe that P must contain a spur or shell with outerpath along its interior. Indeed, at most one spur or shell is broken by the endpoint of P, so the other lies along its interior. (Similarly, the case that D is a single 2-cell is handled by treating its outerpath as P.) If this shell were missing then the algorithm had not yet terminated.

This solves the membership problem which is equivalent to being able to build arbitrarily large parts of the based covering space $\widehat{X}$ associated to H. To be more explicit, let $g \in \pi_1 X$, and let $\sigma_o \to X$ represent g. Define $Y_1 = \vee_0^r \sigma_i$. Now σ_0 maps to a closed path in Y_t if and only if $g \in H$. Indeed, if $g \in H$, then $g = W(h_1, \ldots, h_r)$ for some word W in $\{h_1^{\pm 1}, \ldots, h_r^{\pm 1}\}$, and so there is a disk diagram D whose boundary path equals $\sigma_0^{-1} W(\sigma_1, \ldots, \sigma_r)$. At each stage, some path that is path-homotopic to it would give a fold or missing shell, until this path is closed at the end of the algorithm.   □

Remark 8.22 (Complexity). The number of steps in the algorithm is at most quadratic in the sum of the lengths of the input elements and universal constants depending on X.

Theorem 8.23 (Quasi-isometrically embedded subgroups). *Let X be a compact 2-complex. Suppose that for each diagram $D \to X$ with $\partial_{\mathsf{p}} D = P_1 P_2$, and $P_1 \to \widetilde{X}$ a geodesic (with respect to the graph metric on the 1-skeleton) either there is a spur or perimeter reducing shell whose outerpath is on P_2, or else $P_1 \subset \mathcal{N}_J(P)$ within D, for some $J = J(X)$.*

Then every f.g. subgroup $H \subset \pi_1 X$ is quasi-isometrically embedded.

Proof. Choose $Y \to X$ to be a combinatorial immersion with Y compact, with $\pi_1 Y$ mapping to H, and with $\mathbf{P}(Y)$ minimal among all such choices. Consider the map $\widetilde{Y} \to \widetilde{X}$. It is injective by Theorem 8.20 whose hypothesis is covered by the case where P_1 is trivial.

Consider a geodesic $\gamma \to \widetilde{X}$ whose endpoints lie in $\widetilde{Y}$. We claim that $\gamma \subset \mathcal{N}_J(\widetilde{Y})$. To see this, consider a disk diagram $D \to \widetilde{X}$ with $\partial_{\mathsf{p}} D = \gamma \lambda$ where $\lambda \to \widetilde{Y}$ is a path, and D is chosen to be minimal among all choices (D, λ) where λ is allowed to vary in $\widetilde{Y}$.

By hypothesis, either $\gamma \subset \mathcal{N}_J(\lambda)$ in D, and hence $\gamma \subset \mathcal{N}_J(\widetilde{Y})$, or else, D has a perimeter reducing shell R whose outerpath lies on λ. The complex $Y \cup_Q R$ which is the quotient of $\widetilde{Y} \cup HR$ by H, provides a new map with $\mathbf{P}(Y \cup_Q R) < \mathbf{P}(Y)$, and so we obtain a lower $\mathbf{P}$ immersion after folding.   □

Remark 8.24. A more careful analysis actually shows that for the coherence and core results above it suffices to assume the shells are weakly perimeter reducing. We refer to [MW05].

Problem 8.25. Let X be a compact nonnegatively weighted 2-complex. Suppose $\mathbf{P}(D) < \mathbf{P}(\partial_{\mathsf{p}} D)$ for every minimal area disk diagram $D \to X$. Does it follow that $\pi_1 X$ is coherent?

8.f   The spelling theorem and one-relator groups with torsion

Theorem 8.26 (Newman spelling theorem). *Let $G = \langle x_1, \dots \mid W^n \rangle$ be a one-relator group with torsion. Let U be a freely and cyclically reduced word representing the identity of G. Then U contains a subword Q such that Q is a subword of the cyclic word W^n and $|Q| > |W^{n-1}|$.*

Newman's spelling theorem was announced in [New68], and details can be found in [LS77, IV.5.5]. Theorem 8.26 is most naturally proven as a consequence of the following which we state for staggered 2-complexes with torsion (see Section 14.a):

Theorem 8.27 (Two external 2-cells). *Let X be a staggered 2-complex and let $D \to X$ be a reduced disk diagram. Suppose that D has no spurs, and D contains* *at least two 2-cells. Then D contains two distinct 2-cells C_1 and C_2 such for each i there is a subpath Q_i of the attaching path of C_i and $|Q_i| > |W|^{n-1}$.*

When S is a proper subword of W, using the unit perimeter, we have $\mathbf{P}(S) < n\mathbf{Wt}(W)$. Consequently:

Theorem 8.28. *Let W be a cyclically reduced word and let $G = \langle a_1, \dots \mid W^n \rangle$. If $n \geq |W| - 1$, then G is coherent.*

The following "ladder theorem" for staggered groups with torsion is proven in [HW01, Thm 5.2]. A more exacting statement is described in [LW13].

The 2-cell R of the disk diagram D is λ-*external* if there is a subpath Q of $\partial_{\mathsf{p}} R$ such that:

1. Q is a subpath of $\partial_{\mathsf{p}} D$,
2. $|Q| > \lambda |\partial_{\mathsf{p}} R|$.

Theorem 8.29 (Ladder). *Let X be a staggered 2-complex and suppose that there are constants n and k such that each 2-cell of X is attached along a path U^n, where U is a non-periodic path of length k.*

Let $D \to X$ be a reduced disk diagram with no spurs. Then either:

1. *D consists of a single 0-cell;*
2. *D is a ladder with an $\left(\frac{n-1}{n}\right)$-external 2-cell at each end; or*
3. *D has at least three $\left(\frac{n-2}{n}\right)$-external 2-cells.*

Assuming each $\left(\frac{n-2}{n}\right)$-external 2-cell is perimeter reducing yields local-quasiconvexity for such staggered groups by Theorem 8.23. (Note that they are word-hyperbolic [HW01].) In particular:

Theorem 8.30. *The group $G = \langle a, \dots \mid W^n \rangle$ is locally-quasiconvex provided that $n + 1 > 2|W|$.*

8.g   Small-cancellation theory

We now review the basic notions of small-cancellation theory. The reader is referred to [MW02] for a rigorous development of these notions that is consistent with their use here. The classical reference is the book [LS77], and Strebel gives a very nice introduction to small-cancellation theory in [Str90].

Definition 8.31 (Piece). *Let X be a combinatorial 2-complex. Intuitively, a piece of X is a path which is contained in the boundaries of the 2-cells of X in at least two distinct ways. More precisely, a nontrivial path $P \to X$ is a piece of X if there are 2-cells R_1 and R_2 such that $P \to X$ factors as $P \to R_1 \to X$ and as $P \to R_2 \to X$ but there does not exist a homeomorphism $\partial R_1 \to \partial R_2$ such that there is a commutative diagram*

$$
\begin{array}{ccc}
P & \to & \partial R_2 \\
\downarrow & \nearrow & \downarrow \\
\partial R_1 & \to & X.
\end{array}
$$

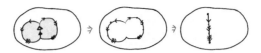

Figure 7: Removing a cancellable pair.

Excluding commutative diagrams of this form ensures that P occurs in ∂R_1 and ∂R_2 in essentially distinct ways.

Definition 8.32 (Cancellable pair and reduced). *A cancellable pair in a disk diagram $D \to X$ in a 2-complex X is a pair of 2-cells R_1, R_2 that meet along a nontrivial arc A, and so that $\partial_p R_1 = AB_1$ and $\partial_p R_2 = AB_2$ and B_1, B_2 map to the exact same path in X. A disk diagram with no cancellable pair is reduced. A piece is a nontrivial arc A that appears in the overlap between two 2-cells in a reduced diagram.*

A disk diagram $D \to X$ with minimal area among all diagrams with the same boundary path must be reduced. Indeed, we can "remove a cancellable pair" as in Figure 7.

Definition 8.33 ($C(p)$-$T(q)$-complex and $C'(\alpha)$-complex). *An arc in a diagram is a path whose internal vertices have valence 2 and whose initial and terminal vertices have valence ≥ 3. The arc is internal if its interior lies in the interior of D, and it is a boundary arc if it lies entirely in ∂D.*

A 2-complex X satisfies the $T(q)$ condition if for every minimal area disk diagram $D \to X$, each internal 0-cell of D has valence 2 or valence $\geq q$. Similarly, X satisfies the $C(p)$ condition if the boundary path of each 2-cell in D either contains a nontrivial boundary arc, or is the concatenation of at least p nontrivial internal arcs. Finally, for a fixed positive real number α, the complex X satisfies $C'(\alpha)$ provided that for each 2-cell $R \to X$, and each piece $P \to X$ which factors as $P \to R \to X$, we have $|P| < \alpha|\partial R|$. Note that if X satisfies $C'(\frac{1}{n})$ then X satisfies $C(n+1)$.

When $D \to X$ is minimal area, each nontrivial arc in the interior of D is a piece in the sense of Definition 8.31. See also Definition 8.32.

When $\frac{1}{p} + \frac{1}{q} \leq \frac{1}{2}$, nontrivial minimal area diagrams in X always contain certain features of positive curvature that we now discuss.

Definition 8.34 (*i*-shells and spurs). *Let D be a diagram. An i-shell of D is a 2-cell $R \hookrightarrow D$ whose boundary cycle ∂R is the concatenation $P_0 P_1 \cdots P_i$ where $P_0 \to D$ is a boundary arc, the interior of $P_1 \cdots P_i$ maps to the interior of D, and $P_j \to D$ is a nontrivial interior arc of D for all $j > 0$. The path P_0 is the outerpath of the i-shell.*

A 1-cell e in ∂D which is incident with a valence 1 0-cell v is a spur. In analogy with the outerpath of an i-shell, we regard the length 2 path (either ee^{-1} or $e^{-1}e$) that passes through v as the outerpath of the spur.

Illustrated from left to right in Figure 8 are disk diagrams containing a spur, a 0-shell, a 1-shell, a 2-shell, and a 3-shell. In each case, the 2-cell R is shaded, and the boundary arc P_0 is $\partial R \cap \partial D$.

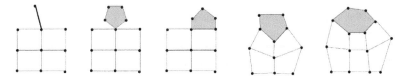

Figure 8: Spur and i-shells.

Figure 9: The trichotomy of Theorem 8.35.

The classical result underpinning small-cancellation theory is "Greendlinger's Lemma" (see [LS77, Thm V.4.5]). The following strengthening of Greendlinger's Lemma was proven in [MW02, Thm 9.4].

Theorem 8.35. *If D is a $C(4)$-$T(4)$ $[C(6)$-$T(3)]$ disk diagram, then one of the following holds:*

1. *D consists of a single 0-cell or a single 2-cell.*
2. *D is a ladder of width ≤ 1, and hence has a spur, 0-shell or 1-shell at each end.*
3. *D contains at least three spurs and/or i-shells with $i \leq 2$ $[i \leq 3]$.*

The utility of Theorem 8.35 is that when X is $C(6)$ then any reduced disk diagram $D \to X$ is also $C(6)$. Likewise for $C(4)$-$T(4)$.

8.h   Injectivity and quasiconvexity using small-cancellation

Definition 8.36 (No missing shells). *Let $Y \to X$ be a combinatorial immersion. Let R be a shell in a reduced disk diagram $D \to X$ and let Q be its outerpath. Let $Q \to Y$ be a lift of $Q \to X$, in the sense that $Q \to D \to X$ equals $Q \to Y \to X$. We say R is* missing *at $Q \to Y$, if there is no lift $R \to Y$ of $R \to X$, so that the path $Q \to Y \to X$ equals $Q \to R \to X$.*

We say $Y \to X$ has no missing shells *if there is no shell R in a disk diagram $D \to X$ that is missing along some lift $Q \to Y$ of its outerpath.*

The following serves as the main interface between small-cancellation and coherence in [MW05]. It has useful further generalizations to cubical small-cancellation theory.

Lemma 8.37 (Injectivity). *Let X be a $C(6)$ $[or\ C(4)$-$T(4)]$ complex. Let $Y \to X$ be a combinatorial immersion with no missing shells and with Y connected. Then $\widetilde{Y} \to \widetilde{X}$ is injective. Hence $\pi_1 Y \to \pi_1 X$ is injective.*

Proof. It suffices to prove the first statement. Let $p, q \in \widetilde{Y}$ be distinct points that map to the same point in $\widetilde{X}$. Let $P \to \widetilde{Y}$ be an immersed path in $\widetilde{Y}$ from p to q. Let $D \to \widetilde{X}$ be a disk diagram for P. Moreover, choose (D, P) so that D has a minimal number of cells among all such possible choices with p, q fixed.

Consider the three possibilities for D given by Theorem 8.35. If D is trivial then $p = q$. If $D = R$ is a single 2-cell, then $\partial_p R = P$ is a single outerpath with trivial innerpath, so R lifts to $\widetilde{Y}$, and so again $p = q$. If D has at least two spurs and/or shells, then at least one of these has outerpath that is a subpath of P. The case of a spur implies that D can be made smaller by removing a backtrack from P, the case of a shell R with $\partial_p R = QS$ and Q a subpath of P, implies by no missing shells, that R lies in $\widetilde{Y}$. Hence there is a small diagram D' obtained by removing R and obtaining P' from P by replacing Q by S. $\square$

Definition 8.38. *The $C(6)$ complex [or $C(4)$-$T(4)$] X has* short innerpaths *if $|S| < |Q|$ for each i-shell R with outerpath Q and innerpath S and $i \leq 3$ [and $i \leq 2$].*
For instance, $C'(\frac{1}{6})$ and $C'(\frac{1}{4})$-$T(4)$ complexes have short innerpaths.

Theorem 8.39 (Quasiconvexity). *Suppose X is $C(6)$ [or $C(4) - T(4)$] and has short innerpaths. Let $Y \to X$ be a combinatorial immersion with no missing shells. Then $\widetilde{Y} \to \widetilde{X}$ is κ-quasiconvex where $\kappa = \max_{R \in 2-cells}(|\partial_p R|)$.*

Proof. Let $\gamma \to \widetilde{X}$ be a combinatorial geodesic with endpoints on $\widetilde{Y}$. Let D be a disk diagram between γ and a path $\sigma \to \widetilde{Y}$. Moreover, choose (D, σ) such that the number of cells in D is minimal among all such choices with γ fixed. By Theorem 8.35, either D is trivial or a single 2-cell, or D is a ladder, or D has at least three shells and/or spurs.

No spur lies on γ since γ is a geodesic, and no spur lies on σ by minimality. No shell has outerpath on γ since short innerpaths would contradict that γ is a geodesic. No shell has outerpath on σ since $\widetilde{Y}$ has no missing shells, and so D could be made smaller by absorbing the shell into $\widetilde{Y}$.

Thus there are at most two spurs and/or shells—with one at each end of D at the transition points of $\partial_p D = \gamma \sigma^{-1}$. Hence D is a ladder. Thus $\gamma \subset \mathcal{N}_\kappa(\sigma) \subset \mathcal{N}_\kappa(\widetilde{Y})$. $\square$

Corollary 8.40. *Let X be a $C(6)$-complex or $C(4)$-$T(4)$ complex with short innerpaths. Let $Y \to X$ be a combinatorial immersion with no missing shells. If Y is compact, then $\pi_1 Y \to \pi_1 X$ is quasi-isometrically embedded.*

8.i   Applications to small-cancellation complexes

We now state the applications to coherence properties of $C(6)$ or $[C(4)$-$T(4)]$ small-cancellation complexes. An i-shell R with $i \leq 3$ [or $i \leq 2$] is *perimeter reducing* if $r^n = \partial_p R = QS$ where the innerpath S is the concatenation of i pieces and satisfies $\mathbf{P}(S) < n\mathbf{Wt}(r)$.

Theorem 8.41 (Compact core). *Let X be a locally-finite $C(6)$ [or $C(4)$-$T(4)$] complex. Suppose each i-shell is perimeter reducing. Then X has the compact core property. Specifically, let $\widehat{X} \to X$ be a finite cover, and let $C \subset \widehat{X}$ be a compact subcomplex. Then there is a compact connected subcomplex Y with $C \subset Y$, and such that the inclusion $Y \to \widehat{X}$ is a π_1-isomorphism.*

Proof. The criterion of Theorem 8.20 is satisfied by Theorem 8.35.   □

Corollary 8.42 (Coherence using perimeter). *Let X be a locally-finite $C(6)$-complex [or $C(4)$-$T(4)$-complex]. Suppose each i-shell is perimeter reducing. Then $\pi_1 X$ is coherent.*

Proof. This is a consequence of Theorem 8.41.   □

Theorem 8.43 (Compact core). *Let X be a locally-finite $C(6)$[or $C(4)$-$T(4)$] complex. Suppose each i-shell is perimeter reducing. There is a quadratic time algorithm to produce a finite presentation for a subgroup given a finite set of generators. There is a quadratic time algorithm to solve the membership problem in a f.g. subgroup.*

Proof. The criterion of Theorem 8.21 is satisfied by Theorem 8.35.   □

Theorem 8.44. *Let X be a compact $C(6)$ or $C(4)$-$T(4)$ complex with short inner-paths. Suppose each i-shell is perimeter reducing. Then $\pi_1 X$ is locally quasi-isometrically embedded.*

Proof. The criterion of Theorem 8.23 is satisfied by Theorem 8.35.   □

9   INCOHERENT EXAMPLES

9.a   Bieri's restriction on f.g. normal subgroups

A common source of non-f.p. subgroups are subgroups that are f.g. and normal. A very useful tactic in this vein comes from the following theorem from [Bie81] which also has a higher dimensional form:

Theorem 9.1 (Bieri). *Let N be a f.g. normal subgroup of a f.p. group G of cohomological dimension ≤ 2. Then N is f.p. if and only if either N is trivial, or N is of finite index, or N is free and G/N is virtually free.*

In many cases it is convenient to couple Theorem 9.1 with the following observation which was applied in this context by Stallings, and which holds since euler characteristic is multiplicative for fibrations.

Theorem 9.2. *Consider a short exact sequence $1 \to N \to G \to Q \to 1$. Suppose that χ is defined for each factor. Then $\chi(G) = \chi(N)\chi(Q)$.*

We will frequently use the following:

Corollary 9.3. *Let G be π_1 of a compact aspherical complex, and suppose $\chi(G) \neq 0$. Suppose G has a finite index subgroup G' with a f.g. normal subgroup N with $G'/N \cong \mathbb{Z}$. Then N is not f.p. Hence G is not coherent.*

Proof. By Theorem 9.1, if N is f.p. then N is free. But then $\chi(G') = \chi(N)\chi(\mathbb{Z}) = 0$ by Theorem 9.2. Since $\chi(G) = [G:G']\chi(G')$, this contradicts that $\chi(G) \neq 0$.   □

9.b   Rips's construction

It was unknown whether every $C(6)$ small-cancellation group is coherent until Rips gave the following construction in [Rip82].

Theorem 9.4. *Let Q be a f.p. group, then there exists a f.p. $C'(\frac{1}{6})$ small-cancellation group G with a subgroup $N = \langle x, y \rangle$ such that there is a short exact sequence:*

$$1 \to N \to G \to Q \to 1.$$

Sketch. Let Q be presented by:

$$Q = \langle q_1, \ldots, q_n \mid R_1, \ldots, R_m \rangle.$$

Then we construct G as follows:

$$G = \langle x, y, q_1, \ldots, q_n \mid R_i W_i : 1 \leq i \leq n, \ x^{q_i} = X_{+i}, \ x^{q_i^{-1}} = X_{-i}, \ y^{q_i} = Y_{+i}, \ y^{q_i^{-1}} = Y_{-i} \rangle$$

where $\{W_i, X_{\pm i}, Y_{\pm i}\}$ are sufficiently long small-cancellation words. For instance we can choose them to be sufficiently long disjoint subwords of the infinite word $xyxy^2xy^3\cdots$.

   The conjugation relations ensure $N = \langle x, y \rangle$ is normal. And clearly $G/N \cong Q$.   □

Rips used Theorem 9.4 by beginning with an arbitrary f.p. group Q with some pathological property, and lifting this property to an associated pathology in G. For instance, if Q has a f.g. subgroup H that is not f.p., the preimage K of H is f.g. but not f.p. since if it were, then by adding two generators of N as relations, we would obtain a finite presentation for $K/N \cong H$.

   The presentation for G is $C'(\frac{1}{6})$ and its relators have no proper powers, and so its standard 2-complex is aspherical [LS77].

   There are several ways to see that N is not f.p. unless Q is finite. Rips observed that N has 2 generators, and that a relation of the form $x = (x^{q_i})^{q_i^{-1}}$ arising from a generator q_i, yields a relation, so N cannot be free, since it is noncyclic.

   Since Q has n generators and m relations, G has $n + 2$ generators and $2n + m$ relators, so $\chi(G) > 0$. Hence N is not f.p. through the more general approach encapsulated by Corollary 9.3.

   Rips's construction was varied in [Wis98] so that G is instead π_1 of a compact negatively curved 2-complex, and it was later strengthened in [Wis03b] so that G is $C'(\frac{1}{6})$ and residually finite. (We now know that in this case G is always residually finite since it is virtually special.)

Remark 9.5. Note that $F_2 \times F_2$ is a $C(4)$-$T(4)$ group as it is π_1 of the product of two graphs. However, there is no embedding of $F_2 \times F_2$ into any $C(6)$ group [BW13].

9.c   Failed ascending HNN extensions

A simpler construction for producing incoherent groups was given in [Wis98].

   Consider the following presentation where W_i and R are words in $a_1^{\pm 1}, \ldots, a_m^{\pm 1}$:

$$\langle a_1, \ldots, a_m, t \mid R = 1, \ a_1^t = W_1, \ \ldots, \ a_m^t = W_m \rangle. \tag{3}$$

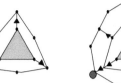

Figure 10: The diagrams D_n.

Theorem 9.6. *Let X denote the standard 2-complex of Presentation (3). Let $A \subset \pi_1 X$ denote the subgroup generated by $\{a_1, \ldots, a_m\}$. If X is aspherical then $\mathsf{H}_2(A) \cong \mathbb{Z}^\infty$ and so A is not f.p.*

Sketch. The hypothesis that X is aspherical implies that R is not freely equivalent to the identity. Let $\widehat{X}$ denote the based cover of X corresponding to the subgroup $A \subset G$. Let $B \subset \widehat{X}$ denote the subcomplex consisting of the basepoint and its neighboring a_i 1-cells. Note that $\mathsf{H}_2(A) = \mathsf{H}_2(\widehat{X})$.

Let r denote the 2-cell of X associated with the relation $R = 1$. For each $n \geq 0$ there is a disk diagram D_n whose boundary path is a word in the $a_i^{\pm 1}$ that equals R^{t^n}. The sequence of diagrams from left to right in Figure 10 are meant to suggest D_0, D_1 and D_2. The shaded triangle r_n at the center of each of the diagrams corresponds to an r 2-cell of X. The remaining 2-cells correspond to relations of the form: $a_i^t = W_i$. The edges with a black arrow correspond to t letters. The basepoint of each diagram is the large black vertex at its lower-left corner.

The based lift of D_n to $\widehat{X}$ has ∂D_n mapped to B. But since $t^n \neq 1$ for $n \neq 0$, the central 2-cells r_n lift to distinct 2-cells in $\widehat{X}$. It follows that there is a non-f.g. subgroup of the 2-dimensional cellular chain group $C_2(\widehat{X})$ which maps to the f.g. subgroup $C_1(B) \subset C_1(\widehat{X})$. Thus $\mathsf{H}_2(\widehat{X}) \cong \mathbb{Z}^\infty$.   □

Example 9.7 (Incoherent small-cancellation group). Since standard 2-complexes of torsion-free small cancellation presentations are aspherical [LS77], we can apply Theorem 9.6 to see that the following group is incoherent.

$$\langle a, b, \, t \, | \, a^1 b^1 a^2 b^2 \cdots a^{30} b^{30} = 1, \, a^t = ab^{10}ab^{11} \cdots ab^{30}, \, b^t = ba^{10}ba^{11} \cdots ba^{30} \rangle$$

Theorem 9.6 was applied in [Wis98] to give elementary examples of incoherent two-relator groups. For instance, the following $C(6)$ group is incoherent, and is also a two-relator group, as we can replace c by $b^t b^{-2}$ and then replace b by $a^t a^{-2}$. Similar examples yield hyperbolic groups with the same property.

$$\langle a, b, c, t \, | \, ab^{-1}ca^{-1}bc^{-1}, a^t = aab, b^t = bbc, c^t = cca \rangle$$

On the other hand, it is known that Thompson's group F has a two-relator presentation, but $\mathbb{Z} \wr \mathbb{Z} \hookrightarrow F$, so F is incoherent by Example 7.3. For proofs of these two statements, see, e.g., [Bur19].

There are embeddings of $F_2 \times F_2$ into a two-relator group. However $F_2 \times F_2$ cannot embed in a one-relator group (see, e.g., [Bag76], [LS77, Thm 5.4]).

Theorem 9.6 leads to the following problem which is an important test case for the rationale behind Problem 12.12:

Problem 9.8. Let X be the mapping torus of an endomorphism of a f.g. free group. Let Y be obtained from X by adding a single 2-cell, and suppose that Y is aspherical. Is $\pi_1 Y$ incoherent?

9.d Free × free, and free-by-free and tree-by-tree

The group $F_2 \times F_2$ was first observed to be incoherent in [BBN59]. Baumslag-Roseblade proved that $F_2 \times F_2$ has uncountably many non-isomorphic f.g. subgroups that are not f.p. [BR84]. Moreover, they proved the following striking result:

Theorem 9.9. *Every f.p. subgroup of $F_2 \times F_2$ is either free or virtually $F_m \times F_n$ for some m and n.*

The Baumslag-Roseblade proof is homological, but a geometric proof is given in [BW99]. Further proofs and generalizations are given in [Sho01], [Mil02], [BHMS02].

An elegant explanation of the incoherence of $F_2 \times F_2$ resides in the kernel of a certain homomorphism. Let

$$F_2 \times F_2 = \langle a_1, b_1 \rangle \times \langle a_2, b_2 \rangle$$

and define $\phi_2 : F_2^2 \to \mathbb{Z}$ by $a_i \mapsto 1$ and $b_i \mapsto 1$. As explained in Example 9.22, $\ker(\phi_2)$ is f.g. but not f.p. A very fruitful line of generalizations of the incoherence of $F_2 \times F_2$ was begun by Stallings who showed that for the analogous map $\phi_3 : F_2^3 \to \mathbb{Z}$ the subgroup $\ker(\phi_3)$ is of type $\mathcal{F}_2$ but not of type $\mathcal{F}_3$ [Sta63]. This was subsequently generalized by Bieri who proved that $\ker(\phi_n)$ was of type $\mathcal{F}_{n-1}$ but not of type $\mathcal{F}_n$ [Bie81]. This line of research culminated in the "Morse theory" of Bestvina-Brady that is reviewed in Section 9.d and is a very effective tool for producing incoherent groups. Their criterion detects finiteness properties of kernels of analogous homomorphisms to $\mathbb{Z}$. Their method works especially well for right-angled Artin groups, and was used in [Bra99] to produce an example of a word-hyperbolic group with a f.p. subgroup that is not word-hyperbolic because it failed the $\mathcal{F}_3$ finiteness property. Brady's example replaces a similar but incorrect example of Gromov who used a branched cover of a 5-torus along generic codimension-2 subtori [Gro87].

Remark 9.10. Gersten observed in [Ger81] that an amalgamated free product obtained by doubling a rank ≥ 2 free group along a subgroup of finite index ≥ 3 is incoherent. In fact, such a double is commensurable with $F_2 \times F_2$ see [Wis02b]. Many other doubles $F_2 *_H F_2$ are incoherent because they contain such a "Gersten double." However, it is not immediately clear exactly which doubles are coherent.

More generally:

Problem 9.11. Which finite graphs of f.g. free groups are coherent?

In particular:

Problem 9.12. Is every group of the form $F_n \rtimes F_m$ incoherent if $n, m \geq 2$?

For instance, as a small step towards Problem 9.12 we have:

Lemma 9.13. *Let G be $F_m \rtimes_\phi F_n$, with $m, n \geq 2$ and suppose the homomorphism $F_n \to \mathrm{Out}(F_n)$ is not injective. Then G is not coherent.*

Proof. Suppose $x \in F_n$ conjugates F_m by an inner automorphism of F_m, so for some $y \in F_m$, we have $x^{-1} f x = y^{-1} f y$ for each $f \in F_m$. Then conjugation by $t = xy^{-1}$ is the identity on F_m. We may thus rechoose the splitting so that F_n contains a nontrivial element t centralizing F_m. Choose $s \in F_n$ such that $\langle s, t \rangle$ is a rank 2 free group, and observe that $t' = sts^{-1}$ also centralizes F_m and the subgroup $\langle t, t' \rangle$ is isomorphic to F_2. Hence we have an embedding $F_m \times F_2 \hookrightarrow F_m \rtimes F_n$. Thus it is incoherent by Example 9.22. $\square$

Following Problem 9.12, I also expect the following source of incoherence:

Problem 9.14. Is every group of the form $\pi_1 S \rtimes F_m$ incoherent, when $m \geq 2$ and S is a closed surface with $\chi(S) < 0$?

A *complete square complex* (CSC) is a square complex X such that for each $x \in X^0$, link(x) is a complete bipartite graph. In addition, X is *nonelementary* if link(x) is nonplanar for some $x \in X^0$.

Problem 9.15. Is $\pi_1 X$ incoherent for each compact nonelementary CSC X?

The product $A \times B$ of two graphs is a CSC, and if $A \times B$ is nonelementary then $\pi_1(A \times B)$ is commensurable with $F_2 \times F_2$. Moreover, like the universal cover of $A \times B$, the universal cover $\tilde{X}$ of a CSC X is isomorphic to the product of two trees.

If Problem 9.15 has a negative solution, then this solves Problem 12.12, as well as the following:

Problem 9.16. Give an example of a coherent group that is quasi-isometric to an incoherent group.

One expects that there are many examples resolving Problem 9.16. However, since many of the most natural conditions implying coherence are very strong geometric conditions, it is likely that while the conditions themselves may not be closed under quasi-isometry, all the groups quasi-isometric to groups satisfying these conditions are coherent.

We return for a moment back to the case of a double $G = F_2 *_H F_2$. It seems overly optimistic to hope that either its standard 2-complex has nonpositive immersions, or G is incoherent because of an application of Theorem 9.19 to some immersed subcomplex. Alternatively, one might perform a $b_1^{(2)} = 0$ computation to a compact immersed subcomplex with $\chi > 0$, and then apply Theorem 17.1. So perhaps this is a good test case for Problem 9.11, even when H is malnormal in which case the word-hyperbolicity of G is an easy special case of [BF95]. We mention in passing the following related problem:

Problem 9.17. Give an algorithm which decides if $G = F_2 *_H F_2$ has the f.g.i.p. relative to H (in which case G is coherent).

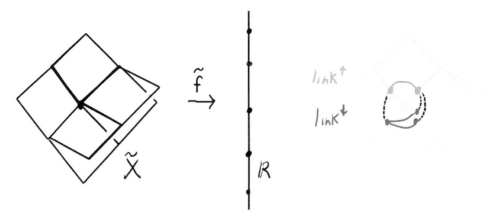

Figure 11: On the left is a Morse function $\tilde{f} : \widetilde{X} \to \mathbb{R}$. On the right is a depiction of $\mathrm{link}^{\uparrow}(v)$ and $\mathrm{link}^{\downarrow}(v)$.

9.e   Bestvina-Brady Morse theory

Definition 9.18. *A Morse function $\tilde{f} : \widetilde{X} \to \mathbb{R}$ on a piecewise-Euclidean complex is a map that restricts to a rank-1 affine map on each n-cell with $n \geq 1$. The ascending link with respect to $\tilde{f}$ at v is the subgraph $\mathrm{link}^{\uparrow}(v) \subset \mathrm{link}(v)$ consisting of cells associated to corners of n-cells having a minimum at v. Similarly, the descending link $\mathrm{link}^{\downarrow}(v)$ consists of the subcomplex of corners having a maximum at v. See Figure 11.*

The following is proven in [BB97]:

Theorem 9.19 (Bestvina-Brady). *Let X be a compact complex. Let $f : X \to S^1$ be a map such that $\tilde{f} : \widetilde{X} \to \widetilde{S}^1$ is a Morse function. Suppose $\mathrm{link}^{\uparrow}(v)$ and $\mathrm{link}^{\downarrow}(v)$ are connected for each v. Then $K = \ker(f : \pi_1 X \to \pi_1 S^1)$ is f.g.*

More generally, if each $\mathrm{link}^{\updownarrow}(v)$ is $(n-1)$-connected then K is of type $\mathcal{F}_n$, meaning that $K = \pi_1 Y$ where Y is an aspherical complex with Y^n compact.

Sketch. We indicate how to prove the first statement in Theorem 9.19. The main point of the argument is to show that there is a path-connected K-cocompact subcomplex of $\widetilde{X}$. Let $m \in \mathbb{Z}$ be an upper bound on $\frac{1}{2}\mathrm{diameter}(\tilde{f}(r))$ where r varies over all 2-cells of X. Let $\widetilde{Y}$ be the complex consisting of all cells that lie in $\tilde{f}^{-1}[-m, m]$.

Consider the following notion of complexity for a path σ.

$$\mathrm{Comp}(\sigma) = \big(\ldots, v_{m+2}, v_{m+1}, v_m\big)$$

where v_p equals the number of vertices in $\tilde{f}^{-1}\big([p, p-1) \cup (-p+1, -p]\big)$ that σ passes through. Declare $\mathrm{Comp}(\sigma) > \mathrm{Comp}(\sigma')$ if there exists r such that $v_s = v'_s$ for all $s > r$, but $v_r > v'_r$.

We claim that $\widetilde{Y}$ is path-connected. Indeed, suppose $p, q \in \widetilde{Y}^0$. Let $\sigma \to \widetilde{X}$ be a combinatorial path joining p, q. Suppose that $\sigma \not\subset \widetilde{Y}$. Let x be a vertex of σ such that $\tilde{f}(x)$ is maximal and $\tilde{f}(x) > m$. (A similar argument using ascending links works for "minimal and $\tilde{f}(x) < m$.") Let e, e' be the edges of σ at x. By hypothesis, the vertices v, v' in $\mathrm{link}(x)$ that are associated to e, e' are connected by a path

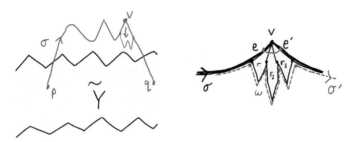

Figure 12: The assumption on link$^\uparrow$ and link$^\downarrow$ ensures that combinatorial paths with endpoints in $\widetilde{Y}$ can be homotoped to lie entirely within the K-cocompact subcomplex $\widetilde{Y}$.

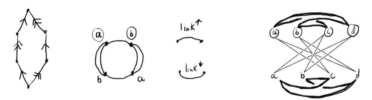

Figure 13: link$^\uparrow$ and link$^\downarrow$ for $\langle a,b\,|\,aba=bab \rangle$ is on the left, and that of $F_2 \times F_2$ is on the right. We use ⓔ and e to denote the outgoing and incoming ends of the edge e.

$a_1 a_2 \cdots a_t$ in link$^\downarrow(x)$. The edges of this path correspond to a sequence of 2-cells $r_1, r_2, \cdots, r_t$, and the path ee' can be replaced by a path ω along the bottom of the concatenation of these 2-cells. See Figure 12. This allows us to replace σ by a path σ' having the same endpoints but such that $\mathrm{Comp}(\sigma') < \mathrm{Comp}(\sigma)$. We conclude that there exists a path joining p, q in $\widetilde{Y}$.   □

Example 9.20. Let $G = \langle a, b\,|\,aba = bab \rangle$ and let X be the associated 2-complex. The map $G \to \mathbb{Z}$ is induced by a map $f : X \to S^1$ where $\tilde{f}$ is a Morse function. The ascending and descending links are contractible, and so the kernel is f.p. In this case, one finds that $G \cong F \rtimes \mathbb{Z}$ where F is free.

Remark 9.21. Bestvina-Brady applied their theory to RAAGs, leading to many remarkable examples with exotic finiteness properties of K.

Example 9.22. Let $F_2 \times F_2$ be the product of two free groups presented by $\langle a, b, c, d\,|\,[a,c,], [a,d], [b,c], [b,d] \rangle$. Consider the homomorphism $F_2 \times F_2 \to \mathbb{Z} = \langle z \rangle$ induced by $\langle a \mapsto z, b \mapsto z, c \mapsto z, d \mapsto z \rangle$. See Figure 13. Then Theorem 9.19 applies, to see its kernel is f.g. thus giving another proof that $F_2 \times F_2$ is not f.p., since the kernel K cannot be f.p. by Theorem 9.1.

Definition 9.23 (Artin group). *Let (x, y, n) be the length n initial subword of $(xy)^n$. Let M be a symmetric matrix with entries in $\{2, 3, \ldots, \infty\}$. An Artin group A is given by a presentation of the form:*

$$\langle x_1, x_2, \ldots \mid (x_i, x_j, m_{ij}) = (x_j, x_i, m_{ji}) \quad : \quad i \neq j \rangle.$$

Customarily, the data is described on a simplicial graph Γ whose vertices correspond to the generators, and where the weight m_{ij} is assigned to the edge joining i, j. The edges with weight ∞ are often omitted.

A is a right-angled Artin group *or RAAG if each entry is 2 or ∞. In that case $A = A(\Gamma)$ is entirely determined by the graph, and the presentation merely asserts that some generators commute.*

When the relator x_i^2 is added for each i, then one obtains the Coxeter group associated to A. We refer to [McC17] for a survey of Artin groups focusing on geometric aspects.

Example 9.22 generalizes to see that:

Proposition 9.24. *Let Γ be a cycle such that each edge has a weight in $\{2, 3, \ldots\}$, (i.e., no ∞). Suppose Γ is not a 3-cycle with weights 2-3-3, 2-3-4, 2-3-5, or 2-4-4. Then the Artin group $A = A(\Gamma)$ is not coherent. Likewise, the same holds if Γ contains such an induced cycle.*

Remark 9.25. The common denominator here is that the associated Coxeter group is finite. The case 2-2-n is omitted from this list, as the group $A(2\text{-}2\text{-}n)$ is commensurable with a group of the form $\mathbb{Z} \times (F \rtimes \mathbb{Z})$ which is coherent by Theorem 6.1 and Lemma 1.3.

Sketch. We focus on the case of a cycle. One verifies that the 2-complex X for the presentation is aspherical. And hence $\chi(A) = 1$. The quotient $A \to \langle z \rangle = \mathbb{Z}$ induced by $x_i \mapsto z$ is represented by a map $f : X \to S^1$ such that $\tilde{f}$ is a Morse function. The ascending and descending links are isomorphic to Γ.

Consequently, the kernel K of $A \to Z$ is f.g. by Theorem 9.19. Hence it is not f.p. by Corollary 9.3.

The general incoherence statement holds since the subgroup of an Artin group generated by some of its generators is itself an Artin group.   □

The above observation shows that Theorem 9.19 explains the coherence of almost all Artin groups. The few Artin groups that are coherent enjoy this property because of a repeated application of Theorem 5.1, and are recognizable since their graphs can be built by repeatedly gluing together other graphs along cliques. This was addressed in the right-angled case in [Dro87], and then resolved for general Artin groups by Gordon [Gor04], with an exceptional incoherent case handled in [Wis13]. We emphasize that the main tool for verifying incoherence in every case is Theorem 9.19, sometimes after a bit of preparation.

9.f   Incoherent Bourdon and Coxeter groups

Definition 9.26. *A Bourdon building of type $I_{p,r}$ has each chamber a p-gon, the thickness of each edge equal to r, and moreover the link of each 0-cell isomorphic to the complete bipartite graph $K(r, r)$. We refer to [Bou97]. We also note that for $p > 4$, Haglund proved the commensurability of any two groups acting properly discontinuously and cocompactly on the same Bourdon building [Hag06].*

We proved the following in [Wis11]:

Theorem 9.27. *Let G act properly and cocompactly on a Bourdon building $I_{p,r}$. For each $p \geq 6$, there exists R such that if $r \geq R$ then G has a finite index subgroup G' such that G' contains a f.g. normal subgroup N that is not f.p., and where $G'/N \cong \mathbb{Z}$.*

Note that Theorem 9.27 is revisited and amplified in Theorem 17.18.

The method is to study the cube complex X that is dual to a natural wallspace for $I_{p,r}$. We pass to an appropriate finite cover $\widehat{X}$, and then choose a Morse map $\widehat{X} \to S^1$ by randomly orienting the hyperpanes. It is shown that when r is sufficiently large, each ascending and descending link is nonempty and connected but not simply-connected. Accordingly the kernel has the desired properties by the Bestvina-Brady theory. (The proof there avoids mentioning the dual cube complex, but it can be thought of in these terms.) A very similar argument shows the following proven in [JW16]:

Theorem 9.28. *Let $G(r, m)$ be the Coxeter group with the following presentation:*

$$\langle a_1, \ldots, a_r \mid a_i^2, (a_i a_j)^m \rangle.$$

For each $r \geq 3$, there exists R such that if $m \geq R$ then G is incoherent.

Define the euler characteristic $\chi(G)$ to be $\frac{1}{[G:K]} \chi(K)$ where K is a torsion-free finite index subgroup. It is conjectured that the following holds:

Conjecture 9.29. *Let G be a Coxeter group of virtual cohomological dimension 2. Then G is coherent if and only if each rank ≥ 3 Coxeter subgroup $H \subset G$ satisfies $\chi(H) \leq 0$.*

In another direction, in [JNW] we examined a targeted finite cover to show that the Bestvina-Brady theory can be applied to many right-angled Coxeter groups to obtain noncoherence. Moreover, we showed that it cannot directly be applied in some cases (without perhaps using a different cube complex from the one provided by the reflection group). The results are briefly described in Section 17.a.

Conjecture 9.30. *For every n, there is a right-angled hyperbolic reflection group G_n, having a finite index subgroup G'_n, such that $G'_n = \pi_1 M$ where M is a closed pseudo-manifold, and where G'_n is incoherent.*

9.g Hyperbolic 4-manifolds and higher

M. Kapovich-Potyagailo constructed the first examples of hyperbolic groups of dimension ≥ 4 that are f.g. but not f.p. [KP91]. Their examples contain parabolic elements, but shortly thereafter two constructions accomplished this without parabolics. Potyagailo gave an example of a closed hyperbolic 4-manifold group that is incoherent [Pot94]. His group has a f.g. normal subgroup that is not f.p. Using Lemma 10.1, Bowditch-Mess produced an example of a subgroup of a discrete cocompact group of isometries of real hyperbolic 4-space which is f.g. but not f.p.

[BM94]. This discrete cocompact group is the group generated by reflections in the faces of a right regular 120-cell.

M. Kapovich constructed examples of incoherent groups which are fundamental groups of closed complex-hyperbolic surfaces. He did this by showing that fundamental groups of complex surfaces which are f.p.-by-surface groups are actually surface-by-surface, and this cannot occur in the complex-hyperbolic case [Kap98]. He also refers to [Hil00] as obtaining similar results.

M. Kapovich showed the following:

Theorem 9.31. *Every simple-type arithmetic hyperbolic manifold M^n with $n \geq 4$ has $\pi_1 M^n$ incoherent.*

It suffices to prove this for $n = 4$ since an intersection of regular hyperplanes provides such a π_1-injective submanifold. His method capitalizes on the virtual fibering of all simple-type (arithmetic) hyperbolic 3-manifolds [Kap13].

Further investigation of hyperbolic manifolds with incoherent π_1 is given in [JNW]. These examples are drawn from right-angled reflection groups, and in particular the reflection group in the 120-cell is shown to have a finite index subgroup with a f.g. normal subgroup that is not f.p. See Section 17.a.

One is led to the following conjecture:

Conjecture 9.32. *Let M be a finite-volume hyperbolic manifold of dimension ≥ 4. Then $\pi_1 M$ is not coherent.*

More recently, in view of Corollary 17.5, we now understand that there are f.g. virtually normal subgroups within a greater context, and this informs Conjecture 9.32 and the results described above.

Finally, we note that results in Section 10 can often be used to obtain incoherent groups.

10   FINITELY GENERATED INTERSECTION PROPERTY

We have already mentioned that local-quasiconvexity implies the f.g.i.p. in Proposition 1.10, which is certainly the greatest source of groups with the f.g.i.p. On the other hand, Theorem 5.3 explains the way the relative f.g.i.p. of edge groups in a splitting of G can support a proof of the coherence of G. There is an interesting interplay here between coherence and the f.g.i.p. which can transfer information in both directions, and be used to provide incoherent examples.

The following goes back to Bernhard Neumann and we refer to [BM94] for a short proof:

Lemma 10.1. *Let $G = A *_C B$ split as an amalgamated product. Suppose G is f.p. and A and B are f.g. Then C is f.g.*

A useful variant of the contrapositive of Lemma 10.1 is:

Lemma 10.2. *Let $G = A *_C B$ be an amalgamated product. Let $A' \subset A$ and $B' \subset B$ be f.g. subgroups such that $A' \cap C = C' = B' \cap C$. If C' is not f.g. then G is not coherent.*

Proof. This follows from Lemma 10.1 since $\langle A', B' \rangle$ is f.g. but not f.p.   □

We note the following special case of Lemma 10.1:

Lemma 10.3. *Suppose H is a subgroup of G such that $H \cap K$ is not f.g. for some f.g. subgroup K of G. Then the double $G *_{H=\bar{H}} \bar{G}$ is not coherent.*

Proof. The subgroup $\langle K, \bar{K} \rangle$ is f.g. but not f.p. Indeed, one can check that its H_2 is not f.g.   □

Remark 10.4. In view of Lemma 10.3 and Theorem 5.3, one might be led to believe that for a splitting of G, some local conditions on the vertex groups with respect to the edge groups might imply coherence of G. However, while there might be some interesting possibilities in this direction, it is going to be subtle. In particular, coherence of G does not follow from coherence of its vertex groups and f.g.i.p. of the vertex relative to their edge groups. As noted in Remark 9.10, a double of a f.g. free group along a finite index subgroup is commensurable with the product of two f.g. free groups, and is thus incoherent when $\chi > 0$.

A hyperbolic example with coherent vertex groups that have the f.g.i.p. relative to their edge groups is described in Example 10.6.

Proposition 10.5. *Let $F = \langle x_1, x_2 \rangle$ be free group with a rank 4 subgroup $\langle W_{11}, W_{12}, W_{21}, W_{22} \rangle$ such that:*

1. *each W_{ij} is a positive nontrivial word in $\{x_1, x_2\}$,*
2. *$|W_{12}| = |W_{21}|$,*
3. *W_{ij} starts with x_i and ends with x_j.*

Then the following amalgamated product is incoherent:

$$G = F *_{W_{11}=W_{11}', W_{12}=W_{21}', W_{21}=W_{12}', W_{22}=W_{22}'} F'.$$

Proof. We use the standard 2-complex of the presentation:

$$\langle x_1, x_2, x_1', x_2' \mid W_{11} = W_{11}', W_{12} = W_{21}', W_{21} = W_{12}', W_{22} = W_{22}' \rangle.$$

Let S^1 be a directed loop labelled by z. Then the map $X \to S^1$ induced by $x_i \mapsto z$ satisfies the criterion of Theorem 9.19.   □

Example 10.6. For instance, the above condition is satisfied for

$$W_{11} = x_1 x_2 x_1 x_2^2 \cdots x_1 x_2^{100} x_1, \qquad W_{12} = x_1^2 x_2^2 \cdots x_1^{100} x_2^{100},$$
$$W_{22} = x_2 x_1 x_2 x_1^2 \cdots x_2 x_1^{100} x_2, \qquad W_{21} = x_2^1 x_1^2 \cdots x_2^{100} x_1^{100}.$$

Note that $M \subset F$ is malnormal provided $\langle x_1, x_2 \mid W_1, W_2, W_3 \rangle$ satisfies the $C'(\frac{1}{10})$ condition [Wis01, Thm 2.14]. In particular, $M \subset F$ is malnormal for the above choice. Thus G is a hyperbolic group [BF92] that is an amalgamated product of free groups along a malnormal subgroup.

Lemma 10.3 gives a quick way to produce incoherent groups. For instance, returning to $F_2 \times F_2$ we have:

Example 10.7. Let $G = \langle a, b, t \mid [a, t], [b, t] \rangle \cong F_2 \times \mathbb{Z}$, and let $H = a, b$. Then the double of G along H is isomorphic to $F_2 \times F_2$. This double is incoherent since, letting $K = \langle at, bt \rangle$, the group $K \cap H = \ker(\phi)$ where $\phi : H \to \mathbb{Z}$ is defined by $a \mapsto 1, b \mapsto 1$. And K is not f.g. since it is an infinite index nontrivial normal subgroup of F_2.

The mode of proof in Example 10.7 works for the doubles of many other mapping tori along their base groups.

Let us now illustrate how to use the contrapositive of Lemma 10.3 to glean relative f.g.i.p information from coherence.

Proposition 10.8. *Let M be a 3-manifold, and let S be a subsurface of a component of ∂M. Then $\pi_1 M$ has the f.g.i.p. relative to the image of $\pi_1 S$ in $\pi_1 M$.*

Proof. Let $\bar{M} = M \cup_S M$ be the 3-manifold obtained by doubling M along S. Then $\pi_1 \bar{M}$ is coherent by Theorem 4.3. Let $G = \pi_1 M$ and let H be the image of $\pi_1 S$ in G. Then $\pi_1 \bar{M}$ itself splits as a double $G *_H G$. Hence G has the f.g.i.p. relative to H by Lemma 10.3. □

Proposition 10.9. *Suppose every staggered group is coherent. Then each one-relator group has the f.g.i.p. relative to each of its Magnus subgroups.*

Proof. Doubling the one-relator group G along the Magnus subgroup M, and observe that $G *_M G$ is a staggered group. The result now follows from Lemma 10.3. □

While Proposition 10.9 is conditional, the same scheme gives the following, which was proven in [MW05] under the additional assumption that $n > |W|$.

Theorem 10.10. *Let $G = \langle a, b, \ldots \mid W^n \rangle$ be a one-relator group with torsion, and let $M \subset G$ be a Magnus subgroup. Then $M \subset G$ has the f.g.i.p.*

Proof. The double of G along M is staggered with torsion, and thus coherent by Corollary 14.13 and Theorem 12.18. Hence M has the f.g.i.p. by Lemma 10.3. □

Theorem 10.11. *Let $\phi : F \to F$ be an endomorphism, and let $G = \langle F, t \mid f^t = \phi(f) : f \in F \rangle$ be the mapping torus of ϕ. Let $J = \langle E, t \rangle$ where $E \subset F$ is a ϕ-invariant free factor. Then G has the f.g.i.p. relative to J.*

Proof. Double G along J to obtain an ascending HNN extension of a free group. It is coherent by Theorem 6.1. The result holds by Lemma 10.3. □

It is possible to show the following for which we omit the proof:

Proposition 10.12. *Let G be an ascending HNN extension of a free group of rank $2 \leq n < \infty$. Then G does not have the f.g.i.p.*

We give one further example illustrating the theme linking coherence to the relative f.g.i.p., but we now employ notions from Section 11.

Lemma 10.13. *Let X be an angled 2-complex, and let $Y \subset X$ be a subcomplex. Let Z be the double of X along Y. Suppose Z has negative sectional curvature, where the angle assignments are given by the angles in the two copies of X. Then $\pi_1 X$ has the f.g.i.p. relative to the image of $\pi_1 Y$.*

Note that the hypothesis indirectly implies the inclusion map $\pi_1 Y \to \pi_1 X$ is π_1-injective. This is because the angle distance between points p, q in $\text{link}_X(y)$ outside of $\text{link}_Y(y)$ is always $\geq \pi$, for otherwise there would be a planar section in Z with positive curvature. When all angles are ≥ 0, it then follows that $Y \to X$ is π_1-injective since it is a local isometry. Indeed it holds by Lemma 11.13 in this combinatorial setting:

Similarly, the analog of Lemma 10.13 would hold if Conjecture 12.11 holds.

Remark 10.14. Let X be a compact piecewise-Euclidean 2-complex whose angles satisfy negative sectional curvature. Then $\pi \widehat{X}_1 \cap \pi_1 \widehat{X}_2$ can be computed by taking the fiber-product of cores $C_1 \otimes C_2$, that have been thickened to contain all closed geodesics.

Generalizing the argument used to establishing the $\Theta = 2$ weakened form of the SHNC in Equation (9), there is a formula:

$$\sum \chi(\pi_1 \widehat{X}_1^g \cap \widehat{X}_2) \leq K \chi(C_1) \chi(C_2)$$

where $K = K(X)$.

Then by Lemma 10.3, $\pi_1 X$ has the f.g.i.p. relative to a subgroup $\pi_1 \widehat{X}$, provided $\pi_1 \widehat{X}$ has a core $C \subset \widehat{X}$ such that $\widehat{X}$ has nonpositive curvature relative to C.

Such an approach works more generally for X if there is a class of normal forms which are ensured to exist in some uniformly thick core.

11   SECTIONAL CURVATURE

In [Wis04a] we introduced a notion of sectional curvature for 2-complexes and examined some of its consequences. Instead of considering only ordinary planar sections, we consider more general sections, and measure the curvatures of these sections by utilizing the curvature measure arising in the generalized combinatorial Gauss-Bonnet theorem. Since we shall demand that these general sections have negative curvature, our spaces are considerably more restrictive than spaces with negative planar sectional curvature. However, we are then able to prove a variety of theorems about such 2-complexes and their fundamental groups. Many of these theorems are either false for fundamental groups of spaces with negative planar sectional curvature, or they are more difficult to prove. The following summarizes the results:

Theorem 11.1. *Let X be a compact space with negative sectional curvature. Then the following hold:*

1. *Let $\widehat{X} \to X$ be a covering space with $\pi_1 \widehat{X}$ f.g. Then every compact subcomplex of $\widehat{X}$ is contained in a compact core of $\widehat{X}$.*

2. $\pi_1 X$ *is coherent.*
3. *For each r, there are finitely many conjugacy classes of noncyclic indecomposable r-generated subgroups of $\pi_1 X$.*
4. $\pi_1 X$ *is 2-free.*
5. $\pi_1 X$ *is cohopfian if it is noncyclic and freely indecomposable.*

A group is *cohopfian* if it is not isomorphic with a proper subgroup. Torsion-free noncyclic freely indecomposable word-hyperbolic groups were shown to be cohopfian by Sela [Sel97].

A group is *n-free* if every subgroup generated by at most n elements is free. See Section 19. There are a variety of results about n-free groups, for instance [BS89, FGRS95, AO96, Bum01]. Our approach places the n-free condition in a broader context by showing that the number of relators is bounded by a linear function of the number of generators.

Following Gersten and Pride [Ger87, Pri88] our notion of negative sectional curvature is defined for a 2-complex with angles assigned to the corners of its 2-cells, and we do not presume an actual metric of nonpositive curvature. However, when such a metric exists, as it does in many natural cases, we are able to prove the following stronger result:

Theorem 11.2. *Let X be a nonpositively curved piecewise-Euclidean 2-complex. If X has negative sectional curvature then $\pi_1 X$ is locally-quasiconvex.*

Recall that a subgroup H of the group $G = \langle g_1, \ldots, g_n \rangle$ is *quasiconvex* if there exists $K > 0$, such that for any geodesic γ in the Cayley graph of G, if γ starts and ends on vertices of H then γ is contained in a K-neighborhood of H (see [Sho91] for a clear account). When G is word-hyperbolic the notion of quasiconvexity does not depend on the choice of finite generating set. Similarly, a group is *locally-quasiconvex* if each of its f.g. subgroups is quasiconvex. Among the many reasons for being interested in locally-quasiconvex groups is that they have solvable generalized word problem.

In fact we actually prove a more general result than stated in Theorem 11.2 depending upon the existence of "quasi-geodesic straight lines." A different class consisting primarily of small-cancellation groups satisfying a "perimeter condition" were shown to be locally-quasiconvex in [MW05]. It will be interesting to see how these classes of groups are eventually integrated. I believe they will ultimately be merged under the rubric of "negative immersions" as in Conjecture 12.17.

Thus far we have focused on results concerning 2-complexes with negative sectional curvature. However, we now describe a connection providing a substantial reason for being interested in nonpositive sectional curvature. A 2-complex X has *nonpositive immersions* if for every immersion $Y \to X$ with Y compact and connected, either $\chi(Y) \leq 0$ or $\pi_1 Y = 1$. We show that every 2-complex with nonpositive sectional curvature has nonpositive immersions. In a certain sense, the notion of nonpositive sectional curvature is a strong local form of the nonpositive immersion property. We proved in [Wis18] that if X has nonpositive immersions, then $\pi_1 X$ is locally indicable (see Lemma 12.6). We thus obtain the following consequence:

Theorem 11.3. *If X has nonpositive sectional curvature then $\pi_1 X$ is locally indicable.*

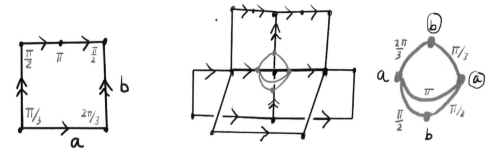

Figure 14: The 2-complex associated to $\langle a, b \mid baab^{-1}a^{-1} \rangle$ is made into an angled 2-complex by assigning an angle to each corner of the 2-cell. The assignment is indicated on the left, and the associated assignment to corresponding edges of the link of the 0-cell is indicated on the right.

The list of properties described above appear to suggest that 2-complexes with nonpositive or negative sectional curvature are quite restrictive. Nevertheless, many of the examples that have traditionally occupied the center stage in geometric group theory are fundamental groups of 2-complexes with nonpositive or negative sectional curvature. Here are some examples.

1. Negatively curved 2-complexes whose vertices have spherical links have negative sectional curvature.
2. Finite-volume cusped hyperbolic 3-manifolds have spines with nonpositive sectional curvature.
3. 2-complexes satisfying *Sieradski's asphericity test*: See Definition 11.10.
4. Ascending HNN extensions of free groups arising from immersion endomorphisms.
5. Coxeter groups with sufficiently large exponents.

11.a   Combinatorial Gauss-Bonnet theorem

Definition 11.4 (Combinatorial complexes). *We will work in the category of combinatorial 2-complexes. A map $X \to Y$ between CW-complexes is combinatorial if its restriction to each open cell of X is a homeomorphism onto an open cell of Y. A CW-complex is combinatorial provided that the attaching map of each of its cells is a combinatorial map (after a suitable subdivision).*

We now restrict ourselves to 2-dimensional combinatorial complexes—as in most of this text.

Definition 11.5 (Angled 2-complex). *The* link *of $v \in X^0$ is the graph* $\text{link}(v)$ *corresponding to the "epsilon sphere" about v in X. The* corners *of 2-cells of X at v are in one-to-one correspondence with the edges in $\text{link}(v)$. We say X is an angled 2-complex provided that for each $v \in X^0$, every corner c of X at v has been assigned a real number $\angle c$ called the* angle *at c. See Figure 14. We say X is [nonnegatively] positively angled if all these angles are [nonnegative] positive. We shall always assume that immersions between angled 2-complexes preserve angles.*

Definition 11.6 (Curvature). *Let X be an angled 2-complex. We use the notation $|\partial_{\mathsf{p}} f|$ for the length of the attaching map of a 2-cell f in X. The* curvature *of f is defined to be the sum of the angles assigned to its corners minus $(|\partial_{\mathsf{p}} f| - 2)\pi$ (the expected Euclidean angle sum). In symbols we have:*

$$\kappa(f) \;=\; \left(\sum_{c \in\, Corners(f)} \measuredangle c \right) - (|\partial_{\mathsf{p}} f| - 2)\pi. \tag{4}$$

If v is a 0-cell of X then the curvature *of v is defined in Equation (5). If X embeds in the plane and v is an interior 0-cell then $\mathrm{link}(v)$ is a circle and the curvature measures the difference between the expected Euclidean angle sum of 2π and the actual angle sum.*

$$\kappa(v) \;=\; 2\pi - \pi \cdot \chi(\mathrm{link}(v)) - \left(\sum_{c \in\, Corners(v)} \measuredangle c \right). \tag{5}$$

Letting $\mathrm{def}(c) = \pi - \measuredangle(c)$, these are equivalent to:

$$\kappa(f) = 2\pi \;-\; \sum_{c \in\, Corners(f)} \mathrm{def}(c). \tag{6}$$

$$\kappa(v) = (2 - v)\pi + \sum_{c \in\, Corners(v)} \mathrm{def}(c). \tag{7}$$

We can now state the combinatorial Gauss-Bonnet theorem. This was first published in [BB96], and later rediscovered in [MW02]. Ballman informs me that it has a much longer history, which I have not investigated.

Theorem 11.7 (Combinatorial Gauss-Bonnet). *If X is an angled 2-complex then the sum of the 2-cell curvatures and the 0-cell curvatures is 2π times the Euler characteristic of X.*

$$\sum_{f \in\, 2\text{-}cells(X)} \kappa(f) + \sum_{v \in\, 0\text{-}cells(X)} \kappa(v) \;=\; 2\pi \cdot \chi(X). \tag{8}$$

11.b   General sections and planar sections

Definition 11.8 (Sections). *A* section *of X at the 0-cell x is a based immersion $(S, s) \to (X, x)$. The section is* regular *if $\mathrm{link}(s)$ is nonempty, compact, connected, and has $\deg(u) \geq 2$ for each vertex u. Pull back the angles at corners of 2-cells of X at x to corners of 2-cells of S at s. The* curvature of the section *$(S, s) \to (X, x)$ to be $\kappa(s)$.*

We say X has sectional curvature *$\leq \alpha$ at x, if all regular sections of X at x have curvature $\leq \alpha$. X has* sectional curvature *$\leq \alpha$ if for each 0-cell X has sectional curvature $\leq \alpha$, and if $\kappa(f) \leq \alpha$ for each 2-cell f.*

A section is planar *if $\mathrm{link}(s)$ is a circle. We say X has* planar sectional curvature *$\leq \alpha$ if all planar sections of X at 0-cells have curvature $\leq \alpha$, and $\kappa(f) \leq \alpha$ for each 2-cell f.*

The reason for considering regular sections above is:

Lemma 11.9. *Let X be an angled 2-complex. Suppose each regular section at $x \in X^0$ has nonpositive curvature. Then for each section $(S, s) \to (X, x)$ with $\text{link}_S(s)$ nonempty, compact, and having no degree 0 or 1 vertices, we have $\kappa(S, s) \leq 0$.*

Definition 11.10 (Sieradski). *An angled 2-complex satisfies* Sieradski's weight test *if it has an angle assignment such that all angles are either 0 or π, and each 2-cell has at least two corners with 0 angles, and each cycle in each link has at least two edges with π angles. Sieradski used this condition to verify asphericity—one essentially applies Theorem 11.7 to an angled 2-sphere [Sie83]. His condition was subsequently generalized to the "Gersten-Pride weight test"; see Section 11.e.*

It is shown in [Wis04a] that Sieradski's weight test implies nonpositive sectional curvature.

Example 11.11. Standard 2-complexes of Adian's "cycle-free" presentations satisfy Sieradski's weight test. These are presentations of the form:

$$\langle a_1, a_2, \ldots \mid U_1 = V_1, U_2 = V_2, \ldots, \rangle$$

where each U_i, V_i is a nontrivial positive word in $\{a_1, a_2, \ldots\}$, and the following two graphs are acyclic: The *ascending* [*descending*] *link* has a vertex for each generator a_i and has an edge joining the initial [terminal] letters of U_j and V_j for each j.

The angle assignment places an angle of 0 at the two corners of each relator $U_j = V_j$, and an angle of π at each other corner.

11.c Nonpositive sectional curvature implies nonpositive immersions

Nonpositive sectional curvature implies the following condition which is slightly stronger than the nonpositive immersion property of Definition 12.1.

Theorem 11.12. *Let X be a 2-complex with nonpositive sectional curvature. Let $Y \to X$ be an immersion where Y is collapsed, connected, and compact. Then either $\chi(Y) = 0$ or Y consists of a single 0-cell.*

In particular X has nonpositive immersions.

Proof. The first statement of the theorem follows from Theorem 11.7 and the hypothesis that X has nonpositive sectional curvature. Indeed, if Y is not a single 0-cell, then $\text{link}(y)$ is nonempty for each $y \in Y^0$, and the hypothesis that Y is compact and collapsed implies that $\text{link}(y)$ is finite, nonempty, and has no degree 0 or 1 vertex. Therefore Lemma 11.9 implies that $\kappa(y) \leq 0$.

To see the second part, observe that Y collapses onto a collapsed subcomplex Y' having the same homotopy type. Now by the first part of the theorem, either Y' consists of a single 0-cell, in which case $\pi_1 Y = \pi_1 Y'$ is trivial, or $\chi(Y) = \chi(Y') \leq 0$. $\square$

We close by mentioning the following corollary of Theorem 11.7 which is in the same spirit as Lemma 8.37.

Lemma 11.13. *Let $Y \to X$ be an angle-preserving combinatorial immersion between angled 2-complexes with nonnegative angles and planar nonpositive sectional curvature. Suppose that for each 0-cell y mapping to a 0-cell x, the embedding $\mathrm{link}(y) \hookrightarrow \mathrm{link}(x)$ has the property that the angular distance between p and q in $\overline{\mathrm{link}(x) - \mathrm{link}(y)}$ is $\geq \pi$. Then $Y \to X$ is π_1-injective.*

Proof. Choose an essential path $P \to Y$ bounding a reduced disk diagram $D \to X$, and assume D is minimal among all possible such (D, P). By Theorem 11.7, there is a 0-cell $v \in \partial D$ with $\kappa(v) > 0$. Consider the arc A in $\mathrm{link}(v)$ between the vertices p, q of $\mathrm{link}(v)$ corresponding to the edges of D at v. Then p, q are vertices of $\mathrm{link}(y)$, and the arc A joins them but does not lie entirely in $\mathrm{link}(y)$, so since its angular length is $< \pi$ there is a contradiction (perhaps associated to a subarc). $\square$

11.d Negative sectional curvature

Definition 11.14. *The angled 2-complex X has* negative sectional curvature *if each 2-cell has negative curvature, and each regular section has negative curvature.*

When X is compact, by decreasing possibly the angles slightly, it suffices to have negative curvature for sections and nonpositive curvature for 2-cells.

Similarly, by possibly increasing the angles slightly, it suffices to have nonpositive curvature for sections and negative curvature for 2-cells.

As for nonpositive sectional curvature, a natural source of examples that give intuition are (p, q, r)-complexes where r is sufficiently small compared to p, q. See Theorem 13.5.

Theorem 11.15. *Let X have negative sectional curvature. Then $\pi_1 X$ is coherent.*

For a complex Z, we use the notation ∂Z for the subgraph of Z^1 consisting of all edges that are either isolated from 2-cells or are free faces of 2-cells. We let $|\partial Z|$ denote the number of edges in ∂Z.

Sketch. Let H be a f.g. subgroup of $\pi_1 X$. Choose a combinatorial immersion $Y \to X$ with Y compact and $\pi_1 Y_1 \twoheadrightarrow H$.

For each i, if $Y_i \to X$ is injective, then we stop as this provides a finite presentation. Otherwise we obtain a new immersion $Y_{i+1} \to X$ such that $Y_i \to X$ factors as $Y_i \to Y_{i+1} \to X$ where $Y_i \to Y_{i+1}$ is obtained by forming $Y_i^+ = Y_i \cup_P D$ by attaching a reduced disk diagram along an essential path, and then folding to obtain an immersion. Hence $\partial Y_{i+1} \subset \mathrm{image}(\partial Y_i)$.

In this way we obtain a sequence $Y_1 \to Y_2 \to \cdots$. This sequence must terminate since $-b_1(H) \leq \chi(Y_i) \leq \kappa F_i + \kappa V_i + \delta |\partial Y_1|$ where F_i and V_i are the numbers of 2-cells and 0-cells. And $\kappa < 0$ is an upper bound on the curvatures of nonboundary sections. And $\delta > 0$ is an upper bound on the curvature of boundary sections.

There are finitely many possible immersions $Y_i \to X$ with $|\partial Y_i|$ uniformly bounded, and hence there cannot be an infinite sequence. $\square$

The key property of negative sectional curvature is generalized as "negative immersions" in Definition 12.16. It is a quick consequence of Theorem 11.7 that if a compact angled 2-complex X has negative sectional curvature, then X has negative immersions. Furthermore, as described in [Wis18], the proof of Theorem 11.15 generalizes to this broader framework. See Section 12.e.

11.e Spherical links

A 2-complex X satisfies the Gersten-Pride *weight test* if each 2-cell has nonpositive curvature, and for each 0-cell x, each immersed cycle in $\operatorname{link}(x)$ has length $\geq 2\pi$. For instance, the angled 2-complex in Figure 14 satisfies the weight test. Similarly, X satisfies the *negative weight test* if negative curvature and/or $> 2\pi$ holds. See [Ger87, Pri88].

An intriguing class of examples arises from the following, which is a result of a simple but surprising computation [Wis04a, Thm 10.5].

Say X has *spherical links* if $\operatorname{link}(x) \subset S^2$ for each $x \in X^0$.

Theorem 11.16. *Let X be an angled 2-complex satisfying the nonpositive [negative] weight test. Suppose X has spherical links. Then X has nonpositive [negative] sectional curvature.*

The following is a natural instance of a 2-complex with spherical links [Wis04a, Thm 12.1]:

Theorem 11.17. *Let M be a cusped hyperbolic 3-manifold. Then M has a spine X that is an angled 2-complex satisfying the weight test and with angles in $[0, \pi]$. Hence X has nonpositive sectional curvature since X has spherical links.*

11.f Local quasiconvexity

In this section we sketch a proof of the following:

Theorem 11.18. *Let X be a compact piecewise-Euclidean 2-complex with negative sectional curvature at each vertex. Then $\pi_1 X$ is locally-quasiconvex.*

Sketch. We use that any two points of the CAT(0) space $\widetilde{X}$ are joined by a geodesic [BH99]. Let H be a f.g. subgroup of $\pi_1 X$, and by Theorem 11.1.(11.1), let $\widetilde{Y} \subset \widetilde{X}$ be a H-cocompact core such that $Y = H \backslash \widetilde{Y}$ is collapsed. Let $\tilde{\gamma} \to \widetilde{X}$ be a CAT(0) geodesic whose endpoints lie on $\widetilde{Y}$. Let X' be a subdivision of X such that γ lies in the 1-skeleton of $\widetilde{X}'$. Let $D \to \widetilde{X}'$ be a reduced disk diagram between $\tilde{\gamma}$ and a path $\tilde{\sigma} \to \widetilde{Y}$. We refer to Figure 15. Let $\widetilde{Z} = \widetilde{Y} \cup HD$.

Each point $\tilde{y} \in \tilde{\gamma}$ is connected to a point of $\widetilde{Y}$ by a path $\tilde{\tau} = e_1 \cdots e_m$ in D and hence in $\widetilde{Z}$ such that e_1 lies in a single original 2-cell or 1-cell, and e_i is an original 1-cell of $\widetilde{X}$ for $i > 1$, and such for $1 < i < m$, the endpoint v_i of e_i has the property that $\operatorname{link}_{v_i}(\widetilde{Z})$ contains a cycle. Hence the same statement holds for the image $y \in Z$ of $\tilde{y}$. Let τ' be a minimal length such path. We claim that $m \leq K$ for some $K = K(X)$. As K is independent of the choice of $\tilde{\gamma}$, a lift of τ' to $\widetilde{X}$ shows that $\gamma \subset \mathcal{N}_{E(K+1)}(\widetilde{Y})$ where E is an upper bound on the diameters of 1-cells of $\widetilde{X}$. Hence H is quasiconvex.

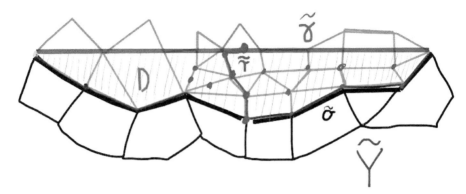

Figure 15: The diagram D between a geodesic $\tilde{\gamma}$ and a path $\tilde{\sigma} \to \widetilde{Y}$. The path $\tilde{\tau} \to D$ connects a point of $\tilde{\gamma}$ to $\tilde{\sigma} \subset \widetilde{Y}$.

By compactness, there are constants $\alpha, \beta > 0$ depending on X such that $\kappa(S, s) \leq \alpha$ for an arbitrary section $(S, s) \to (X, x)$ and such that $\kappa(S, s) \leq -\beta$ for any section $(S, s) \to (X, x)$ such that $\mathrm{link}_S(s)$ contains a cycle.

Each 0-cell of $\widetilde{Z}$ is either in $\widetilde{Y}$, or is in the interior of hD for some $h \in H$, or lies on $h\tilde{\gamma}$ for some $h \in H$. Accordingly,

$$2\pi\chi(Y) = 2\pi\chi(Z) \leq \sum_{z \in \mathrm{Vertices}} \kappa(z)$$

$$\leq \sum_{z \in Y^0} \kappa(z) \; + \sum_{\substack{\mathrm{link}(z) \text{ has cycle} \\ z \notin Y^0}} \kappa(z) \; + \sum_{\substack{z \in \mathrm{im}(\tilde{\gamma}) \\ \mathrm{link}(z) \text{ acyclic and } z \notin Y^0}} \kappa(z).$$

As each $z \in \mathrm{im}(\tilde{\gamma})$ has $\kappa(z) \leq 0$ we have:

$$2\pi\chi(Y) \; \leq \; \alpha|Y^0| + \sum_{\mathrm{link}(z) \text{ has cycle}} \kappa(z) + 0 \; \leq \; \alpha|Y^0| + \sum_{\mathrm{link}(z) \text{ has cycle}} (-\beta).$$

Since each internal vertex of τ has a cycle in its link, and we may assume τ is embedded and only touches Y at its final point we have:

$$2\pi\chi(Y) \leq |Y^0|\alpha + |\tau|(-\beta).$$

Rearranging the terms, and using $|\tau| = m$ we have:

$$m \leq \frac{1}{\beta}\big(|Y^0|\alpha - 2\pi\chi(Y)\big). \qquad \square$$

Problem 11.19. Generalize the negative sectional curvature result so that it works for orbihedra with negative sectional curvature and noetherian vertex stabilizers. In particular, this should work with finite stabilizers. Negative sectional curvature plays the role of the graph in Theorem 5.3.

Conjecture 11.20. *If X is compact and has negative sectional curvature then $\pi_1 X$ is linear and hence residually finite.*

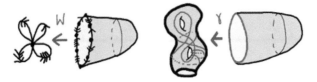

Figure 16: The ordinary 2-complex associated to a presentation on the left, and the surface presentation on the right.

The following should be helpful, both towards the above conjecture, and for somehow "classifying" these complexes:

Problem 11.21. Show that if X is compact and has nonpositive sectional curvature [nonpositive immersions?] then X has a "hierarchy" consisting of a finite sequence of graphs of groups, where each term is built from the previous terms by forming an amalgamated free product or HNN extension along a f.g. (free?) subgroup, or by taking an ascending HNN extension of a free group. In the negative sectional curvature case, we should be able to avoid HNN extensions.

A potentially useful point is the result of [BS90] that if G is f.g. and G is indicable then either G is an ascending HNN extension of a f.g. subgroup, or G splits as an amalgamated free product.

In the first case, we can use the coherence and the result of [GMSW01] to see that the base group must be free. In the second case, we should hope for a splitting over a f.g. free group. One hopes this holds for negative immersions using a $\mathbb{Z}^2$ variant. Otherwise, look at graphs of abelian groups to find a counterexample in the nonpositive immersion case.

11.g   Generic negative sectional curvature

In this subsection we describe a "superficial" approach towards showing that a group G is π_1 of a 2-complex with negative generalized sectional curvature.

Construction 11.22 (Surface presentation). Let $G = \langle a_1, \ldots, a_s \mid w_1, \ldots, w_t \rangle$ be a f.p. group. Choose a hyperbolic surface S with $\pi_1 S \cong \langle a_1, \ldots, a_s \rangle$. For each w_i, let $\gamma_i \to S$ be the closed geodesic representing w_i. Let X be the 2-complex obtained from S by attaching a disk D_i along γ_i for each i. Hence $G \cong \pi_1 X$.

Subdivide S so that each γ_i lies in $S^1 = X^1$. For generic G, the complement $S - \cup \gamma_i$ consists of the union of polygons and a crown for each cusp/boundary circle of S.

For generic G, there are no triple intersection points for the curves $\{\gamma_1, \ldots, \gamma_t\}$. Hence each 0-cell v of X has the property that $\mathrm{link}_X(v) \cong K(4)$, where there is a 4-cycle corresponding to $\mathrm{link}_S(v)$ together with an edge attaching opposite vertices for each corner of a disk at v.

Conjecture 11.23. *For generic G with $t < s$, the 2-complex X has nonpositive generalized sectional curvature. By Theorem 11.16, this is the same as asserting that X satisfies the nonpositive weight test.*

For generic G with $t < s$, the 2-complex X has a metric of nonpositive curvature.

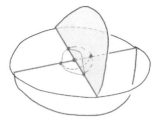

Figure 17: The link of a generic vertex in $X = S \cup_\gamma D$ is a copy of $K(4)$.

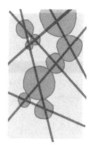

Figure 18: The balls B_{ij} that are centered at intersection points.

Computer experiments give evidence that Conjecture 11.23 is correct when $t < s - 1$. Interestingly, the experiments carried out by Della Veccia suggest the genus of S has an impact when $t = s - 1$ [Vec16].

A geometric rationale for believing in Conjecture 11.23 comes from an expectation about the way a set of random long geodesics are positioned within a surface. To simplify matters, we work with a closed surface S_g of genus g (otherwise we would also have to consider a "crown" for each circle in ∂S_g in the construction below). Let $\{\gamma_1, \ldots, \gamma_k\}$ be closed geodesics chosen uniformly at random from all closed geodesics of length at most L. For simplicity, assume all noninjectivity is at double points, as triple and higher self-intersections generically do not arise. Assume the curves are sufficiently dense that $S_g - \cup \mathrm{im}(\sqcup_i(\gamma_i))$ consists of a set of contractible 2-cells, and note that fixing g, this too will generically be the case for sufficiently large L. For each i, let $p_{i1}, \ldots, p_{ik_i}$ be the points in γ_i where $\sqcup_i^k \gamma_i \to S_g$ is not injective. Let B_{ij} be balls in S_g that are centered at p_{ij}, and assume the $\{B_{ij}\}$ are disjoint from each other. Finally, choose B_{ij} so that $\sum \mathsf{Area}(B_{ij})$ is maximized among all such choices.

Conjecture 11.24. *For each k there exists g_k, such that for $g \geq g_k$, as $L \mapsto \infty$, for each i we have:* $\sum_j \mathsf{Area}(B_{ij}) \geq 4\pi$.

The connection between Conjecture 11.24 and Conjecture 11.23 follows: We can use the area of each ball B centered at a vertex v to guide us in enlarging angles at the four corners in S at v while simultaneously decreasing the angles at the corners of disks attached at v. Specifically, if the angles at v are θ, θ' and the corresponding areas of B are ϵ, ϵ' then we add ϵ and ϵ' to the angles at the corners of S at v, and we assign angles of $\pi - \epsilon - \epsilon'$ to the two corners of disks at v. See Figure 19 which depicts the case of a single attached geodesic. Nonpositive curvature of each 2-cell D_i with $\partial_\mathsf{p} = \gamma_i$ holds since $\kappa(D_i) = 2\pi - \sum_j (\epsilon_{ij} + \epsilon'_{ij}) \leq 0$, where the equality is

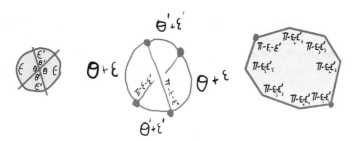

Figure 19: Adjusting angles in S, and assigning angles at corners of each D.

Equation (6), and the inequality would hold by Conjecture 11.24. Similarly, negative curvature would hold if the inequality in Conjecture 11.24 were strict.

We have explored Conjecture 11.23 in [FW16], and reached some conditional positive results depending on the conjectured behavior of random geodesics in a hyperbolic surface.

11.h   Coxeter groups

Let G be a f.g. Coxeter group with presentation

$$\langle a_i \mid a_i^2, (a_i a_j)^{m_{ij}} \rangle$$

where the $(a_i a_j)^{m_{ij}}$ relators are included for pairs $i < j$ with $2 \leq m_{ij} < \infty$.

As Coxeter groups are virtually torsion-free (they are linear and have finitely many conjugacy classes of finite-order elements), we can choose a torsion-free finite index subgroup, and pass to the corresponding finite based covering space Y of the standard 2-complex of the presentation above. Let X_G denote the 2-complex obtained from Y by removing duplicate 2-cells and identifying bigons to edges.

There is a natural angle assignment for X_G where the 2-cell corresponding to $(a_i a_j)^{m_{ij}}$ is assigned angles $\frac{2m_{ij}-2}{2m_{ij}}\pi = (1 - \frac{1}{m_{ij}})\pi$.

Definition 11.25. *For a Coxeter group H presented by:*

$$\langle \ h_1, \ldots, h_n \ \mid \ h_i^2 : 1 \leq i \leq n, \ (h_i h_j)^{m_{ij}} : 1 \leq i < j \leq n \ \rangle$$

we define $\bar{\chi}(H) = 1 - n + \sum \frac{1}{m_{ij}}$.

Definition 11.26. *A Coxeter subgroup H of G is the subgroup generated by a subset of the generators of G. In fact, H is presented by those generators together with all relators in those generators within the presentation of G.*

Theorem 11.27. *The following are equivalent:*

1. X_G has nonpositive sectional curvature,
2. $\bar{\chi}(H) \leq 0$ for each Coxeter subgroup $H \subset G$ with $\mathrm{rank}(H) \geq 2$.

Proof. Observe that for a subgraph of the link with a fixed set of vertices, the curvature is maximal if we include all possible edges, in the sense that we use the subgraph induced by the chosen vertices. Indeed, this holds in general provided $\measuredangle \leq \pi$ for all angles, since then $\mathrm{def}(e) \geq 0$. Because $\kappa = 2\pi - \pi\chi \, \mathrm{link} - \sum \measuredangle = (2 - v)\pi + \sum \mathrm{def}(e)$ where $\mathrm{def}(e) = \pi - \measuredangle(e)$.

Thus, if there were a nontrivial positive curvature section, then we would obtain a Coxeter subgroup H with $\bar{\chi}(H) > 0$. $\qquad\qquad\qquad\qquad\qquad\square$

Problem 11.28. Suppose X_G has nonpositive planar sectional curvature, but $\bar{\chi}(G) > 0$ (and G has ≥ 3 generators). Is it true that $\pi_1 G$ is incoherent?

One hopes that Theorem 9.19 and Corollary 9.3 can be applied to an appropriate finite index subgroup. If the question had an affirmative answer then it would prove the following:

Conjecture 11.29. *If X_G has nonpositive planar sectional curvature then the following are equivalent:*

1. *G is coherent,*
2. *X_G has nonpositive sectional curvature.*

Partial results towards Problem 11.28 are given in Theorem 9.28 and we expect there is more to come in this direction.

12   NONPOSITIVE IMMERSIONS

Definition 12.1 (Nonpositive immersions). *A 2-complex X has* nonpositive immer- *sions* provided that for each immersion $Y \to X$ with Y compact and connected either $\chi(Y) \leq 0$ or Y is contractible.

There are several useful variants of Definition 12.1. For instance, we could weaken "or Y is contractible" to "or $\pi_1 Y = 1$," or we can strengthen it to "or Y is collapsible." Another general variant appears in Theorem 12.8.

Problem 12.2. Does there exist an algorithm to recognize whether or not a compact 2-complex has nonpositive immersions?

12.a   Towers

We recall some background on towers due to Howie [How81].

Definition 12.3 (Tower). *A map $A \to B$ of connected CW-complexes is a* tower *if it is the composition of maps that are alternately inclusions of subcomplexes and covering maps as follows:*

$$A = B_n \hookrightarrow \widehat{B}_{n-1} \to B_{n-1} \hookrightarrow \cdots \hookrightarrow \widehat{B}_2 \to B_2 \hookrightarrow \widehat{B}_1 \to B_1 = B.$$

Let $C \to B$ be a map of connected CW-complexes. A map $C \to A$ is a tower lift of $C \to B$ if there is a tower $A \to B$ such that the following diagram commutes:

$$
\begin{array}{ccc}
 & & A \\
 & \nearrow & \downarrow \\
C & \to & B
\end{array}
$$

The tower lift $C \to A$ is maximal if, for any tower lift $C \to D$ of $C \to A$, the map $D \to A$ is an isomorphism.

The main fact about towers we need is contained in Lemma 12.4. See Howie's proof in [How81] which generalizes from the combinatorial to the cellular case in [Wis18].

Lemma 12.4. *Let S be a compact connected CW-complex and $S \to K$ be a cellular map. Then $S \to K$ has a maximal tower lift.*

12.b Local indicability

Lemma 12.5. *Let X be a based 2-complex. Let $H \subset \pi_1 X$ be a f.g. subgroup. There exists a based 2-complex Y, and a based cellular map $\phi: Y \to X$ such that $\phi_*(\pi_1 Y) = H$ and the map $H_1(Y) \to H_1(H)$ is an isomorphism.*

Note that by replacing $Y \to X$ with $T \to X$ where $Y \to T$ is a maximal tower lift, we can assume the map is not just cellular, but is actually a combinatorial immersion.

Sketch. Let $\{h_1, \ldots, h_r\}$ be based combinatorial paths in X generating H. Let $K^1 = \vee_1^r h_i$. There are finitely many relations among the generators of H that determine $H_1(H)$, and we attach corresponding 2-cells to K^1 to form K. They map to disk diagrams in X, which might be singular, however the maps can be made cellular by subdividing. $\square$

Theorem 12.6. *Let X have nonpositive immersions. Then $\pi_1 X$ is locally indicable.*

Proof. Let $H \subset \pi_1 X$ be a f.g. subgroup with finite abelianization. By Lemma 12.5, let $K \to X$ be a based cellular map such that K is compact, $\pi_1 K$ maps surjectively to H, and $\pi_1 K$ has finite abelianization. By Lemma 12.4, $K \to X$ has a maximal tower lift $K \to T$ where $T \to X$ is a tower. Because K has these properties, T is compact and $\pi_1 T$ has finite abelianization. Thus $\chi(T) \geq 1$, so $\pi_1 T$ and hence H is trivial. $\square$

Theorem 12.6 does not hold for subgroups that are not f.g. Indeed:

Example 12.7. $\mathbb{Q}$ is isomorphic to the fundamental group of a 2-complex with nonpositive immersions, but $\mathbb{Q}$ has no infinite cyclic quotient. Indeed, $\mathbb{Q} = \cup_n \frac{1}{n!}\mathbb{Z}$ is the ascending union of cyclic groups, and the associated infinite mapping telescope provides a 2-complex with nonpositive immersions whose π_1 equals $\mathbb{Q}$.

Note that $\mathbb{Q}$ embeds in an ascending HNN extension of a free group. Baumslag has pointed out examples of one-relator groups with a nontrivial perfect subgroup (that is not f.g.). For instance

$$\langle\, a, b \ \mid \ a = [[a, b], [a, b^2]] \,\rangle.$$

12.c   Asphericity

We close this section by noting that while the 2-sphere satisfies the "$\chi(Y) \leq 0$ or $\pi_1 Y = 1$" version of nonpositive immersions, in practice, 2-complexes with nonpositive immersions tend to be aspherical because of the following [Wis18]:

Theorem 12.8. *Suppose that $\chi(Y) \leq 1$ for any immersion $Y \to X$ with Y compact and connected. Then X is aspherical.*

Proof. Let $S^2 \to X$ be a map from a 2-sphere to X, and after subdividing and homotoping, we assume it is cellular. By Lemma 12.4, let $S^2 \to Y$ be a maximal tower lift of $S^2 \to Y$, where $Y \to X$ is a tower map. As Y is compact, $\chi(Y) \leq 1$. Since the tower lift is maximal, $\pi_1 Y = 1$, so $b_1(Y) = 0$. Thus $b_2(Y) = \chi(Y) - \beta_0(Y) + \beta_1(Y) \leq 1 - 1 + 0 = 0$. Thus $\pi_2(Y) = \mathsf{H}_2(Y) = 0$ by the Hurewicz theorem [Hat02]. A nullhomotopy of $S^2 \to Y$ projects to a nullhomotopy of $S^2 \to X$.   □

Because of the asphericity result in [Wis18], the following would imply Whitehead's asphericity conjecture which posits that a subcomplex of an aspherical 2-complex is itself aspherical [Whi41, Bog93].

Conjecture 12.9. *Let X be a contractible 2-complex. Then X has nonpositive immersions.*

More generally:

Problem 12.10. Is the nonpositive immersion property a (simple) homotopy invariant of 2-complexes?

12.d   Coherence

The main conjecture in this text is the following:

Conjecture 12.11. *If X has nonpositive immersions then $\pi_1 X$ is coherent.*

Intuition. Starting with an immersion $Y_1 \to X$ with Y_1 compact, we repeat the following process: If $Y_i \to X$ is π_1-injective we stop. Otherwise, find a closed essential path $P \to Y_i$ whose projection $P \to X$ is nullhomotopic, and thus bounds a disk diagram $D \to X$. Let $Y_i^+ = Y_i \cup_P D$, and note that $\chi(Y_i^+) = \chi(Y_i) + 1$. We obtain Y_{i+1} from Y_i^+ by folding or passing to a maximal tower lift, and note that Y_{i+1} has the same π_1-image as Y_i. The erroneous but plausible expectation is that $\chi(Y_{i+1}) \geq \chi(Y_i^+)$. Consequently, nonpositive immersions imply this process must eventually terminate.   □

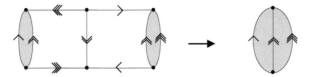

Figure 20: The above immersion is not a tower map.

It may very well be that in the 2-dimensional case, coherence is the same as non-positive immersions. However, the rationale in the sketch above should persist under the assumption that $\chi < c$ for some uniform $c > 0$, or even a more flexible assumption such as $\chi < cb_1$. However, it seems difficult to imagine 2-complexes satisfying such a condition without actually having $\chi(Y) \leq 1$ for each immersion.

Problem 12.12. Is there a compact aspherical 2-complex X such that $\chi(X) \geq 1$ and $\pi_1 X \neq 1$ but $\pi_1 X$ is coherent?

I thought that I had proven Conjecture 12.11 in 2002, but the proof was flawed. The argument utilized the following result of Howie [How81]:

Theorem 12.13 (Locally indicable Freiheitssatz). *Let $\pi_1 Y$ be locally indicable, and suppose Y is a subcomplex of a 2-complex X with $\mathsf{H}_2(X, Y) = 0$. Then $\pi_1 Y \to \pi_1 X$ is injective.*

Definition 12.14. X *has* nonpositive towers, *if for each tower map $Y \to X$ with Y compact and connected, either $\chi(Y) \leq 0$ or $\pi_1 Y$ is trivial.*

Generally speaking, statements that hold when X has nonpositive immersions also hold under the weaker hypothesis that X has nonpositive towers. In practice however, it seems that it is more natural to prove that a 2-complex has the more restrictive nonpositive immersion property. See also Remark 16.3. While nonpositive immersions obviously implies nonpositive towers, I do not know if the following holds:

Problem 12.15. Does the nonpositive tower property imply the nonpositive immersion property?

Tower maps and immersions are exactly the same for 1-complexes, since any immersion is the top of a length 1 tower. However, in Figure 20 we illustrate an immersion of 2-complexes that is not a tower map.

12.e Negative immersions

Definition 12.16. *A 2-complex X has* negative immersions *if there is a constant $c > 0$ such that for each immersion $Y \to X$ where Y is compact and has no free faces, either Y is a single vertex, or $\chi(Y) \leq -c\mathsf{Area}(Y)$ where $\mathsf{Area}(Y)$ denotes the number of 2-cells in Y. A similar definition requires that $\chi(Y) \leq -c|Y^0|$ when Y is not a single vertex and has no free face or cut point. I do not know if negative immersions is equivalent to the assumption that $\chi(Y) < 0$ whenever Y is nontrivial, has no free faces, and has no cut point.*

The ideas and proofs about negative sectional curvature can usually be repeated for negative immersions with little change. In particular:

Conjecture 12.17. *Let X be a compact 2-complex with negative immersions. Then: $\pi_1 X$ is word-hyperbolic; $\pi_1 X$ is locally-quasiconvex; $\pi_1 X$ has a hierarchy.*

Theorem 12.18. *Let X be a compact 2-complex with negative immersions. Then $\pi_1 X$ is coherent.*

Proof. Since free products of coherent groups are coherent, we can give the proof by induction on the minimal number of generators of a subgroup H. (In [Wis18] we handled the possibility of isolated edges occurring further down in the sequence by invoking Proposition 4.1 to choose a f.p. group K, and a homomorphism $K \to H \subset \pi_1 X$, such that for any $K \to \bar{K} \to H$ with $K \to \bar{K}$ surjective, the group $\bar{K}$ is indecomposable.)

By Lemma 12.5, there is an immersion $Y \to X$ such that $\pi_1 Y \to \pi_1 X$ maps to H, with $\mathsf{H}_1(Y) \to \mathsf{H}_1(H)$ an isomorphism. And moreover, Y is compact, collapsed, and has no isolated edges. (Indecomposability of H allows us to avoid a nontrivial free decomposition associated to an isolated edge, and changing the basepoint of X and conjugating allows us to avoid isolated edges at the basepoint of Y.)

If $Y \to X$ is not a π_1-injection, then we can find an essential closed path $P \to Y$ whose image $P \to X$ bounds a reduced disk diagram $D \to X$. We then form $Y^+ = Y \cup_P D$, and consider a maximal tower lift $Y^+ \to T$ of $Y \to X$ by Lemma 12.4. Isolated edges in T must occur in Y^+ but there are none. Collapsible edges in T must have preimages in Y^+ that are either collapsible edges (none), or cancellable pair edges within D (none since reduced). Thus T has no collapsible or isolated edges.

Apply this procedure to $Y_i \to X$, and let $Y_{i+1} = T$, and we thus obtain a sequence of immersions $Y = Y_1 \to Y_2 \to \cdots$ such that: each $Y_i \to X$ factors as $Y_i \to Y_{i+1} \to X$, and each $\pi_1 Y_i \to \pi_1 X$ has image H. If the sequence terminates, then we obtain a finite presentation for H. Otherwise, we have the situation that $b_1(Y_i) = b_1(H)$ for each i. Hence $-b_1(H) \leq \chi(Y_i) \leq -c\mathsf{Area}(Y_i)$ and so there is an upper bound on the number of 2-cells in Y_i. However, there are finitely many distinct immersed complexes $Y_i \to X$ with a fixed number of 2-cells and no isolated 1-cell. Thus some $Y_i \to X$ is the same as $Y_j \to X$ for $i < j$. However, the path $P_i \to Y_i$ is essential in Y_i but nullhomotopic in Y_{i+1} and hence in Y_j. So this is impossible. $\square$

One can attempt to generalize some of the methods we have mentioned and the proof of Theorem 12.18 to cover groups of the following form:

$$\langle a, b \mid W_1^{n_1}, \ldots, W_k^{n_k} \rangle$$

where $\sum_{i=1}^k \frac{1}{n_i} \leq 1$. A similar statement holds with r generators and $\sum \leq \frac{1}{r-1}$, but one must be careful that each generator appears in each relator.

In a similar vein, other examples to pursue are the nonpositively curved *generalized triangle groups* [BMS87], and *nonpositively curved generalized tetrahedral groups* (see [HK06] and the references therein). One expects a version of negative immersion in the negatively curved case.

While in many cases it is possible to sidestep torsion by passing to a finite cover as in Construction 14.12, it seems worthwhile to tackle the situation directly. For

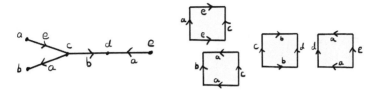

Figure 21: The LOT on the left is associated to the presentation $\langle a, b, c, d, e \mid ae = ec, ca = ab, cb = bd, ea = ad \rangle$ whose relators are indicated on the right.

instance, when a 2-cell has exponent n, then a counting mechanism would give it an χ-value of $+\frac{1}{n}$, and there should be a more permissive notion of collapsibility.

Problem 12.19. Generalize negative/nonpositive immersions and its conclusions to work for orbihedra.

12.f LOTs

Definition 12.20 (LOT). *Let T be a tree where each edge e is directed and labelled by an element $\ell(e) \in Vertices(T)$. Let $\iota(e), \tau(e)$ denote the initial and terminal vertices of e. Define a presentation by:*

$$\langle \; v \in Vertices(T) \; \mid \; \iota(e)\ell(e) = \ell(e)\tau(e) : e \in Edges(T) \; \rangle.$$

Let X be the 2-complex associated to this presentation. See Figure 21.

We refer to [How85], for background on these labelled oriented trees *(LOTs) and the connection to ribbon disk complements. We refer to Harlander-Rosebrock for the latest progress on the asphericity of LOTs [HR17].*

Given Theorem 12.8, substantial progress on LOTs would be achieved by resolving the following:

Conjecture 12.21. *Let X be the standard 2-complex of a labelled oriented tree [How85], then X has nonpositive immersions.*

Example 12.22. Let $a_1, \ldots, a_n \in \{\pm 1, \pm 2, \ldots, \pm n\}$. Let $[a_1, a_2, \ldots, a_n]$ denote the labelled oriented interval $[0, n]$, whose edge $[i-1, i]$ is labelled by $|a_i|$ and oriented forward or backward according to whether a_i is positive or negative. The presentation corresponding to this labelled oriented tree is:

$$\langle 1, \ldots, n \mid ia_i = a_i(i+1) \; : \; (1 \le i < n) \; \rangle.$$

See Figure 22 for an angle assignment on the 2-complex of $[5, 4, 1, 2]$ having nonpositive generalized sectional curvature.

Curiously, if $|a_1|, |a_2|, \ldots, |a_4|$ are distinct then the 2-complex of $[a_1, \ldots, a_4]$ has nonpositive sectional curvature for an appropriate angle assignment. However, this fails for $[a_1, \ldots, a_5]$. A Maple computer computation showed that the following LOTs do not have nonpositive sectional curvature: The 2-complex of $[-5, -4, 5, -1]$ fails to have nonpositive planar sectional curvature. The 2-complex of $[3, 5, 1, 3]$ has a spherical link but fails to have nonpositive planar sectional curvature.

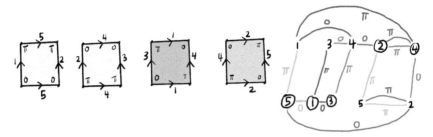

Figure 22: The LOT $[5412]$ satisfies Sieradski's weight test, and hence has nonpositive generalized sectional curvature.

13 (p, q, r)-COMPLEXES

I believe the following types of complexes offer good test cases for the various methods of pursuing coherence.

Definition 13.1. *The 2-complex X is a (p, q, r)-complex provided the attaching map of each 2-cell has length $\geq p$, each cycle in the link of each 0-cell has length $\geq q$, and each 1-cell occurs $\leq r$ times among the attaching maps.*

The perimeter approach in Section 8 places the focus on the ratio $\frac{r}{p}$.

From the viewpoint of (p, q, r)-complexes, the perimeter method appears to be inferior to the methods involving sectional curvature and L^2-betti numbers. Following the results described in Section 8.i we have:

Example 13.2. The perimeter reduction hypothesis holds for a (p, q, r)-complex:

1. for $\frac{r}{p} \leq \frac{1}{3}$ when $q = 3$ and $p \geq 6$,
2. for $\frac{r}{p} \leq \frac{1}{2}$ when $q \geq 4$ and $p \geq 4$.

Hence coherence holds for π_1 of such (p, q, r) complexes. Similarly, local-quasiconvexity holds when $\frac{r}{p} < \frac{1}{3}$ or $\frac{r}{p} \leq \frac{1}{2}$ respectively.

In [MW08] we explore matters further using the perimeter approach, coupled with more elaborate features of positive curvature ("fans" instead of "shells") to reduce the perimeter. The following theorem which is a special case of [MW08, Thm 7.8 and 7.9] allows us to probe the relationship between p and r more deeply. It is conditional on a residual finiteness assumption which is now known in the hyperbolic case. It shows that as $q \to \infty$, the upper bound required on $\frac{r}{p}$ to ensure coherence/local-quasiconvexity gets arbitrarily close to 1.

Theorem 13.3. *Let X be a (p, q, r)-complex. Coherence holds when*

1. $\frac{r}{p} \leq \frac{q-3}{q-2}$ *and* $p \geq 4, q \geq 4$,
2. $\frac{r}{p} \leq \frac{(q-2)(p-3)-1}{(q-1)(p-3)-1}$ *and* $p \geq 6, q \geq 3$.

And local-quasiconvexity holds when the initial inequalities are strict.

For low values of p, q, the other methods we describe are superior.

The following is proven in [Wisa] and discussed in Section 16 where "supertrees" are defined.

Theorem 13.4. *X has supertrees if it satisfies any of the following:*

1. *X is a $(p, 3, p-4)$-complex with $p \geq 6$.*
2. *X is a $(p, 4, p-2)$-complex with $p \geq 4$.*
3. *X is a $(p, 5, p-1)$-complex with $p \geq 4$.*

The following is proven in [Wis08, Thm 2.2 and 2.4]:

Theorem 13.5. *If X satisfies one of the following then X has nonpositive sectional curvature:*

1. *X is a $(p, 3, p-3)$-complex with $p \geq 6$.*
2. *X is a $(p, 4, p-2)$-complex with $p \geq 4$.*
3. *X is a $(p, 5, p-1)$-complex with $p \geq 4$.*

Furthermore, X has negative sectional curvature in the above cases, except for the cases $(4, 4, 2)$ and $(6, 3, 3)$.

Theorem 13.5 is optimal in the following sense: Examples were given in [Wis08] of compact noncontractible $(p, 3, p-2)$-complexes, $(p, 4, p-1)$-complexes, and $(p, 5, p)$-complexes with positive euler characteristic.

The next milestone is to construct $(p, 3, p-2)$-complexes and $(p, 4, p-1)$-complexes with incoherent fundamental group (for arbitrarily large p). Moreover, it would be interesting to know if every nontrivial such example has incoherent fundamental group. I expect this will be resolved by proving $b_1^{(2)} = 0$ for an immersed subcomplex with $\pi_1 \neq 1$ and then applying Theorem 17.1 and virtual specialness.

To get some feel for the intuition of Theorem 13.5, we describe the following diminished special case:

Lemma 13.6. *Let X be a $(4, 12, 3)$-complex. Then X has nonpositive sectional curvature, and thus X has nonpositive immersions.*

Proof. We assign an angle of $\frac{\pi}{2}$ at the corner of each 2-cell, so each 2-cell has nonpositive curvature. Consider a section with vertex x, and where $\text{link}(x)$ has v vertices and e edges. Then $\kappa(x) = 2\pi - \pi\chi(\text{link}(x)) - \sum_e \angle(e) = 2\pi - \pi(v - e) - \frac{\pi}{2}e = \pi(2 - v + \frac{e}{2})$. Thus $\kappa(x) \leq 0$ provided that:

$$v \geq \frac{2}{3}e \geq 2 + \frac{e}{2}.$$

The first inequality holds since each vertex in $\text{link}(x)$ has degree ≤ 3, and so $3v \geq 2e$. The second inequality holds for any section since $\frac{1}{6}e \geq 2$ because the girth is ≥ 12 so any regular section has at least 12 edges. $\qquad\square$

Example 13.7. Theorem 13.5 fails for $(4, 4, 3)$-complexes, as the following example shows. Let Θ denote the graph with two vertices joined by three edges. Then $X = \Theta \times \Theta$ is a $(4, 4, 3)$-complex with $\chi(X) = 1$. Moreover $\pi_1 X \cong F_2 \times F_2$ is incoherent, as in Examples 9.22 and 10.7.

Theorem 13.5 fails for $(4, 5, 4)$-complexes, since $(4, q, 4)$-complexes with incoherent fundamental group exist for all $q \geq 4$ [MW08]. Moreover, a similar argument produces (p, q, p)-complexes with incoherent fundamental group for arbitrarily large p and q.

14   ONE-RELATOR GROUPS

Much of the research described in this survey is motivated by:

Problem 14.1 (Baumslag [Bau74]). Is every one-relator group coherent?

A *Magnus subgroup* of a one-relator group is a subgroup generated by a subset of the generators that omits a generator occurring in the relator. The famous "Freiheitssatz" of Magnus which underpins the theory of one-relator groups, is the following. See [LS77, IV.5.1],[How87].

Theorem 14.2. *A Magnus subgroup is freely generated by its generators.*

While one-relator groups have an inductive definition as a sequence of HNN extensions associating Magnus subgroups [LS77] (see also [Wisc, Const 19.5]) Theorem 5.3 cannot be applied to obtain coherence. Indeed, consider the following example:

Example 14.3. The one-relator group $G \cong \langle a, t \mid at^{-1}at^{-1}at^2 \rangle$ can be constructed as a HNN extension of the one-relator group $H \cong \langle a, b, c \mid abc \rangle$ by adding the stable letter t which conjugates the subgroup $\langle a, b \rangle$ to the subgroup $\langle b, c \rangle$ according to $a^t = b$ and $b^t = c$.

However, $G \cong F_2 \rtimes \mathbb{Z}$ does not have the f.g.i.p. relative to $\langle a, b \rangle$, so Theorem 5.3 cannot be applied to prove coherence. See Proposition 10.12.

It is concievable that Theorem 5.3 can sometimes be applied, or that some strategic transitional steps can be employed before applying it in general.

As discussed in Section 8.f, the perimeter method together with Theorem 8.26 gave a quick proof of Theorem 14.4 that $\langle a_1, a_2, \ldots \mid W^n \rangle$ is coherent for $n \geq |W| - 1$.

The following is proven in [Wis04a]:

Theorem 14.4. *Let X be the standard 2-complex of a one-relator group $\langle a_1, a_2, \ldots \mid W \rangle$. Suppose that X is an angled 2-complex with nonpositive [negative] planar sectional curvature, and each angle is $\leq \pi$. Then X has nonpositive [negative] sectional curvature.*

For instance, this theorem holds for a one-relator group all of whose pieces have length 1. Indeed, if $|W| \geq 6$ then assign an angle of $\frac{2\pi}{3}$ to each corner, and the finitely many remaining possibilities can be handled on a case-by-case basis.

My approach to affirmatively prove Problem 14.1 consisted of proving Conjecture 12.11, and then establishing the following whose proof is discussed in Section 15:

Theorem 14.5. *Let X be the 2-complex associated to a one-relator presentation $\langle a, b, \ldots \mid W \rangle$. Assume that W is cyclically reduced and is not a proper power. Then X has nonpositive immersions.*

A version of this holds for the case where W is a proper power and is discussed in Section 14.b. Furthermore, the proofs of Theorem 14.5 generalize naturally to staggered 2-complexes, which are intimately connected to one-relator groups and are discussed in Section 14.a.

14.a Staggered 2-complexes

Definition 14.6 (Staggered). *Let X be a 2-complex such that the attaching map of each 2-cell is an immersion. We say X is* staggered *if there are linear orderings on the 2-cells and on a subset $\mathcal{O}$ of the 1-cells (called the* ordered 1-*cells) such that*

1. *for each 2-cell α, at least one ordered 1-cell is traversed by $\partial_p \alpha$, and*
2. *for 2-cells α and β, if $\alpha < \beta$ then $(\min \alpha) < (\min \beta)$ and $(\max \alpha) < (\max \beta)$ where $\min \alpha$ and $\max \alpha$ are respectively the least and greatest ordered 1-cells traversed by $\partial_p \alpha$.*

A group presentation is staggered *if its standard 2-complex is staggered.*

The notion of *staggered* defined above appears in [LS77, p. 152] and was implicit in Magnus's original proof of Theorem 14.2 for one-relator groups [Mag30]. Howie illuminated the theory of one-relator groups by explicitly developing the interplay between towers and staggered 2-complexes [How87].

Definition 14.7. *A staggered 2-complex X is* positive *if there is an orientation on the 1-cells in $\mathcal{O}$, such that for each 2-cell R of X, the attaching map $\partial_p R \to X^1$ consistently traverses $\max(R)$ in the same direction, and consistently traverses $\min(R)$ in the same direction.*

A staggered presentation is positive *if its standard 2-complex is positive.*

Example 14.8. The following staggered presentation is positive where $a < b < c < d$ are the ordered 1-cells, and the 2-cells are ordered as they appear in the presentation.

$$\langle a, b, c, d, e \mid aeb^{-1}ae^{-1}b^{-1}c^2, bebececedede \rangle$$

The motivating case is a one-relator group, or more precisely, a one-relator presentation, whose relator is a positive word in the generators. More generally, $\langle a_1, a_2, \ldots \mid W \rangle$ is a positive staggered presentation in the sense of Definition 14.7 if and only if for some i, the generator a_i^{+1} appears in W but a_i^{-1} does not, or vice versa.

The following was proven in [Wisb]:

Theorem 14.9. *Let X be a compact positive staggered 2-complex, then X has a finite cover $\widehat{X}$ with a π_1-isomorphic subcomplex $\bar{X}$ such that $\bar{X}$ has nonpositive immersions.*

The same statement is now known without the "positive" hypothesis, and is sketched in Section 15. The proof in [Wisb] has a different style that uses a sequence of cyclic covering spaces, and an argument by contradiction.

Conjecture 14.10. *Every group with a staggered presentation is coherent.*

A different approach, taken in [Wis05], employs a counting argument engaged with Equation (9) below to prove:

Theorem 14.11. *Let X be the standard 2-complex of $\langle a_1, a_2, \ldots \mid W^n \rangle$. Let $\bar{X}$ be the complex obtained from a torsion-free finite index subgroup in Construction 14.12. Then $\bar{X}$ has nonpositive immersions if $n \geq 2$.*

Note that Theorem 14.11 was conditional on the Strengthened Hanna Neumann Conjecture (SNHC), which has since been proven, and we refer to Section 15.b.

14.b   Torsion

The nonpositive immersion hypothesis of Definition 12.1 does not directly work in conjunction with torsion. In fact, let X be the standard 2-complex of the one-relator group $\langle a \mid a^n \rangle$. Then $\chi(\widetilde{X}) = n$, and similarly, the 2-complex of $\langle a, b \mid a^n \rangle$ admits immersions with arbitrarily positive χ. Fortunately, one-relator groups are virtually torsion-free [LS77], which facilitates the following:

Construction 14.12. Given a one-relator group G presented by $\langle a, b, \cdots \mid W^n \rangle$ where W is a freely and cyclically reduced word that is not a proper power, let X be its standard 2-complex. It is convenient to pass to a certain subcomplex of a finite covering space to remove a degree of redundancy of the 2-cells when $n \geq 2$. Using the fact that free groups are potent, let $\widehat{X}$ be the covering space corresponding to a finite quotient $\bar{G}$ of G such that $\bar{W}$ has order n. The 2-cells of $\widehat{X}$ are partitioned into equivalence classes according to whether they have the same attaching map, and there are n 2-cells in each equivalence class. Let $\bar{X}$ be the subcomplex of $\widehat{X}$ obtained by including exactly one 2-cell from each equivalence class. Note that $\pi_1 \bar{X} \cong \pi_1 \widehat{X}$, and when $n = 1$ we may let $\bar{X} = X$.

The following is a consequence of the proof of Theorem 15.1:

Corollary 14.13. *$\bar{X}$ has negative immersions.*

We refer to Definition 12.16 for the definition of negative immersions.

Note that a version of Corollary 14.13 has also been noticed more recently by Louder-Wilton [LW18b] (I had not circulated my results).

Corollary 14.14. *$\langle a, b \mid W^n \rangle$ is coherent when $n \geq 2$.*

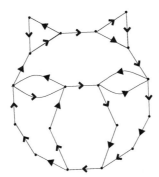

Figure 23: The a-edges in the graph above are indicated by arrows and the b-edges by solid triangles.

Proof. Since coherence is a commensurability invariant, the coherence of $\pi_1 X$ follows from the coherence of $\pi_1 \bar{X}$. The latter holds by Corollary 14.13 and Theorem 12.18. □

Remark 14.15. Versions of Corollary 14.13 and Corollary 14.14 also hold for staggered 2-complexes *with torsion* which have the additional property that all the attaching maps of 2-cells are proper powers.

15 COUNTING W-CYCLES

Let Γ be a connected finite graph whose edges are directed and labelled from some alphabet set $\{a_1, \ldots, a_n\}$. Suppose no vertex has two incoming edges with the same label, and no vertex has two outgoing edges with the same label. A combinatorial path $P \to \Gamma$ starting at some vertex v, gives rise to a word in $\{a_1^{\pm 1}, \ldots, a_n^{\pm 1}\}$. Conversely, a word W in $\{a_1^{\pm 1}, \ldots, a_n^{\pm 1}\}$ determines at most one path P starting at a vertex v. Let W be a word that is not periodic and that is reduced and cyclically reduced. A *based W-cycle* is a closed based path corresponding to the word W^n for some $n \geq 1$. We regard two based W-cycles as *equivalent* if there is a path of the form W^m between their basepoints for some $m \in \mathbb{Z}$. A *W-cycle* is an equivalence class of based W-cycles. For example, letting $W = aab$, there are seven based W-cycles in Figure 23 but there are only three W-cycles.

We use the notation $\#_W(\Gamma)$ for the number of W-cycles in Γ.

The nonpositive immersion property for one-relator groups that is stated in Theorem 14.5 is a consequence of the proof of the following:

Theorem 15.1. *Let W be a reduced cyclically reduced word. Then $\#_W(\Gamma) \leq \beta_1(\Gamma)$. In particular, if Γ is connected then $\#_W(\Gamma) \leq 1 + e - v$, where e and v are the numbers of edges and vertices in Γ.*

Remark 15.2. Theorem 15.1 was conjectured in [Wis03a], and there are several partial results on it: In [Wisb], Theorem 15.1 is proven when W is a positive staggered 2-complex. In [Wis05], an entirely different approach is used to prove the following weakened form of Theorem 15.1 that is parallel to Theorem 14.11:

Proposition 15.3. $\#_W(\Gamma) \le 4\beta_1(\Gamma)$ and $\#_W(\Gamma) \le 2\beta_1(\Gamma)$ if the Strengthened Hanna Neumann Conjecture holds.

Theorem 15.1 was recently proven in general in [LW17] and [HW16], and we discuss this in Section 15.a.

15.a Proofs of the W-cycle conjecture:

We describe two proofs of Theorem 15.1. As indicated by Theorem 14.11 and Proposition 15.3, there is a connection between the w-cycle conjecture and the Strengthened Hanna Neumann Conjecture, and one should view Theorem 15.1 as a "rank-1" variant of the SHNC. Thus, resolution of the SHNC by Friedman and Mineyev [Fri15, Min12b] leads one to examine their approach. At this time, it is unknown how Friedman's approach might be utilized. However, Dicks abstracted the essence of Mineyev's proof and explained the precise way its counting mechanism interacted with orderability [Dic11, Min12a]. Helfer-Wise adapted the Dicks-Mineyev proof so that it could be applied in this case, and we describe that argument in Section 15.c. Louder-Wilton created another argument predicated on an ingenious counting mechanism that also hinges upon orderability, and we sketch their argument in Section 15.d. We now know that the Louder-Wilton condition "good stacking" implies the Helfer-Wise condition "bi-slim structure" but the converse is still unclear [BCGW19].

The proofs generalize from one-relator complexes to *staggered 2-complexes* which are discussed in Definition 14.6. There is a "slim" variant of the notion of "bislim" in [HW16] which also implies nonpositive immersions, and covers the more general *reducible 2-complexes* [How81].

15.b The Strengthened Hanna Neumann Conjecture

This section reviews the SHNC and the allied notion of fiber-product of graphs. See [Sta83, Neu90].

For a free group H, let $\widetilde{\mathrm{rk}}(H) = \max\{\mathrm{rank}(H) - 1, 0\}$. The Hanna Neumann Conjecture states that if H_1 and H_2 are subgroups of a free group F, then

$$\widetilde{\mathrm{rk}}(H_1 \cap H_2) \ \le \ \widetilde{\mathrm{rk}}(H_1)\, \widetilde{\mathrm{rk}}(H_2).$$

The SHNC asserts the following stronger inequality where the summation is taken over distinct double coset representatives. Note that only finitely many of these intersections are nontrivial when H_1 and H_2 are f.g.

$$\sum_{H_1 g H_2 \ \in \ H_1 \backslash F / H_2} \widetilde{\mathrm{rk}}(g^{-1} H_1 g \cap H_2) \ \le \ \widetilde{\mathrm{rk}}(H_1)\, \widetilde{\mathrm{rk}}(H_2).$$

Let $\phi_1 \colon A_1 \to B$ and $\phi_2 \colon A_2 \to B$ be combinatorial immersions of graphs. Their *fiber-product* $A_1 \otimes A_2$ is the graph whose vertices are pairs of vertices in A_1 and A_2 mapping to the same vertex of B, and whose edges are pairs of edges in A_1 and A_2 mapping to the same edge in B. More precisely, each vertex is a pair (v_1, v_2) where $\phi_1(v_1) = \phi_2(v_2)$, and each edge is a pair (e_1, e_2) where $\phi_1(e_1) = \phi_2(e_2)$. The edge (e_1, e_2) has initial vertex $(\iota(e_1), \iota(e_2))$ and terminal vertex $(\tau(e_1), \tau(e_2))$, where

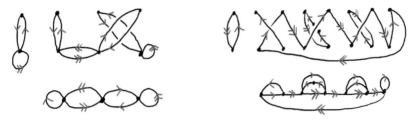

Figure 24: The fiber-product on the left has the property that all of its components are covers of A_1, since $A_2 \to B$ is a covering map. The fiber-product on the right has the property that its components are all intervals or cycles, since A_1 is a cycle.

$\iota(e_i)$ and $\tau(e_i)$ are the initial and terminal vertices of e_i in A_i. There is a natural map $A_1 \otimes A_2 \to B$. See Figure 24.

For a connected graph A, let $\widetilde{\mathrm{rk}}(A) = \max\{b_1(A) - 1, 0\}$, so $\widetilde{\mathrm{rk}}(A) = \widetilde{\mathrm{rk}}(\pi_1 A)$. As observed in [Neu90], the following inequality is readily deduced with $\Theta = 2$:

$$\sum_{E \in \text{Components}(A_1 \otimes A_2)} \widetilde{\mathrm{rk}}(E) \ \leq \ \Theta \, \widetilde{\mathrm{rk}}(A_1) \, \widetilde{\mathrm{rk}}(A_2). \tag{9}$$

As explained in [Neu90], the SHNC is equivalent to the statement that Equation (9) holds with $\Theta = 1$. Indeed, for each i, the subgroup H_i can be identified with $\pi_1 A_i$ for an immersed based graph $A_i \to B$. The non-tree components of $A_1 \otimes A_2$ are in correspondence with the double cosets $H_1 g H_2$ such that $g^{-1} H_1 g \cap H_2$ is nontrivial.

Thus the Strengthened Hanna Neumann Conjecture says:

Theorem 15.4. *Let $A_1 \to B$ and $A_2 \to B$ be immersions of graphs. Then:*

$$\sum_{E \in \text{Components}(A_1 \otimes A_2)} \widetilde{\mathrm{rk}}(E) \ \leq \ \widetilde{\mathrm{rk}}(A_1) \, \widetilde{\mathrm{rk}}(A_2). \tag{10}$$

15.c Bislim structure

A *preorder* $\preceq$ on a set $\mathcal{E}$ is a reflexive, transitive relation on $\mathcal{E}$. As usual, $a \prec b$ means $(a \preceq b) \wedge \neg(b \preceq a)$. An element $a \in S$ is *minimal* in $S \subseteq \mathcal{E}$ if there is no $s \in S$ with $s \prec a$. The reader should keep in mind the special case of a total ordering.

Definition 15.5 (Bislim). *A combinatorial 2-complex X is* bislim *if:*

1. *there is a $\pi_1 X$-invariant preorder on the edges of $\widetilde{X}$;*
2. *∂r has distinguished edges r^+ and r^- for each 2-cell r of $\widetilde{X}$; moreover, r^+ is traversed exactly once by $\partial_{\mathsf{p}} r$;*
3. *$g(r^{\pm}) = (gr)^{\pm}$ for each $g \in \pi_1 X$ and each 2-cell r;*
4. *if r_1 and r_2 are distinct 2-cells in $\widetilde{X}$ and $r_1^+ \subset \partial r_2$ then $r_1^+ \prec r_2^+$;*
5. *if r_1 and r_2 are distinct 2-cells in $\widetilde{X}$ and $r_2^- \subset \partial r_1$ then $r_1^+ \prec r_2^+$.*

Remark 15.6. The definition of bislim in [HW16] states the additional requirement that r^+ has the property that $r^+ \npreceq e$ for any other edge e in ∂r, and moreover that

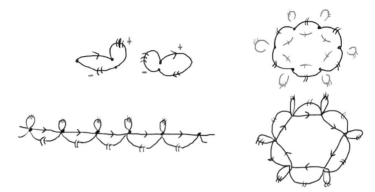

Figure 25: Let X be the standard 2-complex of $\langle a, b, c \mid abc, ac^{-1}b \rangle$. Let $\widehat{X} \to X$ denote the $\mathbb{Z}$-covering of the bouquet of three circles illustrated on the left. Consider the $\mathbb{Z}$-invariant ordering where $e \prec e'$ if the rightmost vertex of e is to the left of the rightmost vertex of e', and if their rightmost vertices are the same, then use $b < a < c$. The preorder on the 1-cells of $\widetilde{X}$ is induced by the order on the 1-cells of $\widehat{X}$. A bislim structure is obtained on $\widetilde{X}$: Let $(abc)^+$ be its b-edge, and $(abc)^-$ be its c-edge. Let $(ac^{-1}b)^+$ be its c-edge and $(ac^{-1}b)^-$ be its b-edge. On the right is an isle-decomposition for Y^1 where $Y \to X$ is a degree 6 cover.

r^+ is the unique such edge in ∂r. This "unique strictly maximal" property does not get used in the proofs of the properties of bislim complexes. In particular it is not used in the definition of the pre-widges there, nor in the proof of [HW16, Thm 4.1].

A bislim structure leads to nonpositive immersions using the following:

Construction 15.7. Let X have a bislim structure. Let $Y \to X$ be an immersion. A *widge* of Y^1 is an edge traversed by an r_i^+ for some cycle r_i^n. The *isles* of Y^1 are components obtained by removing the open widges. See Figure 25.

Let $\mathcal{W}$ denote the set of widges. Let $\mathcal{I}$ denote the set of isles. Let $\mathcal{R}$ denote the set of 2-cells in Y. Observe that $\chi(Y^1) = \sum_{I \in \mathcal{I}} \chi(I) - |\mathcal{W}|$.

Hypothesis (15.5) ensures that if $r_i \neq r_j$ then $r_i^+ \neq r_j^+$. Hence there is an injection $\mathcal{R} \to \mathcal{W}$. Consequently, if $\chi(I) \leq 0$ for each $I \in \mathcal{I}$ then:

$$\chi(Y) = \chi(Y^1) + |\mathcal{R}| \leq \chi(Y^1) + |\mathcal{W}| = \sum_{I \in \mathcal{I}} \chi(I) \leq 0.$$

Nonpositive immersions follow by ensuring that each $\chi(I) \leq 0$. The order properties of the bislim structure are utilized to prove the following in [HW16]:

Lemma 15.8 (No tree isle). *Suppose Y is compact and does not collapse along a free face. If some isle is a tree, then Y is a single 0-cell.*

15.d   Good stackings

We now describe the beautiful ideas introduced by Louder-Wilton [LW17]. Let X be a 2-complex with 2-cells $\{r_i\}$ having boundary paths $\partial_{\mathsf{p}} r_i = w_i$, and whose

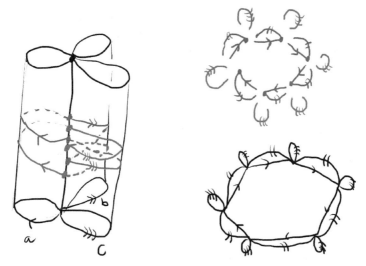

Figure 26: On the left is a good stacking of the 2-complex associated to the presentation $\langle a, b, c \mid abc, ac^{-1}b \rangle$. On the right is the arc decomposition of Y^1 where $Y \to X$ is a degree 6 covering map.

attaching maps $\{\phi_i : w_i \to X^1\}$ are closed immersed paths. Let $W = \sqcup w_i$ and let $\Phi : W \to X^1$ be induced by $\{\phi_i\}$. Let ρ_X and $\rho_{\mathbb{R}}$ be the projections of $X^1 \times \mathbb{R}$ onto X^1 and $\mathbb{R}$.

A *stacking* of X is an injection $\Psi : W \to X^1 \times \mathbb{R}$ so that $\Phi = \rho_X \circ \Psi$. The stacking is *good* if for each i, there exist $h_i, \ell_i \in w_i$ so that: $\rho_{\mathbb{R}}(\Psi(\ell_i))$ is lowest in $\rho_{\mathbb{R}}(\Psi(\Phi^{-1}(\Phi(\ell_i))))$ and $\rho_{\mathbb{R}}(\Psi(h_i))$ is highest in $\rho_{\mathbb{R}}(\Psi(\Phi^{-1}(\Phi(h_i))))$. See Figure 26 for a simple example where the 2-complex X has only two 2-cells.

The proof that one-relator 2-complexes (without torsion) have a good stacking is related to right-orderability. A group G is *right-orderable* if there is a right-invariant total order on G. *Left-orderability* is defined similarly. It is easy to see that G is left-orderable if and only if it is right-orderable. Locally indicable groups are right-orderable [BH72]. Thus every one-relator group without torsion $\langle a, b, \ldots \mid w \rangle$ is right-orderable, since it is locally indicable [Bro84, How82]. Another fundamental fact is that each proper nontrivial subword of w represents a nontrivial element of $\pi_1 X$ [Wei72].

Given the right-orderability and the nontrivialness of proper subwords, good stackings exist for one-relator 2-complexes without torsion as a consequence of the following characterization of Farrell [Far76]. Indeed, an embedding of a single 2-cell boundary path $w \subset \Upsilon$ provides the good stacking $w \hookrightarrow B \times \mathbb{R}$.

Theorem 15.9. *Let G be a group with Cayley graph Υ, so there is a covering map $\Upsilon \to B$ where B is a bouquet of circles corresponding to the generators of G. Then G is right-orderable if and only if there is an embedding $\Upsilon \hookrightarrow B \times \mathbb{R}$ such that the following diagram commutes:*

$$
\begin{array}{ccc}
\Upsilon & \hookrightarrow & B \times \mathbb{R} \\
 & \searrow & \downarrow \\
 & & B
\end{array}
$$

We now use the properties of one-relator groups above to directly sketch that a good stacking exists for a one-relator group without torsion $\langle a_1, a_2, \ldots \mid w \rangle$. A similar argument works for staggered 2-complexes.

Construction 15.10. For simplicity assume there are two generators. Let X^1 be the bouquet of circles on $\{a, b\}$. Each vertex v_i is associated to an initial subword w_i. Suppose $w_{i+1} = w_i c$ where $c \in \{a^{\pm 1}, b^{\pm 1}\}$. Position an edge travelling above the c-edge of X^1 linearly from v_i to v_{i+1}. Finally, right-orderability ensures that two edges cannot cross. Indeed, crossing edges must be of the same type. However for vertices u, v, u', v': if $uc = v$ and $u'c = v'$, then $u < v$ iff $u' < v'$. Thus those two edges cannot cross.

An *open-arc* A is the union of n distinct edges $e_1, \ldots, e_n$ and $n - 1$ distinct vertices $v_1, \ldots, v_{n-1}$ where e_i and e_{i+1} are concatenated at v_i. Note that the subspace $A = \bigcup v_i \cup \bigcup e_i$ is homeomorphic to $(0, 1)$.

An *open-arc decomposition* of a graph Γ is a decomposition of Γ into the disjoint union of open arcs $A_1, \ldots, A_m$. Note that $\chi(\Gamma) = -m$.

The counting mechanism in [LW17] takes the following form:

Lemma 15.11. *Let X be a 2-complex with a good stacking. Let $Y \to X$ be an immersion such that Y has no free faces. Then there is an open-arc decomposition $\mathcal{A} = \{A_1, \ldots, A_m\}$ of Y. Moreover, there is an injection $\mathcal{W} \to \mathcal{A}$. Consequently, if Y is compact we have: $\chi(Y) = |\mathcal{W}| - m \leq 0$.*

Proof. The good stacking of X induces a good stacking of Y. Each arc of $\mathcal{A}$ is the part of a 2-cell visible from the top. Since each cycle in the good stacking is highest somewhere and lowest somewhere else (there are no free faces) we see that there is at least one open arc per 2-cell. $\square$

16 L^2-BETTI NUMBERS

In this section we describe an approach to detect the nonpositive immersions property. The method is to use L^2-betti numbers, and we refer to [Lüc02] for a comprehensive account of this theory. I am grateful to Misha Gromov for suggesting to explore the connection with L^2-betti numbers when I asked him for a coarse generalization of the nonpositive immersion property in 2002.

16.a L^2-betti numbers

We now briefly describe L^2-homology. For a complex X, we use the L^2-chain-complex $\mathcal{C}^{(2)}(\widetilde{X})$, where $C_n^{(2)}(\widetilde{X})$ consists of square-summable chains with $\mathbb{R}$-coefficients, instead of ordinary $\mathbb{R}$-chains. The boundary of a square-summable chain is again square-summable, and one has $\partial_k \circ \partial_{k+1} = 0$ as usual. One defines $\mathsf{H}_n^{(2)}(X) = Z_n^{(2)} / \overline{B_n^{(2)}}$ where $Z_n^{(2)} = \ker(\partial_n)$ and $\overline{B_n^{(2)}}$ is the closure of $\mathrm{im}(\partial_{n+1})$. Although $\mathsf{H}_n(\widetilde{X})$ may not be finite dimensional over $\mathbb{R}$, there is a notion of *von Neumann dimension* of $\mathsf{H}_n(\widetilde{X})$ as a module over the "group von Neumann algebra" $\mathcal{N}(\pi_1 X)$, and when X is compact, this provides a finite measure of its size called the *n-th L^2-betti number* of X. We use the notation $b_n^{(2)}(X)$ for the n-th L^2-betti number and $b_n(X)$ for the n-th ordinary betti number. Surprisingly, when X and Y are compact aspherical complexes with $\pi_1 X \cong \pi_1 Y$, we have $b_n^{(2)}(X) = b_n^{(2)}(Y)$. Our discussions here are mostly about these L^2-betti numbers, and we will only require a few simple properties. See [Lüc02, Thm 1.35].

Lemma 16.1. *Let X be a finite complex. Then:*

1. $\sum_n (-1)^n b_n(X) = \chi(X) = \sum_n (-1)^n b_n^{(2)}(\widetilde{X})$;
2. *if X is 2-dimensional then:* $\left(\mathsf{H}_n^{(2)}(\widetilde{X}) = 0\right) \ \Leftrightarrow \ \left(b_n^{(2)}(\widetilde{X}) = 0\right)$;
3. $b_0^{(2)}(X) = \frac{1}{|\pi_1 X|}$.

16.b Supertrees

We now relate L^2-betti numbers and nonpositive immersions.

Lemma 16.2. *Let K be a category of compact 2-complexes that is closed under immersions. Suppose each 2-complex $X \in K$ has $\mathsf{H}_2^{(0)}(\widetilde{X}) = 0$. Then each 2-complex in K has nonpositive immersions.*

Sketch. Let $Y \to X$ be an immersion with Y compact. By Lemma 16.1, we have $\chi(Y) = b_0^{(2)} - b_1^{(2)} + b_2^{(2)} \leq b_0^{(2)} = \frac{1}{|\pi_1 X|}$. Consequently, as $\chi(Y) \in \mathbb{Z}$, either $\chi(Y) \leq 0$ or $\pi_1 Y = 1$. In the latter case one then deduces that Y is contractible from Theorem 12.8. □

Remark 16.3. Recall the notion of nonpositive towers from Definition 12.14. If we knew that $\mathsf{H}_2^{(2)} = 0$ is preserved by subcomplexes of 2-complexes (in some category), then the computation in the proof of Lemma 16.2 will show that nonpositive tower maps follow from $\mathsf{H}_2^{(2)}(X) = 0$. See Problem 12.15.

Definition 16.4 (Bipartite dual and supertrees). *The bipartite dual D of $\widetilde{X}$ has a face-vertex for each 2-cell of $\widetilde{X}$ and an edge-vertex for each 1-cell of $\widetilde{X}$. The edges of D correspond to the incidence relation between 1-cells and 2-cells of $\widetilde{X}$. Namely, for a 2-cell whose attaching map is a length n combinatorial path, there are n edges from the corresponding face-vertex to the various edge-vertices corresponding to 1-cells in the attaching map.*

An s-supertree is a subgraph T of D such that T is a tree, each face-vertex of T has valence $\geq s$ and each edge-vertex has valence $\leq s$. More generally, T is a supertree if the valence of each edge-vertex is $\leq$ the valences of its adjacent face vertices.

A supertree T in D is associated to a *cactus* which is a subcomplex of $\widetilde{X}$ whose 2-cells correspond to the face-vertices of T and where 2-cells meet along 1-cells corresponding to edge-vertices of T. See Figure 27.

Theorem 16.5. *Let D be the bipartite dual of $\widetilde{X}$. Suppose each face-vertex of D is contained in a supertree. Then $\mathsf{H}_2^{(2)}(\widetilde{X}) = 0$.*

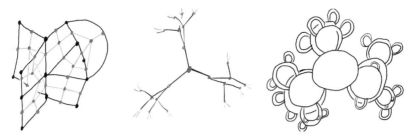

Figure 27: A bipartite dual on the left, a 3-supertree in the middle, and its associated caetus on the right.

Sketch. The idea of the proof of Theorem 16.5 is simple. Suppose $\mathsf{H}_2^{(2)}(\widetilde{X}) \neq 0$. Consider a nonzero square summable cycle $\sum_{c \in 2\text{-cells}} \alpha_c c \in Z_2^{(2)}$. As the cycle is nonzero, the coefficient $\alpha_{c_o} \neq 0$ for some 2-cell c_o. By hypothesis, the face-vertex associated to c_o lies in a supertree T in the bipartite dual. Consider the associated cactus subcomplex of $\widetilde{X}$, which has a collection of 2-cells joined together at 1-cells, so that the amount of branching at these 1-cells is bounded above by the number of branches around each 2-cell. Viewing the vertex dual to c_o as the root of T, and giving each edge of T length $\frac{1}{2}$, consider the family of vertices at each integer distance i from c_o, This decomposes the 2-cells into rings of 2-cells corresponding to the radius-i spheres in T. A computation using the hypothesized relationship between the branching at face-vertices and edge-vertices of T implies that the square sum of the coefficients of the 2-cells in the $(i+1)$-th ring is bounded below by the square sum in the i-th ring. And all of these are bounded below by $\alpha_{c_o}^2$. This contradicts that the cycle is square-summable. $\qquad\square$

The property of each face-vertex of $\widetilde{X}$ being contained in a supertree is closed under immersions to X. Consequently, by combining Theorem 16.5 and Lemma 16.2 we obtain:

Theorem 16.6. *Suppose that for each 2-cell in $\widetilde{X}$, its dual face-vertex lies in a supertree in the bipartite dual. Then X has nonpositive immersions.*

We close by mentioning the connection between small-cancellation complexes and supertrees [Wisa].

Theorem 16.7. *Let X be a $C(p)$-$T(q)$ complex.*

1. *If $p \geq 6$ and $q \geq 3$ and the perimeter of each 1-cell is $\leq p-4$, then the dual of each 2-cell of $\widetilde{X}$ lies in a $(p-4)$-supertree.*
2. *If $p \geq 4$ and $q \geq 4$ and the perimeter of each 1-cell is $\leq p-2$, then the dual of each 2-cell of $\widetilde{X}$ lies in a $(p-2)$-supertree.*
3. *If $p \geq 4$ and $q \geq 5$ and the perimeter of each 1-cell is $\leq p-1$, then the dual of each 2-cell of $\widetilde{X}$ lies in a $(p-1)$-supertree.*

We thus obtain the following which was stated earlier as Theorem 13.4:

Corollary 16.8. *X has supertrees and thus nonpositive immersions if it satisfies any of the following:*

1. X is a $(p, 3, p-4)$-complex with $p \geq 6$.
2. X is a $(p, 4, p-2)$-complex with $p \geq 4$.
3. X is a $(p, 5, p-1)$-complex with $p \geq 4$.

Corollary 16.8 motivated a more careful inspection of the nonpositive sectional curvature for (p, q, r)-complexes, and as reported in Section 13, a careful analysis showed that nonpositive sectional curvature holds under the analogous hypothesis.

17 VIRTUAL ALGEBRAIC

A group G *virtually algebraically fibers* if G has a finite index subgroup G', such that G' has a normal subgroup N such that G'/N is infinite cyclic.

As observed by Stallings [Sta62], if $G = \pi_1 M$ for a compact irreducible 3-manifold M, then G virtually algebraically fibers precisely if M virtually fibers, in the sense that M has a finite cover $\widehat{M}$ that is a compact surface bundle over a circle. A group G is *RFRS* if every nontrivial $g \in G$ lies outside a finite index subgroup which is obtained by finitely many replacements of G by G', where G' is the kernel of a homomorphism $G \to A \to F$ where A is torsion-free abelian and F is finite. For instance, it is easy to check that every right-angled Artin group is RFRS. Agol proved that if M is a compact 3-manifold with $\chi(M) = 0$ and $\pi_1 M$ is RFRS then $\pi_1 M$ virtually fibers.

Kielak proved the following generalization [Kie19]:

Theorem 17.1. *Let G be a RFRS group, and suppose $b_1^{(2)}(G) = 0$. Then G virtually algebraically fibers.*

This section is primarily concerned with consequences of Theorem 17.1. But we first interpose another virtual algebraic fibering discussion:

17.a Right-angled Coxeter groups and legal systems

In [JNW], we gave a condition ensuring the virtual algebraic fibering of a right-angled Coxeter groups using Bestvina-Brady Morse theory. In particular, we showed that there are closed and finite-volume hyperbolic 4-manifolds whose fundamental groups virtually algebraically fiber. We conjectured as well that every finite-volume hyperbolic manifold of dimension ≥ 4 virtually algebraically fibers. See Conjecture 9.32. The condition in [JNW] is expressed in terms of a game involving colorings of the graph Γ which we now describe:

Definition 17.2 (Legal systems). *Let Γ be a simplicial graph with vertices V and edges E. A state is a subset $S \subset V$. The state S is legal if Γ_S and Γ_{V-S} are connected and nonempty, where Γ_A is the full subgraph induced by A. A move at the vertex v is a subset $M_v \subset V$ such that $v \in M_v$, but $u \notin M_v$ when $\{u, v\} \in E$.*

Identify 2^V with $\mathbb{Z}_2^V$ so, e.g., $\{2, 5\} \leftrightarrow (0, 1, 0, 0, 1)$. Let $\mathcal{M} = \{M_v : v \in V\}$ be a collection of moves. The subgroup $\langle \mathcal{M} \rangle$ acts on 2^V. Let S be a state. The orbit $\langle \mathcal{M} \rangle S$ is legal if each state in $\langle \mathcal{M} \rangle S$ is legal. We then say $(\mathcal{M}, S)$ is a legal system.

The main result in [JNW] is:

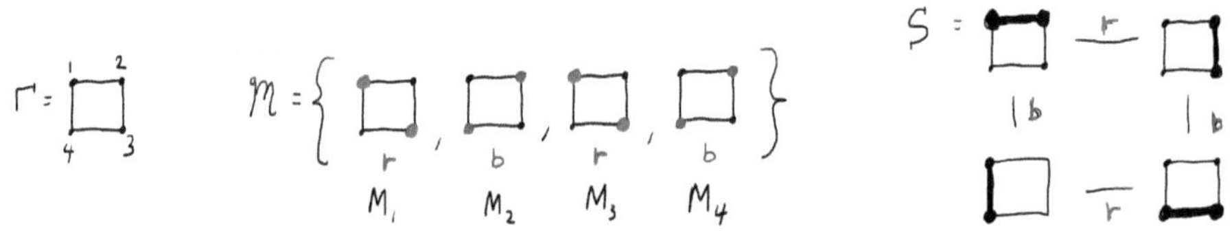

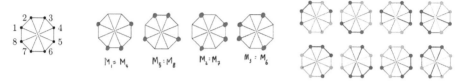

Figure 28: A legal system for the graph Γ on the left is determined by the collection $\mathcal{M}$ of 4 moves in the center (there are repeats); a legal orbit is indicated on the right.

Figure 29: A legal system for Wagner's graph Γ on the left is determined by the 4 different types of moves in the middle (each move is associated to two vertices); a legal orbit is indicated on the right.

Theorem 17.3. *If the finite graph Γ has a legal system, then the associated right-angled Coxeter group $G(\Gamma)$ virtually algebraically fibers.*

The proof of Theorem 17.3 is a straightforward application of Theorem 9.19 applied to a cube complex that is the (orbihedral) cover associated to the homomorphism $G \to \mathbb{Z}^2$ induced by $v \mapsto M_v$. The legal orbit $\langle \mathcal{M} \rangle S$ determines a way of directing all the edges in the cover, so that the associated Morse function has connected ascending and descending links. These correspond to the hypothesized pair of connected induced subgraphs associated to each state in the legal orbit. The nonemptiness condition is to ensure that the homomorphism to $\mathbb{Z}$ is surjective.

Building on Theorem 17.3, Fiz Pontiveros-Glebov-Karpas [PGK17] showed:

Theorem 17.4. *Let Γ be random finite graph on n vertices and with an edge included with probability $\rho(n)$. Then the right-angled Coxeter group $G(\Gamma)$ almost surely virtually algebraically fibers provided:*

$$\frac{\log(n) + \log\log(n) + \omega(1)}{n} \leq \rho(n) \leq 1 - \omega\left(\frac{1}{n^2}\right). \tag{11}$$

As noted in [PGK17], Condition (11) is precisely the probability ensuring the random graph Γ almost surely has degree ≥ 2 for each of its vertices.

17.b Hyperbolic manifolds

It is a classical result that $b_1^{(2)}(M) = 0$ for any finite-volume hyperbolic manifold of dimension ≥ 3. See [Dod79] and [Lue03, Exmp 6.13]. Given this, an immediate consequence of Kielak's theorem is the following:

Corollary 17.5 (Kielak). *Every virtually special finite-volume hyperbolic manifold of dimension ≥ 3 virtually algebraically fibers.*

Although virtual specialness is conjectured (by me, but some disagree) for all finite-volume hyperbolic manifolds, there are still relatively few positive results. A hyperbolic n-manifold is *arithmetic* if the commensurator of $\pi_1 M$ is dense in the isometry group of $\mathbb{H}^n$. It is of *simple-type* if there exists an isometric copy $\mathbb{H}^{n-1} \subset \mathbb{H}^n$ such that $\mathsf{Stab}_{\pi_1 M}(\mathbb{H}^{n-1})$ acts on $\mathbb{H}^{n-1}$ with finite-volume quotient. The following was proven in [BHW11], and covers roughly $\frac{3}{4}$ of all arithmetic hyperbolic manifolds. (There are some exceptional cases in dimensions 3 and 7, but besides those, all finite-volume arithmetic hyperbolic manifolds, and "half" of the closed arithmetic hyperbolic manifolds are of simple-type.)

Theorem 17.6. *Every simple-type arithmetic lattice is virtually special.*

In particular, Kielak's result gives:

Corollary 17.7 (Kielak). *Every arithmetic hyperbolic manifold of simple-type virtually algebraically fibers.*

I expect that the f.g. kernels associated to Kielak's algebraic virtual fiberings of higher dimensional hyperbolic manifolds are not f.p. Hence none of these hyperbolic manifolds have coherent π_1, in a manner bolstering Theorem 9.31. This was the case for the 4-manifolds in [JNW].

17.c   Baumslag's virtual algebraic fibering conjecture

Baumslag suggested the possibility that each one-relator group G has a finite index subgroup H that is free-by-cyclic. This attractive possibility cannot hold in general, since H and thus G would be residually finite [Bau71], and this does not hold for all one-relator groups [BS62].

Conjecture 17.8. *Let G be a hyperbolic one-relator group. Then G is virtually free-by-cyclic.*

It is almost surely the case that Conjecture 17.8 holds for one-relator groups with torsion, which was Baumslag's original conjecture. In the torsion-free hyperbolic case, Conjecture 17.8 is now a consequence of virtual specialness of hyperbolic one-relator groups which was conjectured in [Wis14].

However, Corollary 17.12 shows that Conjecture 17.8 is generically true, as it holds for $C'(\frac{1}{6})$ one-relator groups, and generic f.g. one-relator groups embed in 2-generator $C''(\frac{1}{6})$ one-relator groups.

There have been some interesting investigations about the proportion of two-generator one-relator groups that algebraically fiber. Borisov-Sapir report on separate computational experiments of Sapir and Schupp indicating that the vast majority of two-generator one-relator groups are free-by-cyclic [BS05]. Although it seems that while most do, a definite proportion do not algebraically fiber as shown by Dunfield-Thurston [DT06]. However, as in Conjecture 17.15, we expect they generically virtually algebraically fiber.

The following result was proven by Dicks-Linnell [DL07] (see [Lue03, Exmp 6.12]).

Theorem 17.9. *Let G be a one-relator group. Then $b_2^{(2)}(G) = 0$.*

and so:

Corollary 17.10. *If G is a 2-generator one-relator group without torsion then $b_1^{(2)}(G) = 0$.*

Proof. This holds from the following:

$$0 = 1 - 2 + 1 = \chi(G) = b_0^{(2)} - b_1^{(2)} + b_2^{(2)} = 0 - b_1^{(2)} + 0 = b_1^{(2)}$$

where $b_2^{(2)} = 0$ by Theorem 17.9, and $\chi(G) = b_2^{(2)} - b_1^{(2)} + b_2^{(2)}$ holds for G of finite type, and $b_2^{(2)} = 0$ for $|G| = \infty$. See Lemma 16.1. □

As a consequence of Corollary 17.10 and Theorem 17.1 we have:

Theorem 17.11. *Every virtually special two-generator one-relator group without torsion has a finite-index subgroup that is f.g. free-by-cyclic.*

Theorem 17.11 has the following consequence:

Corollary 17.12. *Generic one-relator groups are virtually free-by-cyclic.*

Note that since $\chi < 0$, the free normal subgroup will not be f.g. when the one-relator group has more than 2 generators.

Proof. A random group G with a fixed number of generators and relators is readily seen to be $C'(\alpha)$ for each $\alpha > 0$, since the maximal length of a piece grows logarithmically in terms of the lengths of the relators. Hence, with overwhelming probability, H is $C'(\frac{1}{7})$. Lemma 17.13 embeds H in a 2-generator one-relator group G that is $C'(\frac{1}{6})$. The group G acts freely on a CAT(0) cube complex [Wis04b] and so G is virtually special [Ago13]. Hence Theorem 17.11 provides a finite index subgroup $G' \subset G$ that is f.g. free-by-cyclic, and letting $H' = G' \cap H$, we see that H' is free-by-cyclic. □

Lemma 17.13. *Every f.g. one-relator group H is a subgroup of a 2-generator one-relator group G. Moreover, if H is a one-relator group with torsion, then so is G, and if H is $C'(\frac{1}{q+1})$ then G is $C'(\frac{1}{q})$.*

Proof. The first explanation shows that a f.g. one-relator group with torsion embeds in a 2-generator one-relator group with torsion. This well-known procedure uses the Magnus-Moldavanskii HNN extension in reverse [LS77] (see also [Wisc, Const 19.5]).

Let $H = \langle a_1, \ldots, a_m \mid W^n \rangle$ be a one-relator group, and suppose for simplicity that each a_i appears in $W^{\pm 1}$. Let t be a new generator, and consider the HNN extension: $G = H *_{M^t = M'}$ where $M = \langle a_1, \ldots, a_{m-1} \rangle$ and $M' = \langle a_2, \ldots, a_m \rangle$ are rank $r - 1$ free subgroups of H by Theorem 14.2, and $a_i^t = a_{i+1}$. Then $G = \langle a_1, t \mid (W')^n \rangle$ is a one-relator group with the same torsion exponent.

We now discuss the second statement which uses small-cancellation theory. For each $\alpha > 0$ there is a set of words $\{A_1, \ldots, A_m\}$ in x, y such that $\langle x, y \mid A_1, \ldots, A_m \rangle$

satisfies $C'(\alpha)$, and moreover each $|A_i|$ has the same length ℓ. Choose these so that $\alpha < \frac{1}{6} - \frac{|P|}{|A_i|}$ for any piece P.

Let W' be the word obtained from $W(A_1, \ldots, A_m)$ by reducing and cyclically reducing. That is, W' is obtained by substituting A_i for each a_i in W, and then removing all backtracks. Note that each backtrack has length $< \alpha|A_i|$. Hence $|W'| \geq (1 - 2\alpha)\ell|W|$. Let $G = \langle x, y \mid W' \rangle$.

Each piece in the presentation for G is either the concatenation of one or two pieces in $\langle x, y \mid A_1, \ldots, A_m \rangle$, or it is of the form $BP'C$ where P is a piece of $\langle a_1, \ldots, a_m \mid W \rangle$ and P' is a subword of the word obtained from $P(A_1, \ldots, A_m)$ by removing all backtracks, and B, C are pieces of $\langle x, y \mid A_1, \ldots, A_m \rangle$. The first type of piece has length $< \alpha\ell + \alpha\ell = 2\alpha\ell$. The second type of piece has length $< \alpha\ell + \ell\frac{1}{q+1}|W| + \alpha\ell$. Consequently, the presentation for G satisfies $C'(\frac{1}{q})$ provided $(2\alpha + \frac{1}{q+1}|W|)\ell < \frac{1}{q}(1 - 2\alpha)\ell|W|$, which holds when $(\frac{2\alpha q}{|W|} + \frac{q}{q+1}) < (1 - 2\alpha)$, and hence holds for $\alpha < \frac{1}{4(q+1)^2}$.

Let Y be the 2-complex associated to the presentation for H, and let X be the 2-complex associated to that for G. Subdivide each a_i 1-cell of Y and replace it by A_i. Now fold the result to obtain a complex Z, and note that the map $Y \to Z$ is a π_1-isomorphism, since the folding is highly limited by the small-cancellation hypothesis on $\{A_1, \ldots, A_m\}$. We thus obtain a combinatorial immersion $Z \to X$.

We now verify that $Z \to X$ has no missing shells in the sense of Definition 8.36. Observe that Z^1 equals the result of folding the wedge $\vee_{i=1}^m A_i$. Let A_i' denote the subpath of A_i obtained by removing the initial and terminal subpath of length $\lceil \alpha|A_i| \rceil$. Observe that for $i \neq j$, each A_i' is disjoint from each A_j in Z^1, since at most a piece of length $< \alpha|A_i|$ is folded at each end of A_i. The only way that A_i' can appear in the boundary path of relator $R = W(A_1, \ldots, A_m)$, is as a subpath of its original occurrence within some A_i arising from a_i in the construction of the relator. Consequently, the only way a long subpath of W can appear as a path in Z^1 is the way that the boundary path of the 2-cell of Z appears. We conclude that $Z \to X$ is π_1-injective by Lemma 8.37. $\qquad\square$

17.d   Noncoherence at $\chi > 0$

While Theorem 17.1 is about virtual algebraic fibering, I expect it can be moved towards noncoherence as follows:

Conjecture 17.14. *Let $G = \langle a_1, \ldots, a_r \mid w_1, \ldots, w_s \rangle$ be a random presentation where r and s are fixed, and where each w_i is drawn uniformly at random from the set of length $\leq \ell$ cyclically reduced words in $\{a_1^{\pm 1}, \ldots, a_r^{\pm 1}\}$. Then the following hold with overwhelming probability as $\ell \to \infty$:*

1. If $r \leq s + 1$ then $b_1^{(2)} = 0$.
2. If $r \geq s + 1$ then $b_2^{(2)} = 0$.

Conjecture 17.14.(17.14) would lead to the following consequence:

Conjecture 17.15. *Let $G = \langle a_1, \ldots, a_r \mid w_1, \ldots, w_s \rangle$ be a random group with r, s fixed, and the relators having length $\leq \ell$. Then with overwhelming probability, as $\ell \to \infty$ we have:*

1. G is incoherent when $s \geq r$;
2. G is virtually f.g. free-by-cyclic when $s = r - 1$;
3. G is virtually free-by-cyclic when $s < r - 1$.

Explanation when $\chi \geq 0$. A random group G with a fixed number of generators and relators is readily seen to be $C'(\frac{1}{6})$ since the maximal length of a piece grows logarithmically in terms of the lengths of the relators. Hence, with overwhelming probability, G is hyperbolic and acts on a CAT(0) cube complex [Wis04b] and so G is virtually special [Ago13].

Assuming Conjecture 17.14 holds, when $s \geq r - 1$ with overwhelming probability $b_2^{(2)}(G) = 0$, and so Theorem 17.1 implies that G virtually algebraically fibers. When $s > r - 1$, the kernel is f.g. but not f.p. by Theorem 9.1, as if the kernel were free then $\chi(G) = 0$. But the presentation 2-complex of G is aspherical since it is $C'(\frac{1}{6})$ [LS77], and almost surely torsion-free since with overwhelming probability, no relator is a proper power. $\qquad\square$

While $b_1 = 0$ is preserved by epimorphisms of groups, the following shows that $b_1^{(2)} = 0$ is not. Hence adding additional relators could increase $b_1^{(2)}$, so one cannot directly deduce information about Conjecture 17.14 from Corollary 17.10.

Example 17.16. Let M be a finite-volume hyperbolic 3-manifold with an epimorphism $\pi_1 M \to F_2$ to a rank 2 free group. Then $b_1^{(2)}(\pi_1 M) = 0$ but $b_1^{(2)}(F_2) = 1$.

Another example is $F_2 \times F_n \to F_2$ where $n \geq 1$, since $b_1^{(2)}(F_2) = 1$ but $b_1^{(2)}(F_2 \times F_n) = 0$ by the Künneth Formula [Lüc02, Thm 1.35.(4)].

17.e   Bourdon buildings

We return to Bourdon buildings (Definition 9.26). Let $X = X_{p,q}$ be the right-angled 2-dimensional hyperbolic building, with chamber a p-gon and thickness q along every edge. Here $p \geq 5$ and $q \geq 2$.

In [Dym04] (see also [BP03] for this computation), it is proven that:

Theorem 17.17. *If $q < p - 1$ then $b_2^{(2)}(X) = 0$. If $q \geq p - 1$ then $b_1^{(2)}(X) = 0$. Thus $b_2^{(2)} = 0$ when $\chi(X) < 0$ and $b_1^{(2)} = 0$ when $\chi(X) > 0$.*

The following resolves a problem that has troubled me since [Wis11] where Theorem 9.27 is proven. (My most recent attempts used a computer to prove it when $p = 6$ and $r \geq 7$ by finding a Bestvina-Brady Morse function on a finite covering map.)

Theorem 17.18. *Let $X_{p,r}$ be a Bourdon building, then $\pi_1(X_{p,r})$ is incoherent if and only if $r \geq p - 1$.*

Proof. We add a 0-cell at the center of each p-gon to subdivide each p-gon into p squares. The typing of the building ensures that no hyperplane intersects its translates. Hence X is virtually special by [Wisc]. When $r \geq p - 1$ we have $b_1^{(2)}(X_{p,r}) = 0$ by Theorem 17.17. Hence $\pi_1 X$ virtually algebraically fibers by Theorem 17.1. Since $\chi(X) > 0$, and X is aspherical since it is nonpositively curved, the kernel is f.g. but not f.p. by Theorem 9.1.

When $r < p - 1$, we have $r \leq p - 2$, so $X_{p,r}$ is a $(p, q, p - 2)$-complex with $p \geq 5$ and $q = 4$. Then $X_{p,r}$ has nonpositive generalized sectional curvature when $r < p - 1$ by Theorem 13.5. Hence $\pi_1 X_{p,r}$ is locally-quasiconvex by Theorem 18.3.   □

18   LOCAL QUASICONVEXITY

Definition 18.1 (Locally-quasiconvex). *A word-hyperbolic group G is locally-quasiconvex if every f.g. subgroup H of G is quasiconvex, or equivalently, quasi-isometrically embedded.*

A first source of locally-quasiconvex word-hyperbolic groups comes from the following result of Thurston's about compact atoroidal 3-manifolds with $\chi < 0$. We refer to [Can94], [Mor84, Prop 7.1].

Theorem 18.2 (Thurston). *Let N be a geometrically finite hyperbolic manifold such that $\partial \mathrm{Core}(N) \neq \emptyset$. Then N' is geometrically finite whenever $N' \to N$ is a covering space with $\pi_1 N'$ f.g.*

Swarup showed that when N has no cusps, geometrically finite subgroups are quasiconvex [Swa93]. This has been explained in general in [Hru10].

A second source comes from small-cancellation theory as in Theorem 8.10. For the most part, the theorems utilizing perimeter provide local-quasiconvexity when perimeter reductions are strict.

A third source is the following result from [Wis04a, MPW13] which was stated in Theorem 11.18.

Theorem 18.3. *Let X be a compact piecewise-Euclidean 2-complex. Suppose X has negative generalized sectional curvature using its induced angle assignment. Then $\pi_1 X$ is locally-quasiconvex.*

The idea of the proof of Theorem 18.3 is that for a f.g. subgroup $H \subset \pi_1 X$ represented by an immersion $Y \to X$ with Y compact, there is a uniform narrowness of the "band" between a geodesic $\widetilde{\gamma}$ and $\widetilde{Y} \subset \widetilde{X}$. This is achieved by knowing that the vertices along $\widetilde{\gamma}$ have a nonpositive contribution to the curvature, but the vertices in the interior of the band provide a negative contribution.

It is almost surely the case that Theorem 18.3 holds equally well for arbitrary compact 2-complexes with negative generalized sectional curvature. However, even the following is unknown:

Conjecture 18.4. *Let X be an angled 2-complex with negative generalized sectional curvature. Then $\pi_1 X$ is word-hyperbolic.*

Most of the ideas and proofs about negative generalized sectional curvature hold in the somewhat more general context of negative immersions, and we refer to Section 12.e.

Heuristic arguments support the following generalization of Theorem 8.14. Moreover, when $G = \pi_1 X$ with X a nonpositively curved 2-complex with negative sectional curvature, then [MPW13, Thm 9.1] suggests it is true for generic g.

Conjecture 18.5. *Let G be a locally-quasiconvex word-hyperbolic group. Let $g \in G$ be an infinite order element. Then $G/\langle\langle g^n \rangle\rangle$ is locally-quasiconvex for $n \gg 0$.*

It is easy to see that free products of locally-quasiconvex groups are locally-quasiconvex. More generally, we refer to [Kap01] for the following which is parallel to Theorem 5.1:

Theorem 18.6. *Let G be a word-hyperbolic group that splits as a graph of groups with cyclic edge groups. If the vertex groups are locally-quasiconvex then G is locally-quasiconvex.*

There are few known sources of locally-quasiconvex groups. A striking fact about locally-quasiconvex word-hyperbolic groups is the following result of Kapovich-Weidmann [KW].

Theorem 18.7. *Let G be a locally-quasiconvex word-hyperbolic group. Then G has finitely many indecomposable subgroups of each rank.*

As mentioned in Section 3, local-quasiconvexity appears to be a 2-dimensional phenomenon, and we refer to Problem 3.2 about the existence of higher dimensional negatively curved manifolds with locally quasiconvex π_1.

A typical way that a subgroup fails to be quasiconvex is when it is conjugated properly into itself.

Problem 18.8. Is there a context in which whenever $H \subset G$ is a f.g. nonquasiconvex subgroup, then H contains a f.g. subgroup K such that $K^g \subsetneq K$ for some $g \in G$?

See Section 21 Problem (21).

A salient property of quasiconvex subgroups of hyperbolic groups regards their "height." A subgroup $H \subset G$ has *finite height* if there is a finite upper bound on the number of distinct cosets $\{g_i H\}$ such that $\cap_i g_i H g_i^{-1}$ is infinite. This generalizes the notion of H being *malnormal* which means that $gHg^{-1} \cap H = 1$ whenever $g \notin H$.

The following is proven in [GMRS98], and we refer to [HW09] for some relatively hyperbolic generalizations.

Theorem 18.9. *If $H \subset G$ is quasiconvex then H has finite height.*

A major problem is to understand the potential for a converse to Theorem 18.9. In particular:

Problem 18.10. Suppose $M \subset G$ is a malnormal f.g. subgroup of a hyperbolic group. Is M quasiconvex?

Some progress was made on Problem 18.10 for right-angled Coxeter groups in [Tra17].

19 n-FREE

We now describe several results providing n-free groups. Conditions for 2-freeness of small-cancellation groups have been studied for a while; see, e.g., [Gub86]. The following is a rough restatement of the main result in [AO96]. A similar but weaker result was obtained in [Bum01].

Theorem 19.1. *Let $\mathcal{P}(m, n, d)$ consist of all presentations of the form $\langle a_1, \ldots, a_m \mid w_1, \ldots, w_n \rangle$ for fixed m, n and satisfying $\sum |w_i| \leq d$. Let G be the group of a presentation randomly chosen from $\mathcal{P}(m, n, d)$. Then G is $(m-1)$-free with probability converging to 1 as $d \to \infty$.*

As mentioned in Theorem 11.1.(11.1) we have the following which was stated in [Wis04a, Thm 7.5] (but the proof there is flawed).

Theorem 19.2. *Let X have negative immersions. Then $\pi_1 X$ is 2-free.*

Note that the proof doesn't actually use a core, or that X is compact, or that H is f.p. Instead it uses that any compact immersion $Y \to X$ that H factors through will have $\pi_1 Y$ locally indicable and $\chi(Y) < 0$.

Proof. By the proof of Theorem 12.18, the subgroup H is conjugate to $\pi_1 Y$ with Y compact and $Y \to X$ a combinatorial immersion. By collapsing, we can assume Y has no free faces, and by assuming H is indecomposable, we can assume Y has no isolated edges.

Let B be a bouquet of two circles, and let $\psi : B \to Y$ be a π_1-surjective combinatorial map with $\pi_1 B$ mapping to a rank 2 subgroup H. Let C be the mapping cylinder of ψ, so $C = (B \times [0, 1] \sqcup Y)/\{(b, 1) \sim \psi(b) : b \in B\}$.

Consider the long exact sequence for the homology of (C, B):

$$\mathsf{H}_2(C) \to \mathsf{H}_2(C, B) \to \mathsf{H}_1(B) \to \mathsf{H}_1(C).$$

If Y is a graph then H is free. (Likewise one could simplify things if Y had a cut vertex.) Otherwise, observe that $-1 \geq \chi(Y) = b_0(Y) - b_1(Y) + b_2(Y) = 1 - b_1(Y) + b_2(Y)$, so $2 = b_1(B) \geq b_1(Y) \geq 2 + b_2(Y)$, since $B \to Y$ is π_1-surjective. Thus $b_2(Y) = 0$ and $b_1(Y) = 2$. Since C deformation retracts to Y we deduce that $\mathsf{H}_2(C) = 0$, and that $b_1(C) = 2$. Hence $\mathsf{H}_1(B) \to \mathsf{H}_1(C)$ is injective, since it is surjective as $\pi_1 B \to \pi_1 Y$ is surjective, and a surjection $\mathbb{Z}^2 \to \mathbb{Z}^2$ must be injective. Finally, from exactness, we deduce that $\mathsf{H}_2(C, B) = 0$.

Since $\pi_1 C \cong \pi_1 Y$ and $\pi_1 Y$ is locally indicable by Theorem 12.6, we can apply Theorem 12.13 to see that: The map $\pi_1 B \to \pi_1 C = \pi_1 Y$ is injective. As we already knew surjectivity, we conclude that H is free. $\square$

Theorem 19.2 was revisited in [LW18a], who proved it in a setting that ostensibly relaxes the negative immersions property, as its proof only uses that $\chi(Y) < 0$ for each compact collapsed immersion $Y \to X$.

n-freeness is often implicated in the study of limit groups (see the end of Section 5). For instance, a limit group is 2-free if it does not contain $\mathbb{Z}^2$. Indeed, if $[\bar{x}, \bar{y}] \neq 1$ in a free quotient, then $\langle \bar{x}, \bar{y} \rangle$ is a rank 2 free group, and hence so is $\langle x, y \rangle$.

Figure 30: The universal cover $\widetilde{X}$ of $X = \langle p, t \mid ptpp^{-1}t^{-1} \rangle$.

See [FGRS95] for more on 3-freeness in the context of limit groups. In [BS89], it is examined how 3-freeness is preserved by certain amalgams along cyclic subgroups. Finally, we refer to [CS12] and the references therein for a study of k-freeness in certain hyperbolic manifolds.

20   COMPACT CORE

Definition 20.1. *X has the* compact core *property if, for each $\widehat{X}$ with $\pi_1\widehat{X}$ f.g., each compact subspace $S \subset \widehat{X}$ is contained in a compact subspace $C \subset \widehat{X}$ whose inclusion induces a π_1-isomorphism.*

Theorem 20.2. *The compact core property holds in the following contexts:*

1. *X is a 3-manifold [Sco73a].*
2. *X satisfies the nonpositive perimeter reduction hypothesis (along the lines of Theorem 8.20).*
3. *X has negative sectional curvature. (See Theorem 11.1.(11.1).)*

Remark 20.3. I expect the compact core property to hold when X is the mapping torus of an immersion from a graph to itself. However, in general, one needs to assume that certain pathologies are avoided. Specifically, Example 20.4 shows that it can fail. I didn't check the exact requirements.

The compact core property does not always hold under the hypothesis of nonpositive sectional curvature, and hence nonpositive immersions. We now describe a 2-complex X without the compact core property [Wis02a]. It was revisited in [Wis04a] where it was shown to have nonpositive sectional curvature.

Example 20.4. Consider the 2-complex $\widetilde{X}$ in Figure 30. It is the universal cover of the 2-complex X associated to the presentation $\langle p, t \mid pt^{-1}p^{-1}pt \rangle$, where t is the triangle arrow and p is the simple arrow. There does not exist a π_1-isomorphically embedded finite subcomplex of $\widetilde{X}$ that contains a p-circle. The point is that the tail of each 2-cell forms a circle, which is nullhomotopic using the 2-cell to its right. By identifying p with a more serious relation, we can create a more interesting example, for instance: Let Y be the 2-complex associated to the presentation $\langle a, b, t \mid [a, b]t^{-1}[a, b][a, b]^{-1}t \rangle$. Now the subgroup $\langle a, b \rangle$ which is isomorphic to $\mathbb{Z}^2$ has the property that the associated based covering space $\widehat{Y}$ does not have a compact core.

It is possible to put an angle assignment with nonpositive generalized sectional curvature, however some of the angles are negative.

Currently, all known examples X with nonpositive sectional curvature, but without the compact core property, require negative angles.

Problem 20.5. Does the compact core property hold for X if X has nonpositive sectional curvature with nonnegative or positive angles?

The compact core property is helpful for proving separability. For instance, Schupp [Sch03] runs the perimeter algorithm and then uses the resulting core to prove separability. This facilitates the proof of subgroup separability of certain Coxeter groups that are not right-angled. More substantially, it is used to prove the separability of quasiconvex subgroups of special groups in [HW08].

Here are some related problems stated in [Wis02a] that seem interesting:

Problem 20.6. Let X satisfy one of the following.

1. X is a nonpositively curved 2-complex.
2. X is a small-cancellation complex.
3. X is the standard 2-complex of a cyclically reduced one-relator presentation.

Is it always true that a covering space $\widehat{X}$ has a compact core if $\pi_1 \widehat{X}$ is f.p.?

21 PROBLEM LIST

1. (Baumslag) Is every one-relator group coherent?
2. Let $G = H * F / \langle\!\langle W \rangle\!\rangle$ where H is coherent, F is free, and H embeds in G. Prove that G is coherent. More generally, let $G = A * B / \langle\!\langle W \rangle\!\rangle$ where A and B are coherent and $A, B \subset G$. Is G coherent?
3. (Serre) Is $SL_2(\mathbb{Z}[\frac{1}{p}])$ incoherent for p prime?
4. (Serre) Is $SL_3(\mathbb{Z})$ incoherent?

   Note: This is stated in [Wal79] together with the previous problem on $SL_2(\mathbb{Z}[\frac{1}{p}])$. Note that $SL_2(\mathbb{Z})$ is coherent since it is virtually free, and $F_2 \times F_2 \hookrightarrow SL_2 \times SL_2 \hookrightarrow SL_n(\mathbb{Z})$ for $n \geq 4$, so these groups are incoherent.
5. Does there exist an infinite coherent group with Property-T?

   Note: $SL_3(\mathbb{Z})$ has *Property (T)*, which means that every isometric action on a Hilbert space has a global fixed point [CCJ+01].
6. Prove that generically, $\langle a_1, \ldots, a_r \mid W_1, \ldots, W_s \rangle$ is coherent for $r > s$, is incoherent for $r < s$, and is virtually f.g. free-by-$\mathbb{Z}$ for $r = s$.

   Note: It suffices to restrict to the case when $C'(\frac{1}{6})$ holds. This problem, originally raised in [Wisa], has been elucidated by recent advances. See Conjecture 17.15.

   For non-metric small-cancellation groups, Rips has observed that it is not even known whether:
7. (Rips) Is there a f.p. $C(6)$ group that isn't a surface group, but such that every f.g. subgroup is either free or of finite index?

   Note: *Tarski-monsters* constructed by Olshanskii [Ol'84] and Rips (unpublished) have the property that while they are not f.p., all their proper subgroups are cyclic and thus f.p.
8. Let $\langle a_1, \ldots, a_r \mid w_1, \ldots, w_s \rangle$ be a random presentation. Show that it is generically π_1 of a 2-complex with negative [nonpositive] generalized sectional curvature when $r < s - 1$ [when $r < s$].

9. Suppose X has nonpositive immersions. Prove that $\pi_1 X$ is coherent.

10. Find an aspherical 2-complex X with $\pi_1 X$ coherent and $\chi(X) > 1$.

11. Suppose X is an aspherical 2-complex with $\pi_1 X$ coherent. Does X have non-positive immersions?

12. Suppose X is compact and has negative immersions. Prove that $\pi_1 X$ is hyperbolic. Prove that $\pi_1 X$ is locally-quasiconvex.

13. Is there an algorithm to detect negative immersions or nonpositive immersions (within some category)?

14. Let X be a 2-complex with $b_2^{(2)}(X) = 0$. Does X have nonpositive immersions?

15. Let X be a 2-complex with $b_2^{(2)}(X) = 0$. Let $Y \subset X$ be a subcomplex. Show that $b^{(2)}(Y) = 0$.

16. Suppose X has nonpositive immersions/nonpositive sectional curvature. Show that $b_2^{(2)}(X) = 0$.

17. Show that a hyperbolic free-by-$\mathbb{Z}$ group fails to be locally-quasiconvex if and only if it contains a F_n-by-$\mathbb{Z}$ subgroup.

18. Let X have negative immersions/negative sectional curvature. Show that $\pi_1 X$ has a quasiconvex hierarchy. Hence, show that $\pi_1 X$ is virtually compact special.

19. Let X be a contractible 2-complex. Show that X has nonpositive immersions.

20. Show that every LOT presentation has nonpositive immersions.

21. Is π_1 of a compact complete square complex incoherent?

22. Show that $A *_C B$ is incoherent whenever A and B are f.g. free groups of rank ≥ 2, and C is a finite index subgroup of each factor.

23. Let G split as a graph of free groups. Characterize when G is coherent.

24. Prove that every F_n-by-F_m group is incoherent when $n, m \geq 2$.

25. Prove that $\pi_1 S \rtimes F_2$ is incoherent when S is a closed orientable surface of genus ≥ 2.

26. Is Higman's group coherent or incoherent?

27. Let G act on a contractible 2-complex $\tilde{X}$ with nonpositive (negative) generalized sectional curvature, and with coherent (polycyclic?) vertex stabilizers and finite edge stabilizers. Is G coherent?

28. Does there exist a torsion-free hyperbolic coherent group with $\mathsf{cd}(G) \geq 4$.

29. Does there exist a torsion-free locally-quasiconvex hyperbolic group with $\mathsf{cd}(G) = 3$.

30. Let G be locally-quasiconvex [and cubulated], and let $g \in G$ be of infinite order. Show that $G/\langle\!\langle g^n \rangle\!\rangle$ is locally-quasiconvex for $n \gg 0$.

31. Suppose G is word-hyperbolic and coherent and $\mathsf{cd}(G) = 2$. Suppose $\chi(G) = 0$. Let $g \in G$. Is $G/\langle\!\langle g^n \rangle\!\rangle$ incoherent for all $n \gg 0$?

32. Does coherence imply Tits' alternative? Does nonpositive immersions imply Tits' alternative?

33. Does every contractible 2-complex have nonpositive immersions?

34. Is having nonpositive immersions an Andrews-Curtis invariant?

35. Does there exist a coherent G_1 and a noncoherent G_2 such that G_1 and G_2 are quasi-isometric?

36. Does coherence imply a solvable word problem?

37. Does coherence imply a solvable membership problem for each f.g. subgroup? A solvable word problem? For instance, I expect this holds for free-by-cyclic groups.

38. Let X be a coherent group with $\pi_1 X$ hyperbolic. Is there an algorithm to compute a presentation for a f.g. subgroup?

39. Let G be a hyperbolic group. Suppose the conformal dimension of ∂G is at most 2. Is G coherent? Suppose the conformal dimension of ∂G is strictly less than 2. Is G locally-quasiconvex?

40. Is every f.p. subgroup separable group coherent?

   Note: G is *subgroup separable* if every f.g. subgroup H of G is the intersection of finite index subgroups. The circumstantial connection between subgroup separability and coherence is that proofs of separability often rely on a compact core, and these are often provided by a proof of coherence. Grigorchuk's group is not subgroup separable [GW03], which suggests the problem has a negative solution.

41. Give a "language theoretic" characterization of coherence of G in terms of the kernel of a surjection $F \to G$ from a free group.

   Note: Few results follow this (implausible?) direction: e.g., regular languages for finite groups, and context free languages for virtually free groups.

Acknowledgments

I am grateful to my many students and colleagues who collaborated with me on different aspects of coherence. Life would have been incoherent without them! Thanks to: Macarena Arenas, Jacob Bamberger, David Carrier, Antony Della Veccia, David Futer, Jonah Gaster, Joseph Helfer, Chris Hruska, Joseph Lauer, Eduardo Martinez Pedroza, Jon McCammond, and James Requeima. And many thanks to the referee for helpful corrections.

REFERENCES

[Ago13] Ian Agol. The virtual Haken conjecture. *Doc. Math.*, 18:1045–1087, 2013. With an appendix by Agol, Daniel Groves, and Jason Manning.

[AO96] G. N. Arzhantseva and A. Yu. Ol'shanskiĭ. Generality of the class of groups in which subgroups with a lesser number of generators are free. *Mat. Zametki*, 59(4):489–496, 638, 1996.

[Bag76] G. H. Bagherzadeh. Commutativity in one-relator groups. *J. London Math. Soc. (2)*, 13(3):459–471, 1976.

[Bau71] Gilbert Baumslag. Finitely generated cyclic extensions of free groups are residually finite. *Bull. Austral. Math. Soc.*, 5:87–94, 1971.

[Bau74] Gilbert Baumslag. Some problems on one-relator groups. In *Proceedings of the Second International Conference on the Theory of Groups (Australian Nat. Univ., Canberra, 1973)*, volume 372 of *Lecture Notes in Math.*, pages 75–81. Springer, Berlin, 1974.

[BB96] W. Ballmann and S. Buyalo. Nonpositively curved metrics on 2-polyhedra. *Math. Z.*, 222(1):97–134, 1996.

[BB97] Mladen Bestvina and Noel Brady. Morse theory and finiteness properties of groups. *Invent. Math.*, 129(3):445–470, 1997.

[BBN59] Gilbert Baumslag, W. W. Boone, and B. H. Neumann. Some unsolvable problems about elements and subgroups of groups. *Math. Scand.*, 7:191–201, 1959.

[BCGW19] Jacob Bamberger, David Carrier, Jonah Gaster, and Daniel T. Wise. Good stackings, bislim structures, and equivariant staggerings. Unpublished, pages 1–10, 2020.

[BF92] M. Bestvina and M. Feighn. A combination theorem for negatively curved groups. *J. Differential Geom.*, 35(1):85–101, 1992.

[BF95] Mladen Bestvina and Mark Feighn. Stable actions of groups on real trees. *Invent. Math.*, 121(2):287–321, 1995.

[BH72] R. G. Burns and V.W.D. Hale. A note on group rings of certain torsion-free groups. *Canad. Math. Bull.*, 15:441–445, 1972.

[BH99] Martin R. Bridson and André Haefliger. *Metric spaces of non-positive curvature*. Springer-Verlag, Berlin, 1999.

[BHMS02] Martin R. Bridson, James Howie, Charles F. Miller III, and Hamish Short. The subgroups of direct products of surface groups. *Geom. Dedicata*, 92:95–103, 2002. Dedicated to John Stallings on the occasion of his 65th birthday.

[BHW11] Nicolas Bergeron, Frédéric Haglund, and Daniel T. Wise. Hyperplane sections in arithmetic hyperbolic manifolds. *J. Lond. Math. Soc. (2)*, 83(2):431–448, 2011.

[Bie81] Robert Bieri. *Homological dimension of discrete groups*. Queen Mary College Department of Pure Mathematics, London, second edition, 1981.

[BM94] B. H. Bowditch and G. Mess. A 4-dimensional Kleinian group. *Trans. Amer. Math. Soc.*, 344(1):391–405, 1994.

[BMS87] Gilbert Baumslag, John W. Morgan, and Peter B. Shalen. Generalized triangle groups. *Math. Proc. Cambridge Philos. Soc.*, 102(1):25–31, 1987.

[Bog93] William A. Bogley. J.H.C. Whitehead's asphericity question. In *Two-dimensional homotopy and combinatorial group theory*, volume 197 of *London Math. Soc. Lecture Note Ser.*, pages 309–334. Cambridge Univ. Press, Cambridge, 1993.

[Bou97] M. Bourdon. Immeubles hyperboliques, dimension conforme et rigidité de Mostow. *Geom. Funct. Anal.*, 7(2):245–268, 1997.

[BP03] Marc Bourdon and Hervé Pajot. Cohomologie l_p et espaces de Besov. *J. Reine Angew. Math.*, 558:85–108, 2003.

[BR84] Gilbert Baumslag and James E. Roseblade. Subgroups of direct products of free groups. *J. London Math. Soc. (2)*, 30(1):44–52, 1984.

[Bra99] Noel Brady. Branched coverings of cubical complexes and subgroups of hyperbolic groups. *J. London Math. Soc. (2)*, 60(2):461–480, 1999.

[Bro84] S. D. Brodskiĭ. Equations over groups, and groups with one defining relation. *Sibirsk. Mat. Zh.*, 25(2):84–103, 1984.

[BS62] Gilbert Baumslag and Donald Solitar. Some two-generator one-relator non-Hopfian groups. *Bull. Amer. Math. Soc.*, 68:199–201, 1962.

[BS79] Robert Bieri and Ralph Strebel. Soluble groups with coherent group rings. In *Homological group theory (Proc. Sympos., Durham, 1977)*, pages 235–240. Cambridge Univ. Press, Cambridge, 1979.

[BS89] Gilbert Baumslag and Peter B. Shalen. Groups whose three-generator subgroups are free. *Bull. Austral. Math. Soc.*, 40(2):163–174, 1989.

[BS90] Gilbert Baumslag and Peter B. Shalen. Amalgamated products and finitely presented groups. *Comment. Math. Helv.*, 65(2):243–254, 1990.

[BS05] Alexander Borisov and Mark Sapir. Polynomial maps over finite fields and residual finiteness of mapping tori of group endomorphisms. *Invent. Math.*, 160(2):341–356, 2005.

[Bum01] Inna Bumagin. On small cancellation k-generated groups with $(k-1)$-generated subgroups all free. *Internat. J. Algebra Comput.*, 11(5):507–524, 2001.

[Bur19] José Burillo. *Introduction to Thompson's group F*. 2019.

[BW99] Martin R. Bridson and Daniel T. Wise. $\mathcal{VH}$ complexes, towers and subgroups of $F \times F$. *Math. Proc. Cambridge Philos. Soc.*, 126(3):481–497, 1999.

[BW03] Inna Bumagin and Daniel T. Wise. Coherence of word-hyperbolic by cyclic groups. Preprint, 2003.

[BW13] Hadi Bigdely and Daniel T. Wise. C(6) groups do not contain $F_2 \times F_2$. *J. Pure Appl. Algebra*, 217(1):22–30, 2013.

[Can94] Richard D. Canary. Covering theorems for hyperbolic 3-manifolds. In *Low-dimensional topology (Knoxville, TN, 1992)*, Conf. Proc. Lecture Notes Geom. Topology, III, pages 21–30. Internat. Press, Cambridge, MA, 1994.

[CCJ+01] Pierre-Alain Cherix, Michael Cowling, Paul Jolissaint, Pierre Julg, and Alain Valette. *Groups with the Haagerup property, Gromov's a-T-menability*, volume 197 of *Progress in Mathematics*. Birkhäuser Verlag, Basel, 2001.

[CS12] Marc Culler and Peter B. Shalen. 4-free groups and hyperbolic geometry. *J. Topol.*, 5(1):81–136, 2012.

[Deh11] M. Dehn. Über unendliche diskontinuierliche Gruppen. *Math. Ann.*, 71(1):116–144, 1911.

[Del95] T. Delzant. L'image d'un groupe dans un groupe hyperbolique. *Comment. Math. Helv.*, 70(2):267–284, 1995.

[Dic11] Warren Dicks. Simplified Mineyev. http://mat.uab.es/~dicks/Simpli
          fiedMineyev.pdf, 2011.

[DL07]   Warren Dicks and Peter A. Linnell. L^2-Betti numbers of one-relator
          groups. *Math. Ann.*, 337(4):855–874, 2007.

[Dod79] Jozef Dodziuk. L^2 harmonic forms on rotationally symmetric Rieman-
          nian manifolds. *Proc. Amer. Math. Soc.*, 77(3):395–400, 1979.

[Dro87] Carl Droms. Graph groups, coherence, and three-manifolds. *J. Alge-
          bra*, 106(2):484–489, 1987.

[DT06]   Nathan M. Dunfield and Dylan P. Thurston. A random tunnel number
          one 3-manifold does not fiber over the circle. *Geom. Topol.*, 10:2431–
          2499, 2006.

[Dym04] Jan Dymara. L^2-cohomology of buildings with fundamental class.
          Proc. Amer. Math. Soc., 132(6):1839–1843, 2004.

[Far76] F. Thomas Farrell. Right-orderable deck transformation groups. *Rocky
          Mountain J. Math.*, 6(3):441–447, 1976.

[FGRS95] Benjamin Fine, Anthony M. Gaglione, Gerhard Rosenberger, and Den-
          nis Spellman. n-free groups and questions about universally free groups.
          In *Groups '93 Galway/St. Andrews, Vol. 1 (Galway, 1993)*, volume 211
          of *London Math. Soc. Lecture Note Ser.*, pages 191–204. Cambridge
          Univ. Press, Cambridge, 1995.

[FH99]   Mark Feighn and Michael Handel. Mapping tori of free group auto-
          morphisms are coherent. *Ann. of Math. (2)*, 149(3):1061–1077, 1999.

[Fri15] Joel Friedman. Sheaves on graphs, their homological invariants, and
          a proof of the Hanna Neumann conjecture. *Mem. Amer. Math. Soc.*,
          233(1100):xii+106, 2015. With an appendix by Warren Dicks.

[FW16]   David Futer and Daniel T. Wise. Generic curves and sectional curva-
          ture. pages 1–17, 2016. In preparation.

[Ger81] S. M. Gersten. Coherence in doubled groups. *Comm. Algebra*, 9(18):
          1893–1900, 1981.

[Ger87] S. M. Gersten. Reducible diagrams and equations over groups. In
          Essays in group theory, pages 15–73. Springer, New York-Berlin,
          1987.

[GMRS98] Rita Gitik, Mahan Mitra, Eliyahu Rips, and Michah Sageev. Widths
          of subgroups. *Trans. Amer. Math. Soc.*, 350(1):321–329, 1998.

[GMSW01] Ross Geoghegan, Michael L. Mihalik, Mark Sapir, and Daniel T. Wise.
          Ascending HNN extensions of finitely generated free groups are Hop-
          fian. *Bull. London Math. Soc.*, 33(3):292–298, 2001.

[Gor04] C. McA. Gordon. Artin groups, 3-manifolds and coherence. *Bol. Soc.
          Mat. Mexicana (3)*, 10(Special Issue):193–198, 2004.

[Gro78]   J.R.J. Groves. Soluble groups in which every finitely generated sub-
          group is finitely presented. *J. Austral. Math. Soc. Ser. A*, 26(1):115–
          125, 1978.

[Gro87]   M. Gromov. Hyperbolic groups. In *Essays in group theory*, volume 8
          of *Math. Sci. Res. Inst. Publ.*, pages 75–263. Springer, New York,
          1987.

[Gub86]   V. S. Guba. Conditions under which 2-generated subgroups in small
          cancellation groups are free. *Izv. Vyssh. Uchebn. Zaved. Mat.*, (7):12–
          19, 87, 1986.

[GW03]   R. I. Grigorchuk and J. S. Wilson. A structural property concerning
          abstract commensurability of subgroups. *J. London Math. Soc. (2)*,
          68(3):671–682, 2003.

[Hag06]   Frédéric Haglund. Commensurability and separability of quasiconvex
          subgroups. *Algebr. Geom. Topol.*, 6:949–1024 (electronic), 2006.

[Hat02]   Allen Hatcher. *Algebraic topology.* Cambridge University Press, Cam-
          bridge, 2002.

[Hil00]   Jonathan A. Hillman. Complex surfaces which are fibre bundles.
          Topology Appl., 100(2–3):187–191, 2000.

[HK06]   James Howie and Natalia Kopteva. The Tits alternative for generalized
          tetrahedron groups. *J. Group Theory*, 9(2):173–189, 2006.

[How81]   James Howie. On pairs of 2-complexes and systems of equations over
          groups. *J. Reine Angew. Math.*, 324:165–174, 1981.

[How82]   James Howie. On locally indicable groups. *Math. Z.*, 180(4):445–461,
          1982.

[How85]   James Howie. On the asphericity of ribbon disc complements. *Trans.
          Amer. Math. Soc.*, 289(1):281–302, 1985.

[How87]   James Howie. How to generalize one-relator group theory. In S. M.
          Gersten and John R. Stallings, editors, *Combinatorial group theory and
          topology*, pages 53–78. Princeton Univ. Press, Princeton, N.J., 1987.

[HR17]   Jens Harlander and Stephan Rosebrock. Injective labeled oriented
          trees are aspherical. *Math. Z.*, 287(1–2):199–214, 2017.

[Hru10]   G. Christopher Hruska. Relative hyperbolicity and relative quasicon-
          vexity for countable groups. *Algebr. Geom. Topol.*, 10(3):1807–1856,
          2010.

[HW01]   G. Christopher Hruska and Daniel T. Wise. Towers, ladders and
          the B. B. Newman spelling theorem. *J. Aust. Math. Soc.*, 71(1):53–69,
          2001.

[HW08]   Frédéric Haglund and Daniel T. Wise. Special cube complexes. *Geom.
          Funct. Anal.*, 17(5):1551–1620, 2008.

[HW09] G. Christopher Hruska and Daniel T. Wise. Packing subgroups in relatively hyperbolic groups. *Geom. Topol.*, 13(4):1945–1988, 2009.

[HW16] Joseph Helfer and Daniel T. Wise. Counting cycles in labeled graphs: the nonpositive immersion property for one-relator groups. *Int. Math. Res. Not. IMRN*, (9):2813–2827, 2016.

[JNW]   Kasia Jankiewicz, Sergei Norine, and Daniel T. Wise. *Journal of the Institute of Mathematics of Jussieu*, pages 1–30. To appear.

[JW16] Kasia Jankiewicz and Daniel T. Wise. Incoherent Coxeter groups. *Proc. Amer. Math. Soc.*, 144(5):1857–1866, 2016.

[Kap98] Michael Kapovich. On normal subgroups in the fundamental groups of complex surfaces. Preprint, 1998.

[Kap01] Ilya Kapovich. The combination theorem and quasiconvexity. *Internat. J. Algebra Comput.*, 11(2):185–216, 2001.

[Kap02] Ilya Kapovich. Subgroup properties of fully residually free groups. *Trans. Amer. Math. Soc.*, 354(1):335–362 (electronic), 2002.

[Kap13] Michael Kapovich. Noncoherence of arithmetic hyperbolic lattices. *Geom. Topol.*, 17(1):39–71, 2013.

[Kie19] Dawid Kielak. Residually finite rationally-solvable groups and virtual fibring. 2019.

[KM99] O. Kharlampovich and A. Myasnikov. Description of fully residually free groups and irreducible affine varieties over a free group. In *Summer School in Group Theory in Banff, 1996*, volume 17 of *CRM Proc. Lecture Notes*, pages 71–80. Amer. Math. Soc., Providence, RI, 1999.

[KP91] M. È. Kapovich and L. D. Potyagaĭlo. On the absence of finiteness theorems of Ahlfors and Sullivan for Kleinian groups in higher dimensions. *Sibirsk. Mat. Zh.*, 32(2):61–73, 212, 1991.

[KS70] A. Karrass and D. Solitar. The subgroups of a free product of two groups with an amalgamated subgroup. *Trans. Amer. Math. Soc.*, 150:227–255, 1970.

[KW]   Ilya Kapovich and Richard Weidmann. Freely indecomposable groups acting on hyperbolic spaces. Internat. J. Algebra Comput., 14(2):115–171, 2004.

[LS77] Roger C. Lyndon and Paul E. Schupp. *Combinatorial group theory.* Springer-Verlag, Berlin, 1977. Ergebnisse der Mathematik und ihrer Grenzgebiete, Band 89.

[Lüc02] Wolfgang Lück. *L²-invariants: theory and applications to geometry and K-theory.* Springer-Verlag, Berlin, 2002.

[Lue03] Wolfgang Lueck. *L²*-invariants from the algebraic point of view, 2003.

[LW13] Joseph Lauer and Daniel T. Wise. Cubulating one-relator groups with torsion. *Math. Proc. Cambridge Philos. Soc.*, 155(3):411–429, 2013.

[LW17] Larsen Louder and Henry Wilton. Stackings and the W-cycles conjecture. *Canad. Math. Bull.*, 60(3):604–612, 2017.

[LW18a] Larsen Louder and Henry Wilton. Negative immersions for one-relator groups, 2018.

[LW18b] Larsen Louder and Henry Wilton. One-relator groups with torsion are coherent, 2018.

[Mag30] Wilhelm Magnus. Über diskontinuierliche Gruppen mit einer definierenden Relation (Der Freiheitssatz). *J. Reine Angew. Math.*, 163:141–165, 1930.

[McC17] Jonathan P. McCammond. The mysterious geometry of artin groups. Pages 1–44, 2017.

[Mil02] Charles F. Miller III. Subgroups of direct products with a free group. *Q. J. Math.*, 53(4):503–506, 2002.

[Min12a] Igor Mineyev. Groups, graphs, and the Hanna Neumann conjecture. *J. Topol. Anal.*, 4(1):1–12, 2012.

[Min12b] Igor Mineyev. Submultiplicativity and the Hanna Neumann conjecture. *Ann. of Math. (2)*, 175(1):393–414, 2012.

[Mor84] John W. Morgan. On Thurston's uniformization theorem for three-dimensional manifolds. In *The Smith conjecture (New York, 1979)*, volume 112 of *Pure Appl. Math.*, pages 37–125. Academic Press, Orlando, FL, 1984.

[MPW13] Eduardo Martínez-Pedroza and Daniel T. Wise. Coherence and negative sectional curvature in complexes of groups. *Michigan Math. J.*, 62(3):507–536, 2013.

[MW02] Jonathan P. McCammond and Daniel T. Wise. Fans and ladders in small cancellation theory. *Proc. London Math. Soc. (3)*, 84(3):599–644, 2002.

[MW05] J. P. McCammond and D. T. Wise. Coherence, local quasiconvexity, and the perimeter of 2-complexes. *Geom. Funct. Anal.*, 15(4):859–927, 2005.

[MW08] Jonathan P. McCammond and Daniel T. Wise. Locally quasiconvex small-cancellation groups. *Trans. Amer. Math. Soc.*, 360(1):237–271 (electronic), 2008.

[Neu90] Walter D. Neumann. On intersections of finitely generated subgroups of free groups. In *Groups—Canberra 1989*, pages 161–170. Springer, Berlin, 1990.

[New68] B. B. Newman. Some results on one-relator groups. *Bull. Amer. Math. Soc.*, 74:568–571, 1968.

[Ol'84]   A. Yu. Ol'šanskiĭ.   On a geometric method in the combinatorial group theory. In *Proceedings of the International Congress of Mathematicians, Vol. 1, 2 (Warsaw, 1983)*, pages 415–424, Warsaw, 1984. PWN.

[PGK17]   Gonzalo Fiz Pontiveros, Roman Glebov, and Ilan Karpas. Virtually fibering random right-angled Coxeter groups, 2017.

[Pot94]   L. Potyagaĭlo. Finitely generated Kleinian groups in 3-space and 3-manifolds of infinite homotopy type. *Trans. Amer. Math. Soc.*, 344(1): 57–77, 1994.

[Pri88]   Stephen J. Pride. Star-complexes, and the dependence problems for hyperbolic complexes. *Glasgow Math. J.*, 30(2):155–170, 1988.

[PW18]   Piotr Przytycki and Daniel T. Wise. Mixed 3-manifolds are virtually special. *J. Amer. Math. Soc.*, 31(2):319–347, 2018.

[Rip82]   E. Rips. Subgroups of small cancellation groups. *Bull. Lond. Math. Soc.*, 14(1):45–47, 1982.

[Rol90]   Dale Rolfsen. *Knots and links*, volume 7 of *Mathematics Lecture Series*. Publish or Perish Inc., Houston, TX, 1990. Corrected reprint of the 1976 original.

[Sch03]   Paul E. Schupp. Coxeter groups, 2-completion, perimeter reduction and subgroup separability. *Geom. Dedicata*, 96:179–198, 2003.

[Sco73a]   G. P. Scott. Compact submanifolds of 3-manifolds. *J. London Math. Soc. (2)*, 7:246–250, 1973.

[Sco73b]   G. P. Scott. Finitely generated 3-manifold groups are finitely presented. *J. London Math. Soc. (2)*, 6:437–440, 1973.

[Sel97]   Zlil Sela. Structure and rigidity in (Gromov) hyperbolic groups and discrete groups in rank 1 Lie groups. II. *Geom. Funct. Anal.*, 7(3):561–593, 1997.

[Sel01]   Zlil Sela. Diophantine geometry over groups. I. Makanin-Razborov diagrams. *Publ. Math. Inst. Hautes Études Sci.*, (93):31–105, 2001.

[Sho91]   Hamish Short. Quasiconvexity and a theorem of Howson's. In É. Ghys, A. Haefliger, and A. Verjovsky, editors, *Group theory from a geometrical viewpoint (Trieste, 1990)*, pages 168–176. World Sci. Publishing, River Edge, NJ, 1991.

[Sho01]   Hamish Short. Finitely presented subgroups of a product of two free groups. *Q. J. Math.*, 52(1):127–131, 2001.

[Sie83]   Allan J. Sieradski. A coloring test for asphericity. *Quart. J. Math. Oxford Ser. (2)*, 34(133):97–106, 1983.

[Sta62]   John Stallings. On fibering certain 3-manifolds. In *Topology of 3-manifolds and related topics (Proc. Univ. of Georgia Institute, 1961)*, pages 95–100. Prentice-Hall, Englewood Cliffs, N.J., 1962.

[Sta63] John Stallings. A finitely presented group whose 3-dimensional integral homology is not finitely generated. *Amer. J. Math.*, 85:541–543, 1963.

[Sta77] John Stallings. Coherence of 3-manifold fundamental groups. In *Séminaire Bourbaki, Vol. 1975/76, 28 ème année, Exp. No. 481*, pages 167–173. Lecture Notes in Math., Vol. 567. Springer, Berlin, 1977.

[Sta83] John R. Stallings. Topology of finite graphs. *Invent. Math.*, 71(3):551–565, 1983.

[Str90] Ralph Strebel. Appendix. Small cancellation groups. In *Sur les groupes hyperboliques d'après Mikhael Gromov (Bern, 1988)*, volume 83 of *Progr. Math.*, pages 227–273. Birkhäuser Boston, Boston, MA, 1990.

[Swa73] G. Ananda Swarup. Finding incompressible surfaces in 3-manifolds. *J. London Math. Soc. (2)*, 6:441–452, 1973.

[Swa93] G. A. Swarup. Geometric finiteness and rationality. *J. Pure Appl. Algebra*, 86(3):327–333, 1993.

[Swa04] Gadde A. Swarup. Delzant's variation on Scott complexity, 2004.

[Tra17] Hung Cong Tran. Malnormality and join-free subgroups in right-angled Coxeter groups, 2017.

[Vec16] Antony Della Veccia. Sectional curvature of one-relator groups. Master's thesis, McGill University, 2016.

[Wal79] List of problems. In C.T.C. Wall, editor, *Homological group theory (Proc. Sympos., Durham, 1977)*, pages 369–394. Cambridge Univ. Press, Cambridge, 1979.

[Wei72] C. M. Weinbaum. On relators and diagrams for groups with one defining relation. *Illinois J. Math.*, 16:308–322, 1972.

[Whi41] J.H.C. Whitehead. On adding relations to homotopy groups. *Ann. of Math. (2)*, 42:409–428, 1941.

[Wisa] Daniel T. Wise. On the vanishing of the 2nd L^2 betti number. Available at http://www.math.mcgill.ca/wise/papers., pp. 1–21. Submitted.

[Wisb] Daniel T. Wise. Positive one-relator groups are coherent. Available at http://www.math.mcgill.ca/wise/papers., pp. 1–19.

[Wisc] Daniel T. Wise. *The structure of groups with a quasiconvex hierarchy.* Annals of Mathematics Studies, to appear.

[Wis98] Daniel T. Wise. Incoherent negatively curved groups. *Proc. Amer. Math. Soc.*, 126(4):957–964, 1998.

[Wis01] Daniel T. Wise. The residual finiteness of positive one-relator groups. *Comment. Math. Helv.*, 76(2):314–338, 2001.

[Wis02a] Daniel T. Wise. A covering space with no compact core. *Geom. Ded.*, 92(1):59–62, 2002.

[Wis02b] Daniel T. Wise. The residual finiteness of negatively curved polygons of finite groups. *Invent. Math.*, 149(3):579–617, 2002.

[Wis03a] Daniel T. Wise. Nonpositive immersions, sectional curvature, and subgroup properties. *Electron. Res. Announc. Amer. Math. Soc.*, 9:1–9 (electronic), 2003.

[Wis03b] Daniel T. Wise. A residually finite version of Rips's construction. *Bull. London Math. Soc.*, 35(1):23–29, 2003.

[Wis04a] D. T. Wise. Sectional curvature, compact cores, and local quasiconvexity. *Geom. Funct. Anal.*, 14(2):433–468, 2004.

[Wis04b] Daniel T. Wise. Cubulating small cancellation groups. *GAFA, Geom. Funct. Anal.*, 14(1):150–214, 2004.

[Wis05] Daniel T. Wise. The coherence of one-relator groups with torsion and the Hanna Neumann conjecture. *Bull. London Math. Soc.*, 37(5):697–705, 2005.

[Wis08] Daniel T. Wise. Nonpositive sectional curvature for (p, q, r)-complexes. *Proc. Amer. Math. Soc.*, 136(1):41–48 (electronic), 2008.

[Wis11] Daniel T. Wise. Morse theory, random subgraphs, and incoherent groups. *Bull. Lond. Math. Soc.*, 43(5):840–848, 2011.

[Wis13] Daniel T. Wise. The last incoherent Artin group. *Proc. Amer. Math. Soc.*, 141(1):139–149, 2013.

[Wis14] Daniel T. Wise. The cubical route to understanding groups. In *Proceedings of the International Congress of Mathematicians—Seoul 2014. Vol. II*, pages 1075–1099. Kyung Moon Sa, Seoul, 2014.

[Wis18] Daniel T. Wise. Coherence, local-indicability, and nonpositive immersions. 2018. Submitted.

[Wis19] Daniel T. Wise. A note on the coherence of 3-manifold groups. pages 1–4, 2019.

A Decade of Thurston Stories

Dennis Sullivan

In December of 1971 a dynamics seminar ended at Berkeley with the solution to a thorny problem in the plane which had a nice application in dynamics. The solution purported to move N distinct points to a second set of "epsilon near" N distinct points by a motion which kept the points distinct and only moved while staying always "epsilon prime near." The senior dynamicists in the front row were upbeat because the dynamics application up to then had only been possible in dimensions at least three where this matching problem is obvious by general position. Now the dynamics theorem also worked in dimension two. But then a graduate student in the back of the room stood up and said he thought the algorithm of the proof didn't work. He went shyly to the blackboard and drew two configurations of about seven points each and started applying to these the method from the end of the lecture. Little paths started emerging and getting in the way of other emerging paths which to avoid collision had to get longer and longer. The algorithm didn't work at all for this quite involved diagrammatic reason. I had never seen such comprehension and then the creative construction of a counterexample done so quickly. This combined with my awe at the sheer complexity of the geometry that emerged.

A couple of days later the grad students invited me (I was also heavily bearded with long hair) to paint math frescoes on the corridor wall separating their offices from the elevator foyer. While milling around before painting, that same grad student came up to ask, "Do you think this is interesting to paint?" It was a complicated maze-like looking smooth one-dimensional object encircling three points in the plane. I asked, "What is it?" and was astonished to hear, "It is a simple closed curve." I said, "You bet it's interesting!" So we proceeded to spend several hours painting this curve on the wall. It was a great learning and bonding experience. For such a curve drawing to look good it has to be drawn in sections of short, parallel, slightly curved strands (like the flow boxes of a foliation) which are subsequently smoothly spliced together. When I asked how he got such curves, he said by successively applying to a given simple curve a pair of Dehn twists along intersecting curves. The "wall curve painting," two meters high and four meters wide (see AMS *Notices* cover "Cave Drawings"), dated and signed, lasted on that Berkeley wall with periodic restoration for almost four decades before finally being painted over a few years ago.

That week in December 1971 I was visiting Berkeley from MIT to give a series of lectures on differential forms and the homotopy theory of manifolds. Since foliations and differential forms were appearing everywhere, I thought to use the one forms that emerged in my story describing the lower central series of the fundamental group to construct foliations. Leaves of these foliations would cover graphs of maps of the manifold to arithmetic nilmanifolds associated to all the higher nilpotent quotients of the fundamental group. These would begin with and generalize Abel's map to the torus associated with the first homology torus. Being uninitiated in Lie theory, I had asked all the differential geometers at MIT and Harvard about this possibility but couldn't make myself understood. It was too vague and too much algebraic topology. I presented the discussion in my first lecture at Berkeley and to Bill privately without much hope because of the weird algebra/geometry mixture. However, the next day Bill came with a complete solution and a full explanation. For him it was elementary and really only involved actually understanding the basic geometric meaning of the Jacobi relation in the form due to Élie Cartan, namely, $dd = 0$.

In between the times of the first two stories above I had spoken to old friend Mo Hirsch about this grad student, Bill Thurston, who was working with Mo and was finishing in his fifth year after an apparently slow start. Mo or someone else told how Bill's oral exam was a slight problem because when asked for an example of the universal cover of a space Bill chose the surface of genus two and started drawing awkward octagons with many (eight) coming together at each vertex. This exposition quickly became an unconvincing mess on the blackboard. I think Bill might have been the only one in the room who had thought about this nontrivial kind of universal cover. Mo then said, "Lately, Bill has started solving thesis level problems at the rate of about one every month." Some years later I heard from Bill that his first child Nathaniel didn't like to sleep at night so Bill was sleep deprived "walking the floor with Nathaniel" for about a year of grad school.

That week of math at Berkeley was life-changing for me. I was very grateful to be able to seriously appreciate the Mozart-like phenomenon I had been observing; and I had a new friend. Upon returning to MIT after the week in Berkeley I related my news to the colleagues there, but I think my enthusiasm was too intense to be believed "I have just met the best graduate student I have ever seen or ever expect to see." It was arranged for Bill to give a talk at MIT which evolved into a plan for him to come to MIT after going first to IAS in Princeton. It turned out he did come to MIT for just one year (1973–74). That year I visited IHES where I ultimately stayed for 20 odd years while Bill was invited back to Princeton to be professor at the university.

IAS PRINCETON, 1972–73

When I visited the environs of Princeton from MIT in 72–73 I had chances to interact more with Bill Thurston. One day walking outside towards lunch at IAS, I asked Bill what a horocycle was. He said, "You stay here," and he started walking away into the institute meadow. After some distance he turned and stood still saying, "You are on the circumference of a circle with me as center." Then he turned, walked much further away, turned back, and said something which I couldn't hear because of the distance. After shouting back and forth to the amusement of the members, we realized he was saying the same thing, "You are on the circumference of a circle with me as center." Then he walked even further away, just a small figure in the distance and certainly out of hearing, whereupon he turned and started shouting presumably the same thing again. We got the idea what a horocycle was. The next day Atiyah asked some of us topologists if we knew if flat vector bundles had a classifying space (he had constructed some new characteristic classes for flat bundles). We knew a classifying space existed from Brown's representability characterization but we didn't know how to construct it explicitly. The next day Atiyah said he asked Thurston this question who did it by what was then a shocking construction: take the Lie structure group of the vector bundle as an abstract group with the discrete topology and form its classifying space. Later I heard about Thurston drawing Jack Milnor a picture proving any dynamical pattern for any unimodal map appears in the quadratic family $x \mapsto x^2 + c$. Since I was studying dynamics I planned to spend a semester with Bill at Princeton to learn about the celebrated Milnor-Thurston universality paper that resulted from this drawing.

PRINCETON UNIVERSITY, FALL 1976

I expected to learn about one-dimensional dynamics upon arriving in Princeton in September 1976 but Thurston had already developed a new theory of surface transformations. During the first few days he expounded on this in a wonderful three-hour extemporaneous lecture at the institute. Luckily for me, the main theorem about limiting foliations was intuitively clear because of the painstaking wall painting experience mentioned above. At the end of the stay that semester Bill told me he believed the mapping torus of these very twisting surface transformations carried 3D hyperbolic metrics. When I asked why, he told me he couldn't explain it to me because I didn't understand enough differential geometry. A few weeks after I left Princeton, with more time to work—without my constant distractions, for example—Bill essentially understood the proof of the existence of hyperbolic metrics for appropriate Haken manifolds. The mapping torus case took two more years as discussed below.

During that fall of '76 grad course that Bill gave, the grad students and I learned several key ideas: (1) We learned the quasi analogue of "hyperbolic geometry at infinity becomes conformal geometry on the sphere at infinity." (A notable memory here is the feeling that Bill conveyed about really being inside hyperbolic space rather than being outside and looking at a particular model. For me this made a psychological difference.) (2) We learned about the intrinsic geometry of convex surface boundaries away from the extreme points: Bill came into class one day and for several silent minutes rolled a paper contraption he had made around and around in the lecturer's table without saying a word until we felt the flatness. (3) We learned about the thick-thin decomposition of hyperbolic surfaces. I remember how Bill drew a 50-meter-long thin part winding all around the wide blackboard near the common room, and suddenly everything was clear, including geometric convergence to the points of the celebrated DM compactification of the space of Riemann surfaces.

During that fall of '76 semester stay at Princeton, Bill and I discussed understanding the Poincaré conjecture by trying to prove a general theorem about all closed three-manifolds based on the idea that three is a relatively small dimension. We included in our little paper on "canonical coordinates" the sufficient for PC candidate possibility that all closed three-manifolds carried special coordinate atlases, e.g., the conformally flat class is closed under connected sum and contains many prime three-manifolds. However, an undergrad who was often around—Bill

Goldman—disproved this a few years later for the nilmanifold prime. We decided to try to spend a year together in the future.

MEETING IN THE ALPS, SPRING 1978

In the next period, Bill developed limits of quasi-Fuchsian Kleinian groups and pursued the mapping torus hyperbolic structure in Princeton while I pursued the Ahlfors limit set measure problem in Paris. After about a year Bill had made substantial positive progress, e.g., closing the cusp, and I had made substantial negative progress (showing all known ergodic methods coupled with all known Kleinian group information were inadequate: there was too much potential nonlinearity). We met in the Swiss Alps at the Plans-sur-Bex conference and compared notes. His mapping torus program was positively finished but very complicated while my negative information actually revealed a rigidity result extending Mostow's to infinite volume growth strictly less than hyperbolic space itself. This allowed a simplification of Bill's mapping tour proof (linear volume growth). (See my Bourbaki report of 1980 on Bill's mapping torus theorem.)

THE STONYBROOK MEETING, SUMMER 1978

There was a big conference on Kleinian groups at Stonybrook and Bill was in attendance but not as a speaker. Gromov and I got him to give a lengthy impromptu talk outside the schedule. It was a wonderful trip out into the end of a hyperbolic three-manifold, combined with convex hulls, pleated surfaces and ending laminations. During the lecture, Gromov leaned over and said watching Bill made him feel like "this field hasn't officially started yet."

COLORADO, JUNE 1980 TO AUGUST 1981

Bill and I shared the Stanislaus Ulam visiting chair at Boulder and ran two seminars, a big one drawing together all the threads for the full hyperbolic theorem and a smaller one on the dynamics of Kleinian groups and dynamics in general. All aspects of the hyperbolic proof passed in review with many grad students in attendance. One day in the other seminar Bill was late and Dan Rudolph was very energetically explaining in just one hour a new shorter version of an extremely complicated proof. The theorem promoted an orbit equivalence to a conjugacy between two ergodic transformations if the discrepancy from conjugation of the orbit equivalence was controlled, e.g., AE bounded, or even summable, which could be reduced to the bounded case rather quickly. The new proof was due to a subset I don't recall of the triumvirate Katznelson, Weiss and Ornstein and was notable because it could be explained in one hour whereas the original proof took a minicourse to explain. Thurston at last came in and asked me to bring him up to speed, which I did. The lecture continued to the end with Bill wondering in loud whispers what the difficulty was and with me shushing him out of respect for the context. Finally at the end, Bill said just imagine a bi-infinite string of beads on a wire with finitely

many missing spaces and just slide them all to the left, say. Up to some standard
bookkeeping this gave a new proof. Later that day an awestruck Dan Rudolph said
to me he never realized before then just how smart Bill Thurston really was.

LA JOLLA AND PARIS, END OF SUMMER 1981

The Colorado experience was very good, relaxing in the big Thurston seminar with
geometry (one day we worked out the eight geometries and another day we voted on
terminology—"manifolded" or "orbifold"—and writing several papers of my own
on Hausdorff dimension, dynamics and measures on dynamical limit sets. Later,
closer to Labor Day, I was flying from Paris to La Jolla to give a series of AMS lectures on the dynamics stuff when I changed the plan and decided instead to try to
expose the entire Hyperbolic Theorem "for the greater good" and as a self-imposed
Colorado final exam. I managed to come up with a one-page sketch while on the
plane. There were to be two lectures a day for four or five days. The first day would
be ok, I thought, just survey things and then try to improvise for the rest, but I
needed a stroke of luck. The stroke of luck came, big time. There is a nine-hour time
difference between California and Paris and the first day I awoke around midnight
local time and went to my assigned office to prepare. After a few hours I had generated many questions and fewer answers about the hyperbolic argument. I noticed a
phone on the desk that miraculously allowed long distance calls and by then it was
around 4:00 a.m. California time and 7:00 a.m. in Princeton. I called Bill's house
and he answered. I posed my questions. He gave quick responses, I took notes, and
he said call back after he dropped kids at school and he got to his office. I gave my
objections to his answers around 9:30 a.m. his time and he responded more fully.
We ended up with various alternate routes that all-in-all covered every point. By
8:00 a.m. my time I had a pair of lectures prepared. The first day went well: lecture,
lunch/beach/swim, second lecture, dinner then goodbye to colleagues and back to
bed. This took some discipline but as viewers of the videos will see the audience
was formidable (Ahlfors, Bott, Chern, Kirby, Siebenmann, Edwards, Rosenberg,
Freedman, Yau, Maskit, Kra, Keen, Dodziuk, ...) and I was motivated. Bill and I
repeated this each day, perfecting the back and forth, so that by 8:00 a.m. California
time each day I had my two lectures prepared and they were getting the job done.
The climax came when presenting Bill's delicious argument that controls the length
of a geodesic representing the branching locus of a branched pleated surface by the
dynamical rate of chaos or entropy created by the geodesic flow on the intrinsic
surface. One knows this rate is controlled by the area growth of the universal cover
of the branched surface, this because negative curvature is controlled by the volume
growth of the containing hyperbolic three space which grows like $\exp(2r)$. QED.

There was, in addition, Bill's beautiful example showing the estimate was quali-
tatively sharp. This splendid level of lecturing was too much for Harold Rosenberg,
my astute friend from Paris, who was in the audience. He came to me afterwards
and asked frustratingly, "Dennis, do you keep Thurston locked up in your office
upstairs?" The lectures were taped by Michael Freedman and I have kept my lips
sealed until now. The URLs of the taped (Thurston)-Sullivan lectures can be found
in two locations: https://drive.google.com/file/d/11xWsIggCAQ0FT1ZkkKhG6nx
u9txiTtYK/edit and http://www.math.stonybrook.edu/Videos/Einstein/.

THURSTON IN PARIS, FALL 1981

Bill visited me in Paris and I bought a comfy sofa bed for my home office where he could sleep. He politely asked what would I have talked about in La Jolla had I not changed plans for the AMS lectures, and in particular, what had I been doing in detail in Colorado beyond the hyperbolic seminar. There were about six papers to tell him about. One of the most appealing ideas I had learned from him. Namely, the visual Hausdorf f-dimensional measure of a set on the sphere at infinity as viewed from a point inside defines a positive eigenfunction for the hyperbolic three space Laplacian with eigenvalue $f(2-f)$. I started going through the ideas and statements. I made a statement, and he either immediately gave the proof or I gave the idea of my proof. We went through all the theorems in the six papers in one session with either him or me giving the proof. There was one missing result that the bottom eigenfunction when f was >1 would be represented by a normalized eigenfunction whose square integral norm was estimated by the volume of the convex core. Bill lay back for a moment on his sofa bed, his eyes closed, and immediately proved the missing theorem. He produced the estimate by diffusing geodesics transversally and averaging. Then we went out to walk through Paris from Porte d'Orléans to Porte de Clignancourt. Of course we spoke so much about mathematics that Paris was essentially forgotten, except maybe the simultaneous view of Notre Dame and the Conciergerie as we crossed over the Seine.

PRINCETON-MANHATTAN, 1982–83

I began splitting time between IHES and the CUNY grad center where I started a thirteen-year-long Einstein chair seminar on dynamics and quasiconformal homeomorphisms (which changed then to quantum objects in topology) while Bill continued developing a cadre of young geometers to spread the beautiful ideas of negatively curved space. Bill delayed writing a definitive text on the hyperbolic proof in lieu of letting things develop along many opening avenues by his increasingly informed cadre of younger and older geometers. He wanted to avoid what had happened when his basic papers on foliations "tsunamied" the field in the early '70s. Once we planned to meet in Manhattan when I came in from IHES to discuss holomorphic dynamics in one variable and its analogies with which I had been preoccupied—hyperbolic geometry and Kleinian groups. We were not disciplined and began talking about other things at the apartment and finally got around to our agenda about thirty minutes before he had to leave for his train back to Princeton. I sketched the general analogy: Poincaré's limit set, domain of discontinuity, deformations, rigidity, classification, Ahlfors finiteness theorem...the work of Ahlfors-Bers...to be compared with Julia set, Fatou set, deformations, rigidity, classification, no-wandering-domain theorem...the work of Hubbard-Douady, etc., which he perfectly and quickly absorbed until he had to leave for the train. Two weeks later we heard about his reformulation of a holomorphic dynamical system as a fixed point on Teichmüller space analogous to part of his hyperbolic theorem. There were many new results including those of Curt McMullen some years later and the subject of holomorphic dynamics was raised to another higher level.

THE END

POSTSCRIPT

Thurston and I met again at Milnor's 80th fest at Banff after essentially thirty years and picked up where we had left off (I admired his checked green shirt the second time it appeared and he presented it to me the next day).

We sat side by side during the Jeremy Kahn lecture on the beautiful proof by Kahn and Markovic of the hyperbolic subsurface conjecture. As each step was revealed, I remembered when some aspect of an analogous device was introduced by Bill more than thirty years before and then later taught to his protégés at Princeton. This, except for one crucial step.

After the talk Bill mentioned he had not thought of the offset step in their proof which prevents everything from slipping off to infinity.

We promised to try to attack together a remaining big hole in the Kleinian group/holomorphic dynamics dictionary—"the invariant line field conjecture." It was a good idea but unfortunately turned out to be impossible.

For me personally, "What's Next?" means the following question, given that three manifolds have this very specific form predicted and developed by Bill Thurston almost forty years ago:

Why are three manifolds the way that they are?

Namely, one wonders what unexpected and beautiful significance and/or applications does this very specific structure mean for other parts of mathematics, physics, or science in general.

For one attempt, see my paper "Geometry on All Prime Three Manifolds" (with A. Kwon, arXiv:1906.10820).

Currently, I personally am trying to complete this example of a uniform texture for prime three manifolds to include connected sums in the context of 4D Lorentz structures and general relativity (GR). This will be reported at the conference "Geometry and Duality," Max Planck Institute for Gravitational Physics (Albert Einstein Institute), Potsdam, December 2–6, 2019.

List of Contributors

Ian AGOL, University of California, Berkeley, 970 Evans Hall #3840, Berkeley, CA 94720, USA
ianagol@berkeley.edu

Hyungryul BAIK, Department of Mathematical Sciences, KAIST, 291 Daehak-ro Yuseong-gu, Daejeon, 34141, South Korea
hrbaik@kaist.ac.kr

Michel BOILEAU, Aix-Marseille Université, CNRS, Centrale Marseille, I2M, Marseille, France
michel.boileau@univ-amu.fr

Martin R. BRIDSON, Mathematical Institute, Andrew Wiles Building, University of Oxford, Oxford OX2 6GG, UK
bridson@maths.ox.ac.uk

Danny CALEGARI, Department of Mathematics, University of Chicago, 5734 S University Ave., Chicago, IL 60637, USA
dannyc@math.uchicago.edu

Vincent DELECROIX, LaBRI, Domaine Universitaire, 351 Cours de la Libération, 33405 Talence, France
20100.delecroix@gmail.com

Nathan M. DUNFIELD, Department of Mathematics, MC-382, University of Illinois, 1409 W. Green St., Urbana, IL 61801, USA
nathan@dunfield.info

Benson FARB, Department of Mathematics, University of Chicago, 5734 S University Ave., Chicago, IL 60637, USA
farb@math.uchicago.edu

John FRANKS, Department of Mathematics, Northwestern University, Evanston, IL 60208, USA
j-franks@northwestern.edu

Stefan FRIEDL, Fakultät für Mathematik, Universität Regensburg, Germany
sfriedl@gmail.com

GAO Yan, Department of Mathematics, Sichuan University, Chengdu, 610065, China
gyan@scu.edu.cn

Misha GROMOV, IHÉS, Université Paris-Saclay, Le Bois-Marie, 35 rte. de Chartres, 91440 Bures-sur-Yvette, France and Courant Institute of Mathematical Sciences, New York University, 251 Mercer St., New York, NY 10012, USA
gromov@ihes.fr

John H. HUBBARD, Department of Mathematics, Cornell University, Ithaca, NY 14853, USA
jhh8@cornell.edu

Richard KENYON, Department of Mathematics, Yale University, 442 Dunham Lab, 10 Hillhouse Ave., New Haven, CT 06511, USA
richard.kenyon@yale.edu

Sarah KOCH, Department of Mathematics, University of Michigan, 530 Church St., Ann Arbor, MI 48109, USA
kochsc@umich.edu

Kathryn A. LINDSEY, Department of Mathematics, University of Chicago, 5734 S University Ave., Chicago, IL 60637, USA
lindseka@bc.edu

Yi LIU, Beijing International Center for Mathematical Research, Peking University, Beijing 100871, China
liuyi@bicmr.pku.edu.cn

Vladimir MARKOVIC, Mathematics 253-37, California Institute of Technology, Pasadena, CA 91125, USA
markovic@caltech.edu

Yi NI, Mathematics 253-37, California Institute of Technology, Pasadena, CA 91125, USA
yini@caltech.edu

Alan W. REID, Department of Mathematics, Rice University, Houston, TX 77005, USA
alan.reid@rice.edu

Dennis SULLIVAN, CUNY Graduate Center, 365 Fifth Ave., New York, NY 10016, USA and Mathematics Department, SUNY Stony Brook, Stony Brook, NY 11794, USA
sullivan0212@gmail.com

TAN Lei, late of Université d'Angers, Angers, France

Dylan P. THURSTON, Department of Mathematics, Indiana University–Bloomington, Bloomington, IN 47405, USA
dpthurst@indiana.edu

William P. THURSTON, late of Cornell University, Ithaca, NY 14853, USA

Daniel T. WISE, Department of Mathematics, McGill University, Montreal, QC H3A 2K6, Canada
wise@math.mcgill.ca

Xingru ZHANG, Department of Mathematics, University at Buffalo, Buffalo, NY 14260, USA
xinzhang@buffalo.edu

Anton ZORICH, Center for Advanced Studies, Skoltech; Institut de Mathématiques de Jussieu—Paris Rive Gauche, Case 7012, 8 Place Aurlie Nemours, 75205 Paris Cedex 13, France
anton.zorich@imj-prg.fr